Surveys in Differential Geometry

Vol. 1: Lectures given in 1990,
edited by S.-T. Yau and H. Blaine Lawson

Vol. 2: Lectures given in 1993,
edited by C.C. Hsiung and S.-T. Yau

Vol. 3: Lectures given in 1996,
edited by C.C. Hsiung and S.-T. Yau

Vol. 4: Integrable systems,
edited by Chuu Lian Terng and Karen Uhlenbeck

Vol. 5: Differential geometry inspired by string theory,
edited by S.-T. Yau

Vol. 6: Essays on Einstein manifolds,
edited by Claude LeBrun and McKenzie Wang

Vol. 7: Papers dedicated to Atiyah, Bott, Hirzebruch, and Singer

Vol. 8: Papers in honor of Calabi, Lawson, Siu, and Uhlenbeck

Vol. 9: Eigenvalues of Laplacians and other geometric operators,
edited by A. Grigor'yan and S.-T. Yau

Vol. 10: Essays in geometry in memory of S.S. Chern,
edited by S.-T. Yau

Vol. 11: Metric and comparison geometry,
edited by Jeffrey Cheeger and Karsten Grove

Vol. 12: Geometric flows,
edited by Huai-Dong Cao and S.-T. Yau

Volume X

Surveys in Differential Geometry

Essays in geometry in memory of S.S. Chern

edited by
Shing-Tung Yau

International Press

Surveys in Differential Geometry, Vol. 10

International Press
P.O. Box 43502
Somerville, MA 02143
www.intlpress.com

2000 Mathematics Subject Classification. 03-02

Essays in geometry in memory of S.S. Chern
Shing-Tung Yau, editor

ISBN 978-1-57146-122-3

Originally published in 2006 under ISBN 978-1-57146-116-2.
This larger-format reissue includes a correction to the table of contents, a revised preface, and an updated series listing at the front of the book.

Typeset using the LaTeX system.
Printed in the USA on acid-free paper.

Shiing-Shen Chern
1911 - 2004

Preface

A venerated leader in the field of geometry and a beloved teacher, Professor Shiing-Shen Chern passed away in 2004, at the age of 93.

In the spring of 2005, the Journal of Differential Geometry Conference, held annually at Harvard University, was dedicated in his honor, and attended by many of his friends, his former students, and even his son, who came from far away to be present at the conference. There were several talks related to his lifelong work in geometry and topology, as well as lectures given on new contributions to these fields.

The conference was a great success, and Professor Chern would have loved to see most of his students and friends gathered in memory of him.

Contents

On the space-time monopole equation

Bo Dai, Chuu-Lian Terng, and Karen Uhlenbeck

ABSTRACT. The space-time monopole equation is obtained from a dimension reduction of the anti-self dual Yang-Mills equation on $\mathbb{R}^{2,2}$. A family of Ward equations is obtained by gauge fixing from the monopole equation. In this paper, we give an introduction and a survey of the space-time monopole equation. Included are alternative explanations of results of Ward, Fokas-Ioannidou, Villarroel and Zakhorov-Mikhailov. The equations are formulated in terms of a number of equivalent Lax pairs; we make use of the natural Lorentz action on the Lax pairs and frames. A new Hamiltonian formulation for the Ward equations is introduced. We outline both scattering and inverse scattering theory and use Bäcklund transformations to construct a large class of monopoles which are global in time and have both continuous and discrete scattering data.

1. Introduction

The self-dual Yang-Mills equations in $\mathbb{R}^4$ and their reduction to monopole equations in $\mathbb{R}^3$ have become central topics of study and useful tools in modern geometry. The same self-dual equations in the case of a different signature of $\mathbb{R}^{2,2}$ are not of the general type to be used much in geometry. However, their dimensional reduction to the *space-time monopole equation* in $\mathbb{R}^{2,1}$ yields an extremely interesting system of non-linear wave equations which deserve to be better known. These equations can be encoded in a Lax pair. Moreover, with a mild additional assumption and a gauge fixing they can be rewritten for a map from $\mathbb{R}^{2,1}$ into a Lie group. These equations differ only slightly from the usual wave map equation.

This article is meant to be an introduction to and a survey of the literature on the space-time monopole equations. We also give a construction of the inverse scattering of the monopole equations via loop group factorizations. These equations form a hyperbolic system for a connection and a Higgs field, and hence have a gauge symmetry. A simple restriction and

The research of the second author was supported in part by NSF grant DMS-0529756. The research of the third author was supported in part by Sid Richardson Regents' Chair Funds, University of Texas system and NSF grant DMS-0305505.

coordinate change produces the equation for a map into the gauge group. This last equation was introduced by Richard Ward, who studied them using a version of Riemann-Hilbert problem and twistor theory. He produced the basic examples and a number of interesting papers [**22, 23, 24**]. Hence the equation for the map is referred to as either Ward's equation, or in his original language, the modified chiral model. Additional work on the equations is due to T. Ioannidou, W. Zakrewski [**10, 11, 12**], Manakov and Zakharov [**13**], A. K. Fokas and Ioannidou [**7**] and Villaroel [**21**]. The last three references present both the continuous scattering theory and the inverse scattering transform. The construction of a complete set of soliton solutions has been carried out by the first two authors in a previous paper [**6**].

The plan of the paper is as follows. We derive the monopole equations with their Lax pairs, paying special attention to the difference between monopole equations in space and in space-time in Section 2. A family of Ward equations for maps into groups is constructed in Section 3. Section 4 describes the action of the Lorentz group on the Lax pair system and on frames. We make use of Lorentz boosts in the construction of solitons and of the spacial rotation group in deriving estimates in the appendix. Next we list special classes of solutions, so we can continue the discussion with a lot of examples in mind. Section 6 contains a very brief Hamiltonian formulation for the family of Ward equations. In Section 7, we introduce the transform which produces the continuous scattering data as well as the inverse scattering transform. Since the inverse scattering transform always exists, this produces many global solutions to the equations that are decaying at spacial infinity. The details of the fixed point theorem which yields continuous scattering data for small initial data are in the appendix. In Section 8 to 10, we review Bäcklund transformations, and use these transformations to construct soliton monopoles and monopoles with both continuous and discrete scattering data.

Due to soliton theory, Bäcklund transformations and the inverse scattering transform, we discover a very large class of solutions which are global in time. This is in contrast to the closely related wave map equation from $\mathbb{R}^{2,1}$ to G. It is a difficult theorem first of T. Tao [**15**], extended by D. Tataru [**16**], to show that small initial data results in solutions for all time. Whether the difference is entirely due to integrability, or whether there is a deeper analytic theory or more examples to be found, remains open.

2. The Monopole equations

The curvature $F_A = \sum_{i,j} F_{ij}\, dx_i \wedge dx_j$ of a $u(n)$-valued connection 1-form $A = \sum_{i=1}^4 A_i dx_i$ on $\mathbb{R}^4$ is

$$F_{ij} = [\nabla_i, \nabla_j] = -\partial_{x_i}(A_j) + \partial_{x_j}(A_i) + [A_i, A_j],$$

where

$$\nabla_i = \partial_{x_i} - A_i.$$

The connection A is *anti-self-dual Yang-Mills* (ASDYM) on $\mathbb{R}^4$ if

$$*F_A = -F_A,$$

where $*$ is the Hodge star operator with respect to the Euclidean metric $\sum_{i=1}^4 dx_i^2$ on $\mathbb{R}^4$. The ASDYM on $\mathbb{R}^4$ written in coordinates is

$$F_{12} = -F_{34}, \quad F_{13} = -F_{42}, \quad F_{14} = -F_{23}. \tag{2.1}$$

The ASDYM has a Lax pair formulation. The term "Lax pair" refers to any equation which is written as a "zero curvature" equation for a connection, or a portion of a connection. This connection contains an additional complex parameter μ which is variously interpreted as a "spectral", "twistor", or "Riemann-Hilbert parameter". Set

$$z = x_1 + ix_2, \quad w = x_3 + ix_4,$$

$\nabla_z = \frac{1}{2}(\nabla_1 - i\nabla_2) = \frac{\partial}{\partial z} - A_z$, $\nabla_{\bar z} = \frac{1}{2}(\nabla_1 + i\nabla_2) = \frac{\partial}{\partial \bar z} - A_{\bar z}$, and $\nabla_w, \nabla_{\bar w}$ similarly. Since $A_i \in u(n)$, $A_{\bar z} = -A_z^*$ and $A_{\bar w} = -A_w^*$, where $B^* = \bar{B}^t$. The equation

$$[\nabla_{\bar w} + \mu\nabla_z, \nabla_w - \mu^{-1}\nabla_{\bar z}] = 0 \tag{2.2}$$

is equivalent to the ASDYM (2.1) on $\mathbb{R}^4$. This is because (2.2) holds for all $\mu \in \mathbb{C} \setminus \{0\}$ if and only if the coefficients of $\mu, 1$ and μ^{-1} of (2.2) are zero, which is (2.1).

If we assume the ASDYM connection A is independent of x_4, then $A_w = \frac{1}{2}(A_t - i\Phi)$ and $A_{\bar w} = \frac{1}{2}(A_t + i\Phi)$, where $\Phi = A_4$ is the Higgs field, $A = A_t\, dt + A_z\, dz + A_{\bar z}\, d\bar z$ is a connection 1-form on $\mathbb{R}^3$. Then (A, Φ) satisfies the Euclidean monopole equation (cf. [**3**])

$$D_A\Phi = *F_A,$$

where $*$ is the Hodge star operator with respect to the metric $dx^2 + dy^2 + dt^2$. The Euclidean monopole equation is an important equation in both geometry and physics (cf. [**8, 3**]).

The ASDYM on $\mathbb{R}^{2,2}$ is again $*F_A = -F_A$, but $*$ is the Hodge star operator with respect to the metric $dx_1^2 + dx_2^2 - dx_3^2 - dx_4^2$. In coordinates, the ASDYM is

$$F_{12} = F_{34}, \quad F_{13} = -F_{42}, \quad F_{14} = -F_{23}. \tag{2.3}$$

This equation has a Lax pair similar to the ASDYM on $\mathbb{R}^4$,

$$[\nabla_{\bar w} + \mu\nabla_z,\ \nabla_w + \mu^{-1}\nabla_{\bar z}] = 0 \tag{2.4}$$

(the only difference with (2.2) is that the second operator is taken with a plus sign). In other words, A satisfies (2.3) if and only if (2.4) holds for all $\mu \in \mathbb{C} \setminus \{0\}$.

We say $E(\dots, \mu)$ is a *frame* of the Lax pair $[D_1(\mu), D_2(\mu)] = 0$ if $D_1E = D_2E = 0$.

If $[D_1(\mu), D_2(\mu)] = 0$ is a Lax pair of a PDE with spectral parameter μ, then $[\tilde{D}_1(\lambda), \tilde{D}_2(\lambda)] = 0$ is also a Lax pair of the same PDE with parameter

$\mu = \frac{a\lambda+b}{c\lambda+d}$, where $\tilde{D}_i(\lambda) = \sum_{j=1}^{2} f_{ij}(\mu)D_j(\mu)$ and f_{ij} are meromorphic functions. Moreover, if E is a frame for the Lax pair $[D_1(\mu), D_2(\mu)] = 0$, then $F(\ldots, \lambda) = E(\ldots, \frac{a\lambda+b}{c\lambda+d})$ is a frame for the Lax pair $[\tilde{D}_1(\lambda), \tilde{D}_2(\lambda)] = 0$. We say these Lax pairs are *equivalent*.

If there is a g such that $D_1(\mu_0)g = D_2(\mu_0)g = 0$, then $g^{-1}[D_1(\mu), D_2(\mu)]g = 0$ is also a Lax pair for the same PDE, which is called the Lax pair obtained from the Lax pair $[D_1(\mu), D_2(\mu)] = 0$ by *fixing the gauge* at $\mu = \mu_0$. So Lax pairs of a PDE come in many forms. We can choose an equivalent Lax pair and fix a gauge to make our computation easier or more transparent. For example, this is what we will do in later sections when we compute the $SO(2,1)$-actions on frames and discuss the scattering and inverse scattering of the monopole equations.

By making a linear fractional transformations in μ, we obtain an equivalent Lax pair. Set $\mu = \frac{\lambda-i}{\lambda+i}$. Then

$$\begin{cases} 2(\lambda+i)(\nabla_{\bar{w}} + \mu\nabla_z) = L_1 + iL_2, \\ 2(\lambda-i)(\nabla_w + \mu^{-1}\nabla_{\bar{z}}) = L_1 - iL_2, \end{cases} \tag{2.5}$$

where

$$\begin{cases} L_1 = \lambda(\nabla_{x_3} + \nabla_{x_1}) - (\nabla_{x_4} + \nabla_{x_2}), \\ L_2 = \lambda(\nabla_{x_4} - \nabla_{x_2}) + (\nabla_{x_3} - \nabla_{x_1}). \end{cases}$$

It follows that (2.4) holds if and only if $[L_1, L_2] = 0$, i.e.,

$$[\lambda(\nabla_{x_3} + \nabla_{x_1}) - (\nabla_{x_4} + \nabla_{x_2}),\ \lambda(\nabla_{x_4} - \nabla_{x_2}) + (\nabla_{x_3} - \nabla_{x_1})] = 0 \tag{2.6}$$

holds for all $\lambda \in \mathbb{C} \setminus \{i, -i\}$. So (2.6) is an equivalent Lax pair for the ASDYM on $\mathbb{R}^{2,2}$. The use of μ in the section on scattering theory, and λ in the rest of the literature on the Ward equation, is unfortunately confusing, but necessary.

If we assume the connection A is independent of x_4, then

$$A_w = \frac{1}{2}(A_t - i\phi), \qquad A_{\bar{w}} = \frac{1}{2}(A_t + i\phi)$$

where $A_{x_4} = \phi$ is now the Higgs field for our new *space-time monopole equation*. A calculation shows that we can write the space-time monopole equation in the same form as the more familiar Euclidean monopole equation for a connection $A = A_t\,dt + A_z\,dz + A_{\bar{z}}\,d\bar{z}$ and a Higgs field ϕ (cf. [**25**]):

$$D_A\phi = *F_A. \tag{2.7}$$

Here we have used the $*$ operator for the Lorentz metric $dx^2 + dy^2 - dt^2$.

In this paper, we always assume (A, ϕ) decays at spacial infinity.

Equation (2.7) has a Lax pair induced from the Lax pair (2.4) for ASDYM:

$$\left[\frac{1}{2}\nabla_t - \frac{i\phi}{2} + \mu\nabla_z,\ \frac{1}{2}\nabla_t + \frac{i\phi}{2} + \mu^{-1}\nabla_{\bar{z}}\right] = 0. \tag{2.8}$$

Set

$$\mu = \frac{\lambda - i}{\lambda + i}, \quad \xi = \frac{t+x}{2}, \quad \eta = \frac{t-x}{2}.$$

Then (2.6) induces an equivalent Lax pair for the monopole equation in (ξ, η, y) coordinates:

$$\big[\lambda\nabla_\xi - \nabla_y + \phi,\ \lambda^{-1}\nabla_\eta - \nabla_y - \phi\big] = 0. \tag{2.9}$$

This is the Lax pair used by Ward. We have:

PROPOSITION 2.1. *The following statements are equivalent for a connection A on $\mathbb{R}^{2,1}$ and a Higgs field ϕ:*

(1) *(A, ϕ) is a solution of the space time monopole equation* (2.7) *on $\mathbb{R}^{2,1}$.*
(2) (2.8) *holds for all $\mu \in \mathbb{C} \setminus \{0\}$.*
(3) *The linear system,*

$$\begin{cases} (\frac{1}{2}\frac{\partial}{\partial t} + \mu\frac{\partial}{\partial z})E = (\frac{1}{2}(A_t + i\phi) + \mu A_z)E, \\ (\frac{1}{2}\frac{\partial}{\partial t} + \mu^{-1}\frac{\partial}{\partial \bar z})E = (\frac{1}{2}(A_t - i\phi) + \mu^{-1}A_{\bar z})E, \end{cases} \tag{2.10}$$

is compatible for complex parameter μ.
(4) (2.9) *holds for all $\lambda \in \mathbb{C} \setminus \{0\}$.*
(5) *The linear system*

$$\begin{cases} (\lambda\frac{\partial}{\partial \xi} - \frac{\partial}{\partial y})F = (\lambda A_\xi - (A_y + \phi))F, \\ (\lambda^{-1}\frac{\partial}{\partial \eta} - \frac{\partial}{\partial y})F = (\lambda^{-1}A_\eta - (A_y - \phi))F \end{cases} \tag{2.11}$$

is compatible for complex parameter λ.

Moreover, if $E(x, y, t, \mu)$ is a frame of (2.8) *(i.e., a solution of* (2.10)*), then*

(a) *$F(x, y, t, \lambda) = E\left(x, y, t, \frac{\lambda - i}{\lambda + i}\right)$ is a frame of* (2.9) *(i.e., a solution of* (2.11)*),*
(b) *E satisfies the reality condition*

$$E(x, y, t, \bar\mu^{-1})^* E(x, y, t, \mu) = \mathrm{I} \tag{2.12}$$

if and only if F satisfies the reality condition

$$F(x, y, t, \bar\lambda)^* F(x, y, t, \lambda) = \mathrm{I}. \tag{2.13}$$

3. The Ward equations

We call solutions of linear systems (2.10) or (2.11) that satisfy the reality condition *monopole frames.* Note that, unlike the case when the Lax pair is a full connection, locally there can be a serious lack of uniqueness in solving for a frame. We resolve this lack of uniqueness away from $\mu \in S^1$ by observing that the spacial part of the Lax pair is a Cauchy-Riemann operator. Frames, if they exist, are unique if we require $E_\mu = \mathrm{I}$ at spacial infinity. We expect the frames to exist at most points $\mu \notin S^1$.

When $\mu = \rho \in S^1$, the existence of frames is more problematic. To obtain the Ward equation, we need extra assumptions, even for small initial data.

DEFINITION 3.1. Let $\rho \in S^1$, and (A, ϕ) be a solution of the space-time monopole equation such that (A, ϕ) decays at spacial infinity. We say (A, ϕ) is *ρ-regular*, if there is a smooth solution $k : \mathbb{R}^{2,1} \to U(n)$ such that

$$\begin{cases} (\frac{1}{2}\partial_t + \rho\partial_z)k = (A_{\bar{w}} + \rho A_z)k, \\ (\frac{1}{2}\partial_t + \rho^{-1}\partial_{\bar{z}})k = (A_w + \rho^{-1}A_{\bar{z}})k, \end{cases} \tag{3.1}$$

and $k - \mathrm{I}$ and the first derivative of k decays as $|z| \to \infty$. (Note that the second equation of (3.1) is the Hermitian transpose of the first.)

Let f be a $U(n)$-valued map. Then

$$f(\partial_x - A)f^{-1} = \partial_x - (fAf^{-1} + (\partial_x f)f^{-1})$$

is the gauge transformation of f on $\frac{\partial}{\partial x} - A$, or

$$f * A = fAf^{-1} + f_x f^{-1}.$$

Suppose (A, ϕ) is ρ-regular and k is the solution of (3.1). We fix the gauge at $\mu = \rho$, i.e., we apply the gauge transformation of k^{-1} to the Lax pair (2.8) to get

$$\left[\frac{1}{2}\partial_t + \mu\partial_z - (\mu - \rho)\tilde{A}_z,\ \frac{1}{2}\partial_t + \mu^{-1}\partial_{\bar{z}} - (\mu^{-1} - \rho^{-1})\tilde{A}_{\bar{z}}\right] = 0, \tag{3.2}$$

where $\tilde{A}_z = k^{-1} * A_z$ and $\tilde{A}_{\bar{z}} = k^{-1} * A_{\bar{z}}$. Or equivalently, the following linear system is compatible for an open subset of parameters μ:

$$\begin{cases} (\frac{1}{2}\frac{\partial}{\partial t} + \mu\frac{\partial}{\partial z})E = (\mu - \rho)\tilde{A}_z E, \\ (\frac{1}{2}\frac{\partial}{\partial t} + \mu^{-1}\frac{\partial}{\partial \bar{z}})E = (\mu^{-1} - \rho^{-1})\tilde{A}_{\bar{z}} E. \end{cases} \tag{3.3}$$

Suppose (A, ϕ) is also $-\rho$-regular. Then there exists $g : \mathbb{R}^{2,1} \to U(n)$ which satisfies (3.3) with $\mu = -\rho$, i.e.,

$$\begin{cases} (\frac{1}{2}\partial_t - \rho\partial_z)g = -2\rho\tilde{A}_z g, \\ (\frac{1}{2}\partial_t - \rho^{-1}\partial_{\bar{z}})g = -2\rho^{-1}\tilde{A}_{\bar{z}} g. \end{cases} \tag{3.4}$$

A computation shows that

$$-(g_t g^{-1})_t + (g_x g^{-1})_x + (g_y g^{-1})_y + [g_t g^{-1}, \cos\theta\, g_x g^{-1} + \sin\theta\, g_y g^{-1}] = 0, \tag{3.5}$$

where $\rho = e^{i\theta}$. This is the one-parameter family of *Ward equations* [**22**]. We then obtain

PROPOSITION 3.2. *Suppose $E(x, y, t, \mu)$ is a frame for the solution of the space-time monopole equation, (i.e., E is a solution of the linear system (2.10)), and $E(x, y, t, \mu)$ are smooth at $\mu = \pm e^{i\theta}$. Then*

$$g(x, y, t) := E(x, y, t, e^{i\theta})^{-1}E(x, y, t, -e^{i\theta})$$

is a solution of the Ward equation (3.5).

The Lax pair (2.9) is equivalent to the following Lax pair:

$$[\lambda\nabla_\xi - \nabla_y + \phi, \quad \lambda(\nabla_y + \phi) - \nabla_\eta] = 0. \tag{3.6}$$

We fix the gauge of (3.6) at $\lambda = \infty$ (equivalent to fixing the gauge of (2.8) at $\mu = 1$), i.e., take the gauge transformation of h^{-1} on (3.6) to get

$$[\lambda\partial_\xi - (\partial_y - \hat{A}), \quad \lambda\partial_y - (\frac{\partial}{\partial\eta} - \hat{B})] = 0, \tag{3.7}$$

where $\partial_y h = (A_y - \phi)h$, $\partial_\xi h = A_\xi h$, $\hat{A} = -2h^{-1}\phi h$ and $\hat{B} = h^{-1} * A_\eta$. So we have the following proposition:

PROPOSITION 3.3. *The following statements are equivalent:*

(1) *Equation* (3.7) *holds for all* $\lambda \in \mathbb{C}$,

(2)

$$\begin{cases} \partial_\xi \hat{B} = \partial_y \hat{A}, \\ [\partial_y - \hat{A}, \partial_\eta - \hat{B}] = 0. \end{cases}$$

(3) *The linear system*

$$\begin{cases} (\lambda\partial_\xi - \partial_y)H = -\hat{A}H, \\ (\lambda\partial_y - \partial_\eta)H = -\hat{B}H, \end{cases} \tag{3.8}$$

is locally solvable for an open subset of $\lambda \in \mathbb{C}$. *Moreover, if* $H(x, y, t, \lambda)$ *is a solution of* (3.8) *and is smooth at* $\lambda = 0$, *then* $g = H(\ldots, 0)$ *satisfies*

$$\frac{\partial}{\partial t}\left(\frac{\partial g}{\partial t} g^{-1}\right) - \frac{\partial}{\partial x}\left(\frac{\partial g}{\partial x} g^{-1}\right) - \frac{\partial}{\partial y}\left(\frac{\partial g}{\partial y} g^{-1}\right) - \left[\frac{\partial g}{\partial t} g^{-1}, \frac{\partial g}{\partial x} g^{-1}\right] = 0, \tag{3.9}$$

i.e., g is a solution of Ward equation (3.5) *with* $\theta = 0$.

As a consequence of the above proposition, we see that to construct solutions of the monopole equation that are $\lambda = \infty$-regular, it suffices to construct $H(\xi, \eta, y, \lambda)$ such that $(\lambda\partial_\xi H - \partial_y H)H^{-1}$ and $(\lambda\partial_y H - \partial_\eta H)H^{-1}$ are independent of λ.

PROPOSITION 3.4. *If a monopole is* $\mu = \pm 1$ *regular (i.e.,* $\lambda = \infty, 0$ *regular), then it is gauge equivalent to a monopole* (A, ϕ) *such that* $A_\xi = 0$, $A_y = \phi$. *Conversely, if* $(\hat{A}, \hat{B})$ *satisfies* (3.7), *then* $A_\xi = 0$, $A_y = \phi = \hat{A}/2$, *and* $A_\eta = \hat{B}$ *is a monopole.*

4. The Action of $SO(2,1)$

The Lorentz group $SO(2,1)$ is the group of all $g \in SL(3, \mathbb{R})$ such that $g^t \mathrm{I}_{2,1} g = \mathrm{I}_{2,1}$, where $\mathrm{I}_{2,1} = \mathrm{diag}(1, 1, -1)$. The group $SO(2,1)$ acts on $\mathbb{R}^{2,1}$ by the standard action $g \cdot p = gp$ (here $p \in \mathbb{R}^{2,1}$ is identified as a 3×1 vector). Given a connection $A = A_1 dx + A_2 dy + A_3 dt$, a Higgs field ϕ, and $g \in SO(2,1)$, the action $g \cdot (A, \phi) = (g \cdot A, g \cdot \phi)$ is defined by $g \cdot A = g \cdot A_1\ dx + g \cdot A_2\ dy + g \cdot A_3\ dt$, where

$$(g \cdot A_i)(p) = A_i(g \cdot p), \quad (g \cdot \phi)(p) = \phi(g \cdot p).$$

The space-time monopole equation is invariant under the Lorentz group $SO(2,1)$, i.e., if (A,ϕ) is a solution then so is $g\cdot(A,\phi)=(g\cdot A, g\cdot\phi)$ for $g\in SO(2,1)$. In order to make the scattering theory estimates tractable and to understand the 1-solitons we need to understand the natural action of $SO(2,1)$ on frames (solutions of linear system (2.11)). Since $SO(2)$ of the xy-plane and $O(1,1)$ of the xt-plane generate $SO(2,1)$, to compute the explicit formula of the action of $SO(2,1)$ on frames, it suffices to compute the action of the following one-parameter subgroups on frames:

$$R(\theta)=\begin{pmatrix}\cos\theta & -\sin\theta & 0\\ \sin\theta & \cos\theta & 0\\ 0 & 0 & 1\end{pmatrix},$$

$$T(s)=\begin{pmatrix}\cosh s & 0 & \sinh s\\ 0 & 1 & 0\\ \sinh s & 0 & \cosh s\end{pmatrix}.$$

We also need the representation $\sigma: SO(2,1)\to SL(2,\mathbb{R})$, whose differential $d\sigma_e$ maps $e_{12}-e_{21}$ to $-\frac{1}{2}(e_{12}-e_{21})$, $e_{13}+e_{31}$ to $\frac{1}{2}(e_{11}-e_{22})$, and $e_{23}+e_{32}$ to $-\frac{1}{2}(e_{12}+e_{21})$. So

$$\sigma(R(\theta))=\begin{pmatrix}\cos\frac{\theta}{2} & \sin\frac{\theta}{2}\\ -\sin\frac{\theta}{2} & \cos\frac{\theta}{2}\end{pmatrix},\quad \sigma(T(s))=\begin{pmatrix}e^{\frac{s}{2}} & 0\\ 0 & e^{-\frac{s}{2}}\end{pmatrix}.$$

The group $SL(2,\mathbb{R})$ acts on the scattering parameter space $\mathbb{C}\cup\{\infty\}$ by the linear fractional transformations:

$$\begin{pmatrix}a & b\\ c & d\end{pmatrix}*\lambda=\frac{a\lambda+b}{c\lambda+d}.$$

THEOREM 4.1. *The group $SO(2,1)$ acts on the Lax pairs and on the frames of the space-time monopole equation. If F is a frame of (A,ϕ), then*

$$(g\cdot F)(p,\lambda)=F(g\cdot p,\sigma(g)*\lambda)$$

is a frame for $g\cdot(A,\phi)$, where $\sigma: SO(2,1)\to SL(2,\mathbb{R})$ is the representation given above and $*$ is the standard action of $SL(2,\mathbb{R})$ on λ via the linear fractional transformation.

PROOF. For the action of $R(\theta)$, it can be checked easily that if E is a solution of (2.10) for (A,ϕ), then $\tilde{E}$ solves (2.10) for $e^{i\theta}\cdot(A,\Phi)$, where

$$\tilde{E}(z,t,\mu)=E(e^{i\theta}z,t,e^{i\theta}\mu).$$

(Here we identify $e^{i\theta}$ with the rotation matrix $R(\theta)$). In other words,

$$R(\theta)\cdot E(p,\mu)=E(R(\theta)p,e^{i\theta}\mu).$$

To compute the action of $R(\theta)$ on solutions F of (2.11), we recall that $\mu=\frac{\lambda-i}{\lambda+i}$, $\lambda=\frac{i(1+\mu)}{1-\mu}$, and

$$F(p,\lambda)=E\left(p,\frac{\lambda-i}{\lambda+i}\right),$$

where E is a solution of (2.10). But for $\tilde{\mu} = e^{i\theta}\mu$, the corresponding $\tilde{\lambda}$ and λ are related by

$$\tilde{\lambda} = \frac{i(1+e^{i\theta})\lambda - (1-e^{i\theta})}{(1-e^{i\theta})\lambda + i(1+e^{i\theta})} = \frac{\cos\frac{\theta}{2}\ \lambda + \sin\frac{\theta}{2}}{-\sin\frac{\theta}{2}\ \lambda + \cos\frac{\theta}{2}} = \begin{pmatrix} \cos\frac{\theta}{2} & \sin\frac{\theta}{2} \\ -\sin\frac{\theta}{2} & \cos\frac{\theta}{2} \end{pmatrix} * \lambda.$$

In other words,

$$(R(\theta)\cdot F)(p,\lambda) = F(R(\theta)p, \sigma(R(\theta)) * \lambda).$$

For the action of $T(s)$, we use the light cone coordinates ξ, η. Then $T(s)$ maps (ξ, η, y) to $(e^s\xi, e^{-s}\eta, y)$. Suppose $F(p,\lambda)$ is a solution of linear system (2.11) for (A, ϕ). Given $s \in \mathbb{R}$, define

$$\hat{F}(\xi,\eta,y,\lambda) = F(T(s)\cdot p, e^s\lambda) = F(e^s\xi, e^{-s}\eta, y, e^s\lambda).$$

It can be checked easily that $\hat{F}$ is a frame for $T(s)\cdot(A,\phi)$. But

$$e^s\lambda = \begin{pmatrix} e^{\frac{s}{2}} & 0 \\ 0 & e^{-\frac{s}{2}} \end{pmatrix} * \lambda = \sigma(T(s)) * \lambda,$$

where $*$ is the standard action of $SL(2,\mathbb{R})$ on $\mathbb{C}$. $\square$

COROLLARY 4.2. *The group $SO(2,1)$ acts on frames E of (2.10) as follows: If E is a frame of (A,Φ), then $(g\cdot E)(p,\mu) = E(g\cdot p, g\sharp\mu)$, where*

$$R(\theta)\sharp\mu = e^{i\theta}\mu,$$
$$T(s)\sharp\mu = \frac{\cosh\frac{s}{2}\ \mu + \sinh\frac{s}{2}}{\sinh\frac{s}{2}\ \mu + \cosh\frac{s}{2}}.$$

5. Special classes of solutions

Many examples of solutions to the monopole equations are obtained by assuming additional conditions. A first set of examples comes from linear solutions of the wave equation. If $H \subset G$ is any abelian subgroup, the monopole equations for maps into H are linear. This provides us with a number of solutions to which we can later apply Bäcklund transformations. Let $\mathcal{H} \subset \mathcal{G}$ be the Lie subalgebra of H. Suppose that $(A,\phi) = (a, a_0, \phi) \in \oplus^2\mathfrak{h}\oplus\mathfrak{h}\oplus\mathfrak{h}$ decays at spacial infinity, where $a_0 = A_t, a = (a_1, a_2) = (A_x, A_y)$, and $\mathfrak{h}$ is the Lie algebra of H. Then the monopole equations (2.7) are written in space-time (here d and $*$ are spacial):

$$(5.1) \qquad \begin{cases} *da = \frac{\partial\phi}{\partial t}, \\ \frac{\partial a}{\partial t} - da_0 = *d\phi. \end{cases}$$

The gauge transformation of e^{-u} appears as

$$(a, a_0, \phi) \mapsto (a - du, a_0 - \frac{\partial u}{\partial t}, \phi).$$

A global way of fixing gauge suitably is to require that $d * a = 0$ (a spacial equation which is easily solved). With this choice, we apply $d*$ to the second equation of (5.1) to obtain

$$\frac{\partial}{\partial t}(d * a) - d * d\, a_0 = d * *d\phi = 0.$$

Since $d * a = 0$, $\triangle a_0 = 0$. We assume a_0 decays at infinity, hence $a_0 = 0$. Finally, we conclude

PROPOSITION 5.1. *The abelian monopole equations are equivalent to the linear wave equation for ϕ:*

$$\Box\phi = (\partial_t^2 - \partial_x^2 - \partial_y^2)\phi = 0,$$

with $*d\, a = \frac{\partial \phi}{\partial t}$, $d * a = 0$, and $a_0 = 0$.

We make special note of the fact that the condition of ρ-regular is not automatic even in the abelian case. It is satisfied if we can find a gauge transformation $g = e^u \in H$ such that for $\rho = e^{i\alpha}$ we have $D_1(\rho)g = D_2(\rho)g = 0$. But

$$D_1(\rho) - D_2(\rho) = i(\sin\alpha\nabla_x - \cos\alpha\nabla_y - \phi),$$

so we have $g^{-1}(D_1(\rho) - D_2(\rho))g = i(\sin\alpha\frac{\partial}{\partial x} - \cos\alpha\frac{\partial}{\partial y})$, which implies that

$$-\sin\alpha(a_1 - \frac{\partial u}{\partial x}) + \cos\alpha(a_2 - \frac{\partial u}{\partial y}) - \phi = 0. \tag{5.2}$$

Here u decays at spacial infinity. But (5.2) is the ODE

$$\frac{du}{d\tau} = a_1\sin\alpha - a_2\cos\alpha + \phi,$$

where $d\tau = \sin\alpha dx - \cos\alpha dy$. The condition that u decays at spacial infinity is given by integral conditions, so it is not automatically true. However, if we can find a decaying solution of (5.2), it is unique. Because all the variables satisfy wave equations, if we find a solution u for (5.2) at time $t = 0$, we can propagate it using the wave equation to any t. It follows that the condition of ρ-regular is a property of the initial data alone.

Our second class of examples is well-known (cf. [**20**]). It involves solutions of the monopole equation which are invariant under time translation.

PROPOSITION 5.2. *Suppose that (A, ϕ) is a solution of the monopole equation, which is fixed under time translation. Then its Lax pair is gauge equivalent to the Lax pair of a harmonic map.*

PROOF. The Lax pair (2.8) becomes

$$\left[\mu\nabla_z - \frac{1}{2}(A_t + i\phi),\ \mu^{-1}\nabla_{\bar z} - \frac{1}{2}(A_t - i\phi)\right] = 0.$$

Since the connection at $\mu = 1$ is now a full connection in $\mathbb{C}$, we fix the gauge at $\mu = 1$, so the Lax pair is now a Lax pair for the harmonic map

$$\left[\frac{\partial}{\partial z} + \frac{1-\mu^{-1}}{2}g^{-1}(A_t + i\phi)g,\ \frac{\partial}{\partial \bar z} + \frac{1-\mu}{2}g^{-1}(A_t - i\phi)g\right] = 0,$$

where $g : \mathbb{C} \to U(n)$ satisfies

$$\begin{cases} \frac{\partial g}{\partial z} g^{-1} = A_z + \frac{1}{2}(A_t + i\phi), \\ \frac{\partial g}{\partial \bar{z}} g^{-1} = A_{\bar{z}} + \frac{1}{2}(A_t - i\phi). \end{cases}$$

Note that this differs from the Euclidean monopole reduction, where the different sign produces Hitchin's self-dual equation ([**9**]) rather than the harmonic map. □

Corollary 5.3. *The Lorentz transformations of the stationary solution corresponding to harmonic maps produce families of solutions to the monopole equations.*

It can be checked that the solutions obtained in the above corollary decay in space.

Special solutions of the monopole equations also come from the reduction to $\mathbb{R}^{1,1}$, which involves the assumption that the field (A, ϕ) is independent of one of the spacial variables, say y. The solution of these equations will not directly yield solutions of the monopole equations which decay in space, but we will see that they do arise in the consideration of radially symmetric solutions.

Assume (A, ϕ) is independent of y. Then the $A_y = \psi$ becomes a second Higgs field and the Lax pair (2.8) reduces to

$$[\nabla_t - i\phi + \mu(\nabla_x + i\psi),\ \nabla_t + i\phi + \mu^{-1}(\nabla_x - i\psi)] = 0.$$

The equations become

$$\begin{cases} [\nabla_t, \nabla_x] = [\phi, \psi], \\ \nabla_t \psi = \nabla_x \phi, \\ \nabla_t \phi = \nabla_x \psi. \end{cases}$$

The usual Lax pair is obtained by restriction to characteristic coordinates. Let $\nabla_t + \nabla_x = \nabla_\xi$, $\nabla_t - \nabla_x = \nabla_\eta$, and $\phi_+ = \phi + \psi$, $\phi_- = \psi - \phi$. We get three equations of ϕ_+ and ϕ_- encoded in the Lax pair

$$[\nabla_\xi + \tau\phi_+,\ \nabla_\eta + \tau^{-1}\phi_-] = 0.$$

The wave map $\phi : \mathbb{R}^{1,1} \to G$ is obtained from this Lax pair in the same manner that the Ward map is obtained from the Lax pair for monopoles.

Finally we are interested in solutions with a radial symmetry. In gauge theory, the symmetry is inspired by requiring that under the pull-back of a space-time symmetry, the fields (A, ϕ) go to gauge transformations of themselves. For example, given $J \in su(n)$ such that $e^{2\pi J} = \mathrm{I}$, then

$$e^{i\theta} \cdot (A, \phi)(z, t) = (e^{J\theta} A(e^{i\theta} z, t) e^{-J\theta},\ e^{J\theta} \phi(e^{i\theta} z, t) e^{-J\theta})$$

is an $SO(2)$-action. If the monopole equation is invariant under this action, then the field

$$(A, \phi) = (e^{J\theta} a e^{-J\theta}, e^{J\theta} a_0 e^{-J\theta}, e^{J\theta} \phi e^{-J\theta}),$$

where a, a_0 and ϕ depend only on r and t. Both a and ϕ must vanish at $r = 0$ unless $J = 0$. We then make a singular gauge transformation by $e^{J\theta}$ to the form

$$(A, \phi) = (a_r, a_\theta - J, a_0, \phi),$$

where J now indicates a singularity at 0 and lack of suitable decay at ∞.

PROPOSITION 5.4. *Fix a representation of $S^1 \to U(n)$ given by $e^{i\theta} \mapsto e^{J\theta}$. The monopole equations for a monopole with radial symmetry induced by J are equivalent to the equations for ∇_r, ∇_t, ϕ and $\psi = \frac{a_\theta - J}{r}$:*

$$\begin{cases} [\nabla_t, \nabla_r] = [\phi, \psi], \\ [\nabla_t, \psi] = [\nabla_r, \phi], \\ [\nabla_t, \phi] = [\nabla_r, \psi] + \frac{1}{r}\,\psi. \end{cases}$$

PROOF. It is useful to notice that

$$\nabla_z = \frac{1}{2}(\nabla_x - i\nabla_y) = \frac{e^{-i\theta}}{2}(\nabla_r - \frac{i}{r}\nabla_\theta),$$

$$\nabla_{\bar{z}} = \frac{1}{2}(\nabla_x + i\nabla_y) = \frac{e^{i\theta}}{2}(\nabla_r + \frac{i}{r}\nabla_\theta).$$

Substitute these expression in the Lax pair (2.8) and compare coefficients of μ, μ^{-1} and the constant term to get

$$[\nabla_t - i\phi, \ \nabla_r + i\psi] = 0,$$

which yields two equations. The third equation come from the equation

$$[\nabla_t - i\phi, \nabla_t + i\phi] + [e^{-i\theta}(\nabla_r - \frac{i}{r}\nabla_\theta), \ e^{i\theta}(\nabla_r + \frac{i}{r}\nabla_\theta)] = 0.$$

This last equation has an extra term which spoils the integrability of the system, and prevents the three equations from being encoded as a Lax pair. □

6. Hamiltonian structures

Let (p, q) denote the standard variables for the cotangent bundle $\mathcal{M}$ of the space of rapidly decaying maps from $\mathbb{R}^2$ to $U(n)$, and $H : \mathcal{M} \to \mathbb{R}$ the functional defined by

$$H = \frac{1}{2}\int\int_{\mathbb{R}^2} ||p||^2 + ||q^{-1}dq||^2 \ dxdy.$$

We introduce a Hamiltonian formulation of the Ward equation, which does not seem to have appeared in the literature before. For each unit direction v of $\mathbb{R}^2$, (i) the symplectic form w^v is the sum of the standard symplectic form on cotangent bundle $\mathcal{M}$ and an extra closed 2-form depending on v, (ii) the Hamiltonian system of H with respect to w^v is the Ward equation with $\rho = v$.

Let Θ be the canonical 1-form on $\mathcal{M}$ defined by

$$\Theta_{(p,q)}(\delta p, q^{-1}\delta q) = p(q^{-1}\delta q).$$

The standard symplectic form on $\mathcal{M}$ is $d\Theta$. We identify the tangent and cotangent spaces in the Lie algebra formulation with the L^2 inner product

$$\langle A, B\rangle = \int_{\mathbb{R}^2} \operatorname{tr}(AB)\, dxdy.$$

Use the Cartan formula

$$d\Theta(\xi_1, \xi_2) = \xi_1(\Theta(\xi_2)) - \xi_2(\Theta(\xi_1)) - \Theta([\xi_1, \xi_2])$$

to compute $d\Theta$ to get

$$\begin{aligned}&d\Theta_{(p,q)}((\delta_1 p, q^{-1}\delta_1 q), (\delta_2 p, q^{-1}\delta_2 q))\\&= \langle \delta_1 p, q^{-1}\delta_2 q\rangle - \langle \delta_2 p, q^{-1}\delta_1 q\rangle - \langle p, [q^{-1}\delta_1 q, q^{-1}\delta_2 q]\rangle.\end{aligned}$$

A computation shows that the following 2-form is closed:

$$\tau^v_{(p,q)}((\delta_1 p, q^{-1}\delta_1 q), (\delta_2 p, q^{-1}\delta_2 q)) = \langle q^{-1}q_v, [q^{-1}\delta_1 q, q^{-1}\delta_2 q]\rangle,$$

where $q_v = dq(v) = q_x \cos\theta + q_y \sin\theta$ if $v = e^{i\theta}$. Then $w^v = d\Theta + \tau^v$ is a symplectic form on $\mathcal{M}$. In fact,

$$\begin{aligned}&w^v_{(p,q)}((\delta_1 p, q^{-1}\delta_1 q), (\delta_2 p, q^{-1}\delta_2 q))\\&= \langle \delta_1 p, q^{-1}\delta_2 q\rangle - \langle \delta_2 p, q^{-1}\delta_1 q\rangle + \langle -p + q^{-1}q_v, [q^{-1}\delta_1 q, q^{-1}\delta_2 q]\rangle.\end{aligned}$$

Since

$$dH_{(p,q)}(\delta p, \delta q) = \langle \delta p, p\rangle - \langle q^{-1}\delta q, d^*(q^{-1}dq)\rangle,$$

a direct computation shows that the Hamiltonian flow for H with respect to the symplectic structure w^v is

$$\begin{cases} q^{-1}q_t = p,\\ p_t + [-p + q^{-1}q_v, q^{-1}q_t] = d^*(q^{-1}dq).\end{cases}$$

So we have proved

PROPOSITION 6.1. *Given a unit vector $v = e^{i\theta}$ in $\mathbb{R}^2$, the 2-form w^v is a symplectic form, and the Hamiltonian system of H with respect to w^v is the Ward equation* (3.5).

In [**12**], Ioannidou and Zakrzewski considered another family of Ward equations, and studied the Lagrangian and Hamiltonian formulations for that family of equations.

7. Scattering theory

Scattering theory for the Ward equations has been treated by a number of authors, including Manakov and Zakharov [**13**], Villarroel [**21**], and Fokas and Ioannidou [**7**]. We include a brief synopsis and interpretation of the results. In particular, we construct the inverse scattering transform via loop group factorizations.

Ward's original analysis of the space-time monopole equations is via twistor theory. We recognize features of this analysis in what follows, although we do not go into the twistor formulation. Recall that

$$D_1(\mu) = \frac{1}{2}\nabla_t - \frac{i\phi}{2} + \mu\nabla_z, \quad D_2(\mu) = \frac{1}{2}\nabla_t + \frac{i\phi}{2} + \mu^{-1}\nabla_{\bar{z}}.$$

We rewrite the Lax pair as the linear system consisting of first a spacial operator

$$D_s(\mu) = D_1(\mu) - D_2(\mu) = \mu\nabla_z - \mu^{-1}\nabla_{\bar{z}} - i\phi \tag{7.1}$$

$$= \frac{(\mu - \mu^{-1})}{2}\nabla_x - i\frac{(\mu + \mu^{-1})}{2}\nabla_y - i\phi. \tag{7.2}$$

The second operator in the Lax pair we write as one of a family of time operators

$$\begin{aligned} D_t(\mu) &= D_1(\mu) + D_2(\mu) + \alpha D_s(\mu) \\ &= \nabla_t + \frac{1}{2}(\mu + \mu^{-1})\nabla_x - \frac{i}{2}(\mu - \mu^{-1})\nabla_y + \alpha D_s(\mu). \end{aligned}$$

The operators can be rescaled (i.e., multiplied by a scalar function of μ), so the points $\mu = 0$ and $\mu = \infty$ are included by changing the scaling factor (for example, if we multiply by μ to the operators, then they are defined at $\mu = 0$, and if we multiply by μ^{-1} to the operators, then they are defined at $\mu = \infty$). Note that $D_1E = D_2E = 0$ if and only if $D_sE = D_tE = 0$.

If $\mu \notin S^1$, then the spacial part of the connection, $D_s(\mu)$, can be thought of as containing a $\bar{\partial}$ operator in the complex structure on $\mathbb{R}^2$ given by the complex coordinate

$$w = \frac{\mu^{-1}z + \mu\bar{z}}{a(\mu)}, \quad \bar{w} = \frac{\bar{\mu}^{-1}\bar{z} + \bar{\mu}z}{\bar{a}(\mu)}.$$

(In fact, $D_s(\mu) = \nabla_{\bar{w}}$). The factor $a(\mu)$ does not change the complex structure. Note at $\mu = \infty$, $w = \bar{z}$, and at $\mu = 0$, $w = z$.

Suppose the Higgs field ϕ and the connection A decay at infinity. For fixed time and every $\mu \in \mathbb{C} \cup \{\infty\} \setminus S^1$ the connection $D_s(\mu)$ determines a bundle holomorphic in the complex parameter $w = w(\mu)$ on $\mathbb{R}^2 \cup \{\infty\} = S^2$. By a theorem of Grothendieck, this bundle is the sum of line bundles

$$L_\mu = L_1 \oplus L_2 \oplus \cdots \oplus L_n$$

with first Chern classes $c(1) \leq c(2) \leq \cdots \leq c(n)$. The reality condition insures that the Chern classes at $\bar{\mu}^{-1}$ are the negative of the Chern classes at μ, since the bundle $L_{\bar{\mu}^{-1}}$ is dual to the bundle L_μ at μ. We call this sequence $\vec{c}(\mu) = (c(1), \ldots, c(n))$ the *Chern vector at* μ.

THEOREM 7.1. *Suppose we have a solution of the monopole equation in a time interval* $[T_1, T_2]$. *Then for every* $\mu \in \mathbb{C} \cup \{\infty\} \setminus S^1$, *the Chern vector of the spacial holomorphic bundle is preserved under the flow in time.*

PROOF. As a warm-up, we first prove this for $\mu = 0$. Then the spacial connection is $\nabla_{\bar{z}}$ and we choose $D_1(0) = \nabla_t - i\phi$ as the evolution operator. We have $[\nabla_t - i\phi, \nabla_{\bar{z}}] = 0$. We may make a complex gauge transformation so that $D_1(0) = \nabla_t - i\phi = h^{-1} \circ \frac{\partial}{\partial t} \circ h$. Hence the gauge equivalent Lax pair gives

$$\frac{\partial}{\partial t}(h \circ \nabla_{\bar{z}} \circ h^{-1}) = 0.$$

The complex structure on the bundle determined by $\nabla_{\bar{z}}$ is carried into a structure which is a gauge equivalent one. Hence they have the same splitting, and so the same Chern vectors.

A similar computation occurs for each μ, where we make the choice of the time direction at μ to be real. For simplicity, let $\mu = e^{i\theta}\rho$ where $\rho \in \mathbb{R}^+$ and rotate variable by $z \mapsto e^{i\theta}z$ so that in this coordinate system $\mu = \rho$ is real. Choose as a suitable evolution operator

$$D_1(\rho) + \rho^2 D_2(\rho) = \frac{1+\rho^2}{2}\nabla_t + \rho\nabla_x - \frac{i(1-\rho^2)\phi}{2}.$$

The derivatives which appear are in the direction of $\frac{\partial}{\partial \tau} = \frac{1+\rho^2}{2}\frac{\partial}{\partial t} + \rho\frac{\partial}{\partial x}$. We can again make a gauge transformation so that

$$\nabla_\tau - \frac{1-\rho^2}{2}\, i\phi = h^{-1} \circ \frac{\partial}{\partial \tau} \circ h,$$
$$\frac{\partial}{\partial \tau}(h \circ D_s(\rho) \circ h^{-1}) = 0.$$

Hence the complex structure of the bundle does not changes under a flow in the τ variable. However, time translation is a translation in the τ direction followed by a translation in the x variable. Translation in S^2 by x is holomorphic and does not change the complex structure. Hence the splitting type of $L_\mu = L_\rho$ does not change under the flow in time. We conclude that the Chern vector $\vec{c}(\mu)$ is preserved. □

What do we expect? Since the space of $\bar{\partial}$ derivatives which lead to non-trivial splitting has codimension at least two, we expect that for most choices of initial data, the non-trivial splitting occurs at isolated points. At these points in $\mathbb{C} \setminus S^1$, we expect to have singularities (poles and zeros). We expect continuous scattering data to be defined as a jump across $\mu \in S^1$.

DEFINITION 7.2. A rapidly decaying spacial pair (A, ϕ) is said to have *continuous scattering data* if the frame, which solves

$$\begin{cases} D_s(\mu)E_\mu = \left(\frac{\mu-\mu^{-1}}{2}\nabla_x - \frac{i(\mu+\mu^{-1})}{2}\nabla_y - i\phi\right)E_\mu = 0, \\ E_\mu(\infty) = \mathrm{I}, \quad E_{\bar{\mu}^{-1}} = (E^*_{\mu})^{-1}, \end{cases}$$

has solutions $E^\pm_\mu$, which are holomorphic in a

$$\mathcal{O}^\pm_\epsilon = \{\mu \in \mathbb{C} \mid 1 < |\mu|^{\pm 1} < 1 + \epsilon\}$$

for some $\epsilon > 0$. Moreover, we assume that the limits

$$\lim_{\mu \in \mathcal{O}^{\pm}, \mu \to e^{i\theta}} E_\mu = S_\theta^{\pm}$$

exist. It follows from the reality condition that $S_\theta^- = (S_\theta^+)^{*^{-1}}$. We call the non-negative Hermitian matrix

$$S_\theta = (S_\theta^-)^{-1} S_\theta^+ = (S_\theta^+)^* S_\theta^+$$

the *scattering matrix*.

Let $W^{2,1}$ denote the space of maps f such that f and first partial derivatives of f are in L^1.

PROPOSITION 7.3. *Assume there is a gauge in which (A, ϕ) is rapidly decaying in spacial variables; moreover, assume (A, ϕ) is small in $W^{2,1}$. Then the Chern vector $\vec{c}(\mu) = 0$ at every $\mu \in \mathbb{C} \cup \{\infty\} \setminus S^1$. Moreover, the continuous scattering matrix S_θ exists, $\mathrm{I} - S_\theta$ decays for each θ, and the scattering matrix S_θ satisfies*

(a) $\mathrm{I} - S_\theta$ *is small in* L^∞,
(b) $S_\theta^* = S_\theta \geq 0$,
(c) $d_{s,\theta} S_\theta = (-\sin\theta \frac{\partial}{\partial x} + \cos\theta \frac{\partial}{\partial y}) S_\theta = 0$.

PROOF. The existence of E_μ away from the circle $\mu \in S^1 \subset \mathbb{C}P^1$ is a straight forward iteration argument involving estimates in L^∞ on E_μ using (A, ϕ) small in $L^1 \cap L^\infty$. We relegate the estimate as $\mu \to e^{i\theta} \in S^1$ to an appendix. Once the basic estimate are in place, proof of regularity and holomorphic dependence on μ is straightforward. We explicitly derive (a) in the appendix. To obtain (b), note that $(E_{\bar{u}^{-1}})^{-1} = E_\mu^*$ due to the reality condition on (A, ϕ). Hence $(S_\theta^-)^{-1} = (S_\theta^+)^*$.

To obtain (c), notice that

$$D_s(e^{i\theta}) = -i(d_{s,\theta} - \psi),$$

where $\psi = -\psi^* = \sin\theta A_x - \cos\theta A_y + \phi$. Since $D_s(\mu) E_\mu = 0$, it follows from the definition of S^+ that $D_s(e^{i\theta}) S^+ = 0$, so $d_{s,\theta} S^+ = \psi S^+$. But

$$d_{s,\theta}((S^+)^*) = (d_{s,\theta} S^+)^* = (\psi S^+)^* = (S^+)^* \psi^* = -(S^+)^* \psi.$$

Now compute

$$\begin{aligned} d_{s,\theta} S &= d_{s,\theta}((S^+)^* S^+) = (d_{s,\theta}(S^+)^*) S^+ + (S^+)^* d_{s,\theta} S^+ \\ &= -(S^+)^* \psi S^+ + (S^+)^* \psi S^+ = 0, \end{aligned}$$

and (c) follows. □

COROLLARY 7.4. *Assume (A, ϕ) is a smooth solution in $\mathbb{R}^2 \times (T_1, T_2)$ and decays in spacial variables, and has a smooth continuous scattering data. Then*

$$0 = \left(\frac{\partial}{\partial t} + \cos\theta \frac{\partial}{\partial x} + \sin\theta \frac{\partial}{\partial y} \right) S_\theta.$$

The corollary assumes that the scattering theory is differentiable in t. Since $D_t(\mu) = D_1(\mu) + D_2(\mu) = \nabla_t + \mu\nabla_z + \mu^{-1}\nabla_{\bar{z}}$ has the property that $[D_s(\mu), D_t(\mu)] = 0$, we conclude that $D_t(\mu)E_\mu = 0$. Now the result follows by the same method as (c) in the proposition.

COROLLARY 7.5. *If (A, ϕ) and $(\tilde{A}, \tilde{\phi})$ are gauge equivalent, then they have the same scattering data.*

PROOF. Suppose $(\tilde{A}, \tilde{\phi})$ is the gauge transformation of (A, ϕ) by a unitary map u. If $E_\mu(x, y, t)$ is the frame for (A, ϕ), then $u(x, y, t)E_\mu(x, y, t)$ is the frame for $(\tilde{A}, \tilde{\phi})$. Hence the limits $\tilde{S}_\theta^+ = uS_\theta^+$ and $\tilde{S}_\theta^- = uS_\theta^-$. But u is unitary, so $\tilde{S} = S$. □

Inverse scattering theory is simpler than scattering theory. To dispose the gauge ambiguity, we need to make a choice somewhere. We choose to do this at $\mu = 1$. First we note that if we have initial scattering data $S_\theta(x, y)$ satisfying $d_{s,\theta}S_\theta(x, y) = 0$, then the scattering data $S_\theta(x, y, t)$ for the solution at time t should satisfy

$$\frac{\partial S_\theta}{\partial t} + \cos\theta\frac{\partial S_\theta}{\partial x} + \sin\theta\frac{\partial S_\theta}{\partial y} = 0.$$

Since $d_{s,\theta}S_\theta(x, y) = 0$, there exists s such that

$$S_\theta(x, y) = s(x\cos\theta + y\sin\theta, \theta).$$

The linear evolution equation for $S_\theta(x, y, t)$ implies that

$$S_\theta(x, y, t) = s(x\cos\theta + y\sin\theta - t, \theta).$$

The problem is now to write

$$S_\mu(x, y, t) = E_\mu^-(x, y, t))^{-1}E_\mu^+(x, y, t), \quad \mu \in S^1,$$

where E_μ^- extends holomorphically to μ inside the unit circle, E_μ^+ extends holomorphically outside the unit circle, and $E_\mu^- = ((E_{\bar{\mu}^{-1}}^+)^*)^{-1}$. We can always do this, and the ambiguity corresponds to the gauge transformations. The condition that $S_1(x, y, t) = \mathrm{I}$ is equivalent to the solution being 1-regular. However, we choose a method of factoring which yields a unique solution for all scattering data. The inverse scattering was given in [**21**] by Villarroel and in [**7**] by Fokas-Ioannidou. We prove it using the Iwasawa loop group factorization theorem of Pressley and Segal [**14**]. Since S_θ is Hermitian symmetric and non-negative, there is a Hermitian symmetric matrix P_μ such that $S_\theta = P_\mu^2$, where $\mu = e^{i\theta}$. The Iwasawa loop group factorization of Pressley and Segal [**14**] says that we can factor

$$P_\mu = U_\mu E_\mu^+$$

uniquely such that U_μ is in $U(n)$, $U_1 = \mathrm{I}$, and E_μ^+ extends holomorphically to outside the unit circle. Set $E_\mu^- = ((E_{\bar{\mu}^{-1}}^+)^*)^{-1}$. Now $S_\theta = (E_\mu^-)^{-1}E_\mu^+$ when $\mu = e^{i\theta}$. Since S_μ is smooth, $E_\mu^\pm$ is smooth.

THEOREM 7.6 (Villarroel [**21**], Fokas-Ioannidou [**7**]). *Let $S_\theta(x,y,t) = s(x\cos\theta + y\sin\theta - t, \theta)$. Then the factorization described above*

$$S_\theta = (E_\mu^-)^{-1}(x,y,t)E_\mu^+(x,y,t),$$

where $\mu = e^{i\theta}$ and $E_\mu^\pm$ can extend holomorphically to $|\mu|^{\pm 1} > 1$ yields smooth frame for a solution of the space-time monopole equation.

PROOF. We need to show that $E_\mu^\pm(x,y,t)$ generates a solution to the monopole equation. To do this, note by construction that $E_\mu^\pm$ is holomorphic in $|\mu|^{\pm 1} > 1$. The operator $d_{s,\theta}$ is a directional derivative, $d_{s,\theta}S = 0$ and $S = (S^-)^{-1}S^+$, so

$$0 = d_{s,\theta}S = (d_{s,\theta}((S^-)^{-1}))S^+ + (S^-)^{-1}d_{s,\theta}S^+.$$

Thus we have

$$(d_{s,\theta}S^+)(S^+)^{-1} = -S^- d_{s,\theta}((S^-)^{-1}) = (d_{s,\theta}S^-)(S^-)^{-1}.$$

Note that

$$d_{s,\mu} = \frac{i(\mu - \mu^{-1})}{2}\frac{\partial}{\partial x} + \frac{\mu + \mu^{-1}}{2}\frac{\partial}{\partial y}$$

is the meromorphic extension of $d_{s,\theta} = d_{s,e^{i\theta}}$. So by meromorphic extension, we obtain the identity on $0 < |\mu| < \infty$

$$(d_{s,\mu}E_\mu^+)(E_\mu^+)^{-1} = (d_{s,\mu}E_\mu^-)E_\mu^{-1}.$$

Since the left hand side is meromorphic in $|\mu| > 1$ with a simple pole at ∞ and the right hand side is meromorphic on $|\mu| < 1$ with a simple pole at 0, both sides are meromorphic in $\mathbb{C}$ with simple poles at 0 and ∞. Let $E_\mu = E_\mu^\pm$ when $|\mu|^\pm > 1$. Then $E_{\bar\mu^{-1}} = (E_\mu^*)^{-1}$. It follows that

$$\xi_\mu := (d_{s,\mu}E_\mu)E_\mu^{-1} = \mu C_1 + \mu^{-1}C_{-1} + C_0$$

for some $C_i(x,y,t)$. But $d^*_{s,\bar\mu^{-1}} = d_{s,\mu}$ and $E_{\bar\mu^{-1}} = (E_\mu^*)^{-1}$, so $\xi^*_{\bar\mu^{-1}} = -\xi_\mu$. Hence $C_{-1} = -C_1^*$ and $C_0^* = -C_0$. Write $C_1 = A_z$, $C_{-1} = A_{\bar z}$ and $C_0 = \phi$; then we have

$$(d_{s,\mu}E_\mu)E_\mu^{-1} = \mu A_z - \mu^{-1}A_{\bar z} + \phi,$$

where $A_z = -(A_{\bar z})^*$ and $\phi^* = -\phi$. In other words, we have proved $(D_1(\mu) - D_2(\mu))E_\mu = 0$.

The proof of the evolution equation, i.e., $D_t(\mu)E_\mu = 0$, is similar and we do not carry it out here. □

Of course, if we start with an initial condition which has only continuous scattering data, we will not necessarily obtain the same initial data by the inverse scattering transform, but obtain a gauge equivalent solution.

COROLLARY 7.7. *Let (A,ϕ) be a solution of the space-time monopole equation rapidly decaying in the spacial variables. Assume in addition that the Chern vector $\vec{c}(\mu) = 0$ for all $\mu \in \mathbb{C} \setminus S^1$. Assume also that the solution has continuous scattering data for all t. Then the solution obtained* *from the scattering data at $t = 0$ agrees with the given solution up to gauge transformation.*

PROOF. We know from the assumption that $S_\theta(x, y, t) = s(x\cos\theta + y\sin\theta - t, \theta)$, since S_θ is unique,

$$S_\theta = \lim_{\mu\to e^{i\theta}, |\mu|>1} (E_{\bar{\mu}^{-1}})^* E_\mu.$$

Hence the frame provides a factorization. This factorization is unique up to a unitary matrix $u = u(x, y, t)$. This unitary matrix gives a gauge transformation between the original solution and the one constructed by inverse scattering. □

8. 1-soliton monopoles

In addition to continuous scattering data, solutions of monopole equation may also have discrete scattering data. We first construct monopoles whose frames have one simple pole, and in later sections we construct frames with multiple poles and show how to combine them with continuous scattering data.

The building blocks of the discrete scattering data are one-solitons, which are easy to describe. We have the harmonic maps $\phi : \mathbb{R}^2 \cup \{\infty\} \to SU(n)$, which yield time-independent solutions to the Ward equation. Among these, we have one-unitons, which come from holomorphic maps into Grassmannians. We also have the Lorentz transformations of these stationary one unitons. This family makes up the one-solitions. It is somewhat more difficult to show that every one-soliton, defined in terms of a single pole for the frame, is of this type.

We need to use another gauge equivalent Lax pair to construct soliton solutions. If a monopole (A, ϕ) is $\lambda = 0$ regular, then we can fix the gauge of (2.9) at $\lambda = 0$ to get

$$[\lambda(\partial_\xi - \tilde{A}_\xi) - \partial_y,\ \lambda(\partial_y + 2\tilde{\phi}) - \partial_\eta] = 0,$$

where $\tilde{\phi} = -\tilde{A}_y$, and $\tilde{A}_\eta = 0$. The above Lax pair is equivalent to

$$[\tau\partial_y - \partial_\xi + \tilde{A}_\xi),\ \tau\partial_\eta - \partial_y - 2\tilde{\phi}] = 0. \tag{8.1}$$

(Here $\tau = \lambda^{-1}$.)

The spectral parameters μ, λ, τ in Lax pairs (2.8), (2.9), and (8.1) are related by

$$\mu = \frac{\lambda - i}{\lambda + i}, \quad \tau = \lambda^{-1}, \quad \text{so } \tau = \lambda^{-1} = \frac{i(\mu - 1)}{\mu + 1}.$$

The above discussion gives the following proposition (cf. [**22**]):

PROPOSITION 8.1. *Suppose there is a smooth $GL(n, \mathbb{C})$-valued $\psi(x, y, t, \tau)$ defined for $(x, y, t) \in \mathbb{R}^{2,1}$ and τ in an open subset of $\mathbb{C}$ such that*

(1) *$P := (\tau\partial_y\psi - \partial_\xi\psi)\psi^{-1}$ and $Q := (\tau\partial_\eta\psi - \partial_y\psi)\psi^{-1}$ are independent of τ,*

(2) $\psi(x,y,t,\bar\tau)^*\psi(x,y,t,\tau)=\mathrm{I}$.

Let A be a connection and ϕ a Higgs field defined by $A_\xi=-P$, $A_\eta=0$, and $A_y=-\phi=-Q/2$. Then (A,ϕ) is a solution of the monopole equation. Conversely, every solution of the monopole equation that is regular at $\lambda=0$ is gauge equivalent to a solution of this type.

DEFINITION 8.2. A map ψ that satisfies (1) and (2) of Proposition 8.1 is called a *Ward frame* if P,Q decay in spacial infinity and $\psi(x,y,t,\infty)=\mathrm{I}$. A Ward frame is a *Ward soliton frame* if ψ is rational in τ.

DEFINITION 8.3. A solution (A,ϕ) of the monopole equation is called a *k-soliton* if it is regular at $\mu=-1$ and has a monopole frame E that is rational in μ with k poles counted with multiplicity.

If $f:S^2\to GL(n,\mathbb{C})$ is rational with one simple pole at $\tau=\alpha$, then it can be checked that (cf. [**19**]) f must be of the form

$$g_{\alpha,\pi}(\tau)=\mathrm{I}+\frac{\alpha-\bar\alpha}{\tau-\alpha}\,\pi^\perp,$$

where $\pi^\perp=\mathrm{I}-\pi$ and π is a Hermitian projection π of $\mathbb{C}^n$. We identify the space of rank m Hermitian projections of $\mathbb{C}^n$ as $\mathrm{Gr}\,(m,\mathbb{C}^n)$ via the map $\pi\mapsto\mathrm{Im}(\pi)$. So a Ward 1-soliton frame must be of the form $g_{\alpha,\pi(x,y,t)}(\tau)=\mathrm{I}+\frac{\alpha-\bar\alpha}{\tau-\alpha}\,\pi^\perp(x,y,t)$ for some constant $\alpha\in\mathbb{C}\setminus\mathbb{R}$ and $\pi:S^2\times\mathbb{R}\to\mathrm{Gr}\,(k,\mathbb{C}^n)$. Note that $(\tau\partial_y\psi-\partial_\xi\psi)\psi^{-1}$ and $(\tau\partial_\eta\psi-\partial_y\psi)\psi^{-1}$ are independent of τ if and only if the residue at $\tau=\alpha$ is zero. This implies

PROPOSITION 8.4. *Given $\alpha\in\mathbb{C}\setminus\mathbb{R}$ a constant and $\pi:\mathbb{R}^{2,1}\to\mathrm{Gr}\,(k,\mathbb{C}^n)$ a smooth map, then $g_{\alpha,\pi}(\tau)=\mathrm{I}+\frac{\alpha-\bar\alpha}{\tau-\alpha}\,\pi^\perp$ is a Ward soliton frame if and only if*

$$\begin{cases}(\alpha\partial_y\pi-\partial_\xi\pi)\pi=0,\\ (\alpha\partial_\eta\pi-\partial_y\pi)\pi=0.\end{cases}\tag{8.2}$$

Moreover, if π is a solution of (8.2), then there exists a holomorphic map $\pi_0:S^2\to\mathrm{Gr}\,(k,\mathbb{C}^n)$ such that $\pi(x,y,t)=\pi_0(y+\alpha\xi+\alpha^{-1}\eta)$.

Note that if $\alpha=\pm i$, then $x+\alpha\xi+\alpha^{-1}\eta=y\pm ix$ and $\pi(x,y,t)=\pi_0(y\pm ix)$. So the 1-soliton $g_{\pm i,\pi}$ is a 1-uniton harmonic map.

The $SO(2,1)$-actions described in Section 4 of 1-unitons are 1-soliton monopoles. In fact, we have

PROPOSITION 8.5. *Given $\alpha=re^{i\theta}\in\mathbb{C}\setminus\mathbb{R}$, let $e^s=r$, $e^c=\csc\theta+\cot\theta=\cot(\theta/2)$, and $h=T(c)R(-\pi/2)T(s)\in SO(2,1)$, where $T(c)$ and $R(\theta)$ are 1-parameter subgroups of $SO(2,1)$ defined in Section 4. Let $\pi_0:S^2\to\mathrm{Gr}\,(k,\mathbb{C}^n)$ be a holomorphic map. Then the action of h on 1-uniton frame g_{i,π_0} gives rise to a monopole solution that is gauge equivalent to the 1-soliton given by g_{α,π_0}. In other words, all 1-solitons monopoles are obtained from the action of $SO(2,1)$ on 1-unitons up to gauge equivalence.*

PROOF. Recall that $\lambda = \tau^{-1}$, and

$$F(x,y,t,\lambda) = g_{i,\pi_0(y+ix)}(\lambda^{-1})$$

is a solution for the linear system (2.11) (Lax pair in λ). Let $(\tilde{x},\tilde{y},\tilde{t})^t = h(x,y,t)^t$. A computation gives

$$\tilde{y} + i\tilde{x} = \frac{i}{\sin\theta}\,(y + \alpha\xi + \alpha^{-1}\eta).$$

Let $\tilde{\lambda} = h * \lambda$. Then $\tilde{\lambda} = \frac{e^c(e^s\lambda - 1)}{e^s\lambda+1}$, so the pole of this expression is when $\tilde{\lambda} = -i$, i.e, when $\frac{e^c(e^s\lambda-1)}{e^s\lambda+1} = -i$. But $r = e^s$ and $e^c = \cot(\theta/2)$, so the pole is at $\lambda = \alpha^{-1}$. This shows that $h \cdot F$ has one simple pole at $\lambda = \alpha^{-1}$. Note that $h \cdot F$ is equal to

$$\begin{aligned}(h \cdot F)(x,y,t,\lambda) &= g_{i,\pi_0(\tilde{y}+i\tilde{x})}((\sigma(h) * \lambda)^{-1}) \\ &= \pi_0(\tilde{y}+i\tilde{x}) - e^{i\theta}\frac{\lambda^{-1}-\bar{\alpha}}{\lambda^{-1}-\alpha}\pi_0^{\perp}(\tilde{y}+i\tilde{x}) \\ &= (\pi(x,y,t) - e^{i\theta}\pi^{\perp}(x,y,t))g_{\alpha,\pi(x,y,t)}(\tau),\end{aligned}$$

where $\pi(x,y,t) = \pi_1(y+\alpha\xi+\alpha^{-1}\eta)$ and $\pi_1(z) = \pi_0(iz/\sin\theta)$ is holomorphic. So $h \cdot F$ is gauge equivalent to the 1-soliton $g_{\alpha,\pi}$. Note since $\pi_0 : S^2 \to \mathrm{Gr}\,(k,\mathbb{C}^n)$ is smooth and $h \in SO(2,1)$, the monopole given by $g_{\alpha,\pi}$ decays at spacial variables. □

9. Bäcklund transformations and construction of soliton monopoles

Multisolitons with simple poles were constructed by Ward [**22**]. Ward, Ioannidou, and Anand ([**24, 10, 2**]) derived methods for computing solitons which have poles with higher multiplicities. These multisolitons have dramatic physical properties. We give here a brief description of a method of "superposing" solitons, which is closely related to the permutability formula for Bäcklund transformations. This technique allowed Dai and Terng [**6**] to construct solitons with an arbitrary number of poles with arbitrary multiplicities.

Intuitively, the permutability formula is based on factoring frames. Given the frames of two solutions ψ_1 and ψ_2 with singularities at different sets S_1 and S_2, $S_1 \cap S_2 = \emptyset$, in $\mathbb{C} \cup \{\infty\}$, we write

$$\psi_3 = \tilde{\psi}_1\psi_2 = \tilde{\psi}_2\psi_1,$$

(i.e., factor $\psi_1\psi_2^{-1} = \tilde{\psi}_2^{-1}\tilde{\psi}_1$). Here ψ_3 has the singularities at $S_1 \cup S_2$, and ψ_j and $\tilde{\psi}_j$ have the same singular set S_j. It is not difficult to see that ψ_3 is a frame for a solution when ψ_1 and ψ_2 are. The details of allowing limiting case where $S_1 \to S_2$ yield the interesting but complex solitons. The converse of factoring solutions is also true, but not completely straightforward. We now go to the details.

THEOREM 9.1 (Algebraic Bäcklund transformation). *Let $\psi(x,y,t,\tau)$ be a Ward frame with $P = (\tau\partial_y\psi - \partial_\xi\psi)\psi^{-1}$ and $Q = (\tau\partial_\eta\psi - \partial_y\psi)\psi^{-1}$, and $g_{\alpha,\pi}$ a 1-soliton Ward frame. Suppose ψ is holomorphic and non-degenerate at $\tau = \alpha$. Let $\tilde{\pi}(x,y,t)$ denote the Hermitian projection of $\mathbb{C}^n$ onto $\psi(x,y,t,\alpha)(\mathrm{Im}(\pi(x,y,t)))$. Then*

(1) *$\tilde{\psi}(x,y,t,\tau) = g_{\alpha,\tilde{\pi}(x,y,t)}(\tau)\psi(x,y,t,\tau)g_{\alpha,\pi(x,y,t)}(\tau)^{-1}$ is holomorphic and non-degenerate at $\tau = \alpha, \bar{\alpha}$,*

(2) *$\psi_1 = g_{\alpha,\tilde{\pi}}\psi = \tilde{\psi}g_{\alpha,\pi}$ is again a Ward frame such that*

$$\begin{cases}(\tau\partial_y\psi_1 - \partial_\xi\psi_1)\psi_1^{-1} = \tilde{P},\\ (\tau\partial_\eta\psi_1 - \partial_y\psi_1)\psi_1^{-1} = \tilde{Q},\end{cases}$$

where $\tilde{P} = P + (\bar{\alpha} - \alpha)\partial_y\tilde{\pi}$ and $\tilde{Q} = Q + (\bar{\alpha} - \alpha)\partial_\eta\tilde{\pi})$.

We will use $g_{\alpha,\pi} * \psi$ to denote ψ_1.

PROOF. We give a sketch the proof. Statement (1) can be proved by computing the residue of $\tilde{\psi}$ at $\tau = \alpha$ and show that it is zero. We use Proposition 8.1 to prove (2). Set $\tilde{g} = g_{\alpha,\tilde{\pi}}$. Since $D = (\tau\partial_y - \partial_\xi)$ is a derivation, we have

$$(D\psi_1)\psi_1^{-1} = (D\tilde{g})\tilde{g}^{-1} + \tilde{g}(D\psi)\psi^{-1}\tilde{g}^{-1} = (D\tilde{g})\tilde{g}^{-1} + \tilde{g}P\tilde{g}^{-1},$$

so it is holomorphic for $\tau \in \mathbb{C}\setminus\{\alpha\}$ and has a simple pole at $\tau = \alpha$. But ψ_1 is also equal to $\tilde{\psi}g$ (here $g = g_{\alpha,\pi}$ is a 1-soliton Ward frame), so

$$(D\psi_1)\psi_1^{-1} = (D\tilde{\psi})\tilde{\psi}^{-1} + \tilde{\psi}(Dg)g^{-1}\tilde{\psi}^{-1}.$$

But $(Dg)g^{-1}$ is independent of τ and $\tilde{\psi}$ is holomorphic and non-degenerate at $\tau = \alpha, \bar{\alpha}$, hence the RHS is holomorphic at $\tau = \alpha$. So $(D\psi_1)\psi_1^{-1}$ is holomorphic in $\mathbb{C}$. But the residue of $(D\psi_1)\psi_1^{-1}$ at $\tau = \infty$ is also zero. Hence it must be independent of τ. Similar argument implies that $(\tau\partial_\eta\psi - \partial_x\psi)\psi^{-1}$ is also independent of τ, so by Proposition 8.1, ψ_1 is a Ward frame.

Set $\tilde{P} = (D\psi_1)\psi_1^{-1} = (D\tilde{g})\tilde{g}^{-1} + \tilde{g}P\tilde{g}^{-1}$. Evaluate the residue at $\tau = \infty$ to get $\tilde{P} = \tilde{P} + (\bar{\alpha} - \alpha)\partial_y\tilde{\pi}$. Similarly, we get the formula for $\tilde{Q}$.

Since P, Q decay at spacial infinity, $\mathrm{I} - \psi(\ldots, \alpha)$ decays at spacial infinity. But $g_{\alpha,\pi}$ is a 1-soliton monopole frame, so $\partial_x\pi, \partial_\eta\pi$ also decay at spacial infinity. Hence $\tilde{P}, \tilde{Q}$ decays at spacial infinity. □

k-soliton monopole frames with only simple poles

Let $\alpha_1, \ldots, \alpha_k$ be distinct complex numbers and $\mathrm{Im}(\alpha_j) > 0$ for all $1 \le j \le k$, $\pi_j^0 : S^2 \to \mathrm{Gr}(k_j, \mathbb{C}^n)$ holomorphic maps, and $\pi_j(x,y,t) = \pi_j^0(y + \alpha_j\xi + \alpha_j^{-1}\eta)$. Then g_{α_j,π_j} is a 1-soliton Ward frame. Apply the Algebraic BT (Theorem 9.1) repeatedly as follows: Set $\psi_1 = g_{\alpha_1,\pi_1}$, and define ψ_j inductively by $\psi_j = g_{\alpha_j,\pi_j} * \psi_{j-1}$ for $2 \le j \le k$. Then ψ_k is a k-soliton Ward frame with k simple distinct poles at $\alpha_1, \ldots, \alpha_k$. These are the same soliton Ward frames constructed by Ward using the solution to the Riemann-Hilbert problem.

k-soliton monopole frames with pole data (α, k)

Ward's limiting construction is as follows: Let f_0, f_1 be rational maps from $\mathbb{C}$ to $\mathbb{C}^2$, and $\pi_{1,\epsilon}$ and $\pi_{2,\epsilon}$ the projections of $\mathbb{C}^2$ onto the complex line spanned by $f_0(w_{i+\epsilon})+\epsilon f_1(w_{i+\epsilon})$ and $f_0(w_{i-\epsilon})-\epsilon f_1(w_{i-\epsilon})$ respectively, where $w_{i\pm\epsilon} = y + (i\pm\epsilon)\xi + (i\pm\epsilon)^{-1}\eta$. Ward showed that

$$\psi = \lim_{\epsilon\to 0} g_{i-\epsilon,\pi_{2,\epsilon}} * g_{i+\epsilon,\pi_{1,\epsilon}}$$

is a 2-soliton Ward frame with a double pole at $\tau = i$ and is not stationary.

Since the algebraic BT is easy to compute, Ward's limiting method can be calculated systematically as follows (for detail see the paper by Dai and Terng [**6**]): Let $\alpha_\epsilon = \alpha + \epsilon$, $a_j : \mathbb{C} \to \mathbb{C}^n$ be rational maps, and $f_{j,\epsilon} = a_0 + a_1\epsilon + \ldots + a_{j-1}\epsilon^{j-1}$. Let $\pi_{j,\epsilon}(x, y, t)$ denote the Hermitian projection of $\mathbb{C}^n$ onto the complex line spanned by $f_{j,\epsilon}(y + \alpha_\epsilon\xi + \alpha_\epsilon^{-1}\eta)$. Set $\psi_1 = \lim_{\epsilon\to 0} g_{\alpha_\epsilon,\pi_{1,\epsilon}}$, which is a 1-soliton frame $g_{\alpha,\pi}$ (here π is the projection onto $\mathbb{C}a_0(y + \alpha\xi + \alpha^{-1}\eta)$). Define ψ_k inductively by $\psi_k = \lim_{\epsilon\to 0} g_{\alpha_\epsilon,\pi_{k,\epsilon}} * \psi_{k-1}$. Then ψ_k is a k-soliton frame with pole data (α, k) (i.e., ψ_k has a single pole at $\tau = \alpha$ with multiplicity k). Note that ψ_k depend on k holomorphic maps from $\mathbb{C}P^1$ to $\cup_{i=1}^{n-1}\mathrm{Gr}\,(i, \mathbb{C}^n)$.

Soliton frames with arbitrary pole data

To get Ward soliton frames with arbitrary pole data, we need a more general BT for adding a k-soliton with pole data (α, k) to an existing Ward frame (cf. [**6**]):

THEOREM 9.2 (Adding a k-soliton with pole data (α, k)). *Suppose ψ is a Ward frame that is holomorphic and non-degenerate at $\tau = \alpha, \bar\alpha$, and ϕ a k-soliton monopole frame with pole data (α, k). Then there exist unique $\tilde\phi$ and $\tilde\psi$ such that $\tilde\phi\psi = \tilde\psi\phi$, $\tilde\phi$ has pole data (α, k), and $\tilde\psi$ is holomorphic and non-degenerate at $\tau = \alpha, \bar\alpha$. Moreover, $\hat\psi = \tilde\phi\psi = \tilde\psi\phi$ is again a Ward frame and $\tilde\phi$ and $\tilde\psi$ are constructed algebraically.*

As a consequence, we see that the two BTs and the limiting method give rise to Ward soliton frames with arbitrary pole data. The following theorem was proved in [**6**].

THEOREM 9.3 ([**6**]). *Algebraic BTs, adding k-soliton BTs, and the limiting method produce all soliton monopoles up to gauge equivalence.*

If $\pi_0 : S^2 \to \mathrm{Gr}\,(m, \mathbb{C}^n)$ is holomorphic, then the limit of the 1-soliton frame,

$$\lim_{||(x,y)||\to\infty} \mathrm{I} + \frac{\alpha-\bar\alpha}{\tau-\alpha}\,\pi_0^\perp(y + \alpha\xi + \alpha^{-1}\eta) = \mathrm{I} + \frac{\alpha-\bar\alpha}{\tau-\alpha}\,\pi_0^\perp(\infty)$$

exists as (x, y) tends to infinity and is independent of t. In other words, the Ward soliton frame tends to a fixed rational map $h(\tau)$ at spacial infinity and is independent of time. It can be checked easily that this property is preserved under the Algebraic BT and limiting method. Hence we have

PROPOSITION 9.4. *If ψ is a Ward soliton frame, then* $\lim_{\|(x,y)\|\to\infty} \psi(x, y, t, \tau)$ *exists and is independent of* t.

10. Monopoles with both continuous and discrete scattering data

The Lax pair (2.9) of the monopole equation is equivalent to

$$[\tau(\nabla_y - \phi) - \nabla_\xi,\ \tau\nabla_\eta - \nabla_y - \phi] = 0. \tag{10.1}$$

The linear system associated to this Lax pair is

$$\begin{cases} (\tau\partial_y - \partial_\xi)\psi = (\tau A_y + \tau\phi - A_\xi)\psi, \\ (\tau\partial_\eta - \partial_y)\psi = (\tau A_\eta + \phi - A_y)\psi. \end{cases} \tag{10.2}$$

The Algebraic BT theorem for the monopole equation can be proved the same way as for the Ward equation. We only state the result:

THEOREM 10.1 (Algebraic BT for Monopoles). *Suppose $\alpha \in \mathbb{C}\setminus\mathbb{R}$ is a constant, and ψ is a frame of the monopole solution (A, ϕ) (i.e., solution of (10.2)), and $\psi(x, y, t, \tau)$ is holomorphic and non-degenerate at $\tau = \alpha$. Let $g_{\alpha,\pi}$ be a 1-soliton Ward frame, $\tilde\pi(x, y, t)$ the Hermitian projection onto $\psi(x, y, t, \alpha)(\mathrm{Im}\pi(x, y, t))$, and*

$$\tilde\psi = g_{\alpha,\tilde\pi}\psi g_{\alpha,\pi}^{-1}.$$

Then

(1) *$\tilde\psi$ is holomorphic and non-degenerate at $\tau = \alpha$,*
(2) *$\psi_1 = g_{\alpha,\tilde\pi}\psi = \tilde\psi g_{\alpha,\pi}$ is a frame for (10.1) with $\tilde A, \tilde\phi$ given by*

$$\begin{cases} \tilde A_\eta = A_\eta, \\ \tilde A_\xi = (1 - \frac{\bar\alpha}{\alpha})(\partial_\xi\tilde\pi)h + h^{-1}A_\xi h, \\ \tilde A_y + \tilde\phi = A_y + \phi, \\ \tilde A_y - \tilde\phi = (1 - \frac{\bar\alpha}{\alpha})(\partial_y\tilde\pi)h + h^{-1}(A_y - \phi)h, \end{cases}$$

where $h = \tilde\pi + \frac{\alpha}{\bar\alpha}\,\tilde\pi^\perp$.

Theorem 9.2 works for normalized monopole frames too.

Suppose $k : \mathbb{C} \to GL(n, \mathbb{C})$ is meromorphic, $k(\infty) = \mathrm{I}$, and $k(\bar\tau)^* k(\tau) = \mathrm{I}$. Then

$$\tilde k(x, y, t, \tau) = k(y + \tau\xi + \tau^{-1}\eta)$$

satisfies

$$(\tau\partial_y - \partial_\xi)\tilde k = 0, \quad (\tau\partial_\eta - \partial_y)\tilde k = 0.$$

So if ψ is a solution of (10.2) for the monopole solution (A, ϕ), then so is $\psi\tilde k$. However, if $\lim_{\|(x,y)\|\to\infty} \tilde k(x, y, t, \tau)$ exists and is independent of t, then k must be the constant map I. Hence we can use this condition to normalize frames to get a unique one:

DEFINITION 10.2. A solution ψ of (10.2) for the monopole (A, ϕ) is called the *normalized monopole frame* if

(1) $\psi(x,y,t,\bar{\tau})^*\psi(x,y,t,\tau) = \mathrm{I}$,
(2) there exists a map $h(\tau)$ such that $\lim_{||(x,y)||\to\infty}\psi(x,y,t,\tau) = h(\tau)$ exists and is independent of t.

By Proposition 9.4, a Ward soliton frame is a normalized monopole frame. By the Inverse scattering Theorem 7.6, given a smooth map $s : \mathbb{R}\times S^1 \to GL(n,\mathbb{C})$ such that $\mathrm{I} - s(\cdot, e^{i\theta})$ decays for each θ and $s^* = s \geq 0$, then there exists a solution $E(x,y,t,\mu)$ of the linear system (2.10) such that

$$((E^-)^{-1}E^+)(x,y,0,e^{i\theta}) = s(x\cos\theta + y\sin\theta, e^{i\theta})$$

$E(x,y,t,\mu) \to \mathrm{I}$ as $||(x,y)|| \to \infty$, and E is holomorphic in $|\mu| \neq 1$. Such E is a normalized monopole frame with only continuous scattering data.

If we apply Algebraic BTs and General Algebraic BTs repeatedly to a normalized monopole frame with only continuous scattering data, then we obtain normalized monopoles frames with both continuous and discrete scattering data. So we get

THEOREM 10.3. *Let $s : \mathbb{R}\times S^1 \to GL(n,\mathbb{C})$ be a smooth map such that $\mathrm{I} - s(\cdot, e^{i\theta})$ decays for each θ and $s^* = s \geq 0$, and ϕ_j a normalized soliton monopole frame with pole data (α_j, n_j) for $j = 1,\ldots,k$. Then there is a unique normalized monopole frame $E(x,y,t,\mu)$ such that*

(1) *E is holomorphic for $\mu \in \mathbb{C}\setminus(S^1\cup\{\alpha_1,\ldots,\alpha_k\})$, has poles at α_j with multiplicity n_j, and*

$$E_\pm(x,y,t,e^{i\theta}) = \lim_{|\mu|^{\pm 1}<1, \mu\to e^{i\theta}} E(x,y,t,\mu)$$

exist and are smooth,

(2) *$(E_-^{-1}E_+)(x,y,t,e^{i\theta}) = s(x\cos\theta + y\sin\theta - t, e^{i\theta})$ is the continuous scattering data of M,*
(3) *$E\phi_j^{-1}$ is holomorphic and non-degenerate at $\mu = \alpha_j$ for $1 \leq j \leq k$.*

Below we want to prove that all normalized monopole frames with only finitely many poles and a jump across S^1 are constructed by the above method.

First we need to recall a factorization result. Let $\mathcal{D}$ denote the group of $f : S^2 = \mathbb{C}\cup\{\infty\} \to GL(n,\mathbb{C})$ that satisfies the following conditions:

(1) $f(\bar{\mu}^{-1})^* f(\mu) = \mathrm{I}$,
(2) $f_\pm(e^{i\theta}) := \lim_{|\mu|^\pm<1,\mu\to e^{i\theta}} f(\mu)$ exist and are smooth,
(3) f is holomorphic in $S^2\setminus S^1$ except with possible finitely many poles.

The following results were proved in [**18**]:

(a) Given $f \in \mathcal{D}$, then there exist uniquely $h_i, g_i \in \mathcal{D}$ such that $f = h_1 g_1 = g_2 h_2$, f_i is holomorphic in $S^2\setminus S^1$, g_i are rational maps, and $g_i(-1) = \mathrm{I}$ for $i = 1,2$.
(b) If $f \in \mathcal{D}$ has a simple pole at α, then there is a unique projection π such that $f g_{\alpha,\pi}^{-1}$ is holomorphic and non-degenerate at α.

Note that if E is a normalized monopole frame then $E(x, y, t, \cdot) \in \mathcal{D}$. Moreover,

(i) E is a soliton frame if and only if E is rational in μ,
(ii) E has only continuous scattering data if and only if $E(x, y, t, \cdot)$ is holomorphic in $S^2 \setminus S^1$.

For general normalized monopole frames with both poles and jumps across S^1, we have

THEOREM 10.4 (Subtracting 1-soliton). *Suppose ψ is a normalized monopole frame, and ψ has a simple pole at $\tau = \alpha$ (may have other singularities as well). Then there exist unique $\tilde{\psi}, \psi_1$ and smooth $\pi, \tilde{\pi} : \mathbb{R}^{2,1} \to \mathrm{Gr}(k, \mathbb{C}^n)$ such that*

(i) $\psi = \tilde{\psi} g_{\alpha,\pi} = g_{\alpha,\tilde{\pi}} \psi_1$,
(ii) *$\tilde{\psi}$ and ψ_1 are holomorphic and non-degenerate at $\tau = \alpha, \bar{\alpha}$,*
(iii) *$g_{\alpha,\pi}$ is a normalized 1-soliton monopole frame, and ψ_1 is a normalized monopole frame.*

PROOF. Statements (i) and (ii) follow from known results stated just before the theorem. To prove (iii) we use residue calculus. Set $D = \tau\partial_y - \partial_\xi$, and let $g := g_{\alpha,\pi}$, $\tilde{g} := g_{\alpha,\tilde{\pi}}$. Then

$$(D\psi)\psi^{-1} = (D\tilde{\psi})\tilde{\psi}^{-1} + \tilde{\psi}(Dg)g^{-1}\tilde{\psi}.$$

Since the LHS is independent of τ, the residue of the RHS at $\tau = \alpha$ must be zero, so

$$-\tilde{\psi}(\ldots, \alpha)(\alpha\partial_y\pi - \partial_\xi\pi)\pi\tilde{\psi}(\ldots, \alpha)^{-1} = 0.$$

But $\tilde{\psi}(\ldots, \alpha)$ is non-degenerate, hence $(\alpha\partial_y\pi - \partial_\xi\pi)\pi = 0$. Similarly, calculate the residue at $\tau = \alpha$ in the expression $(\tau\partial_\eta\psi - \partial_y\psi)\psi^{-1}$ to get $(\alpha\partial_\eta\pi - \partial_y\pi)\pi = 0$. By Proposition 8.4, $g_{\alpha,\pi}$ is a 1-soliton monopole frame. Since $\psi_1 = \tilde{g}^{-1}\psi$,

$$(D\psi_1)\psi_1^{-1} = -\tilde{g}^{-1}D\tilde{g} + \tilde{g}^{-1}(D\psi)\psi^{-1}\tilde{g}.$$

The LHS is holomorphic at $\tau = \alpha, \bar{\alpha}$ and the RHS is holomorphic for all $\tau \neq \alpha, \bar{\alpha}$. So $(D\psi_1)\psi_1^{-1}$ is a degree one polynomial in τ. Similarly, $(\tau\partial_\eta\psi_1 - \partial_x\psi_1)\psi_1^{-1}$ is a degree one polynomial in τ. By Proposition 8.1, ψ_1 is a monopole frame. □

Similar argument gives

THEOREM 10.5 (Subtracting a soliton with pole data (α, k)). *Suppose ψ is a normalized monopole frame, and ψ has a pole at $\tau = \alpha$ with multiplicity k (may have other singularities as well). Then there exists a unique k-soliton monopole frame g with pole data (α, k), a normalized monopole frame ψ_1, and maps $\tilde{\psi}$ and $\tilde{g}$ such that ψ_1 and $\tilde{\psi}$ are holomorphic and non-degenerate at $\tau = \alpha, \bar{\alpha}$ and $\psi = \tilde{\psi}g = \tilde{g}\psi_1$.*

A consequence of the above two theorems is that every normalized monopole frame with continuous scattering data and finitely many poles can be obtained from Theorem 10.3.

11. Appendix: Existence of Continuous Scattering Data for Small Solutions

The over-all details of the scattering data described in Sections 7 and 10 where we find solutions with a combination of a jump across the unit circle and point (singularities) measures do fit into the general scheme proposed by Beals and Coifman [**4**], [**5**]. Beals and Coifman point out that the Self-dual Yang-Mills equations have local scattering data more like the AKNS scattering problem than either type of KP, and Fokas and Ioannidou point out that the Ward equation inherits this similarity. Existence of solutions which have combinations of the two types of scattering data follows from the existence of the inverse scattering transform and the process of adding discrete measures by the described Bäcklund transforms. Papers of Fokas and Ioannidou [**7**] and Villarroel [**21**] discuss the existence of the scattering and inverse scattering transforms.

We find that, as in the AKNS models, small data leads to scattering data which consists only of the continuous part. For simplicity, we assume all the data is rapidly decaying and lies in the Schwartz space. Recall that the spacial operator is

$$D_s(\mu) = D_1(\mu) - D_2(\mu) = \frac{(\mu - \mu^{-1})}{2}\nabla_x - i\frac{(\mu + \mu^{-1})}{2}\nabla_y - i\phi.$$

THEOREM 11.1. *Assume that there exists a gauge transformation such that in the given gauge (A, ϕ) is small in $W^{2,1}$. Then the chern vector $c(\mu)$ is zero for all $\mu \in C \cup \{\infty\} \setminus S^1$. Furthermore there exists continuous scattering data $S_\theta = I + G_\theta$ where G_θ is small in $L^\infty_{(x,y,t)}$.*

COROLLARY 11.2. *Given initial data (A, ϕ) which is small in $W^{2,1}$, there exist global in time solutions of the space-time monopole problem with this initial data. These solutions are unique up to gauge transformation.*

First we explain how the corollary follows from the theorem. The existence of global solutions follows from the existence of scattering data which is purely continuous for the initial data, the known flow of the scattering data under time, and the existence of the inverse scattering transform. Because the scattering data remains small, the solution remains small in time. Suppose there is a time t_θ at which uniqueness fails. Since we presume the solution to be continuous in time, the second solution is small for a short time near t_θ. Hence it has scattering data, and must be identified with the solution constructed by inverse scattering data by a gauge transformation.

To prove the theorem, we note that the proof is rather standard away from the unit circle, although it will follow from the proof we give near the unit circle as well. The difficulty is to prove that the limits exist as

$|\mu| \to 1$. To do this, use the combination of Lorentz and fractional linear transformation so that μ is pure imaginary. Now let

$$\tau = (\mu + \mu^{-1})/(\mu - \mu^{-1}),$$

so $\tau < 1$ is real. We assume $|\mu| < 1$, so $\tau < 0$. For $|\mu| > 1$, then the reality condition implies that $E_\mu = ((E_{\bar{\mu}^{-1}})^*)^{-1}$. The equation for the frame now reads:

$$(\partial_x - i\tau\partial_y)E_\tau = G_\tau$$

where

$$G_\tau(x,y,t) = A_\tau(x,y,t)E_\tau(x,y,t), \quad A_\tau = A_x - i\tau A_y(x,t) - \sqrt{1-\tau^2}\,\Phi.$$

As is usual with solutions with small data, we solve by iteration. Let $E_\tau = \mathrm{I} + \sum_{j\geq 1} Q_{j,\tau}$. Set $Q_{0,\tau} = \mathrm{I}$, $Q_{j,\tau}(-\infty, y) = 0$, and define $Q_{j,\tau}$ iteratively by

$$(\partial_x - i\tau\partial_y)Q_{j,\tau} = G_{j-1,\tau} = A_\tau Q_{j-1,\tau}.$$

Here A_τ is as above.

We now take the Fourier transform in the y variable alone, and denote the transform in y only of an expression Q by $\hat{Q}$. Let ζ be the transform variable of y. We have

$$\partial_x \hat{Q}_{j,\tau} - \zeta\tau\hat{Q}_{j,\tau} = \hat{G}_{j-1,\tau}.$$

But this ODE can be solved explicitly:

$$\hat{Q}_{j,\tau}(x,\zeta,t) = \mathrm{sgn}\,(\zeta\tau)\int_{\zeta\tau(x-r)\leq 0} e^{\zeta\tau(x-r))}\hat{G}_{j-1,\tau}(r,\zeta,t)\;dr.$$

So

$$\int_{-\infty}^{\infty} \hat{Q}_{j,\tau}(x,\zeta,t)d\zeta$$
$$= \int_{-\infty}^{\infty} \mathrm{sgn}\,(\zeta\tau)\int_{\zeta\tau(x-r)\leq 0} e^{\zeta\tau(x-r))}\hat{G}_{j-1,\tau}(r,\zeta,t)\;dr\;d\zeta.$$

If $\tau\zeta > 0$, then the above integral is equal to

$$\int_{-\infty}^{\infty}\int_{-\infty}^{0} e^{\tau\zeta s}\hat{G}_{j-1,\tau}(x-s,\zeta,t)\;dsd\zeta.$$

For $\tau\zeta < 0$, we get a similar but different formula. This explains the jump. So we obtain in a straightforward fashion the estimate independent of τ:

$$||\hat{Q}_{j,\tau}||_1 \leq ||\hat{G}_{j-1,\tau}||_2 = ||\hat{A}_\tau * \hat{Q}_{j-1,\tau}||_2 \leq ||\hat{A}_\tau||_2 ||\hat{Q}_{j-1,\tau}||_1.$$

The norms used are $||\cdot||_1$ as the $L^\infty_x L^1_\zeta$ and $||\cdot||_2$ as the $L^1_{(x,\zeta)}$ norm. Given these estimates, the solution

$$\hat{Q}(x,\zeta,t) = \sum_{j\geq 0}\hat{Q}_j(x,\zeta,t)$$

exists in $L^\infty_x L^1_\zeta$ when $||\hat{A}_\tau||_2 \leq 1$. But it is easy to see that when the $W^{2,1}$ norm of (A,ϕ) is small, this inequality is satisfied. Furthermore, $||\hat{Q}||_1$

bounds $||Q||_\infty$ and existence follows. Iterating these same estimates in the various derivatives of Q will give estimates on the higher norms, although we do not expect or require smallness in the higher derivatives and weights.

References

[1] C.K. Anand, *Ward's solitons*, Geom. Topol. **1** (1997), 9–20.
[2] C.K. Anand, *Ward's solitons* II, *Exact solutions*, Canad. J. Math. **50** (1998), 1119–1137.
[3] M.F. Atiyah and N.J. Hitchin, *The geometry and dynamics of magnetic monopoles*, Princeton University Press, Princeton, New Jersey, 1988.
[4] R. Beals and R.R. Coifman, *Multidimensional inverse scattering and nonlinear partial differential equations*, Proc. Symp. Pure Math. **43** (1985), 45–70.
[5] R. Beals and R.R. Coifman, *Linear spectral problems, non-linear equations and the $\bar{\partial}$-method*, Inverse Problems **5** (1989), 87–130.
[6] B. Dai and C.L. Terng, *Bäcklund transformation, Ward solitons, and unitons*, arXiv:math.DG/0405363.
[7] A.S. Fokas and T.A. Ioannidou, *The inverse spectral theory for the Ward equation and for the* $2+1$ *chiral model*, Comm. Appl. Anal. **5(2)** (2001), 235–246.
[8] N.J. Hitchin, *On the construction of monopoles*, Comm. Math. Phys. **89** (1983), 145–190.
[9] N.J. Hitchin, *The self-duality equations on a Riemann surface*, Proc. London Math. Soc. **55** (1987), 59–126.
[10] T. Ioannidou, *Soliton solutions and nontrivial scattering in an integrable chiral model in* $(2+1)$ *dimensions*, J. Math. Phys. **37** (1996), 3422–3441.
[11] T. Ioannidou and W. Zakrzewski, *Solutions of the modified chiral model in* $(2+1)$ *dimensions*, J. Math. Phys. **39(5)** (1998), 2693–2701.
[12] T. Ioannidou and W. Zakrzewski, *Lagrangian formulation of the general modified chiral model*, Phys. Lett. A **249** (1998), 303–306.
[13] S.V. Manakov and V.E. Zakharov, *Three-dimensional model of relativistic-invariant theory, integrable by the inverse scattering transform*, Lett. Math. Phys. **5** (1981), 247–253.
[14] A. Pressley and G. Segal, *Loop groups*, Oxford University Press, 1986.
[15] T. Tao, *Global regularity of wave maps.* II. *Small energy in two dimensions*, Comm. Math. Phys. **224(2)** (2001), 443–544.
[16] D. Tataru, *The wave maps equation*, Bull. Amer. Math. Soc. **41(2)** (2004), 185–204.
[17] C.L. Terng and K. Uhlenbeck, *Poisson actions and scattering theory for integrable systems.* Surveys in differential geometry: integrable systems, 315–402, Surv. Diff. Geom., IV, International Press, Boston, MA, 1998.
[18] C.L. Terng and K. Uhlenbeck, *Bäcklund transformations and loop group actions*, Comm. Pure Appl. Math. **53** (2000), 1–75.
[19] K. Uhlenbeck, *Harmonic maps into Lie groups (classical solutions of the chiral model)*, J. Differential Geom., **30** (1989), 1–50.
[20] K. Uhlenbeck, *On the connection between harmonic maps and the self-dual Yang-Mills and the sine-Gordon equations*, J. Geom. Phys. **8** (1992), 283–316.
[21] J. Villarroel, *The inverse problem for Ward's system*, Stud. Appl. Math. **83** (1990), 211–222.
[22] R.S. Ward, *Soliton solutions in an integrable chiral model in* $2+1$ *dimensions*, J. Math. Phys. **29** (1988), 386–389.
[23] R.S. Ward, *Classical solutions of the chiral model, unitons, and holomorphic vector bundles*, Comm. Math. Phys. **128** (1990), 319–332.

[24] R.S. Ward, *Nontrivial scattering of localized solutions in a* $(2+1)$*-dimensional integrable systems*, Phys. Letter A **208** (1995), 203–208.
[25] R.S. Ward, *Integrable systems and twistors*, in 'Integrable Systems', Oxf. Grad. Texts Math., **4** (1999), 121-134.
[26] V.E. Zakharov and A.V. Mikhailov, A. *Relativistically invariant two dimensional models of fields theory which are integrable by means of the inverse scattering problem method*, Sov. Phys. JETP **47(6)** (1978), 1017–1027.

LMAM, School of Mathematical Sciences, Peking University, Beijing 100871, P.R. China
E-mail address: daibo@math.pku.edu.cn

Department of Mathematics, University of California at Irvine, Irvine, CA 92697-3875
E-mail address: cterng@math.uci.edu

Department of Mathematics, University of Texas at Austin, Austin, TX 78712
E-mail address: uhlen@math.utexas.edu

The Ehrhart function for symbols

Victor W. Guillemin, Shlomo Sternberg, and Jonathan Weitsman

ABSTRACT. We derive an Ehrhart function for symbols from the Euler-MacLaurin formula with remainder.

In memory of Prof. S.S. Chern, our teacher and mentor.

1. Introduction

Let $\Delta \subset \mathbb{R}^n$ be a convex polytope whose vertices are in $\mathbb{Z}^n$ and such that the origin 0 is in the interior of Δ. Consider the expanded polytope $N \cdot \Delta$.

Ehrhart's theorem [**Ehr**] asserts that for N a positive integer, the number of lattice points in the expanded polytope, i.e.,

$$\#\left((N \cdot \Delta) \cap \mathbb{Z}^n\right), \quad N \in \mathbb{Z}_+$$

is a polynomial in N. More generally, suppose that f is a polynomial, and let

$$p(N, f) := \sum_{\ell \in N \cdot \Delta \cap \mathbb{Z}^n} f(\ell). \tag{1}$$

Then Ehrhart's theorem asserts that $p(N, f)$ is a polynomial in N.

In the case that Δ is a simple polytope (meaning that n edges emanate from each vertex) Ehrhart's theorem is a consequence of the Euler-MacLaurin formula, [**Kh, KP, CS1, CS2, Gu, BV, DR**] and one can be more explicit about the nature of the polynomial $p(N, f)$.

Let us explain how this works in the more restrictive case where Δ is not only simple but is regular, meaning that the local cone at each vertex can be transformed by an integral unimodular affine transformation into a

The work of J.W. is partially supported by National Science Foundation Grant DMS/0405670.

neighborhood of the origin in the standard orthant $\mathbb{R}^n_+$.[1] In this case we can apply the formula of Khovanskii-Pukhlikov [**KP**], which reads as follows: The polytope Δ can be described by a set of inequalities

$$x \cdot u_i + a_i \geq 0, \qquad i = 1, \dots, m$$

where m is the number of facets of Δ, where the u_i are primitive lattice vectors, and where the a_i are positive integers. Then for any positive number t, the expanded polytope $t \cdot \Delta$ is described by the inequalities

$$x \cdot u_i + ta_i \geq 0, \qquad i = 1, \dots, m.$$

Let $\Delta_{t,h}, \;\; h = (h_1, \dots, h_m)$ be the polytope defined by

$$x \cdot u_i + ta_i + h_i \geq 0, \qquad i = 1, \dots, m. \tag{2}$$

Then the function

$$\tilde{p}(t, h, f) := \int_{\Delta_{t,h}} f(x)dx \tag{3}$$

is a polynomial in t and h.

The formula of Khovanskii-Pukhlikov (applied to $N \cdot \Delta$) expresses $p(N, f)$ in terms of a differential operator applied to $\tilde{p}(t, h, f)$: Explicitly, consider the infinite order constant coefficient differential operator

$$\mathrm{Todd}\left(\frac{\partial}{\partial h}\right) = \sum_{\gamma} b_{\gamma} \left(\frac{\partial}{\partial h}\right)^{\gamma} \tag{4}$$

where $\sum_{\gamma} b_{\gamma} x^{\gamma}$ is the Taylor series expansion at the origin of the Todd function

$$\mathrm{Todd}(x) = \prod_{i=1}^{m} \frac{x_i}{1 - e^{-x_i}}.$$

The Khovanskii-Pukhlikov formula says that

$$p(N, f) = \mathrm{Todd}\left(\frac{\partial}{\partial h}\right) \tilde{p}(N, h, f)\Big|_{h=0}.$$

Note that since $\tilde{p}$ is a polynomial in h the right hand side really involves only a finite order differential operator.

For purposes below it will be convenient to write the Khovanskii-Pukhlikov formula in the form

$$p(N, f) - \tilde{p}(N, 0, f) = \left(\mathrm{Todd}\left(\frac{\partial}{\partial h}\right) - \mathrm{Id}\right) \tilde{p}(N, h, f)\Big|_{h=0}. \tag{5}$$

---

[1]Of course, this use of the word "regular" has nothing to do with the term denoting Platonic solids. There are other names in the literature for the property we are describing, such as "smooth", "Delzant" ,"torsionfree", "unimodular" etc.. It is unfortunate that the nomenclature for polytopes with this property has not yet been standardized.

For simple polytopes there is a more general formula due to [**CS2, Gu, BV**]. Our goal in this paper is to prove an analogue of (5) and its generalizations when the polynomial f is replaced by a "symbol", a term whose definition we recall from the theory of partial differential equations: A smooth function $f \in C^\infty(\mathbb{R}^n)$ is called a **symbol of order** r if for every n-tuple of non-negative integers $a := (a_1, \ldots, a_n)$, there exists a constant C_a such that

$$|\partial_1^{a_1} \ldots \partial_n^{a_n} f(x)| \leq C_a(1 + |x|)^{r-|a|}$$

where $|a| = \sum_i a_i$. In particular, a polynomial of degree r is a symbol of order r. Note that if f is a symbol of order r on $\mathbb{R}^n$ then its derivatives of order a are in L^1 if $r < |a| - n$.

For simplicity, we will restrict ourselves in this paper to **polyhomogeneous symbols**, meaning functions $f \in C^\infty(\mathbb{R}^n)$ which admit asymptotic expansions of the form:

$$f(x) \sim \sum_{-\infty}^{r} f_\ell(x) \tag{6}$$

for $\|x\| \gg 0$ where the f_ℓ are homogeneous symbols of degree ℓ meaning that each f_ℓ is a symbol of order ℓ with the property that for $\|x\|$ sufficiently large and $t \geq 1$ we have $f_\ell(tx) = t^\ell f_\ell(x)$. The sum in (6) is over a discrete sequence of numbers tending to $-\infty$. "Asymptotic" means that for any j

$$f(x) - \sum_{\ell=j}^{r} f_\ell(x) = o(\|x\|^j) \tag{7}$$

as $\|x\| \to \infty$. The number r occurring in (6) is again called the **order** of the asymptotic series and the collection of functions satisfying (6) will be called polyhomogeneous symbols of order r and denoted by S^r.

For the sake of exposition in this introduction, we will continue to discuss the case where Δ is regular.

We will show that if f is a polyhomogeneous symbol then the function $p(N, f)$ given by (1) is a polyhomogeneous symbol in N and its asymptotic expansion in powers of N is given by a formula similar to (5) with two key differences:

(1) For symbols, an infinite number of differentiations occur on the right hand side of (5), i.e., the whole Todd operator must be applied. So (5) must be understood as an asymptotic series, not as an equality.
(2) The formula (5) has to be corrected by adding a constant term C to the right hand side, a constant which is zero for the case of a polynomial.

More precisely, we will prove:

THEOREM 1.1. *Let Δ be a regular polytope whose vertices lie in $\mathbb{Z}^n$ with 0 in the interior of Δ. Let $f \in S^r$ and $N \in \mathbb{Z}_+$ and let*

$$p(N,f) := \sum_{\ell \in N\cdot\Delta\cap\mathbb{Z}^n} f(\ell).$$

Let $\tilde{p}(t,h,f)$ be defined by (3) so that

$$\tilde{p}(N,0,f) = \int_{N\cdot\Delta} f(x)dx,$$

Then $p(N,f)-\tilde{p}(N,0,f)$ is a symbol in N and has the asymptotic expansion

$$p(N,f) - \tilde{p}(N,0,f) \sim \left(\mathrm{Todd}\left(\frac{\partial}{\partial h}\right) - \mathrm{Id}\right)\tilde{p}(N,h,f)\bigg|_{h=0} + C$$

where C is a constant.

The constant C is of interest in its own right. It can be thought of as a "regularized" version of the difference

$$\sum_{\ell\in\mathbb{Z}^n} f(\ell) - \int_{\mathbb{R}^n} f(x)dx. \tag{8}$$

Of course there is no reason why either the sum or the integral in (8) should converge. But we can "regularize" both as follows: Define the function $\langle x\rangle$ by

$$\langle x\rangle^2 := 1 + \|x\|^2.$$

For $s \in \mathbb{C}$ let

$$f(x,s) := f(x)\langle x\rangle^s.$$

We will show that

$$C(s) := \sum_{\ell\in\mathbb{Z}^n} f(\ell,s) - \int_{\mathbb{R}^n} f(x,s)dx, \tag{9}$$

which is holomorphic for Re $s << 0$, has an analytic continuation to the entire complex plane and that the missing constant C on the right hand side of (5) is exactly $C(0)$. In particular, the constant C is independent of the particular polytope in question.

This result is related to, and inspired by, a result of Friedlander-Guillemin [**FG**] on "Szego regularizations" of determinants of pseudodifferential operators. In their result, as in ours, there is a missing constant which also has to be computed by the above process of "zeta regularization".

Our result is somewhat insensitive to the mode of regularization. In fact, it can be generalized as follows:Define a "gauged symbol" [**Gugau**] to be a function $f(x,s) \in C^\infty(\mathbb{R}^n \times \mathbb{C})$ which depends holomorphically on s and for fixed s is a symbol of order Re $s + r$. For example, if f is a symbol, the function $f(x)\langle x\rangle^s$ introduced above is such a gauged symbol. We make a similar definition of "gauged polyhomogenous symbols". We will prove that if $f(x,s)$ is a gauged polyhomogenous symbol with

$$f(x) = f(x,0)$$

then the function given by (9) with this more general definition of $f(x, s)$ again extends holomorphically from Re $s << 0$ to the entire plane and $C = C(0)$.

The above results will be proved for the more general case of simple integral polytopes in §2. The proof is largely based on the Euler-MacLaurin formula with remainder as proved in [**KSW**] and motivated by an argument of Hardy on "Ramanujan regularization" [Hardy]. Ramanujan's key idea was to use the classical Euler-MacLaurin formula in one variable to regularize (8) by providing "counter terms" in passing to infinity in the difference between sum and integral in one dimension. The origin of this method goes back to Euler's continuation of the zeta function past the pole at $z = 1$ and his introduction of what is known today as "Euler's constant".

We would like to thank Yael Karshon and Michele Vergne for helpful comments.

2. An Ehrhart formula for simple polytopes

We continue with the notation of §1. So $\Delta \subset \mathbb{R}^n$ is a convex polytope whose vertices are in $\mathbb{Z}^n$ and such that the origin 0 is in the interior of Δ. We assume in this section that Δ is simple, which means that for every vertex p there are exactly n edges emanating from p so they lie on rays

$$p + t\alpha_{i,p} \tag{10}$$

where the $\alpha_{i,p}$, $i = 1, \ldots, n$ form a basis of $\mathbb{R}^n$.

2.1. The Euler-MacLaurin formula for symbols. We want to apply the Euler-MacLaurin formula with remainder, [**KSW**]. In [**KSW**] one dealt with a weighted sum where points in the interior of the polytope are given weight 1, points on the relative interior of a facet are given weight $w(x) := \frac{1}{2}$, and, more generally, points in the relative interior of a face of codimension k are given weights $w(x) := \frac{1}{2^k}$. The weighted sum $p_{\frac{1}{2}}(N, f)$ is then defined as

$$p_{\frac{1}{2}}(N, f) := \sum_{\ell \in (N \cdot \Delta) \cap \mathbb{Z}^n} w(\ell) f(\ell).$$

Theorem 3 of [**KSW**] gives an Euler-MacLaurin formula with remainder for weighted sums of symbols.

More generally, [**AW**] consider the Euler-MacLaurin formula with remainder for more general weightings including the unweighted sum we considered in §1. We refer to equations (28) and (29) in [**AW**] for the definition of a general weighting, w, and we will denote the corresponding weighted sum here by $p_w(N, f)$. They stated their formula with remainder for smooth functions of compact support, but the passage from the case of smooth functions of compact support to that of symbols is exactly the same as in [**KSW**].

To formulate the Euler-MacLaurin formula with remainder we fix a vector $\xi \in \mathbb{R}^n$ such that

$$\alpha_{i,p} \cdot \xi \neq 0 \quad \forall\ p \text{ and } i. \tag{11}$$

We then define

$$\alpha^{\sharp}_{i,p} := \begin{cases} \alpha_{i,p} & \text{if } \alpha_{i,p} \cdot \xi > 0 \\ -\alpha_{i,p} & \text{if } \alpha_{i,p} \cdot \xi < 0 \end{cases} \tag{12}$$

$$(-1)^p := \prod_{i=1}^{n} \frac{\alpha^{\sharp}_{i,p} \cdot \xi}{\alpha_{i,p} \cdot \xi} \tag{13}$$

and

$$C_{p,t} := \left\{ tp + \sum_{i=1}^{n} t_i \alpha^{\sharp}_{i,p}, \quad t_i \geq 0 \right\}. \tag{14}$$

There is a certain infinite order differential operator $\mathbf{M}$ (depending on the weighting) in the variables $h_1, \ldots, h_m$ with constant term 1 whose truncation of order k is given by the sum in equation (89) of [**KSW**] (for weight $\frac{1}{2}$) and the sum in equation (56) in [**AW**] (for general weights) such that for any symbol f of order r and $k > n + r$

(15)

$$p_w(N, f) - \tilde{p}(N, 0, f) = \left((\mathbf{M}^{[k]}) \left(\frac{\partial}{\partial h} \right) - \mathrm{Id} \right) \tilde{p}(N, h, f) \Big|_{h=0} + R^k(f, N)$$

where $\mathbf{M}^{[k]}$ denotes the truncation of $\mathbf{M}$ at order k and

$$R^k(f, N) = \sum_p (-1)^p \int_{C_{p,N}} \left(\sum_{|\gamma|=k}^{|\gamma|=nk} \phi^p_{\gamma,k} D^{\gamma} f \right) dx \tag{16}$$

where the $\phi^p_{\gamma,k}$ are bounded piecewise smooth periodic functions. For an explicit expression for these functions see [**KSW**] or [**AW**]. We will not need this here. If the polytope is regular, and we use unweighted sums, the operator $\mathbf{M}$ is exactly the operator Todd of §1.

Notice that $\mathbf{M}^{[k]} - 1$ has no constant term, so

$$\left((\mathbf{M}^{[k]}) \left(\frac{\partial}{\partial h} \right) - \mathrm{Id} \right) \tilde{p}(N, h, f) \Big|_{h=0}$$

involves integration of derivatives of f over faces of the polytope. These faces are moving out to infinity as $N \to \infty$. A derivative of a homogeneous summand in (6) is itself a homogenous function. So as $N \to \infty$, the integral of this derivative over a face of the polytope is a homogenous function in N whose degree depends on the degree of this derivative and the dimension of the face. So the contributions of the polyhomogenous terms with sufficiently negative degree in (6) will be homogeneous terms of high negative degree in N in (15). By the same token, each homogenous summand in (6) will yield a finite number of terms to each order in (15). So the left hand side of (15) will be polyhomogenous symbol in N.

2.2. The key idea. We want to investigate the behavior of the remainder (16) as $N \to \infty$.

Since the origin 0 is in the interior of the polytope, it is in the interior of the cone generated by the $\alpha_{i,p}$ for any vertex p. This means that when we use the $\alpha_{i,p}$ as a basis of $\mathbb{R}^n$ (based at the origin), the point p has strictly negative coefficients relative to this basis. When we flip the basis from $\alpha_{i,p}$ to $\alpha_{i,p}^\sharp$, the coefficients of those edges which are actually flipped become positive.

Condition (11) implies that the function $x \mapsto \xi \cdot x$ has a unique minimum on the polytope, and that this minimum is achieved at a single vertex, q. At this vertex, no edges are flipped, and at any other vertex $p \neq q$ at least one edge is flipped. So for $p \neq q$ we have

$$p = \sum_{i=1}^{n} a_{i,p} \alpha_{i,p}^\sharp, \tag{17}$$

where at least one $a_{i,p} > 0$. Then the cone $C_{p,t}$ is contained in the half space

$$\left\{ x = \sum_i x_i \alpha_{i,p}^\sharp, \quad x_i \geq a_{i,p} t \right\}$$

and so the p-th summand in (16) tends to zero as $N \to \infty$ for $p \neq q$.

At the vertex q we have we have $q = \sum_i a_{i,q} \alpha_{i,q} = \sum_i a_{i,q} \alpha_{i,q}^\sharp$ with all the $a_{i,q} < 0$ and so the cone $C_{q,t}$ tends to the entire space $\mathbb{R}^n$ as $t \to \infty$. Thus (16) tends to

$$C_k = \int_{\mathbb{R}^n} \left(\sum_{|\gamma|=k}^{|\gamma|=nk} \phi_{\gamma,k}^q D^\gamma f \right) dx. \tag{18}$$

It follows from (15) that this limiting value C_k is independent of k for k sufficiently large. Let us call this common limit $C = C(f)$. It also follows from (15) that $C(f)$ is independent of the choice of the polarizing vector ξ.

We shall interpret this limiting value C using regularization in the next section, and we will find that C is also independent of the particular polytope we are expanding.

If f is a polynomial, so that we choose k to be greater than the degree of f, we see from (18) that $C = 0$, as it must be from the classical Ehrhart theorem.

2.3. Application to polyhomogeneous symbols. Now suppose that f is a polyhomogenous symbol. We can apply the above to each summand in the asymptotic series (6). But notice that if we choose j sufficiently negative, the function

$$g(x) = g_j(x) = f(x) - \sum_{\ell=j}^{r} f_\ell(x)$$

occurring on the left hand side of (7) will have the property that both

$$\sum_{\ell\in\mathbb{Z}^n} g(\ell) \quad \text{and} \quad \int_{\mathbb{R}^n} g(x)dx$$

are absolutely convergent. Furthermore, given any negative number m we can arrange, by choosing j sufficiently negative, that

$$p_w(N,g) - \sum_{\ell\in\mathbb{Z}^n} g(\ell) = o(N^m)$$

and

$$\tilde{p}(N,0,g) - \int_{\mathbb{R}^n} g(x)dx = o(N^m)$$

so

$$[p_w(N,g) - \tilde{p}(N,0,g)] - \left[\sum_{\ell\in\mathbb{Z}^n} g(\ell) - \int_{\mathbb{R}^n} g(x)dx\right] = o(N^m).$$

So if we define

$$C(f) = \sum_{\ell=j}^{r} C(f_\ell) + \left[\sum_{\ell\in\mathbb{Z}^n} g_j(\ell) - \int_{\mathbb{R}^n} g_j(x)dx\right] \tag{19}$$

for j sufficiently negative, then $C(f)$ is independent of the choice of j. Furthermore, we see that if f is a polyhomogeneous symbol, we get an asymptotic expansion of the form

$$p_w(N,f) - \tilde{p}(N,0,f) \sim \sum_{\ell=-\infty}^{r} \left(\mathbf{M}\left(\frac{\partial}{\partial h}\right) - \mathrm{Id}\right) \tilde{p}(N,h,f_\ell)\bigg|_{h=0} + C(f),$$

where each level in the asymptotic expansion in N involves only finitely many f_ℓ. By abuse of language, we shall denote this equation as

$$p_w(N,f) - \tilde{p}(N,0,f) \sim \left(\mathbf{M}\left(\frac{\partial}{\partial h}\right) - \mathrm{Id}\right) \tilde{p}(N,h,f)\bigg|_{h=0} + C(f). \tag{20}$$

2.4. Regularization. Suppose that we replace f by a gauged polyhomogenous symbol $f(x,s)$ with $f(x,0) \in S^r$. Then the remainder term (16) applied to a summand in the asymptotic expansion of $f_s = f(\cdot,s)$ is well defined if $\mathrm{Re}\, s < -r-n+k$. Moreover, if $p \neq q$ so that $a_{i,p} > 0$ for some i the p-th summand on the right of (16) is of order $O(N^{\mathrm{Re}\, s+r+n-k})$.

At the unique vertex q where no edges are flipped q-th summand of (16) differs from the integral

$$\int_{\mathbb{R}^n} \left(\sum_{|\gamma|=k}^{|\gamma|=nk} \phi^q_{\gamma,k} D^\gamma_x f_\ell(x,s)\right) dx \tag{21}$$

by a term of order $O(N^{\mathrm{Re}\, s+\ell+n-k})$. Thus the gauged version of (20) is

$$p_w(N,f_s)-\tilde{p}(N,0,f_s)=\left(\mathbf{M}^{[k]}-\mathrm{Id}\right)\left(\frac{\partial}{\partial h}\right)\tilde{p}(N,h,f_s)\bigg|_{h=0}+C_k(s) \tag{22}$$
$$+O(N^{\mathrm{Re}\, s+r+n-k})$$

for

$$\mathrm{Re}\, s<-n-r+k$$

where $f_s=f(\cdot,s)$ where $C_k(s)$ is(20) with f replaced by f_s, and we have computed $C(s)$ by going out to level k in the Euler-MacLaurin expansion.

All the terms on the right of (22) are holomorphic on the half-plane $\mathrm{Re}\, s<-n-r+k$.

Letting $k\to\infty$ we conclude that on this half-plane we have, in the notation of (20),

$$p_w(N,f_s)-\tilde{p}(N,0,f_s)\sim\left(\mathbf{M}\left(\frac{\partial}{\partial h}\right)-\mathrm{Id}\right)\tilde{p}(N,h,f_s)\bigg|_{h=0}+C(s) \tag{23}$$

where

$$C(s)-C_k(s)=O(N^{\mathrm{Re}\, s+r+n-k}). \tag{24}$$

Since the $C_k(s)$ are holomorphic of the half-plane $\mathrm{Re}\, s<-n-r+k$, it follows that $C(s)$ is holomorphic on the whole plane.

Moreover, in the asymptotic series on the right of (23) all the terms are of order at most $\mathrm{Re}\, s+r+n$. Hence for $\mathrm{Re}\, s<-r-n$ these terms tend to zero and we get

$$\begin{aligned} C(s) &= \lim_{N\to\infty}(p(N,f_s)-\tilde{p}(N,0,f_s)) \\ &= \sum_{\ell\in\mathbb{Z}^n} f(\ell,s)-\int_{\mathbb{R}^n} f(x,s)dx, \end{aligned}$$

and both the sum and the integral converge absolutely. So if we set $s=0$ we obtain

$$p_w(N,f)-\tilde{p}(N,0,f)\sim\left(\mathbf{M}\left(\frac{\partial}{\partial h}\right)-\mathrm{Id}\right)\tilde{p}(N,h,f)\bigg|_{h=0}+C \tag{25}$$

where $f(x)=f(x,0)$ and $C=C(0)$. So we can think of C as a "regularization" of (8). To summarize: We have proved

THEOREM 2.1. *Let Δ be a simple polytope whose vertices lie in $\mathbb{Z}^n$ with 0 in the interior of Δ. Let $f\in S^r$ and $N\in\mathbb{Z}_+$ and let*

$$p_w(N,f):=\sum_{\ell\in N\cdot\Delta\cap\mathbb{Z}^n} w(\ell)f(\ell).$$

Let $\tilde{p}(t,h,f)$ be defined by (3) so that

$$\tilde{p}(N,0,f)=\int_{N\cdot\Delta} f(x)dx,$$

Then $p(N, f) - \tilde{p}(N, 0, f)$ *is a symbol in* N *and has the asymptotic expansion*

$$p_w(N, f) - \tilde{p}(N, 0, f) \sim \left(\mathbf{M}\left(\frac{\partial}{\partial h}\right) - \mathrm{Id}\right) \tilde{p}(N, h, f)\Big|_{h=0} + C$$

where C *is a constant. Furthermore, if* $f(x, s)$ *is a gauged polyhomogenous symbol with* $f(x, 0) = f(x)$ *then* $C = C(0)$ *where* $C(s)$ *is the entire function given by* (23) *and* (24). *For* $\operatorname{Re} s < -r - n$

$$C(s) = \sum_{\ell \in \mathbb{Z}^n} f(\ell, s) - \int_{\mathbb{R}^n} f(x, s) dx.$$

Hence $C(s)$ *and in particular* $C = C(0)$ *is independent of the polytope.*

REMARKS.

1. In the course of the discussion we have proved a similar theorem with polyhomogeneous symbols replaced by symbols and gauged polyhomogenous symbols replaced by gauged symbols.
2. Since the initial posting of this paper we have received the interesting paper [**MP**].

References

[AW] J. Agapito and J Weitsman, *The weighted Euler-MacLaurin formula for a simple integral polytope*, Asian Journal of Mathematics **9** (2005), 199–212.

[BV] M. Brion and M. Vergne, *Lattice points in simple polytopes*, Jour. Amer. Math. Soc. **10** (1997), 371–392.

[CS1] S.E. Cappell and J.L. Shaneson, *Genera of algebraic varieties and counting lattice points*, Bull. AMS **30** (1994), 62–69.

[CS2] S.E. Cappell and J.L. Shaneson, *Euler-Maclaurin expansions for lattices above dimension one*, C.R. Acad. Sci. Paris Sr. I Math. **321** (1995), 885–890.

[DR] R. Diaz and S. Robins, *The Ehrhart Polynomial of a Lattice Polytope*, Ann. Math. (2) **145**(**3**) (1997), 503–518, and *Erratum: "The Ehrhart polynomial of a lattice polytope"*, Ann. Math. (2) **146**(**1**) (1997), 237.

[Ehr] E. Ehrhart, *Sur les polyèdres rationnels homothétiques à n dimensions*, C.R. Acad. Sci. Paris **254** (1962), 616–618.

[FG] L. Friedlander and V. Guillemin, *Determinants of zeroth order operators*, to appear in a special memorial volume of "Surveys in Differential Geometry" dedicated to S.-S. Chern.

[Gugau] V. Guillemin, *Gauged Lagrangian distributions*, Adv. Math. **102**(**2**) (1993), 184–201.

[Gu] V. Guillemin, *Riemann-Roch for toric orbifolds*, J. Differential Geom. **45** (1997), 53–73.

[H] G.H. Hardy, *Divergent Series*, Second Edition, 1991 (unaltered), Chelsea Pub. Col, New-York, NY, especially pp. 326–328.

[KK1] J.M. Kantor and A.G. Khovanskii, *Integral points in convex polyhedra, combinatorial Riemann-Roch and generalized MacLaurin formulae*, Inst. Hautes Études Sci. Publ. Math. (1992) 932–937.

[KK2] J.M. Kantor and A.G. Khovanskii, *Une application du théorème de Riemann-Roch combinatoire au polynôme d'Ehrhart des polytopes entiers de* R^d, C.R. Acad. Sci. Paris Sér. I Math. **317**(**5**) (1993), 501–507.

[KSW] Y. Karshon, S. Sternberg, and J. Weitsman, *Euler-MacLaurin with remainder for a simple integral polytope*, Duke Mathematical Journal **130** 2005, 401–434

[Kh] A.G. Khovanski, *Newton polyhedra and toroidal varieties*, Func. Anal. Appl. **11** (1977), 289–296.

[KP] A.G. Khovanskii and A.V. Pukhlikov, *The Riemann-Roch theorem for integrals and sum of quasipolynomials on virtual polytopes*, Algebra i Analiz **4** (1992), 188–216, translation in St. Petersburg Math. J. **4** (1993), 789–812.

[MP] D. Manchon and S. Paycha, *Shuffle relations for regularized integrals of symbols*, preprint, February 11, 2006.

Department of Mathematics, MIT, Cambridge, MA 02139
E-mail address: `vwg@math.mit.edu`

Department of Mathematics, Harvard University, Cambridge, MA 02138
E-mail address: `shlomo@math.harvard.edu`

Department of Mathematics, University of California,
Santa Cruz, CA 95064
E-mail address: `weitsman@math.UCSC.EDU`

Recent results on the moduli space of Riemann surfaces

Kefeng Liu

ABSTRACT. In this paper we briefly discuss some of our recent results in the study of moduli space of Riemann surfaces. It is naturally divided into two parts, one about the differential geometric aspect, another on the topological aspect of the moduli spaces. To understand the geometry of the moduli spaces we introduced new metrics, studied in detail all of the known classical complete metrics, especially the Kähler-Einstein metric. As a corollary we proved that the logarithmic cotangent bundle of the moduli space is strictly stable in the sense of Mumford. The topological results we obtained were motivated by conjectures from string theory. We will describe in this part our proofs by localization method of the Mariño-Vafa formula, its two partition analogue as well as the theory of topological vertex and the simple localization proofs of the ELSV formula and the Witten conjecture. The applications of these formulas in Gromov-Witten theory and string duality will also be mentioned.

1. Introduction

The study of moduli space and Teichmüller space has a long history. These two spaces lie in the intersections of researches in many areas of mathematics and physics. Many deep results have been obtained in history by many famous mathematicians. Moduli spaces and Teichmüller spaces of Riemann surfaces have been studied for many many years since Riemann, by Ahlfors, Bers, Royden, Deligne, Mumford, Yau, Siu, Thurston, Faltings, Witten, Kontsevich, McMullen, Gieseker, Mazur, Harris, Wolpert, Bismut, Sullivan, Madsen and many others including a young generation of mathematicians. Many aspects of the moduli spaces have been understood, but there are still many unsolved problems. Riemann was the first who considered the space $\mathcal{M}$ of all complex structures on an orientable surface modulo the action of orientation preserving diffeomorphisms. He derived the dimension of this space $\dim_{\mathbb{R}} \mathcal{M} = 6g - 6$, where $g \geq 2$ is the genus of the topological surface.

The author is supported by the NSF and NSFC.

The moduli space appears in many subjects of mathematics, from geometry, topology, algebraic geometry to number theory. For example, Faltings' proof of the Mordell conjecture depends heavily on the moduli space which can be defined over the integer ring. Moduli space also appears in many areas of theoretical physics. In string theory, many computations of path integrals are reduced to integrals of Chern classes on the moduli space. Based on conjectural physical theories, physicists have made several amazing conjectures about generating series of Hodge integrals for all genera and all marked points on the moduli spaces. The mathematical proofs of these conjectures supply strong evidences to their theories.

This article surveys two types of results; the first is on the geometric aspect of moduli spaces, and the second is on the topological aspect, in particular the computations of Hodge integrals. The first part is based on our joint work with X. Sun and S.-T. Yau. The main results are in [**34**], [**35**] and [**36**]. The second part is based on our joint work with C.-C. Liu, J. Zhou, J. Li and Y.-S. Kim. The main results are contained in [**27**], [**28**], [**21**] and [**14**]. Now we briefly describe some background and statements of the main results.

Our goal of the geometric project with Sun and Yau is to understand the geometry of the moduli spaces. More precisely, we want to understand the relationships among all of the known canonical complete metrics introduced in history on the moduli and the Teichmüller spaces, and by using them to derive geometric consequences about the moduli spaces. More importantly, we introduce and study certain new complete Kähler metrics: the Ricci metric and the perturbed Ricci metric. Through a detailed study we proved that these new metrics have very good curvature properties and Poincaré-type growth near the compactification divisor [**34**], [**35**]. In particular we proved that the perturbed Ricci metric has bounded negative Ricci and holomorphic sectional curvature and has bounded geometry. To the knowledge of the authors this is the first known such metric on the moduli space and the Teichmüller space with such good properties. We know that the Weil-Petersson metric has negative Ricci and holomorphic sectional curvature, but it is incomplete and its curvatures are not bounded from below. Also note that one has no control on the signs of the curvatures of the other complete Kähler metrics mentioned above.

We have obtained a series of results in this direction. In [**34**] and [**35**] we have proved that all of these known complete metrics are actually equivalent, and as a consequence we proved two old conjectures of Yau about the equivalence between the Kähler-Einstein metric and the Teichmüller metric and also its equivalence with the Bergman metric. In [**57**] and [**46**], which were both written in early 1980s, Yau raised various questions about the Kähler-Einstein metric on the Teichmüller space. By using the curvature properties of these new metrics, we obtained good understanding of the Kähler-Einstein metric such as its boundary behavior and the strongly bounded geometry.

As one consequence we proved the stability of the logarithmic extension of the cotangent bundle of the moduli space [**35**]. Note that the major parts of our papers were to understand the Kähler-Einstein metrics and the two new metrics. One of our goals is to find a good metric with the best possible curvature property. The perturbed Ricci metric is close to being such a metric. The most difficult part of our results is the study of the curvature properties and the asymptotic behavior of the new metrics near the boundary, only from which we can derive geometric applications such as the stability of the logarithmic cotangent bundle. The comparisons of those classical metrics as well as the two new metrics are quite easy and actually simple corollaries of the study and the basic definitions of those metrics. In particular the argument we used to prove the equivalences of the Bergman metric, the Kobayashi metric and the Carathéodory metric is rather simple from basic definitions and Yau's Schwarz lemma, and is independent of the other parts of our works.

Our results on the topological aspect of the moduli spaces are all motivated by string theory. This project on the topological aspect of the moduli spaces was jointly carried out with C.-C. Liu, J. Zhou, J. Li and Y.-S. Kim. According to string theorists, string theory, as the most promising candidate for the grand unification of all fundamental forces in nature, should be the final theory of the world, and should be unique. But now there are five different-looking string theories. As argued by the physicists, these theories should be equivalent, in a way dual to each other. On the other hand, all previous theories like the Yang-Mills and the Chern-Simons theory should be parts of string theory. In particular their partition functions should be equal or equivalent to each other in the sense that they are equal after certain transformations. To compute partition functions, physicists use localization technique, a modern version of residue theorem, on infinite dimensional spaces. More precisely they apply localization formally to path integrals, which is not well-defined yet in mathematics. In many cases such computations reduce the path integrals to certain integrals of various Chern classes on various finite dimensional moduli spaces, such as the moduli spaces of stable maps and the moduli spaces of vector bundles. The identifications of these partition functions among different theories have produced many surprisingly beautiful mathematical formulas like the famous mirror formula [**24**], as well as the Mariño-Vafa formula [**39**].

The mathematical proofs of these conjectural formulas from string duality also depend on localization techniques on these various finite dimensional moduli spaces. In this part I will briefly discuss the proof of the Mariño-Vafa formula, its generalizations and the related topological vertex theory [**1**]. More precisely we will use localization formulas in various forms to compute the integrals of Chern classes on moduli spaces, and to prove those conjectures from string duality. For the proof of the Mariño-Vafa formula and the theory of topological vertex, we note that many aspects of mathematics are involved, such as the Chern-Simons knot invariants, combinatorics of symmetric groups, representations of Kac-Moody algebras, Calabi-Yau manifolds, geometry and topology of moduli space of stable maps, etc.

We remark that localization technique has been very successful in proving many conjectures from physics, see my ICM 2002 lecture [**31**] for more examples. One of our major tools in the proofs of these conjectures is the functorial localization formula which is a variation of the classical localization formula: it transfers computations on complicated spaces to simple spaces, and connects computations of mathematicians and physicists.

Starting from the proof of the Mariño-Vafa formula [**28**], we have proved a series of results about Hodge integrals on the moduli spaces of stable curves. Complete closed formulas for the Gromov-Witten invariants of open toric Calabi-Yau manifolds are given, and their relationships with equivariant indices of elliptic operators on the moduli spaces of framed stable bundles on the projective plane are found and proved. Simple localization proofs of the ELSV formula and the Witten conjecture are discovered through this project. Here we can only give a brief overview of the results and the main ideas of their proofs. For the details see [**27**], [**22**], [**28**], [**29**], [**30**], [**21**]. While the Marinõ-Vafa formula gives a close formula for the generating series of triple Hodge integrals on the moduli spaces of all genera and any number marked points, the mathematical theory of topological vertex [**21**] gives the most effective ways to compute the Gromov-Witten invariants of any open toric Calabi-Yau manifolds. Recently Pan Peng was able to use our results on topological vertex to give a complete proof of the Gopakumar-Vafa integrality conjecture for any open toric Calabi-Yau manifolds [**45**]. Kim also used our technique to derive new effective recursion formulas for Hodge integrals on the moduli spaces of stable curves [**13**]. Together we were able to give a very simple direct proof of the Witten conjecture by using localization [**14**].

The spirit of our topological results is the duality between gauge theory, Chern-Simons theory and the Calabi-Yau geometry in string theory. One of our observations about the geometric structure of the moduli spaces is the convolution formula which is encoded in the moduli spaces of relative stable maps [**17**], [**18**], and also in the combinatorics of symmetric groups, [**28**], [**21**]. This convolution structure implies the differential equation which we called the cut-and-join equation. The cut-and-join equation arises from both representation theory and geometry. The verification of the cut-and-join equation in combinatorics is a direct computation through character formulas, while its proof in geometry is quite subtle and involves careful analysis of the fixed points on the moduli spaces of relative stable maps, see [**27**]-[**30**] and [**21**] for more details. The coincidence of such a kind of equation in both geometry and combinatorics is quite remarkable.

The mathematical theory of topological vertex was motivated by the physical theory as first developed by the Vafa group [**1**], who has been working on string duality for the past several years. Topological vertex theory is a high point of their work starting from their geometric engineering theory and Witten's conjecture that Chern-Simons theory is a string theory [**51**].

The Gopakumar-Vafa integrality conjecture is a very interesting conjecture in the subject of Gromov-Witten invariants. It is rather surprising that for some cases such invariants can be interpreted as the indices of elliptic operators in gauge theory in [**27**]. A direct proof of the conjecture for open toric Calabi-Yau manifolds was given recently by Peng [**45**], by using the combinatorial formulas for the generating series of all genera and all degree Gromov-Witten invariants of open toric Calabi-Yau. These closed formulas are derived from the theory of topological vertex through the gluing property.

This note is based on my lecture in May 2005, at the Journal of Differential Geometry Conference in memory of the late great geometer Prof. S.-S. Chern. It is essentially a combination of a survey article by Xiaofeng Sun, Shing-Tung Yau and myself on the geometric aspect of the modulis spaces [**37**] with another survey by myself on localization and string duality [**33**]. Through my research career I have been working in geometry and topology on problems related to Chern classes. Twenty years ago, at his Nankai Institute of Mathematics, a lecture of S.-S. Chern on the Atiyah-Singer index formula introduced me to the beautiful subject of geometry and topology. He described Chern classes and the Atiyah-Singer index formula and its three proofs. That is the first seminar on modern mathematics I had ever attended. It changed my life. I would like to dedicate this note to Prof. Chern for his great influence in my life and in my career.

Part I: The Geometric Aspect

2. Basics on Moduli and the Techmüller Spaces

In this section, we recall some basic facts in Teichmüller theory and introduce various notations for the following discussions.

Let Σ be an orientable surface with genus $g \geq 2$. A complex structure on Σ is a covering of Σ by charts such that the transition functions are holomorphic. By the uniformization theorem, if we put a complex structure on Σ, then it can be viewed as a quotient of the hyperbolic plane $\mathbb{H}^2$ by a Fuchsian group. Thus there is a unique Kähler-Einstein metric, or the hyperbolic metric on Σ.

Let $\mathcal{C}$ be the set of all complex structures on Σ. Let $\mathrm{Diff}^+(\Sigma)$ be the group of orientation preserving diffeomorphisms and let $\mathrm{Diff}_0^+(\Sigma)$ be the subgroup of $\mathrm{Diff}^+(\Sigma)$ consisting of those elements which are isotopic to identity.

The groups $\mathrm{Diff}^+(\Sigma)$ and $\mathrm{Diff}_0^+(\Sigma)$ act naturally on the space $\mathcal{C}$ by pull-back. The Teichmüller space is a quotient of the space $\mathcal{C}$

$$\mathcal{T}_g = \mathcal{C}/\mathrm{Diff}_0^+(\Sigma).$$

From the famous Bers embedding theorem, now we know that $\mathcal{T}_g$ can be embedded into $\mathbb{C}^{3g-3}$ as a pseudoconvex domain and is contractible. Let

$$\mathrm{Mod}_g = \mathrm{Diff}^+(\Sigma)/\mathrm{Diff}_0^+(\Sigma)$$

be the group of isotopic classes of diffeomorphisms. This group is called the (Teichmüller) moduli group or the mapping class group. Its representations are of great interest in topology and in quantum field theory.

The moduli space $\mathcal{M}_g$ is the space of distinct complex structures on Σ and is defined to be

$$\mathcal{M}_g = \mathcal{C}/\mathrm{Diff}^+(\Sigma) = \mathcal{T}_g/\mathrm{Mod}_g.$$

The moduli space is a complex orbifold.

For any point $s \in \mathcal{M}_g$, let $X = X_s$ be a representative of the corresponding class of Riemann surfaces. By the Kodaira-Spencer deformation theory and the Hodge theory, we have

$$T_X\mathcal{M}_g \cong H^1(X, T_X) = HB(X)$$

where $HB(X)$ is the space of harmonic Beltrami differentials on X.

$$T_X^*\mathcal{M}_g \cong Q(X)$$

where $Q(X)$ is the space of holomorphic quadratic differentials on X.

Pick $\mu \in HB(X)$ and $\varphi \in Q(X)$. If we fix a holomorphic local coordinate z on X, we can write $\mu = \mu(z)\frac{\partial}{\partial z} \otimes d\bar{z}$ and $\varphi = \varphi(z)dz^2$. Thus the duality between $T_X\mathcal{M}_g$ and $T_X^*\mathcal{M}_g$ is

$$[\mu : \varphi] = \int_X \mu(z)\varphi(z)dzd\bar{z}.$$

By the Riemann-Roch theorem, we have

$$\dim_{\mathbb{C}} HB(X) = \dim_{\mathbb{C}} Q(X) = 3g - 3,$$

which implies

$$\dim_{\mathbb{C}} \mathcal{T}_g = \dim_{\mathbb{C}} \mathcal{M}_g = 3g - 3.$$

3. Classical Metrics on the Moduli Spaces

In 1940s, Teichmüller considered a cover of $\mathcal{M}$ by taking the quotient of all complex structures by those orientation preserving diffeomorphims which are isotopic to the identity map. The Teichmüller space $\mathcal{T}_g$ is a contractible set in $\mathbb{C}^{3g-3}$. Furthermore, it is a pseudoconvex domain. Teichmüller also introduced the Teichmüller metric by first taking the L^1 norm on the cotangent space of $\mathcal{T}_g$ and then taking the dual norm on the tangent space. This is a Finsler metric. Two other interesting Finsler metrics are the Carathéodory metric and the Kobayashi metric. These Finsler metrics have been powerful tools in the study of the hyperbolic property of the moduli and the Teichmüller spaces and the mapping class groups. For example, in the 1970s Royden proved that the Teichmüller metric and the Kobayashi metric are the same, and as a corollary he proved the famous result that the holomorphic automorphism group of the Teichmüller space is exactly the mapping class group.

Based on the Petersson pairing on the spaces of automorphic forms, Weil introduced the first Hermitian metric on the Teichmüller space, the Weil-Petersson metric. It was shown by Ahlfors that the Weil-Petersson metric is Kähler and its holomorphic sectional curvature is negative. The work of Ahlfors and Bers on the solutions of Beltrami equation put a solid foundation of the theory of Teichmüller space and moduli space [**3**]. Wolpert studied in detail the Weil-Petersson metric including the precise upper bound of its Ricci and holomorphic sectional curvature. From these one can derive interesting applications in algebraic geometry. For example, see [**32**].

Moduli spaces of Riemann surfaces have also been studied in detail in algebraic geometry since 1960. The major tool is the geometric invariant theory developed by Mumford. In the 1970s, Deligne and Mumford studied the projective property of the moduli space and showed that the moduli space is quasi-projective and can be compactified naturally by adding in the stable nodal surfaces [**6**]. Fundamental work has been done by Gieseker, Harris and many other algebraic geometers. Note that the compactification in algebraic geometry is the same as the differential geometric compactification by using the Weil-Petersson metric.

The work of Cheng-Yau [**5**] in the early 1980s showed that there is a unique complete Kähler-Einstein metric on the Teichmüller space and is invariant under the moduli group action. Thus it descends to the moduli space. As it is well-known, the existence of the Kähler-Einstein metric gives deep algebraic geometric results, so it is natural to understand its properties like the curvature and the behaviors near the compactification divisor. In the early 1980s, Yau conjectured that the Kähler-Einstein metric is equivalent to the Teichmüller metric and the Bergman metric [**4**], [**57**], [**46**].

In 2000, McMullen introduced a new metric, the McMullen metric, by perturbing the Weil-Petersson metric to get a complete Kähler metric which is complete and Kähler hyperbolic. Thus the lowest eigenvalue of the Laplace operator is positive and the L^2-cohomology is trivial except for the middle dimension [**41**].

So there are many very famous classical metrics on the Teichmüller and the moduli spaces, and they have been studied independently by many famous mathematicians. Each metric has played an important role in the study of the geometry and topology of the moduli and Teichmüller spaces. There are three Finsler metrics: the Teichmüller metric $\|\cdot\|_T$, the Kobayashi metric $\|\cdot\|_K$ and the Carathéodory metric $\|\cdot\|_C$. They are all complete metrics on the Teichmüller space and are invariant under the moduli group action. Thus they descend down to the moduli space as complete Finsler metrics.

There are seven Kähler metrics: the Weil-Petersson metric ω_{WP} which is incomplete, the Cheng-Yau's Kähler-Einstein metric ω_{KE}, the McMullen metric ω_C, the Bergman metric ω_B, the asymptotic Poincaré metric on the moduli space ω_P, the Ricci metric ω_τ and the perturbed Ricci metric $\omega_{\tilde{\tau}}$. The last six metrics are complete. The last two metrics are new metrics studied in details in [**34**] and [**35**].

Now let us give the precise definitions of these metrics and state their basic properties.

The Teichmüller metric was first introduced by Teichmüller as the L^1 norm in the cotangent space. For each $\varphi = \varphi(z)dz^2 \in Q(X) \cong T_X^*\mathcal{M}_g$, the Teichmüller norm of φ is

$$\|\varphi\|_T = \int_X |\varphi(z)|\, dzd\bar{z}.$$

By using the duality, for each $\mu \in HB(X) \cong T_X\mathcal{M}_g$,

$$\|\mu\|_T = \sup\{Re[\mu;\varphi] \mid \|\varphi\|_T = 1\}.$$

It is known that Teichmüller metric has constant holomorphic sectional curvature -1.

The Kobayashi and the Carathéodory metrics can be defined for any complex space in the following way: Let Y be a complex manifold of dimension n. Let Δ_R be the disk in $\mathbb{C}$ with radius R. Let $\Delta = \Delta_1$ and let ρ be the Poincaré metric on Δ. Let $p \in Y$ be a point and let $v \in T_pY$ be a holomorphic tangent vector. Let $\mathrm{Hol}(Y, \Delta_R)$ and $\mathrm{Hol}(\Delta_R, Y)$ be the spaces of holomorphic maps from Y to Δ_R and from Δ_R to Y respectively. The Carathéodory norm of the vector v is defined to be

$$\|v\|_C = \sup_{f\in \mathrm{Hol}(Y,\Delta)} \|f_* v\|_{\Delta,\rho}$$

and the Kobayashi norm of v is defined to be

$$\|v\|_K = \inf_{f\in \mathrm{Hol}(\Delta_R,Y),\ f(0)=p,\ f'(0)=v} \frac{2}{R}.$$

The Bergman (pseudo) metric can also be defined for any complex space Y provided the Bergman kernel is positive. Let K_Y be the canonical bundle of Y and let W be the space of L^2 holomorphic sections of K_Y in the sense that if $\sigma \in W$, then

$$\|\sigma\|_{L^2}^2 = \int_Y (\sqrt{-1})^{n^2} \sigma \wedge \overline{\sigma} < \infty.$$

The inner product on W is defined to be

$$(\sigma, \rho) = \int_Y (\sqrt{-1})^{n^2} \sigma \wedge \overline{\rho}$$

for all $\sigma, \rho \in W$. Let $\sigma_1, \sigma_2, \ldots$ be an orthonormal basis of W. The Bergman kernel form is the non-negative (n,n)-form

$$B_Y = \sum_{j=1}^{\infty} (\sqrt{-1})^{n^2} \sigma_j \wedge \overline{\sigma}_j.$$

With a choice of local coordinates $z_i, \ldots, z_n$, we have

$$B_Y = BE_Y(z,\overline{z})(\sqrt{-1})^{n^2} dz_1 \wedge \cdots \wedge dz_n \wedge d\overline{z}_1 \wedge \cdots \wedge d\overline{z}_n$$

where $BE_Y(z,\overline{z})$ is called the Bergman kernel function. If the Bergman kernel B_Y is positive, one can define the Bergman metric

$$B_{i\overline{j}} = \frac{\partial^2 \log BE_Y(z,\overline{z})}{\partial z_i \partial \overline{z}_j}.$$

The Bergman metric is well-defined and is nondegenerate if the elements in W separate points and the first jet of Y. In this case, the Bergman metric is a Kähler metric.

REMARK 3.1. Both the Teichmüller space and the moduli space are equipped with the Bergman metrics. However, the Bergman metric on the moduli space is different from the metric induced from the Bergman metric of the Teichmüller space. The Bergman metric defined on the moduli space is incomplete due to the fact that the moduli space is quasi-projective and any L^2 holomorphic section of the canonical bundle can be extended over. However, the induced one is complete as we shall see later.

The basic properties of the Kobayashi, Carathéodory and Bergman metrics are stated in the following proposition. Please see [**15**] for the details.

PROPOSITION 3.1. *Let X be a complex space. Then*

(1) $\|\cdot\|_{C,X} \leq \|\cdot\|_{K,X}$;
(2) *Let Y be another complex space and $f : X \to Y$ be a holomorphic map. Let $p \in X$ and $v \in T_pX$. Then* $\|f_*(v)\|_{C,Y,f(p)} \leq \|v\|_{C,X,p}$ *and* $\|f_*(v)\|_{K,Y,f(p)} \leq \|v\|_{K,X,p}$;

(3) *If $X \subset Y$ is a connected open subset and $z \in X$ is a point, then with any local coordinates we have $BE_Y(z) \leq BE_X(z)$;*
(4) *If the Bergman kernel is positive, then at each point $z \in X$, a peak section σ at z exists. Such a peak section is unique up to a constant factor c with norm 1. Furthermore, with any choice of local coordinates, we have $BE_X(z) = |\sigma(z)|^2$;*
(5) *If the Bergman kernel of X is positive, then $\|\cdot\|_{C,X} \leq 2\|\cdot\|_{B,X}$;*
(6) *If X is a bounded convex domain in $\mathbb{C}^n$, then $\|\cdot\|_{C,X} = \|\cdot\|_{K,X}$;*
(7) *Let $|\cdot|$ be the Euclidean norm and let B_r be the open ball with center 0 and radius r in $\mathbb{C}^n$. Then for any holomorphic tangent vector v at 0,*

$$\|v\|_{C,B_r,0} = \|v\|_{K,B_r,0} = \frac{2}{r}|v|,$$

where $|v|$ is the Euclidean norm of v.

The three Finsler metrics have been very powerful tools in understanding the hyperbolic geometry of the moduli spaces, and the mapping class group. It has also been known since the 1970s that the Bergman metric on the Teichmüller space is complete.

The Weil-Petersson metric is the first Kähler metric defined on the Teichmüller and the moduli space. It is defined by using the L^2 inner product on the tangent space in the following way:
Let $\mu, \nu \in T_X\mathcal{M}_g$ be two tangent vectors and let λ be the unique Kähler-Einstein metric on X. Then the Weil-Petersson metric is

$$h(\mu, \nu) = \int_X \mu\bar{\nu}\, dv$$

where $dv = \frac{\sqrt{-1}}{2}\lambda dz \wedge d\bar{z}$ is the volume form. Details can be found in [**34**], [**40**] and [**54**].

The curvatures of the Weil-Petersson metric have been well-understood due to the works of Ahlfors, Royden and Wolpert. Its Ricci and holomorphic sectional curvature are all negative with negative upper bound, but with no lower bound. Its boundary behavior is understood, from which it is not hard to see that it is an incomplete metric.

The existence of the Kähler-Einstein metric was given by the work of Cheng-Yau [**4**]. Its Ricci curvature is -1. Namely,

$$\partial\bar{\partial}\log \omega_{KE}^n = \omega_{KE},$$

where $n = 3g - 3$. They actually proved that a bounded domain in $\mathbb{C}^n$ admits a complete Kähler-Einstein metric if and only if it is pseudoconvex.

The McMullen $1/l$ metric defined in [**41**] is a perturbation of the Weil-Petersson metric by adding a Kähler form whose potential involves the short geodesic length functions on the Riemann surfaces. For each simple closed curve γ in X, let $l_\gamma(X)$ be the length of the unique geodesic in the homotopy class of γ with respect to the unique Kähler-Einstein metric. Then the

McMullen metric is defined as

$$\omega_{1/l} = \omega_{WP} - i\delta \sum_{l_\gamma(X)<\epsilon} \partial\bar{\partial}\mathrm{Log}\frac{\epsilon}{l_\gamma}$$

where ϵ and δ are small positive constants and $\mathrm{Log}(x)$ is a smooth function defined as

$$\mathrm{Log}(x) = \begin{cases} \log x, & x \geq 2, \\ 0, & x \leq 1. \end{cases}$$

This metric is Kähler hyperbolic, which means it satisfies the following conditions:

(1) $(\mathcal{M}_g, \omega_{1/l})$ has finite volume;
(2) The sectional curvature of $(\mathcal{M}_g, \omega_{1/l})$ is bounded above and below;
(3) The injectivity radius of $(\mathcal{T}_g, \omega_{1/l})$ is bounded below;
(4) On $\mathcal{T}_g$, the Kähler form $\omega_{1/l}$ can be written as $\omega_{1/l} = d\alpha$ where α is a bounded 1-form.

An immediate consequence of the Kähler hyperbolicity is that the L^2-cohomology is trivial except for the middle dimension.

The asymptotic Poincaré metric can be defined as a complete Kähler metric on a complex manifold M which is obtained by removing a divisor Y with only normal crossings from a compact Kähler manifold $(\overline{M}, \omega)$.

Let $\overline{M}$ be a compact Kähler manifold of dimension m. Let $Y \subset \overline{M}$ be a divisor of normal crossings and let $M = \overline{M} \setminus Y$. Cover $\overline{M}$ by coordinate charts $U_1, \ldots, U_p, \ldots, U_q$ such that $(\overline{U}_{p+1} \cup \cdots \cup \overline{U}_q) \cap Y = \Phi$. We also assume that for each $1 \leq \alpha \leq p$, there is a constant n_α such that $U_\alpha \setminus Y = (\Delta^*)^{n_\alpha} \times \Delta^{m-n_\alpha}$ and on U_α, Y is given by $z_1^\alpha \cdots z_{n_\alpha}^\alpha = 0$. Here Δ is the disk of radius $\frac{1}{2}$ and Δ^* is the punctured disk of radius $\frac{1}{2}$. Let $\{\eta_i\}_{1\leq i\leq q}$ be the partition of unity subordinate to the cover $\{U_i\}_{1\leq i\leq q}$. Let ω be a Kähler metric on $\overline{M}$ and let C be a positive constant. Then for C large, the Kähler form

$$\omega_p = C\omega + \sum_{i=1}^{p} \sqrt{-1}\partial\bar{\partial}\left(\eta_i \log\log \frac{1}{z_1^i \cdots z_{n_i}^i}\right)$$

defines a complete metric on M with finite volume since on each U_i with $1 \leq i \leq p$, ω_p is bounded from above and below by the local Poincaré metric on U_i. We call this metric the asymptotic Poincaré metric.

The signs of the curvatures of the above metrics are all unknown. We actually only know that the Kähler-Einstein metric has constant negative Ricci curvature and that the McMullen metric has bounded geometry. Also, except the asymptotic Poincaré metric, the boundary behaviors of the other metrics are unknown before our works [**34**], [**35**]. It is interesting that to understand them we need to introduce new metrics.

Now we define the Ricci metric and the perturbed Ricci metric. The curvature properties and asymptotics of these two new metrics are understood by us and will be stated in the following sections. Please also see [**34**] and [**35**] for details.

With the works of Ahlfors, Royden and Wolpert we know that the Ricci curvature of the Weil-Petersson metric has a negative upper bound. Thus we can use the negative Ricci form of the Weil-Petersson metric as the Kähler form of a new metric. We call this metric the Ricci metric and denote it by τ. That is

$$\omega_\tau = \partial\overline{\partial} \log \omega_{WP}^n.$$

Through the careful analysis, we now understand that the Ricci metric is a natural canonical complete Kähler metric with many good properties. However, its holomorphic sectional curvature is only asymptotically negative. To get a metric with good sign on its curvatures, we introduced the perturbed Ricci metric $\omega_{\widetilde{\tau}}$ as a combination of the Ricci metric and the Weil-Petersson metric:

$$\omega_{\widetilde{\tau}} = \omega_\tau + C\omega_{WP}$$

where C is a large positive constant. As we shall see later, the perturbed Ricci metric has desired curvature properties so that we can put it either on the target or on the domain manifold in Yau's Schwarz lemma, from which we can compare the above metrics.

4. The Curvature Formulas

In this section we describe the harmonic lift of a vector field on the moduli space to the universal curve due to Royden, Siu [**48**] and Schumacher [**47**]. Details can also be found in [**34**]. We then use this method to derive the curvature formula for the Weil-Petersson metric, the Ricci metric and the perturbed Ricci metric.

To compute the curvature of a metric on the moduli space, we need to take derivatives of the metric in the direction of the moduli space. However, it is quite difficult to estimate the curvature by using a formula obtained in such a way. The central idea is to obtain a formula where the derivatives are taken in the fiber direction. We can view the deformation of complex structures on a topological surface as the deformation of the Kähler-Einstein metrics.

Let $\mathcal{M}_g$ be the moduli space of Riemann surfaces of genus g, where $g \geq 2$. Let $n = 3g-3$ be the complex dimension of $\mathcal{M}_g$. Let $\mathfrak{X}$ be the total space and let $\pi : \mathfrak{X} \to \mathcal{M}_g$ be the projection map.

Let $s_1, \ldots, s_n$ be holomorphic local coordinates near a regular point $s \in \mathcal{M}_g$ and assume that z is a holomorphic local coordinate on the fiber $X_s = \pi^{-1}(s)$. For holomorphic vector fields $\frac{\partial}{\partial s_1}, \ldots, \frac{\partial}{\partial s_n}$, there are vector fields $v_1, \ldots, v_n$ on $\mathfrak{X}$ such that

(1) $\pi_*(v_i) = \frac{\partial}{\partial s_i}$ for $i = 1, \ldots, n$;

(2) $\overline{\partial} v_i$ are harmonic TX_s-valued $(0,1)$ forms for $i = 1, \ldots, n$.

The vector fields $v_1, \ldots, v_n$ are called the harmonic lift of the vectors $\frac{\partial}{\partial s_1}$, $\ldots, \frac{\partial}{\partial s_n}$. The existence of such harmonic vector fields was pointed out by Siu. Schumacher in his work gave an explicit construction of such lift. We now describe it.

Since $g \geq 2$, we can assume that each fiber is equipped with the Kähler-Einstein metric $\lambda = \frac{\sqrt{-1}}{2}\lambda(z,s)dz \wedge d\overline{z}$. The Kähler-Einstein condition gives the following equation:

$$\partial_z \partial_{\overline{z}} \log \lambda = \lambda. \tag{4.1}$$

For the rest of this paper we denote $\frac{\partial}{\partial s_i}$ by ∂_i and $\frac{\partial}{\partial z}$ by ∂_z. Let

$$a_i = -\lambda^{-1}\partial_i \partial_{\overline{z}} \log \lambda$$

and let

$$A_i = \partial_{\overline{z}} a_i.$$

Then the harmonic horizontal lift of ∂_i is

$$v_i = \partial_i + a_i \partial_z.$$

In particular

$$B_i = A_i \partial_z \otimes d\overline{z} \in H^1(X_s, T_{X_s})$$

is harmonic. Furthermore, the lift $\partial_i \mapsto B_i$ gives the Kodaira-Spencer map $T_s\mathcal{M}_g \to H^1(X_s, T_{X_s})$. Thus the Weil-Petersson metric on $\mathcal{M}_g$ is

$$h_{i\overline{j}}(s) = \int_{X_s} B_i \cdot \overline{B_j}\, dv = \int_{X_s} A_i \overline{A_j}\, dv,$$

where $dv = \frac{\sqrt{-1}}{2}\lambda dz \wedge d\overline{z}$ is the volume form on the fiber X_s.

Let $R_{i\overline{j}k\overline{l}}$ be the curvature tensor of the Weil-Petersson metric. Here we adopt the following notation for the curvature of a Kähler metric:
For a Kähler metric (M, g), the curvature tensor is given by

$$R_{i\overline{j}k\overline{l}} = \frac{\partial^2 g_{i\overline{j}}}{\partial z_k \partial \overline{z}_l} - g^{p\overline{q}} \frac{\partial g_{i\overline{q}}}{\partial z_k} \frac{\partial g_{p\overline{j}}}{\partial \overline{z}_l}.$$

In this case, the Ricci curvature is given by

$$R_{i\overline{j}} = -g^{k\overline{l}} R_{i\overline{j}k\overline{l}}.$$

By using the curvature of the Weil-Petersson metric, we can define the Ricci metric:

$$\tau_{i\overline{j}} = h^{k\overline{l}} R_{i\overline{j}k\overline{l}}$$

and the perturbed Ricci metric:

$$\widetilde{\tau}_{i\overline{j}} = \tau_{i\overline{j}} + C h_{i\overline{j}},$$

where C is a positive constant.

Before we present the curvature formulas for the above metrics, we need to introduce the Maass operators and norms on a Riemann surface [**54**].

Let X be a Riemann surface and let κ be its canonical bundle. For any integer p, let $S(p)$ be the space of smooth sections of $(\kappa \otimes \overline{\kappa}^{-1})^{\frac{p}{2}}$. Fix a conformal metric $ds^2 = \rho^2(z)|dz|^2$. In the following, we will take ds^2 to be the Kähler-Einstein metric although the following definitions work for all metrics.

The Maass operators K_p and L_p are defined to be the metric derivatives $K_p : S(p) \to S(p+1)$ and $L_p : S(p) \to S(p-1)$ given by

$$K_p(\sigma) = \rho^{p-1}\partial_z(\rho^{-p}\sigma)$$

and

$$L_p(\sigma) = \rho^{-p-1}\partial_{\overline{z}}(\rho^p\sigma)$$

where $\sigma \in S(p)$.

The operators $P = K_1K_0$ and $\Box = -L_1K_0$ will play important roles in the curvature formulas. Here the operator $\Box$ is just the Laplace operator. We also let $T = (\Box + 1)^{-1}$ be the Green operator.

Each element $\sigma \in S(p)$ has a well-defined absolute value $|\sigma|$ which is independent of the choice of local coordinate. We define the C^k norm of σ: Let Q be an operator which is a composition of operators K_* and L_*. Denote by $|Q|$ the number of factors. For any $\sigma \in S(p)$, define

$$\|\sigma\|_0 = \sup_X |\sigma|$$

and

$$\|\sigma\|_k = \sum_{|Q|\leq k} \|Q\sigma\|_0.$$

We can also localize the norm on a subset of X. Let $\Omega \subset X$ be a domain. We can define

$$\|\sigma\|_{0,\Omega} = \sup_\Omega |\sigma|$$

and

$$\|\sigma\|_{k,\Omega} = \sum_{|Q|\leq k} \|Q\sigma\|_{0,\Omega}.$$

We let $f_{i\overline{j}} = A_i\overline{A}_j$ and $e_{i\overline{j}} = T(f_{i\overline{j}})$. These functions will be the building blocks for the curvature formulas.

The trick of converting derivatives from the moduli directions to the fiber directions is the following lemma due to Siu and Schumacher:

LEMMA 4.1. *Let η be a relative $(1,1)$-form on the total space $\mathfrak{X}$. Then*

$$\frac{\partial}{\partial s_i}\int_{X_s} \eta = \int_{X_s} L_{v_i}\eta.$$

The curvature formula of the Weil-Petersson metric was first established by Wolpert by using a different method [**52**] and later was generalized by Siu [**48**] and Schumacher [**47**] by using the above lemma:

THEOREM 4.2. *The curvature of the Weil-Petersson metric is given by*

$$R_{i\bar{j}k\bar{l}} = \int_{X_s} (e_{i\bar{j}} f_{k\bar{l}} + e_{i\bar{l}} f_{k\bar{j}})\, dv. \tag{4.2}$$

For the proof, please see [**34**]. From this formula it is rather easy to show that the Ricci and the holomorphic sectional curvature have explicit negative upper bound.

To establish the curvature formula of the Ricci metric, we need to introduce more operators. Firstly, the commutator of the operator v_k and $(\square+1)$ will play an important role. Here we view the vector field v_k as a operator acting on functions. We define

$$\xi_k = [\square + 1, v_k].$$

A direct computation shows that

$$\xi_k = -A_k P.$$

Also we can define the commutator of $\overline{v_l}$ and ξ_k. Let

$$Q_{k\bar{l}} = [\overline{v_l}, \xi_k].$$

We have

$$Q_{k\bar{l}}(f) = \overline{P}(e_{k\bar{l}})P(f) - 2f_{k\bar{l}}\square f + \lambda^{-1}\partial_z f_{k\bar{l}} \partial_{\bar{z}} f$$

for any smooth function f.

To simplify the notation, we introduce the symmetrization operator of the indices. Let U be any quantity which depends on indices $i, k, \alpha, \bar{j}, \bar{l}, \bar{\beta}$. The symmetrization operator σ_1 is defined by taking summation of all orders of the triple (i, k, α). That is

$$\begin{aligned}\sigma_1(U(i,k,\alpha,\bar{j},\bar{l},\bar{\beta})) = {} & U(i,k,\alpha,\bar{j},\bar{l},\bar{\beta}) + U(i,\alpha,k,\bar{j},\bar{l},\bar{\beta}) \\ & + U(k,i,\alpha,\bar{j},\bar{l},\bar{\beta}) + U(k,\alpha,i,\bar{j},\bar{l},\bar{\beta}) \\ & + U(\alpha,i,k,\bar{j},\bar{l},\bar{\beta}) + U(\alpha,k,i,\bar{j},\bar{l},\bar{\beta}).\end{aligned}$$

Similarly, σ_2 is the symmetrization operator of $\bar{j}$ and $\bar{\beta}$ and $\widetilde{\sigma_1}$ is the symmetrization operator of $\bar{j}$, $\bar{l}$ and $\bar{\beta}$.

In [**34**] the following curvature formulas for the Ricci and perturbed Ricci metric were proved:

THEOREM 4.3. *Let $s_1, \ldots, s_n$ be local holomorphic coordinates at $s \in M_g$. Then at s, we have*

$$
\begin{aligned}
(4.3)\quad \widetilde{R}_{i\bar{j}k\bar{l}} = h^{\alpha\bar{\beta}}\Big\{\sigma_1\sigma_2 \int_{X_s} & \Big\{(\Box+1)^{-1}(\xi_k(e_{i\bar{j}}))\bar{\xi}_l(e_{\alpha\bar{\beta}}) \\
& + (\Box+1)^{-1}(\xi_k(e_{i\bar{j}}))\bar{\xi}_\beta(e_{\alpha\bar{l}})\Big\}\, dv\Big\} \\
+ h^{\alpha\bar{\beta}}&\Big\{\sigma_1 \int_{X_s} Q_{k\bar{l}}(e_{i\bar{j}})e_{\alpha\bar{\beta}}\, dv\Big\} \\
- \tau^{p\bar{q}}h^{\alpha\bar{\beta}}h^{\gamma\bar{\delta}}&\Big\{\sigma_1 \int_{X_s} \xi_k(e_{i\bar{q}})e_{\alpha\bar{\beta}}\, dv\Big\}\Big\{\widetilde{\sigma}_1 \int_{X_s} \bar{\xi}_l(e_{p\bar{j}})e_{\gamma\bar{\delta}})\, dv\Big\} \\
+ \tau_{p\bar{j}}h^{p\bar{q}}&R_{i\bar{q}k\bar{l}}
\end{aligned}
$$

and

$$
\begin{aligned}
(4.4)\quad P_{i\bar{j}k\bar{l}} = h^{\alpha\bar{\beta}}\Big\{\sigma_1\sigma_2 \int_{X_s} & \Big\{(\Box+1)^{-1}(\xi_k(e_{i\bar{j}}))\bar{\xi}_l(e_{\alpha\bar{\beta}}) \\
& + (\Box+1)^{-1}(\xi_k(e_{i\bar{j}}))\bar{\xi}_\beta(e_{\alpha\bar{l}})\Big\}\, dv\Big\} \\
+ h^{\alpha\bar{\beta}}&\Big\{\sigma_1 \int_{X_s} Q_{k\bar{l}}(e_{i\bar{j}})e_{\alpha\bar{\beta}}\, dv\Big\} \\
- \widetilde{\tau}^{p\bar{q}}h^{\alpha\bar{\beta}}h^{\gamma\bar{\delta}}&\Big\{\sigma_1 \int_{X_s} \xi_k(e_{i\bar{q}})e_{\alpha\bar{\beta}}\, dv\Big\}\Big\{\widetilde{\sigma}_1 \int_{X_s} \bar{\xi}_l(e_{p\bar{j}})e_{\gamma\bar{\delta}})\, dv\Big\} \\
+ \tau_{p\bar{j}}h^{p\bar{q}}&R_{i\bar{q}k\bar{l}} + CR_{i\bar{j}k\bar{l}}
\end{aligned}
$$

where $R_{i\bar{j}k\bar{l}}$, $\widetilde{R}_{i\bar{j}k\bar{l}}$, and $P_{i\bar{j}k\bar{l}}$ are the curvature of the Weil-Petersson metric, the Ricci metric and the perturbed Ricci metric, respectively.

Unlike the curvature formula of the Weil-Petersson metric, from which we can see the sign of the curvature directly, the above formulas are too complicated and we cannot see the sign. So we need to study the asymptotic behaviors of these curvatures, and first the metrics themselves.

5. The Asymptotics

To compute the asymptotics of these metrics and their curvatures, we first need to find a canonical way to construct local coordinates near the boundary of the moduli space. We first describe the Deligne-Mumford compactification of the moduli space and introduce the pinching coordinate and the plumbing construction according to Earle and Marden. Please see [**40**], [**54**], [**49**] and [**34**] for details.

A point p in a Riemann surface X is a node if there is a neighborhood of p which is isometric to the germ $\{(u,v) \mid uv = 0,\ |u|, |v| < 1\} \subset \mathbb{C}^2$. Let $p_1, \ldots, p_k$ be the nodes on X. X is called stable if each connected component of $X\backslash\{p_1,\dots,p_k\}$ has negative Euler characteristic. Namely, each connected component has a unique complete hyperbolic metric.

Let $\mathcal{M}_g$ be the moduli space of Riemann surfaces of genus $g\geq 2$. The Deligne-Mumford compactification $\overline{\mathcal{M}}_g$ is the union of $\mathcal{M}_g$ and corresponding stable nodal surfaces [**6**]. Each point $y\in\overline{\mathcal{M}}_g\setminus\mathcal{M}_g$ corresponds to a stable noded surface X_y.

We recall the rs-coordinate on a Riemann surface defined by Wolpert in [**54**]. There are two cases: the puncture case and the short geodesic case. For the puncture case, we have a noded surface X and a node $p\in X$. Let a,b be two punctures which are paired to form p.

DEFINITION 5.1. The local coordinate chart (U,u) near a is called rs-coordinate if $u(a)=0$, u maps U to the punctured disc $0<|u|<c$ with $c>0$ and the restriction to U of the Kähler-Einstein metric on X can be written as $\frac{1}{2|u|^2(\log|u|)^2}|du|^2$. The rs-coordinate (V,v) near b is defined in a similar way.

For the short geodesic case, we have a closed surface X, a closed geodesic $\gamma\subset X$ with length $l<c_*$, where c_* is the collar constant.

DEFINITION 5.2. The local coordinate chart (U,z) is called rs-coordinate at γ if $\gamma\subset U$, z maps U to the annulus $c^{-1}|t|^{\frac{1}{2}}<|z|<c|t|^{\frac{1}{2}}$ and the Kähler-Einstein metric on X can be written as

$$\frac{1}{2}\left(\frac{\pi}{\log|t|}\frac{1}{|z|}\csc\frac{\pi\log|z|}{\log|t|}\right)^2|dz|^2.$$

REMARK 5.1. We put the factor $\frac{1}{2}$ in the above two definitions to normalize such that (4.1) holds.

By Keen's collar theorem, we have the following lemma:

LEMMA 5.3. *Let X be a closed surface and let γ be a closed geodesic on X such that the length l of γ satisfies $l<c_*$. Then there is a collar Ω on X with holomorphic coordinate z defined on Ω such that*

(1) *z maps Ω to the annulus $\frac{1}{c}e^{-\frac{2\pi^2}{l}}<|z|<c$ for $c>0$;*
(2) *the Kähler-Einstein metric on X restricted to Ω is given by*

$$(\tfrac{1}{2}u^2r^{-2}\csc^2\tau)|dz|^2, \tag{5.1}$$

where $u=\frac{l}{2\pi}$, $r=|z|$ and $\tau=u\log r$;
(3) *the geodesic γ is given by $|z|=e^{-\frac{\pi^2}{l}}$.*

We call such a collar Ω a genuine collar.

We notice that the constant c in the above lemma has a lower bound such that the area of Ω is bounded from below. Also, the coordinate z in the above lemma is rs-coordinate. In the following, we will keep the notation u, r and τ.

Now we describe the local manifold cover of $\overline{\mathcal{M}}_g$ near the boundary. We take the construction of Wolpert [**54**]. Let $X_{0,0}$ be a noded surface corresponding to a codimension m boundary point. $X_{0,0}$ have m nodes $p_1, \ldots, p_m$. $X_0 = X_{0,0} \setminus \{p_1, \ldots, p_m\}$ is a union of punctured Riemann surfaces. Fix rs-coordinate charts (U_i, η_i) and (V_i, ζ_i) at p_i for $i = 1, \ldots, m$ such that all the U_i and V_i are mutually disjoint. Now pick an open set $U_0 \subset X_0$ such that the intersection of each connected component of X_0 and U_0 is a nonempty relatively compact set and the intersection $U_0 \cap (U_i \cup V_i)$ is empty for all i. Now pick Beltrami differentials $\nu_{m+1}, \ldots, \nu_n$ which are supported in U_0 and span the tangent space at X_0 of the deformation space of X_0. For $s = (s_{m+1}, \ldots, s_n)$, let $\nu(s) = \sum_{i=m+1}^{n} s_i \nu_i$. We assume $|s| = (\sum |s_i|^2)^{\frac{1}{2}}$ is small enough such that $|\nu(s)| < 1$. The noded surface $X_{0,s}$ is obtained by solving the Beltrami equation $\bar{\partial} w = \nu(s) \partial w$. Since $\nu(s)$ is supported in U_0, (U_i, η_i) and (V_i, ζ_i) are still holomorphic coordinates on $X_{0,s}$. Note that they are no longer rs-coordinates. By the theory of Alhfors and Bers [**3**] and Wolpert [**54**] we can assume that there are constants $\delta, c > 0$ such that when $|s| < \delta$, η_i and ζ_i are holomorphic coordinates on $X_{0,s}$ with $0 < |\eta_i| < c$ and $0 < |\zeta_i| < c$. Now we assume $t = (t_1, \ldots, t_m)$ has small norm. We do the plumbing construction on $X_{0,s}$ to obtain $X_{t,s}$. Remove from $X_{0,s}$ the discs $0 < |\eta_i| \leq \frac{|t_i|}{c}$ and $0 < |\zeta_i| \leq \frac{|t_i|}{c}$ for each $i = 1, \ldots, m$. Now identify $\frac{|t_i|}{c} < |\eta_i| < c$ with $\frac{|t_i|}{c} < |\zeta_i| < c$ by the rule $\eta_i \zeta_i = t_i$. This defines the surface $X_{t,s}$. The tuple $(t_1, \ldots, t_m, s_{m+1}, \ldots, s_n)$ are the local pinching coordinates for the manifold cover of $\overline{\mathcal{M}}_g$. We call the coordinates η_i (or ζ_i) the plumbing coordinates on $X_{t,s}$ and the collar defined by $\frac{|t_i|}{c} < |\eta_i| < c$ the plumbing collar.

REMARK 5.2. By the estimate of Wolpert [**53**], [**54**] on the length of short geodesic, the quantity $u_i = \frac{l_i}{2\pi} \sim -\frac{\pi}{\log |t_i|}$.

Now we describe the estimates of the asymptotics of these metrics and their curvatures. The principle is that, when we work on a nearly degenerated surface, the geometry focuses on the collars. Our curvature formulas depend on the Kähler-Einstein metrics of the family of Riemann surfaces near a boundary points. One can obtain an approximate Kähler-Einstein metric on these collars by the graft construction of Wolpert [**54**] which is done by gluing the hyperbolic metric on the nodal surface with the model metric described above.

To use the curvature formulas (4.2), (4.3) and (4.4) to estimate the asymptotic behavior, one also needs to analyze the transition from the plumbing coordinates on the collars to the rs-coordinates. The harmonic Beltrami differentials were constructed by Masur [**40**] by using the plumbing coordinates, and it is easier to compute the integration by using rs-coordinates. Such computation was done in [**49**] by using the graft metric of Wolpert and the maximum principle. A clear description can be found in [**34**]. We have the following theorem:

THEOREM 5.4. *Let (t,s) be the pinching coordinates on $\overline{\mathcal{M}}_g$ near $X_{0,0}$ which corresponds to a codimension m boundary point of $\overline{\mathcal{M}}_g$. Then there exist constants $M, \delta > 0$ and $0 < c < 1$ such that if $|(t,s)| < \delta$, then the j-th plumbing collar on $X_{t,s}$ contains the genuine collar Ω_c^j. Furthermore, one can choose rs-coordinate z_j on the collar Ω_c^j properly such that the holomorphic quadratic differentials $\psi_1, \ldots, \psi_n$ corresponding to the cotangent vectors $dt_1, \ldots, ds_n$ have form $\psi_i = \varphi_i(z_j)dz_j^2$ on the genuine collar Ω_c^j for $1 \le j \le m$ where*

(1) $\varphi_i(z_j) = \frac{1}{z_j^2}(q_i^j(z_j) + \beta_i^j)$ *if* $i \ge m+1$;
(2) $\varphi_i(z_j) = (-\frac{t_j}{\pi})\frac{1}{z_j^2}(q_j(z_j) + \beta_j)$ *if* $i = j$;
(3) $\varphi_i(z_j) = (-\frac{t_i}{\pi})\frac{1}{z_j^2}(q_i^j(z_j) + \beta_i^j)$ *if* $1 \le i \le m$ *and* $i \ne j$.

Here β_i^j and β_j are functions of (t,s), q_i^j and q_j are functions of (t,s,z_j) given by

$$q_i^j(z_j) = \sum_{k<0} \alpha_{ik}^j(t,s)t_j^{-k}z_j^k + \sum_{k>0} \alpha_{ik}^j(t,s)z_j^k$$

and

$$q_j(z_j) = \sum_{k<0} \alpha_{jk}(t,s)t_j^{-k}z_j^k + \sum_{k>0} \alpha_{jk}(t,s)z_j^k$$

such that

(1) $\sum_{k<0} |\alpha_{ik}^j| c^{-k} \le M$ *and* $\sum_{k>0} |\alpha_{ik}^j| c^k \le M$ *if* $i \ne j$;
(2) $\sum_{k<0} |\alpha_{jk}| c^{-k} \le M$ *and* $\sum_{k>0} |\alpha_{jk}| c^k \le M$;
(3) $|\beta_i^j| = O(|t_j|^{\frac{1}{2}-\epsilon})$ *with* $\epsilon < \frac{1}{2}$ *if* $i \ne j$;
(4) $|\beta_j| = (1 + O(u_0))$

where $u_0 = \sum_{i=1}^m u_i + \sum_{j=m+1}^n |s_j|$.

By definition, the metric on the cotangent bundle induced by the Weil-Petersson metric is given by

$$h^{i\bar{j}} = \int_{X_{t,s}} \lambda^{-2} \varphi_i \overline{\varphi}_j \, dv.$$

We then have the following series of estimates, see [**34**]. First by using this formula and taking inverse, we can estimate the Weil-Petersson metric.

THEOREM 5.5. *Let (t,s) be the pinching coordinates. Then*

(1) $h^{i\bar{i}} = 2u_i^{-3}|t_i|^2(1+O(u_0))$ *and* $h_{i\bar{i}} = \frac{1}{2}\frac{u_i^3}{|t_i|^2}(1+O(u_0))$ *for* $1 \le i \le m$;
(2) $h^{i\bar{j}} = O(|t_i t_j|)$ *and* $h_{i\bar{j}} = O(\frac{u_i^3 u_j^3}{|t_i t_j|})$ *if* $1 \le i,j \le m$ *and* $i \ne j$;
(3) $h^{i\bar{j}} = O(1)$ *and* $h_{i\bar{j}} = O(1)$ *if* $m+1 \le i,j \le n$;
(4) $h^{i\bar{j}} = O(|t_i|)$ *and* $h_{i\bar{j}} = O(\frac{u_i^3}{|t_i|})$ *if* $i \le m < j$ *or* $j \le m < i$.

Then we use the duality to construct the harmonic Beltrami differentials. We have

LEMMA 5.6. *On the genuine collar Ω_c^j for c small, the coefficient functions A_i of the harmonic Beltrami differentials have the form:*

(1) $A_i = \frac{z_j}{\bar{z}_j} \sin^2 \tau_j \overline{(p_i^j(z_j) + b_i^j)}$ *if* $i \neq j$;

(2) $A_j = \frac{z_j}{\bar{z}_j} \sin^2 \tau_j \overline{(p_j(z_j) + b_j)}$

where

(1) $p_i^j(z_j) = \sum_{k\leq -1} a_{ik}^j \rho_j^{-k} z_j^k + \sum_{k\geq 1} a_{ik}^j z_j^k$ *if* $i \neq j$;

(2) $p_j(z_j) = \sum_{k\leq -1} a_{jk} \rho_j^{-k} z_j^k + \sum_{k\geq 1} a_{jk} z_j^k$.

In the above expressions, $\rho_j = e^{-\frac{2\pi^2}{l_j}}$ *and the coefficients satisfy the following conditions:*

(1) $\sum_{k\leq -1} |a_{ik}^j| c^{-k} = O(u_j^{-2})$ *and* $\sum_{k\geq 1} |a_{ik}^j| c^k = O(u_j^{-2})$ *if* $i \geq m+1$;

(2) $\sum_{k\leq -1} |a_{ik}^j| c^{-k} = O(u_j^{-2}) O(\frac{u_i^3}{|t_i|})$ *and* $\sum_{k\geq 1} |a_{ik}^j| c^k = O(u_j^{-2}) O(\frac{u_i^3}{|t_i|})$ *if* $i \leq m$ *and* $i \neq j$;

(3) $\sum_{k\leq -1} |a_{jk}| c^{-k} = O(\frac{u_j}{|t_j|})$ *and* $\sum_{k\geq 1} |a_{jk}| c^k = O(\frac{u_j}{|t_j|})$;

(4) $|b_i^j| = O(u_j)$ *if* $i \geq m+1$;

(5) $|b_i^j| = O(u_j) O(\frac{u_i^3}{|t_i|})$ *if* $i \leq m$ *and* $i \neq j$;

(6) $b_j = -\frac{u_j}{\pi \bar{t}_j}(1 + O(u_0))$.

To use the curvature formulas to estimate the Ricci metric and the perturbed Ricci metric, one needs to find accurate estimate of the operator $T = (\square + 1)^{-1}$. More precisely, one needs to estimate the functions $e_{i\bar{j}} = T(f_{i\bar{j}})$. To avoid writing down the Green function of T, we construct approximate solutions and localize on the collars in [**34**]. Pick a positive constant $c_1 < c$ and define the cut-off function $\eta \in C^\infty(\mathbb{R}, [0,1])$ by

$$
\begin{cases} \eta(x) = 1, & x \leq \log c_1, \\ \eta(x) = 0, & x \geq \log c, \\ 0 < \eta(x) < 1, & \log c_1 < x < \log c. \end{cases} \tag{5.2}
$$

It is clear that the derivatives of η are bounded by constants which only depend on c and c_1. Let $\widetilde{e_{i\bar{j}}}(z)$ be the function on X defined in the following way, where z is taken to be z_i on the collar Ω_c^i:

(1) if $i \leq m$ and $j \geq m+1$, then

$$
\widetilde{e_{i\bar{j}}}(z) = \begin{cases} \frac{1}{2}\sin^2 \tau_i \overline{b_i} b_j^i, & z \in \Omega_{c_1}^i, \\ (\frac{1}{2}\sin^2 \tau_i \overline{b_i} b_j^i)\eta(\log r_i), & z \in \Omega_c^i \text{ and } c_1 < r_i < c, \\ (\frac{1}{2}\sin^2 \tau_i \overline{b_i} b_j^i)\eta(\log \rho_i - \log r_i), & z \in \Omega_c^i \text{ and} \\ & c^{-1}\rho_i < r_i < c_1^{-1}\rho_i, \\ 0, & z \in X \setminus \Omega_c^i. \end{cases}
$$

(2) if $i, j \le m$ and $i \neq j$, then

$$\widetilde{e_{i\bar{j}}}(z) = \begin{cases} \frac{1}{2}\sin^2 \tau_i \overline{b_i} b_j^i, & z \in \Omega_{c_1}^i, \\ (\frac{1}{2}\sin^2 \tau_i \overline{b_i} b_j^i)\eta(\log r_i), & z \in \Omega_c^i \text{ and } c_1 < r_i < c, \\ (\frac{1}{2}\sin^2 \tau_i \overline{b_i} b_j^i)\eta(\log \rho_i - \log r_i), & z \in \Omega_c^i \text{ and} \\ & c^{-1}\rho_i < r_i < c_1^{-1}\rho_i, \\ \frac{1}{2}\sin^2 \tau_j \overline{b_i^j} b_j, & z \in \Omega_{c_1}^j, \\ (\frac{1}{2}\sin^2 \tau_i \overline{b_i^j} b_j)\eta(\log r_j), & z \in \Omega_c^j \text{ and } c_1 < r_j < c, \\ (\frac{1}{2}\sin^2 \tau_i \overline{b_i^j} b_j)\eta(\log \rho_j - \log r_j), & z \in \Omega_c^j \text{ and} \\ & c^{-1}\rho_j < r_j < c_1^{-1}\rho_j, \\ 0, & z \in X \setminus (\Omega_c^i \cup \Omega_c^j). \end{cases}$$

(3) if $i \le m$, then

$$\widetilde{e_{i\bar{i}}}(z) = \begin{cases} \frac{1}{2}\sin^2 \tau_i |b_i|^2, & z \in \Omega_{c_1}^i, \\ (\frac{1}{2}\sin^2 \tau_i |b_i|^2)\eta(\log r_i), & z \in \Omega_c^i \text{ and } c_1 < r_i < c, \\ (\frac{1}{2}\sin^2 \tau_i |b_i|^2)\eta(\log \rho_i - \log r_i), & z \in \Omega_c^i \text{ and} \\ & c^{-1}\rho_i < r_i < c_1^{-1}\rho_i, \\ 0, & z \in X \setminus \Omega_c^i. \end{cases}$$

Also, let $\widetilde{f_{i\bar{j}}} = (\Box + 1)\widetilde{e_{i\bar{j}}}$. It is clear that the supports of these approximation functions are contained in the corresponding collars. We have the following estimates:

LEMMA 5.7. *Let $\widetilde{e_{i\bar{j}}}$ be the functions constructed above. Then*

(1) $e_{i\bar{i}} = \widetilde{e_{i\bar{i}}} + O\left(\frac{u_i^4}{|t_i|^2}\right)$ *if* $i \le m$;
(2) $e_{i\bar{j}} = \widetilde{e_{i\bar{j}}} + O\left(\frac{u_i^3 u_j^3}{|t_i t_j|}\right)$ *if* $i, j \le m$ *and* $i \neq j$;
(3) $e_{i\bar{j}} = \widetilde{e_{i\bar{j}}} + O\left(\frac{u_i^3}{|t_i|}\right)$ *if* $i \le m$ *and* $j \ge m+1$;
(4) $\|e_{i\bar{j}}\|_0 = O(1)$ *if* $i, j \ge m+1$.

Now we use the approximation functions $\widetilde{e}_{i\bar{j}}$ in the formulas (4.2), (4.3) and (4.4). The following theorems were proved in [**34**] and [**35**]. We first have the asymptotic estimate of the Ricci metric:

THEOREM 5.8. *Let (t, s) be the pinching coordinates. Then we have*

(1) $\tau_{i\bar{i}} = \frac{3}{4\pi^2}\frac{u_i^2}{|t_i|^2}(1 + O(u_0))$ *and* $\tau^{i\bar{i}} = \frac{4\pi^2}{3}\frac{|t_i|^2}{u_i^2}(1 + O(u_0))$ *if* $i \le m$;
(2) $\tau_{i\bar{j}} = O\left(\frac{u_i^2 u_j^2}{|t_i t_j|}(u_i + u_j)\right)$ *and* $\tau^{i\bar{j}} = O(|t_i t_j|)$ *if* $i, j \le m$ *and* $i \neq j$;
(3) $\tau_{i\bar{j}} = O\left(\frac{u_i^2}{|t_i|}\right)$ *and* $\tau^{i\bar{j}} = O(|t_i|)$ *if* $i \le m$ *and* $j \ge m+1$;
(4) $\tau_{i\bar{j}} = O(1)$ *if* $i, j \ge m+1$.

By the asymptotics of the Ricci metric in the above theorem, we have

COROLLARY 5.1. *There is a constant $C > 0$ such that*

$$C^{-1}\omega_P \leq \omega_\tau \leq \omega_P.$$

Next we estimate the holomorphic sectional curvature of the Ricci metric:

THEOREM 5.9. *Let $X_0 \in \overline{\mathcal{M}_g} \setminus \mathcal{M}_g$ be a codimension m point and let $(t_1, \ldots, t_m, s_{m+1}, \ldots, s_n)$ be the pinching coordinates at X_0 where $t_1, \ldots, t_m$ correspond to the degeneration directions. Then the holomorphic sectional curvature is negative in the degeneration directions and is bounded in the non-degeneration directions. Precisely, there is a $\delta > 0$ such that if $|(t, s)| < \delta$, then*

$$\widetilde{R}_{i\bar{i}i\bar{i}} = \frac{3u_i^4}{8\pi^4|t_i|^4}(1 + O(u_0)) > 0 \tag{5.3}$$

if $i \leq m$ and

$$\widetilde{R}_{i\bar{i}i\bar{i}} = O(1) \tag{5.4}$$

if $i \geq m + 1$.

Furthermore, on $\mathcal{M}_g$, the holomorphic sectional curvature, the bisectional curvature and the Ricci curvature of the Ricci metric are bounded from above and below.

This theorem was proved in [**34**] by using the formula (4.3) and estimating error terms. However, the holomorphic sectional curvature of the Ricci metric is not always negative. We need to introduce and study the perturbed Ricci metric. We have

THEOREM 5.10. *For a suitable choice of positive constant C, the perturbed Ricci metric $\widetilde{\tau}_{i\bar{j}} = \tau_{i\bar{j}} + Ch_{i\bar{j}}$ is complete and comparable with the asymptotic Poincaré metric. Its bisectional curvature is bounded. Furthermore, its holomorphic sectional curvature and Ricci curvature are bounded from above and below by negative constants.*

REMARK 5.3. The perturbed Ricci metric is the first complete Kähler metric on the moduli space with bounded curvature and negatively pinched holomorphic sectional curvature and Ricci curvature.

By using the minimal surface theory and Bers' embedding theorem, we have also proved the following theorem in [**35**]:

THEOREM 5.11. *The moduli space equipped with either the Ricci metric or the perturbed Ricci metric has finite volume. The Teichmüller space equipped with either of these metrics has bounded geometry.*

6. The Equivalence of the Complete Metrics

In this section we describe our arguments that all of the complete metrics on the Teichmüller space and moduli space discussed above are equivalent.

With the good understanding of the Ricci and the perturbed Ricci metrics, the results of this section are quite easy consequences of Yau's Schwarz lemma and the basic properties of these metrics. We first give the definition of equivalence of metrics:

DEFINITION 6.1. Two Kähler metrics g_1 and g_2 on a manifold X are equivalent or two norms $\|\cdot\|_1$ and $\|\cdot\|_2$ on the tangent bundle of X are equivalent if there is a constant $C > 0$ such that

$$C^{-1}g_1 \leq g_2 \leq Cg_1$$

or

$$C^{-1}\|\cdot\|_1 \leq \|\cdot\|_2 \leq C\|\cdot\|_1.$$

We denote this by $g_1 \sim g_2$ or $\|\cdot\|_1 \sim \|\cdot\|_2$.

The main result of this section that we want to discuss is the following theorem proved in [**34**] and [**35**]:

THEOREM 6.2. *On the moduli space $\mathcal{M}_g$ $(g \geq 2)$, the Teichmüller metric $\|\cdot\|_T$, the Carathéodory metric $\|\cdot\|_C$, the Kobayashi metric $\|\cdot\|_K$, the Kähler-Einstein metric ω_{KE}, the induced Bergman metric ω_B, the McMullen metric ω_M, the asymptotic Poincaré metric ω_P, the Ricci metric ω_τ, and the perturbed Ricci metric $\omega_{\tilde{\tau}}$ are equivalent. Namely*

$$\omega_{KE} \sim \omega_{\tilde{\tau}} \sim \omega_\tau \sim \omega_P \sim \omega_B \sim \omega_M$$

and

$$\|\cdot\|_K = \|\cdot\|_T \sim \|\cdot\|_C \sim \|\cdot\|_M.$$

As a corollary we proved the following conjecture of Yau made in the early 1980s [**57**], [**46**]:

THEOREM 6.3. *The Kähler-Einstein metric is equivalent to the Teichmüller metric on the moduli space: $\|\cdot\|_{KE} \sim \|\cdot\|_T$.*

Another corollary was also conjectured by Yau as one of his 120 famous problems [**57**], [**46**]:

THEOREM 6.4. *The Kähler-Einstein metric is equivalent to the Bergman metric on the Teichmüller space: $\omega_{KE} \sim \omega_B$.*

Now we briefly describe the idea of proving the comparison theorem. To compare two complete metrics on a noncompact manifold, we need to write down their asymptotic behavior and compare near infinity. However, if one cannot find the asymptotics of these metrics, the only tool we have is the following Yau's Schwarz lemma [**55**]:

THEOREM 6.5. *Let $f : (M^m, g) \to (N^n, h)$ be a holomorphic map between Kähler manifolds, where M is complete and $Ric(g) \geq -c\,g$ with $c \geq 0$.*

(1) *If the holomorphic sectional curvature of N is bounded above by a negative constant, then $f^*h \leq \tilde{c}\,g$ for some constant $\tilde{c}$.*

(2) *If* $m = n$ *and the Ricci curvature of* N *is bounded above by a negative constant, then* $f^*\omega_h^n \leq \tilde{c}\,\omega_g^n$ *for some constant* $\tilde{c}$.

We briefly describe the proof of the comparison theorem by using Yau's Schwarz lemma and the curvature computations and estimates.

Sketch of proof. To use this result, we take $M = N = \mathcal{M}_g$ and let f be the identity map. We know the perturbed Ricci metric is obtained by adding a positive Kähler metric to the Ricci metric. Thus it is bounded from below by the Ricci metric.

Consider the identity map

$$id : (\mathcal{M}_g, \omega_\tau) \to (\mathcal{M}_g, \omega_{WP}).$$

Yau's Schwarz Lemma implies $\omega_{WP} \leq C_0\omega_\tau$. So

$$\omega_\tau \leq \omega_{\tilde{\tau}} = \omega_\tau + C\omega_{WP} \leq (CC_0 + 1)\omega_\tau.$$

Thus $\omega_\tau \sim \omega_{\tilde{\tau}}$.

To control the Kähler-Einstein metric, we consider

$$id : (\mathcal{M}_g, \omega_{KE}) \to (\mathcal{M}_g, \omega_{\tilde{\tau}})$$

and

$$id : (\mathcal{M}_g, \omega_{\tilde{\tau}}) \to (\mathcal{M}_g, \omega_{KE}).$$

Yau's Schwarz Lemma implies

$$\omega_{\tilde{\tau}} \leq C_0\omega_{KE}$$

and

$$\omega_{KE}^n \leq C_0\omega_{\tilde{\tau}}^n.$$

The equivalence follows from linear algebra.

Thus by Corollary 5.1 we have

$$\omega_{KE} \sim \omega_{\tilde{\tau}} \sim \omega_\tau \sim \omega_P.$$

By using a similar method we have $\omega_\tau \leq C\omega_M$. To show the other side of the inequality, we have to analyze the asymptotic behavior of the geodesic length functions. We showed in [**34**] that

$$\omega_\tau \sim \omega_M.$$

Thus by the work of McMullen [**41**] we have

$$\omega_\tau \sim \omega_M \sim \|\cdot\|_T.$$

The work of Royden showed that the Teichmüller metric coincides with the Kobayashi metric. Thus we need to show that the Carathéodory metric and the Bergman metric are comparable with the Kobayashi metric. This was done in [**35**] by using Bers' Embedding Theorem. The idea is as follows:

By the Bers' Embedding Theorem, for each point $p \in \mathcal{T}_g$, there is a map $f_p : \mathcal{T}_g \to \mathbb{C}^n$ such that $f_p(p) = 0$ and

$$B_2 \subset f_p(\mathcal{T}_g) \subset B_6$$

where B_r is the open ball in $\mathbb{C}^n$ centered at 0 with radius r. Since both Carathéodory metric and Kobayashi metric have the restriction property and can be computed explicitly on balls, we can use these metrics defined on B_2 and B_6 to pinch these metrics on the Teichmüller space. We can also use this method to estimate peak sections of the Teichmüller space at point p. A careful analysis shows

$$\|\cdot\|_C \sim \|\cdot\|_K \sim \omega_B.$$

The argument is quite easy. Please see **[35]** for details. □

7. Bounded Geometry of the Kähler-Einstein Metric

The comparison theorem gives us some control on the Kähler-Einstein Metric. Especially we know that it has Poincaré growth near the boundary of the moduli space and is equivalent to the Ricci metric which has bounded geometry. In this section we sketch our proof that the Kähler-Einstein metric also has bounded geometry. Precisely we have

THEOREM 7.1. *The curvature of the Kähler-Einstein metric and all of its covariant derivatives are uniformly bounded on the Teichmüller spaces, and its injectivity radius has lower bound.*

Now we briefly describe the proof. Please see **[35]** for details.

Sketch of proof. We follow Yau's argument in **[56]**. The first step is to perturb the Ricci metric using Kähler-Ricci flow

$$\begin{cases} \frac{\partial g_{i\bar{j}}}{\partial t} = -(R_{i\bar{j}} + g_{i\bar{j}}), \\ g(0) = \tau \end{cases}$$

to avoid complicated computations of the covariant derivatives of the curvature of the Ricci metric.

For $t > 0$ small, let $h = g(t)$ and let g be the Kähler-Einstein metric. We have

(1) h is equivalent to the initial metric τ and thus is equivalent to the Kähler-Einstein metric,
(2) the curvature and its covariant derivatives of h are bounded.

Then we consider the Monge-Amperé equation

$$\log\det(h_{i\bar{j}} + u_{i\bar{j}}) - \log\det(h_{i\bar{j}}) = u + F$$

where $\partial\bar{\partial}u = \omega_g - \omega_h$ and $\partial\bar{\partial}F = Ric(h) + \omega_h$.

The curvature of $P_{i\bar{j}k\bar{l}}$ of the Kähler-Einstein metric is given by

$$P_{i\bar{j}k\bar{l}} = R_{i\bar{j}k\bar{l}} + u_{p\bar{j}}h^{p\bar{q}}R_{i\bar{q}k\bar{l}} + u_{;i\bar{j}k\bar{l}} - g^{p\bar{q}}u_{;i\bar{q}k}u_{;\bar{j}p\bar{l}}.$$

The comparison theorem implies $\partial\bar{\partial}u$ has C^0-bound and the strong bounded geometry of h implies $\partial\bar{\partial}F$ has C^k-bound for $k \geq 0$. Also, the equivalence of h and g implies $u + F$ is bounded.

So we need the C^k-bound of $\partial\overline{\partial}u$ for $k \geq 1$. Let

$$S = g^{i\bar{j}}g^{k\bar{l}}g^{p\bar{q}}u_{;i\bar{q}k}u_{;\bar{j}p\bar{l}}$$

and

$$V = g^{i\bar{j}}g^{k\bar{l}}g^{p\bar{q}}g^{m\bar{n}}\left(u_{;i\bar{q}k\bar{n}}u_{;\bar{j}p\bar{l}m} + u_{;i\bar{n}kp}u_{;\bar{j}m\bar{l}\bar{q}}\right),$$

where the covariant derivatives of u were taken with respect to the metric h.

Yau's C^3 estimate in [**56**] implies S is bounded. Let $f = (S+\kappa)V$ where κ is a large constant. The inequality

$$\Delta' f \geq Cf^2 + (\text{ lower order terms })$$

implies f is bounded and thus V is bounded. So the curvature of the Kähler-Einstein metric is bounded. The same method can be used to derive boundedness of higher derivatives of the curvature. □

Actually we have also proved that all of these complete Kähler metrics have bounded geometry, which should be useful in understanding the geometry of the moduli and the Teichmüller spaces.

8. Application to Algebraic Geometry

The existence of the Kähler-Einstein metric is closely related to the stability of the tangent and cotangent bundle. In this section we review our results that the logarithmic extension of the cotangent bundle of the moduli space is stable in the sense of Mumford. We first recall the definition.

Definition 8.1. Let E be a holomorphic vector bundle over a complex manifold X and let Φ be a Kähler class of X. The (Φ-)degree of E is given by

$$\deg(E) = \int_X c_1(E)\Phi^{n-1}$$

where n is the dimension of X. The slope of E is given by the quotient

$$\mu(E) = \frac{\deg(E)}{\operatorname{rank}(E)}.$$

The bundle E is Mumford (Φ-)stable if for any proper coherent subsheaf $\mathcal{F} \subset E$, we have

$$\mu(\mathcal{F}) < \mu(E).$$

Now we describe the logarithmic cotangent bundle. Let U be any local chart of $\mathcal{M}_g$ near the boundary with pinching coordinates $(t_1, \ldots, t_m, s_{m+1}, \ldots, s_n)$ such that $(t_1, \ldots, t_m)$ represent the degeneration directions. Let

$$e_i = \begin{cases} \frac{dt_i}{t_i} & i \leq m; \\ ds_i & i \geq m+1. \end{cases}$$

The logarithmic cotangent bundle E is the extension of $T^*\mathcal{M}_g$ to $\overline{\mathcal{M}}_g$ such that on U, $e_1, \ldots, e_n$ is a local holomorphic frame of E. One can write down the transition functions and check that there is a unique bundle over $\overline{\mathcal{M}}_g$ satisfying the above condition.

To prove the stability of E, we need to specify a Kähler class. It is natural to use the polarization of E. The main theorem of this section is the following:

THEOREM 8.2. *The first Chern class* $c_1(E)$ *is positive and* E *is stable with respect to* $c_1(E)$.

We briefly describe here the proof of this theorem. Please see [**35**] for details.

Sketch of the proof. Since we only deal with the first Chern class, we can assume the coherent subsheaf $\mathcal{F}$ is actually a subbundle F.

Since the Kähler-Einstein metric induces a singular metric g^*_{KE} on the logarithmic extension bundle E, our main job is to show that the degree and slope of E and any proper subbundle F defined by the singular metric are finite and are equal to the genuine ones. This depends on our estimates of the Kähler-Einstein metric which are used to show the convergence of the integrals defining the degrees.

More precisely we need to show the following:

(1) As a current, ω_{KE} is closed and represents the first Chern class of E, that is
$$[\omega_{KE}] = c_1(\overline{E}).$$
(2) The singular metric g^*_{KE} on E induced by the Kähler-Einstein metric defines the degree of E
$$\deg(E) = \int_{\mathcal{M}_g} \omega^n_{KE}.$$
(3) The degree of any proper holomorphic sub-bundle F of E can be defined by using $g^*_{KE}\mid_F$:
$$\deg(F) = \int_{\mathcal{M}_g} -\partial\overline{\partial} \log \det \left(g^*_{KE}\mid_F\right) \wedge \omega^{n-1}_{KE}.$$

These three steps were proved in [**35**] by using the Poincaré growth property of the Kähler-Einstein metric together with a special cut-off function. This shows that the bundle E is semi-stable.

To get the strict stability, we proceeded by contradiction. If E is not stable, then E, thus $E\mid_{\mathcal{M}_g}$, split holomorphically. This implies a finite smooth cover of the moduli space splits, which implies a finite index subgroup of the mapping class group splits as a direct product of two subgroups. This is impossible by a topological fact. Again, the detailed proof can be found in [**35**].

Part II: The Topological Aspect

9. The Physics of the Mariño-Vafa Conjecture

Our original motivation to study Hodge integrals was to find a general mirror formula for counting higher genus curves in Calabi-Yau manifolds. To generalize the mirror principle to count the number of higher genus curves, we need to first compute Hodge integrals, i.e., the intersection numbers of the λ classes and ψ classes on the Deligne-Mumford moduli space of stable curves $\overline{\mathcal{M}}_{g,h}$. This moduli space is possibly the most famous and most interesting orbifold. It has been studied since Riemann, and by many Fields medalists for the past 50 years, from many different points of view. Still many interesting and challenging problems about the geometry and topology of these moduli spaces remain unsolved. String theory has motivated many fantastic conjectures about these moduli spaces, including the famous Witten conjecture which is about the generating series of the integrals of the ψ-classes. We start with the introduction of some notations.

Recall that a point in $\overline{\mathcal{M}}_{g,h}$ consists of $(C, x_1, \ldots, x_h)$, a (nodal) curve C of genus g, and n distinguished smooth points on C. The Hodge bundle $\mathbb{E}$ is a rank g vector bundle over $\overline{\mathcal{M}}_{g,h}$ whose fiber over $[(C, x_1, \ldots, x_h)]$ is $H^0(C, \omega_C)$, the complex vector space of holomorphic one forms on C. The λ classes are the Chern Classes of $\mathbb{E}$,

$$\lambda_i = c_i(\mathbb{E}) \in H^{2i}(\overline{\mathcal{M}}_{g,h}; \mathbb{Q}).$$

On the other hand, the cotangent line $T^*_{x_i}C$ of C at the i-th marked point x_i induces a line bundle $\mathbb{L}_i$ over $\overline{\mathcal{M}}_{g,h}$. The ψ classes are the Chern classes:

$$\psi_i = c_1(\mathbb{L}_i) \in H^2(\overline{\mathcal{M}}_{g,h}; \mathbb{Q}).$$

Introduce the total Chern class

$$\Lambda^\vee_g(u) = u^g - \lambda_1 u^{g-1} + \cdots + (-1)^g \lambda_g.$$

The Mariño-Vafa formula is about the generating series of the triple Hodge integrals

$$\int_{\overline{\mathcal{M}}_{g,h}} \frac{\Lambda^\vee_g(1)\Lambda^\vee_g(\tau)\Lambda^\vee_g(-\tau-1)}{\prod_{i=1}^h (1-\mu_i\psi_i)},$$

where τ is considered as a parameter here. Later we will see that it actually comes from the weight of the group action, and also from the framing of the knot. Taking Taylor expansions in τ or in μ_i one can obtain information on the integrals of the Hodge classes and the ψ-classes. The Marinõ-Vafa conjecture asserts that the generating series of such triple Hodge integrals for all genera and any numbers of marked points can be expressed by a close formula which is a *finite* expression in terms of representations of symmetric groups, or Chern-Simons knot invariants.

We remark that the moduli spaces of stable curves have been the source of many interests from mathematics to physics. Mumford has computed some low genus numbers. The Witten conjecture, proved by Kontsevich, is about the integrals of the ψ-classes.

Let us briefly recall the background of the conjecture. Mariño and Vafa [**39**] made this conjecture based on the large N duality between Chern-Simons and string theory. It starts from the conifold transition. We consider the resolution of singularity of the conifold X defined by

$$\left\{ \begin{pmatrix} x & y \\ z & w \end{pmatrix} \in \mathbf{C}^4 : xw - yz = 0 \right\}$$

in two different ways:

(1) Deformed conifold T^*S^3

$$\left\{ \begin{pmatrix} x & y \\ z & w \end{pmatrix} \in \mathbf{C}^4 : xw - yz = \epsilon \right\}$$

where ϵ a real positive number. This is a symplectic resolution of the singularity.

(2) Resolved conifold by blowing up the singularity, which gives the total space

$$\tilde{X} = \mathcal{O}(-1) \oplus \mathcal{O}(-1) \to \mathbf{P}^1$$

which is explicitly given by

$$\left\{ \left([Z_0, Z_1], \begin{pmatrix} x & y \\ z & w \end{pmatrix} \right) \in \mathbf{P}^1 \times \mathbf{C}^4 : \begin{matrix} (x,y) \in [Z_0, Z_1] \\ (z,w) \in [Z_0, Z_1] \end{matrix} \right\}$$

$$\begin{matrix} \tilde{X} & \subset & \mathbf{P}^1 \times \mathbf{C}^4 \\ \downarrow & & \downarrow \\ X & \subset & \mathbf{C}^4 \end{matrix}$$

The brief history of the development of the conjecture is as follows. In 1992 Witten first conjectured that the open topological string theory on the deformed conifold T^*S^3 is equivalent to the Chern-Simons gauge theory on S^3. This idea was pursued further by Gopakumar and Vafa in 1998, and then by Ooguri and Vafa in 2000. Based on the above conifold transition, they conjectured that the open topological string theory on the deformed conifold T^*S^3 is equivalent to the closed topological string theory on the resolved conifold $\tilde{X}$. Ooguri-Vafa only considered the zero framing case. Later Marinõ-Vafa generalized the idea to the non-zero framing case and discovered the beautiful formula for the generating series of the triple Hodge integrals. Recently Vafa and his collaborators systematically developed the theory, and for the past several years, they developed these duality ideas into the most effective tool to compute Gromov-Witten invariants on toric Calabi-Yau manifolds. The high point of their work is the theory of topological vertex. We refer to [**39**] and [**1**] for the details of the physical theory and the history of the development.

Starting with the proof of the Marinõ-Vafa conjecture [**28**], [**29**], we have developed a rather complete mathematical theory of topological vertex [**21**]. Many interesting consequences have been derived during the past year. Now let us see how the string theorists derived mathematical consequence from the above naive idea of string duality. First the Chern-Simons partition function has the form

$$\langle Z(U,V)\rangle = \exp(-F(\lambda,t,V)),$$

where U is the holonomy of the $U(N)$ Chern-Simons gauge field around the knot $K \subset S^3$, and V is an extra $U(M)$ matrix. The partition function $\langle Z(U,V)\rangle$ gives the Chern-Simons knot invariants of K.

String duality asserts that the function $F(\lambda,t,V)$ should give the generating series of the open Gromov-Witten invariants of $(\tilde{X}, L_K)$, where L_K is a Lagrangian submanifold of the resolved conifold $\tilde{X}$ canonically associated to the knot K. More precisely, by applying the t'Hooft large N expansion, and the "canonical" identifications of parameters similar to mirror formula, which at level k are given by

$$\lambda = \frac{2\pi}{k+N}, \quad t = \frac{2\pi i N}{k+N},$$

we get the partition function of the topological string theory on conifold $\tilde{X}$, and then on $\mathbf{P}^1$, which is just the generating series of the Gromov-Witten invariants. This change of variables is very striking from the point of view of mathematics.

The special case when K is the unknot is already very interesting. In non-zero framing it gives the Mariño-Vafa conjectural formula. In this case $\langle Z(U,V)\rangle$ was first computed in the zero framing by Ooguri-Vafa and in any framing $\tau \in \mathbb{Z}$ by Mariño-Vafa [**39**]. Comparing with Katz-Liu's computations of $F(\lambda,t,V)$, Mariño-Vafa conjectured the striking formula about the generating series of the triple Hodge integrals for all genera and any number of marked points in terms of the Chern-Simons invariants, or equivalently in terms of the representations and combinatorics of symmetric groups. It is interesting to note that the framing in the Mariño-Vafa's computations corresponds to the choice of lifting of the circle action on the pair $(\tilde{X}, L_{\mathrm{unknot}})$ in Katz-Liu's localization computations. Both choices are parametrized by an integer τ which will be considered as a parameter in the triple Hodge integrals. Later we will take derivatives with respect to this parameter to get the cut-and-join equation.

It is natural to ask what mathematical consequence we can have for general duality, that is for general knots in general three manifolds; a first naive question is what kind of general Calabi-Yau manifolds will appear in the duality, in place of the conifold. Some special cases corresponding to the Seifert manifolds are known by gluing several copies of conifolds.

10. The Proof of the Mariño-Vafa Formula

Now we give the precise statement of the Mariño-Vafa conjecture, which is an identity between the geometry of the moduli spaces of stable curves and

Chern-Simons knot invariants, or the combinatorics of the representation theory of symmetric groups.

Let us first introduce the geometric side. For every partition $\mu = (\mu_1 \geq \cdots \mu_{l(\mu)} \geq 0)$, we define the triple Hodge integral to be

$$G_{g,\mu}(\tau) = A(\tau) \cdot \int_{\overline{\mathcal{M}}_{g,l(\mu)}} \frac{\Lambda_g^\vee(1)\Lambda_g^\vee(-\tau-1)\Lambda_g^\vee(\tau)}{\prod_{i=1}^{l(\mu)}(1-\mu_i\psi_i)},$$

where the coefficient

$$A(\tau) = -\frac{\sqrt{-1}^{|\mu|+l(\mu)}}{|\operatorname{Aut}(\mu)|}[\tau(\tau+1)]^{l(\mu)-1} \prod_{i=1}^{l(\mu)} \frac{\prod_{a=1}^{\mu_i-1}(\mu_i\tau+a)}{(\mu_i-1)!}.$$

The expressions, although very complicated, arise naturally from localization computations on the moduli spaces of relative stable maps into $\mathbf{P}^1$ with ramification type μ at ∞.

We now introduce the generating series

$$G_\mu(\lambda;\tau) = \sum_{g\geq 0} \lambda^{2g-2+l(\mu)} G_{g,\mu}(\tau).$$

The special case when $g=0$ is given by

$$\int_{\overline{\mathcal{M}}_{0,l(\mu)}} \frac{\Lambda_0^\vee(1)\Lambda_0^\vee(-\tau-1)\Lambda_0^\vee(\tau)}{\prod_{i=1}^{l(\mu)}(1-\mu_i\psi_i)} = \int_{\overline{\mathcal{M}}_{0,l(\mu)}} \frac{1}{\prod_{i=1}^{l(\mu)}(1-\mu_i\psi_i)},$$

which is known to be equal to $|\mu|^{l(\mu)-3}$ for $l(\mu) \geq 3$, and we use this expression to extend the definition to the case $l(\mu) < 3$.

Introduce formal variables $p = (p_1, p_2, \ldots, p_n, \ldots)$, and define

$$p_\mu = p_{\mu_1} \cdots p_{\mu_{l(\mu)}}$$

for any partition μ. These p_{μ_j} correspond to $\operatorname{Tr} V^{\mu_j}$ in the notations of string theorists. The generating series for all genera and all possible marked points are defined to be

$$G(\lambda;\tau;p) = \sum_{|\mu|\geq 1} G_\mu(\lambda;\tau)p_\mu,$$

which encode complete information of the triple Hodge integrals we are interested in.

Next we introduce the representation theoretical side. Let χ_μ denote the character of the irreducible representation of the symmetric group $S_{|\mu|}$, indexed by μ with $|\mu| = \sum_j \mu_j$. Let $C(\mu)$ denote the conjugacy class of $S_{|\mu|}$ indexed by μ. Introduce

$$\mathcal{W}_\mu(\lambda) = \prod_{1\leq a<b\leq l(\mu)} \frac{\sin[(\mu_a-\mu_b+b-a)\lambda/2]}{\sin[(b-a)\lambda/2]} \cdot \frac{1}{\prod_{i=1}^{l(\nu)}\prod_{v=1}^{\mu_i} 2\sin[(v-i+l(\mu))\lambda/2]}.$$

This has an interpretation in terms of quantum dimension in Chern-Simons knot theory.

We define the following generating series

$$R(\lambda;\tau;p)=\sum_{n\geq 1}\frac{(-1)^{n-1}}{n}$$

$$\cdot\sum_{\mu}\left[\sum_{\cup_{i=1}^{n}\mu^i=\mu}\prod_{i=1}^{n}\sum_{|\nu^i|=|\mu^i|}\frac{\chi_{\nu^i}(C(\mu^i))}{z_{\mu^i}}e^{\sqrt{-1}(\tau+\frac{1}{2})\kappa_{\nu^i}\lambda/2}\mathcal{W}_{\nu^i}(\lambda)\right]p_\mu$$

where μ^i are sub-partitions of μ, $z_\mu=\prod_j \mu_j! j^{\mu_j}$ and

$$\kappa_\mu=|\mu|+\sum_i(\mu_i^2-2i\mu_i)$$

for a partition μ which is also standard for representation theory of symmetric groups. There is the relation $z_\mu=|\operatorname{Aut}(\mu)|\mu_1\cdots\mu_{l(\mu)}$.

Finally we can give the precise statement of the Mariño-Vafa formula:

Conjecture: *We have the identity*

$$G(\lambda;\tau;p)=R(\lambda;\tau;p).$$

Before discussing the proof of this conjecture, we first give several remarks. This conjecture is a formula: G : Geometry = R : Representations, and the representations of symmetric groups are essentially combinatorics. We note that each $G_\mu(\lambda,\tau)$ is given by a finite and closed expression in terms of the representations of symmetric groups:

$$G_\mu(\lambda,\tau)=\sum_{n\geq 1}\frac{(-1)^{n-1}}{n}$$

$$\cdot\sum_{\cup_{i=1}^{n}\mu^i=\mu}\prod_{i=1}^{n}\sum_{|\nu^i|=|\mu^i|}\frac{\chi_{\nu^i}(C(\mu^i))}{z_{\mu^i}}e^{\sqrt{-1}(\tau+\frac{1}{2})\kappa_{\nu^i}\lambda/2}\mathcal{W}_{\nu^i}(\lambda).$$

The generating series $G_\mu(\lambda,\tau)$ gives the values of the triple Hodge integrals for moduli spaces of curves of all genera with $l(\mu)$ marked points. Finally we remark that an equivalent expression of this formula is the following non-connected generating series. In this situation we have a relatively simpler combinatorial expression:

$$\begin{aligned}G(\lambda;\tau;p)^\bullet&=\exp\left[G(\lambda;\tau;p)\right]\\&=\sum_{|\mu|\geq 0}\left[\sum_{|\nu|=|\mu|}\frac{\chi_\nu(C(\mu))}{z_\mu}e^{\sqrt{-1}(\tau+\frac{1}{2})\kappa_\nu\lambda/2}\mathcal{W}_\nu(\lambda)\right]p_\mu.\end{aligned}$$

According to Mariño and Vafa, this formula gives values for all Hodge integrals up to three Hodge classes. Lu proved that this is right if we combine with some previously known simple formulas about Hodge integrals.

By taking Taylor expansion in τ on both sides of the Mariño-Vafa formula, we have derived various Hodge integral identities in [**30**].

For examples, as easy consequences of the Mariño-Vafa formula and the cut-and-join equation as satisfied by the above generating series, we have unified simple proofs of the λ_g conjecture by comparing the coefficients in τ in the Taylor expansions of the two expressions,

$$\int_{\overline{\mathcal{M}}_{g,n}} \psi_1^{k_1} \cdots \psi_n^{k_n} \lambda_g = \binom{2g+n-3}{k_1, \dots, k_n} \frac{2^{2g-1}-1}{2^{2g-1}} \frac{|B_{2g}|}{(2g)!},$$

for $k_1+\cdots+k_n = 2g-3+n$, and the following identities for Hodge integrals:

$$\int_{\overline{\mathcal{M}}_g} \lambda_{g-1}^3 = \int_{\overline{\mathcal{M}}_g} \lambda_{g-2}\lambda_{g-1}\lambda_g = \frac{1}{2(2g-2)!} \frac{|B_{2g-2}|}{2g-2} \frac{|B_{2g}|}{2g},$$

where B_{2g} are Bernoulli numbers. And

$$\int_{\overline{\mathcal{M}}_{g,1}} \frac{\lambda_{g-1}}{1-\psi_1} = b_g \sum_{i=1}^{2g-1} \frac{1}{i} - \frac{1}{2} \sum_{\substack{g_1+g_2=g \\ g_1, g_2 > 0}} \frac{(2g_1-1)!(2g_2-1)!}{(2g-1)!} b_{g_1} b_{g_2},$$

where

$$b_g = \begin{cases} 1, & g = 0, \\ \frac{2^{2g-1}-1}{2^{2g-1}} \frac{|B_{2g}|}{(2g)!}, & g > 0. \end{cases}$$

Now let us look at how we proved this conjecture. This is joint work with Chiu-Chu Liu, Jian Zhou. See [**27**] and [**28**] for details.

The first proof of this formula is based on the *Cut-and-Join* equation which is a beautiful match of combinatorics and geometry. The details of the proof is given in [**27**] and [**28**]. First we look at the combinatorial side. Denote by $[s_1, \dots, s_k]$ a k-cycle in the permutation group. We have the following two obvious operations:

Cut: a k-cycle is cut into an i-cycle and a j-cycle:

$$[s,t] \cdot [s, s_2, \dots, s_i, t, t_2, \dots, t_j] = [s, s_2, \dots, s_i][t, t_2, \dots, t_j].$$

Join: an i-cycle and a j-cycle are joined to an $(i+j)$-cycle:

$$[s,t] \cdot [s, s_2, \dots, s_i][t, t_2, \dots t_j] = [s, s_2, \dots, s_i, t, t_2, \dots, t_j].$$

Such operations can be organized into differential equations which we call the cut-and-join equation.

Now we look at the geometry side. In the moduli spaces of stable maps, cut and join have the following geometric meaning: *Cut*: one curve splits into two lower degree or lower genus curves. *Join*: two curves are joined together to give a higher genus or higher degree curve.

The combinatorics and geometry of cut-and-join are reflected in the following two differential equations, which look like a heat equation. It is easy to show that such an equation is equivalent to a series of systems of linear ordinary differential equations by comparing the coefficients on p_μ. These equations are proved either by easy and direct computations in combinatorics or by localizations on moduli spaces of relative stable maps in geometry. In combinatorics, the proof is given by direct computations and was explored in combinatorics in the mid '80s and by Zhou [**27**] for this case. The differential operator on the right hand side corresponds to the cut-and-join operations which we also simply denote by (CJ).

LEMMA 10.1.

$$\frac{\partial R}{\partial \tau} = \frac{1}{2}\sqrt{-1}\lambda \sum_{i,j=1}^{\infty} \left((i+j)p_i p_j \frac{\partial R}{\partial p_{i+j}} + ijp_{i+j} \left(\frac{\partial R}{\partial p_i}\frac{\partial R}{\partial p_j} + \frac{\partial^2 R}{\partial p_i \partial p_j} \right) \right).$$

On the geometry side the proof of such equation is given by localization on the moduli spaces of relative stable maps into the the projective line $\mathbf{P}^1$ with fixed ramifications at ∞:

LEMMA 10.2.

$$\frac{\partial G}{\partial \tau} = \frac{1}{2}\sqrt{-1}\lambda \sum_{i,j=1}^{\infty} \left((i+j)p_i p_j \frac{\partial G}{\partial p_{i+j}} + ijp_{i+j} \left(\frac{\partial G}{\partial p_i}\frac{\partial G}{\partial p_j} + \frac{\partial^2 G}{\partial p_i \partial p_j} \right) \right).$$

The proof of the above equation is given in [**27**]. Together with the following

Initial Value: $\tau = 0$,

$$G(\lambda, 0, p) = \sum_{d=1}^{\infty} \frac{p_d}{2d \sin\left(\frac{\lambda d}{2}\right)} = R(\lambda, 0, p)$$

which is precisely the Ooguri-Vafa formula and which has been proved previously for example in [**58**], we therefore obtain the equality which is the Mariño-Vafa conjecture by the uniqueness of the solution:

THEOREM 10.3. *We have the identity*

$$G(\lambda; \tau; p) = R(\lambda; \tau; p).$$

During the proof we note that the cut-and-join equation is encoded in the geometry of the moduli spaces of stable maps. In fact we later find the convolution formula of the following form, which is a relation for the disconnected version $G^\bullet = \exp G$,

$$G^\bullet_\mu(\lambda, \tau) = \sum_{|\nu|=|\mu|} \Phi^\bullet_{\mu,\nu}(-\sqrt{-1}\tau\lambda) z_\nu K^\bullet_\nu(\lambda)$$

where $\Phi^\bullet_{\mu,\nu}$ is the generating series of double Hurwitz numbers, and z_ν is the combinatorial constant that appeared in the previous formulas. Equivalently this gives the explicit solution of the cut-and-join differential equation with initial value $K^\bullet(\lambda)$, which is the generating series of the integrals of certain Euler classes on the moduli spaces of relative stable maps to $\mathbf{P}^1$. See [**26**] for the derivation of this formula, and see [**29**] for the two partition analogue.

The Witten conjecture as proved by Kontsevich states that the generating series of the ψ-class integrals satisfy an infinite number of differential equations. The remarkable feature of Mariño-Vafa formula is that it gives a finite close formula. In fact, by taking limits in τ and μ_i's one can obtain the Witten conjecture as argued by Okounkov-Pandhrapande. But the combinatorics involved is non-trivial. A much simpler direct proof of the Witten conjecture was obtained recently by Kim and myself. We directly derived the recursion formula which implies both the Virasoro relations and the KdV equations. We will discuss this proof later.

The same argument as our proof of the conjecture gives a simple and geometric proof of the ELSV formula for Hurwitz numbers. It reduces to the fact that the push-forward of 1 is a constant in equivariant cohomology for a generically finite-to-one map. This will also be discussed in a later section. See [**28**] for more details.

We would like to briefly explain the technical details of the proof of the Mariño-Vafa formula. The proof of the combinatorial cut-and-join formula is based on the Burnside formula and various simple results in symmetric functions. See [**58**], [**22**] and [**28**].

The proof of the geometric cut-and-join formula used the functorial localization formula in [**24**] and [**25**]. Here we only state its simple form for manifolds as used in [**24**]; the virtual version of this formula is proved and used in [**25**].

Given X and Y two compact manifolds with torus action. Let $f : X \to Y$ be an equivariant map. Let $F \subset Y$ be a fixed component, and let $E \subset f^{-1}(F)$ denote the fixed components lying inside $f^{-1}(F)$. Let $f_0 = f|_E$; then we have

Functorial Localization Formula: *For $\omega \in H_T^*(X)$ an equivariant cohomology class, we have the identity on F:*

$$f_{0*}\left[\frac{i_E^*\omega}{e_T(E/X)}\right] = \frac{i_F^*(f_*\omega)}{e_T(F/Y)}.$$

This formula, which is a generalization of the Atiyah-Bott localization formula to relative setting, has been applied to various settings to prove many interesting conjectures from physics. It was discovered and effectively used in [**24**]. A virtual version which was first applied to the virtual fundamental cycles in the computations of Gromov-Witten invariants was first proved and used in [**25**].

This formula is very effective and useful because we can use it to push computations on complicated moduli space to simpler moduli space. The moduli spaces used by mathematicians are usually the correct but complicated moduli spaces like the moduli spaces of stable maps, while the moduli spaces used by physicists are usually the simple but wrong ones like the projective spaces. This functorial localization formula has been used successfully in the proof of the mirror formula [**24**], [**25**], the proof of the Hori-Vafa formula [**23**], and the easy proof of the ELSV formula [**28**]. Our first proof of the Mariño-Vafa formula also used this formula in a crucial way.

More precisely, let $\overline{\mathcal{M}}_g(\mathbf{P}^1, \mu)$ denote the moduli space of relative stable maps from a genus g curve to $\mathbf{P}^1$ with fixed ramification type μ at ∞, where μ is a fixed partition. We apply the functorial localization formula to the divisor morphism from the relative stable map moduli space to the projective space,

$$\mathrm{Br}: \ \overline{\mathcal{M}}_g(\mathbf{P}^1, \mu) \to \mathbf{P}^r,$$

where r denotes the dimension of $\overline{\mathcal{M}}_g(\mathbf{P}^1, \mu)$. This is similar to the set-up of mirror principle, only with a different linearized moduli space, but in both cases the target spaces are projective spaces.

We found that the fixed points of the target $\mathbf{P}^r$ precisely label the cut-and-join operations of the triple Hodge integrals. Functorial localization reduces the problem to the study of polynomials in the equivariant cohomology group of $\mathbf{P}^r$. We were able to squeeze out a system of linear equations which implies the cut-and-join equation. Actually we derived a stronger relation than the cut-and-join equation, while the cut-and-join equation we need for the Mariño-Vafa formula is only the very first of such kind of relations. See [**28**] for higher order cut-and-join equations.

As was known in infinite Lie algebra theory, the cut-and-join operator is closely related to and more fundamental than the Virasoro algebras in some sense.

Recently there have appeared two different approaches to the Mariño-Vafa formula. The first one is a direct derivation of the convolution formula which was discovered during our proof of the two partition analogue of the formula [**29**]. See [**26**] for the details of the derivation in this case. The second is by Okounkov-Pandhripande [**44**]; they gave a different approach by using the ELSV formula as initial value, as well as the λ_g conjecture and other recursion relations from localization on the moduli spaces of stable maps to $\mathbf{P}^1$.

11. Two Partition Generalization

The two partition analogue of the Mariño-Vafa formula naturally arises from the localization computations of the Gromov-Witten invariants of the open toric Calabi-Yau manifolds, as explained in [**59**].

To state the formula we let μ^+, μ^- be any two partitions. Introduce the Hodge integrals involving these two partitions:

$$G_{\mu^+,\mu^-}(\lambda;\tau) = B(\tau;\mu^+,\mu^-) \cdot \sum_{g\geq 0} \lambda^{2g-2} A_g(\tau;\mu^+,\mu^-)$$

where

$$A_g(\tau;\mu^+,\mu^-) = \int_{\overline{\mathcal{M}}_{g,l(\mu^+)+l(\mu^-)}} \frac{\Lambda_g^\vee(1)\Lambda_g^\vee(\tau)\Lambda_g^\vee(-\tau-1)}{\prod_{i=1}^{l(\mu^+)}\left(1-\mu_i^+\psi_i\right)\prod_{j=1}^{l(\mu^-)}\tau\left(\tau-\mu_i^-\psi_{j+l(\mu^+)}\right)}$$

and

$$B(\tau;\mu^+,\mu^-)$$
$$= -\frac{(\sqrt{-1}\lambda)^{l(\mu^+)+l(\mu^-)}}{|\operatorname{Aut}(\mu^+)||\operatorname{Aut}(\mu^-)|}[\tau(\tau+1)]^{l(\mu^+)+l(\mu^-)-1}$$
$$\cdot\prod_{i=1}^{l(\mu^+)}\frac{\prod_{a=1}^{\mu_i^+-1}(\mu_i^+\tau+a)}{(\mu_i^+-1)!}\cdot\prod_{i=1}^{l(\mu^-)}\frac{\prod_{a=1}^{\mu_i^--1}(\mu_i^-\frac{1}{\tau}+a)}{(\mu_i^--1)!}.$$

These complicated expressions naturally arise in open string theory, as well as in the localization computations of the Gromov-Witten invariants on open toric Calabi-Yau manifolds.

We introduce two generating series, first on the geometry side,

$$G^\bullet(\lambda;p^+,p^-;\tau)=\exp\left(\sum_{(\mu^+,\mu^-)\in\mathcal{P}^2}G_{\mu^+,\mu^-}(\lambda,\tau)p^+_{\mu^+}p^-_{\mu^-}\right),$$

where $\mathcal{P}^2$ denotes the set of pairs of partitions and $p^\pm_{\mu^\pm}$ are two sets of formal variables associated to the two partitions as in the last section.

On the representation side, we introduce

$$R^\bullet(\lambda;p^+,p^-;\tau)$$
$$=\sum_{|\nu^\pm|=|\mu^\pm|\geq 0}\frac{\chi_{\nu^+}(C(\mu^+))}{z_{\mu^+}}\frac{\chi_{\nu^-}(C(\mu^-))}{z_{\mu^-}}$$
$$\cdot e^{\sqrt{-1}(\kappa_{\nu^+}\tau+\kappa_{\nu^-}\tau^{-1})\lambda/2}\mathcal{W}_{\nu^+,\nu^-}p^+_{\mu^+}p^-_{\mu^-}.$$

Here

$$\mathcal{W}_{\mu,\nu}=q^{l(\nu)/2}\mathcal{W}_\mu\cdot s_\nu(\mathcal{E}_\mu(t))$$
$$=(-1)^{|\mu|+|\nu|}q^{\frac{\kappa_\mu+\kappa_\nu+|\mu|+|\nu|}{2}}\sum_\rho q^{-|\rho|}s_{\mu/\rho}(1,q,\dots)s_{\nu/\rho}(1,q,\dots)$$

in terms of the skew Schur functions s_μ [**38**]. They appear naturally in the Chern-Simons invariant of the Hopf link.

THEOREM 11.1. *We have the identity:*

$$G^\bullet(\lambda;p^+,p^-;\tau)=R^\bullet(\lambda;p^+,p^-;\tau).$$

The idea of the proof is similar to that of the proof of the Mariño-Vafa formula. We prove that both sides of the above identity satisfy the same cut-and-join equation of the following type:

$$\frac{\partial}{\partial\tau}H^\bullet=\frac{1}{2}(CJ)^+H^\bullet-\frac{1}{2\tau^2}(CJ)^-H^\bullet,$$

where $(CJ)^\pm$ denote the cut-and-join operator, the differential operator with respect to the two set of variables $p^\pm$. We then prove that they have the same initial value at $\tau = -1$:

$$G^\bullet(\lambda; p^+, p^-; -1) = R^\bullet(\lambda; p^+, p^-; -1),$$

which is again given by the Ooguri-Vafa formula [**29**], [**59**].

The cut-and-join equation can be written in a linear matrix form, and such equation follows from the convolution formula of the form

$$K^\bullet_{\mu^+,\mu^-}(\lambda) = \sum_{|\nu^\pm|=\mu^\pm} G^\bullet_{\mu^+,\mu^-}(\lambda;\tau) z_{\nu^+} \Phi^\bullet_{\nu^+,\mu^+}(-\sqrt{-1}\lambda\tau) z_{\nu^-} \Phi^\bullet_{\nu^-,\mu^-}\left(\frac{-\sqrt{-1}}{\tau}\lambda\right)$$

where $\Phi^\bullet$ denotes the generating series of double Hurwitz numbers, and K_{μ^+,μ^-} is the generating series of certain integrals on the moduli spaces of relative stable maps. For more details see [**29**].

This convolution formula arises naturally from localization computations on the moduli spaces of relative stable maps to $\mathbf{P}^1 \times \mathbf{P}^1$ with the point (∞, ∞) blown up. So it reflects the geometric structure of the moduli spaces. Such a convolution type formula was actually discovered during our search for a proof of this formula, both on the geometric and the combinatorial side; see [**29**] for the detailed derivations of the convolution formulas in both geometry and combinatorics.

The proof of the combinatorial side of the convolution formula is again a direct computation. The proof of the geometric side for the convolution equation is to reorganize the generating series from localization contributions on the moduli spaces of relative stable maps into $\mathbf{P}^1 \times \mathbf{P}^1$ with the point (∞, ∞) blown up, in terms of the double Hurwitz numbers. It involves careful analysis and computations.

12. Theory of Topological Vertex

When we worked on the Mariño-Vafa formula and its generalizations, we were simply trying to generalize the method and the formula to involve more partitions, but it turned out that in the three partition case, we naturally met the theory of topological vertex. Topological vertex was first introduced in string theory by Vafa et al in [**1**]; it can be deduced from a three partition analogue of the Mariño-Vafa formula in a highly nontrivial way. From this we were able to give a rigorous mathematical foundation for the physical theory. Topological vertex is a high point of the theory of string duality as developed by Vafa and his group for the past several years, starting from Witten's conjectural duality between Chern-Simons and open string theory. It gives the most powerful and effective way to compute the Gromov-Witten invariants for all open toric Calabi-Yau manifolds. In physics it is rare to have two theories agree up to all orders, and topological vertex theory gives a very significant example. In mathematics the theory of topological vertex already has many interesting applications. Here we only briefly sketch the rough idea for the three partition analogue of the Mariño-Vafa formula. For its relation to the theory of topological vertex, we refer the reader to [**21**] for the details.

Given any three partitions $\overrightarrow{\mu} = \{\mu^1, \mu^2, \mu^3\}$, the cut-and-join equation in this case, for both the geometry and representation sides, has the form:

$$\frac{\partial}{\partial \tau} F^\bullet(\lambda;\tau;\mathbf{p}) = (CJ)^1 F^\bullet(\lambda;\tau;\mathbf{p}) + \frac{1}{\tau^2}(CJ)^2 F^\bullet(\lambda;\tau;\mathbf{p}) + \frac{1}{(\tau+1)^2}(CJ)^3 F^\bullet(\lambda;\tau;\mathbf{p}).$$

The cut-and-join operators $(CJ)^1$, $(CJ)^2$ and $(CJ)^3$ are with respect to the three partitions. More precisely they correspond to the differential operators with respect to the three groups of infinite numbers of variables $\mathbf{p} = \{p^1, p^2, p^3\}$.

The initial value for this differential equation is taken at $\tau = 1$, which is then reduced to the formulas of the two partition case. The combinatorial, or the Chern-Simons invariant side is given by $\mathcal{W}_{\overrightarrow{\mu}} = \mathcal{W}_{\mu^1,\mu^2,\mu^3}$ which is a combination of the $\mathcal{W}_{\mu,\nu}$ as in the two partition case. See [**21**] for its explicit expression.

On the geometry side,

$$G^\bullet(\lambda;\tau;\mathbf{p}) = \exp(G(\lambda;\tau;\mathbf{p}))$$

is the non-connected version of the generating series of the triple Hodge integral. More precisely,

$$G(\lambda;\tau;\mathbf{p}) = \sum_{\overrightarrow{\mu}} \left[\sum_{g=0}^{\infty} \lambda^{2g-2+l(\overrightarrow{\mu})} G_{g,\overrightarrow{\mu}}(\tau) \right] p^1_{\mu^1} p^2_{\mu^2} p^3_{\mu^3}$$

where $l(\overrightarrow{\mu}) = l(\mu^1) + l(\mu^2) + l(\mu^3)$ and $G_{g,\overrightarrow{\mu}}(\tau)$ denotes the Hodge integrals of the following form,

$$A(\tau) \int_{\overline{\mathcal{M}}_{g,l_1+l_2+l_3}} \frac{\Lambda_g^\vee(1)\Lambda_g^\vee(\tau)\Lambda_g^\vee(-\tau-1)}{\prod_{j=1}^{l_1}(1-\mu_j^1\psi_j)\prod_{j=1}^{l_2}\tau(\tau-\mu_j^2\psi_{l_1+j})} \cdot \frac{(\tau(\tau+1))^{l_1+l_2+l_3-1}}{\prod_{j=1}^{l_3}(\tau+1)(\tau+1+\mu_j^3\psi_{l_1+l_2+j})},$$

where

$$A(\tau) = \frac{-(\sqrt{-1}\lambda)^{l_1+l_2+l_3}}{|\operatorname{Aut}(\mu^1)||\operatorname{Aut}(\mu^2)||\operatorname{Aut}(\mu^3)|} \prod_{j=1}^{l_1} \frac{\prod_{a=1}^{\mu_j^1-1}(\tau\mu_j^1+a)}{(\mu_j^1-1)!} \cdot \prod_{j=1}^{l_2} \frac{\prod_{a=1}^{\mu_j^1-1}((-1-1/\tau)\mu_j^2+a)}{(\mu_j^2-1)!} \prod_{j=1}^{l_3} \frac{\prod_{a=1}^{\mu_j^1-1}(-\mu_j^3/(\tau+1)+a)}{(\mu_j^3-1)!}.$$

In the above expression, $l_i = l(\mu^i)$, $i = 1,2,3$. Despite its complicated coefficients, these triple integrals naturally arise from localizations on the moduli spaces of relative stable maps into the blow-up of $\mathbf{P}^1 \times \mathbf{P}^1 \times \mathbf{P}^1$ along certain divisors. It also naturally appears in open string theory computations [**1**]. See [**21**] for more details.

One of our results in [**21**] states that $G^\bullet(\lambda;\tau;\mathbf{p})$ has a combinatorial expression $R^\bullet(\lambda;\tau;\mathbf{p})$ in terms of the Chern-Simons knot invariants $W_{\overrightarrow{\mu}}$, and it is a closed combinatorial expression. More precisely it is given by

$$R^\bullet(\lambda;\tau;\mathbf{p}) = \sum_{\overrightarrow{\mu}} \left[\sum_{|\nu^i|=|\mu^i|} \prod_{i=1}^{3} \frac{\chi_{\nu^i}(\mu^i)}{z_{\mu^i}} q^{\frac{1}{2}(\sum_{i=1}^3 \kappa_{\nu^i} \frac{w_{i+1}}{w_i})} W_{\overrightarrow{\nu}}(q) \right] p^1_{\mu^1} p^2_{\mu^2} p^3_{\mu^3}.$$

Here $w_4 = w_1$ and $w_3 = -w_1 - w_2$ and $\tau = \frac{w_2}{w_1}$. Due to the complicated combinatorics in the initial values, the combinatorial expression $W_{\overrightarrow{\mu}}$ we obtained is different from the expression $\mathcal{W}_{\overrightarrow{\mu}}$ obtained by Vafa et al. Actually our expression is even simpler than theirs in some sense. The expression we obtained is more convenient for mathematical applications such as the proof of the Gopakumar-Vafa conjecture for open toric Calabi-Yau manifolds. It should be possible to identify the two combinatorial expressions by using the classical theory of symmetric functions, as pointed out to us by R. Stanley.

THEOREM 12.1. *We have the equality:*

$$G^\bullet(\lambda;\tau;\boldsymbol{p}) = R^\bullet(\lambda;\tau;\boldsymbol{p}).$$

The key point to prove that the above theorem is still the proof of convolution formulas for both sides which imply the cut-and-join equation. The proof of the convolution formula for $G^\bullet(\lambda;\tau;\mathbf{p})$ is much more complicated than the one and two partition cases. See [**21**] for details.

The above theorem is crucial for us to establish the theory of topological vertex in [**21**], which gives the most powerful way to compute the generating series of all genera and all degree Gromov-Witten invariants for open formal Calabi-Yau manifolds. The most useful property of topological vertex is its gluing property induced by the orthogonal relations of the characters of the symmetric group. This is very close to the situation of two dimensional gauge theory. In fact string theorists consider topological vertex as a kind of lattice theory on Calabi-Yau manifolds. By using the gluing formula we can easily obtain closed formulas for generating series of Gromov-Witten invariants of all genera and all degrees, open or closed, for all open toric Calabi-Yau manifolds, in terms of the Chern-Simons knot invariants. Such formulas are always given by finite sum of products of those Chern-Simons type invariants $\mathcal{W}_{\mu,\nu}$'s. The magic of topological vertex is that, by simply looking at the moment map graph of the toric surfaces in the open toric Calabi-Yau, we can immediately write down the closed formula for the generating series for all genera and all degree Gromov-Witten invariants, or more precisely the Euler numbers of certain bundles on the moduli space of stable maps.

Here we only give one example to describe the topological vertex formula for the generating series of the all degree and all genera Gromov-Witten invariants for the open toric Calabi-Yau 3-folds. We write down the explicit close formula of the generating series of the Gromov-Witten invariants for $O(-3) \longrightarrow \mathbf{P}^2$ in terms of the Chern-Simons invariants.

Example: The complete generating series of Gromov-Witten invariants of all degree and all genera for $O(-3) \longrightarrow \mathbf{P}^2$ s given by

$$\exp\left(\sum_{g=0}^{\infty} \lambda^{2g-2} F_g(t)\right)$$

$$= \sum_{\nu_1,\nu_2,\nu_3} \mathcal{W}_{\nu_1,\nu_2}\mathcal{W}_{\nu_2,\nu_3}\mathcal{W}_{\nu_3,\nu_1}(-1)^{\sum_{j=1}^3 |\nu_j|} q^{\frac{1}{2}\sum_{i=1}^3 \kappa_{\nu_i}} e^{t(\sum_{j=1}^3 |\nu_j|)}$$

where $q = e^{\sqrt{-1}\lambda}$. The precise definition of $F_g(t)$ will be given in the next section.

For general open toric Calabi-Yau manifolds, the expressions are just similar. They are all given by finite and closed formulas, which are easily read out from the moment map graphs associated to the toric surfaces, with the topological vertex associated to each vertex of the graph.

In [**1**] Vafa and his group first developed the theory of topological vertex by using string duality between Chern-Simons and Calabi-Yau, which is a physical theory. In [**21**] we established the mathematical theory of the topological vertex, and derived various mathematical corollaries, including the relation of the Gromov-Witten invariants to the equivariant index theory as motivated by the Nekrasov conjecture in string duality [**27**].

13. Gopakumar-Vafa Conjecture and Indices of Elliptic Operators

Let $N_{g,d}$ denote the so-called Gromov-Witten invariant of genus g and degree d of an open toric Calabi-Yau 3-fold. $N_{g,d}$ is defined to be the Euler number of the obstruction bundle on the moduli space of stable maps of degree $d \in H_2(S,\mathbb{Z})$ from genus g curve into the surface base S. The open toric Calabi-Yau manifold associated to the toric surface S is the total space of the canonical line bundle K_S on S. More precisely

$$N_{g,d} = \int_{[\overline{\mathcal{M}}_g(S,d)]^v} e(V_{g,d})$$

with $V_{g,d} = R^1\pi_* u^* K_S$ a vector bundle on the moduli space induced by the canonical bundle K_S. Here $\pi : U \to \overline{\mathcal{M}}_g(S,d)$ denotes the universal curve and u can be considered as the evaluation or universal map. Let us write

$$F_g(t) = \sum_{d\geq 0} N_{g,d}\, e^{-d\cdot t}.$$

The Gopakumar-Vafa conjecture is stated as follows:

Gopakumar-Vafa Conjecture: *There exists an expression:*

$$\sum_{g=0}^{\infty} \lambda^{2g-2} F_g(t) = \sum_{k=1}^{\infty} \sum_{g,d\geq 0} n_d^g \frac{1}{d} (2sin \frac{d\lambda}{2})^{2g-2} e^{-kd\cdot t},$$

such that n_d^g are integers, called instanton numbers.

Motivated by the Nekrasov duality conjecture between the four dimensional gauge theory and string theory, we are able to interpret the above integers n_d^g as equivariant indices of certain elliptic operators on the moduli spaces of anti-self-dual connections [**27**]:

THEOREM 13.1. *For certain interesting cases, these n_d^g's can be written as equivariant indices on the moduli spaces of anti-self-dual connections on* $\mathbf{C}^2$.

For more precise statement, we refer the reader to [**27**]. The interesting cases include open toric Calabi-Yau manifolds when S is Hirzebruch surface. The proof of this theorem is to compare fixed point formula expressions for equivariant indices of certain elliptic operators on the moduli spaces of anti-self-dual connections with the combinatorial expressions of the generating series of the Gromov-Witten invariants on the moduli spaces of stable maps. They both can be expressed in terms of Young diagrams of partitions. We find that they agree up to certain highly non-trivial "mirror transformation", a complicated variable change. This result is not only interesting for the index formula interpretation of the instanton numbers, but also for the fact that it gives the first complete examples that the Gopakumar-Vafa conjecture holds for all genera and all degrees.

Recently P. Peng [**45**] has given the proof of the Gopakumar-Vafa conjecture for all open toric Calabi-Yau 3-folds by using the Chern-Simons expressions from the topological vertex. His method is to explore the property of the Chern-Simons expression in great detail with some clever observation about the form of the combinatorial expressions. On the other hand, Kim in [**13**] has derived some remarkable recursion formulas for Hodge integrals of all genera and any number of marked points, involving one λ-classes. His method is to add marked points in the moduli spaces and then follow the localization argument we used to prove the Mariño-Vafa formula.

14. Simple Localization Proofs of the ELSV Formula

Given a partition μ of length $l(\mu)$, denote by $H_{g,\mu}$ the Hurwitz numbers of almost simple Hurwitz covers of $\mathbb{P}^1$ of ramification type μ by connected genus g Riemann surfaces. The ELSV formula [**8, 10**] states:

$$H_{g,\mu} = (2g-2+|\mu|+l(\mu))! I_{g,\mu}$$

where

$$I_{g,\mu} = \frac{1}{|\operatorname{Aut}(\mu)|} \prod_{i=1}^{l(\mu)} \frac{\mu_i^{\mu_i}}{\mu_i!} \int_{\overline{\mathcal{M}}_{g,l(\mu)}} \frac{\Lambda_g^{\vee}(1)}{\prod_{i=1}^{l(\mu)}(1-\mu_i\psi_i)}.$$

Define generating functions

$$\begin{aligned}
\Phi_\mu(\lambda) &= \sum_{g\geq 0} H_{g,\mu}\frac{\lambda^{2g-2+|\mu|+l(\mu)}}{(2g-2+|\mu|+l(\mu))!},\\
\Phi(\lambda;p) &= \sum_{|\mu|\geq 1} \Phi_\mu(\lambda)p_\mu,\\
\Psi_\mu(\lambda) &= \sum_{g\geq 0} I_{g,\mu}\lambda^{2g-2+|\mu|+l(\mu)},\\
\Psi(\lambda;p) &= \sum_{|\mu|\geq 1} \Psi_\mu(\lambda)p_\mu.
\end{aligned}$$

In terms of generating functions, the ELSV formula reads

$$\Psi(\lambda;p) = \Phi(\lambda;p).$$

It was known that $\Phi(\lambda;p)$ satisfies the following cut-and-join equation:

$$\frac{\partial\Theta}{\partial\lambda} = \frac{1}{2}\sum_{i,j\geq 1}\left(ijp_{i+j}\frac{\partial^2\Theta}{\partial p_i\partial p_j} + ijp_{i+j}\frac{\partial\Theta}{\partial p_i}\frac{\partial\Theta}{\partial p_j} + (i+j)p_ip_j\frac{\partial\Theta}{\partial p_{i+j}}\right).$$

This formula was first proved in [**7**]. Later this equation was reproved by sum formula of symplectic Gromov-Witten invariants [**20**].

The calculations in Section 7 and Appendix A of [**27**] show that

$$\tilde{H}_{g,\mu} = (2g-2+|\mu|+l(\mu))! I_{g,\mu}$$

$$\begin{aligned}
\tilde{H}_{g,\mu} = (2g-3+|\mu|+l(\mu))! \Bigg(&\sum_{\nu\in J(\mu)} I_{g,\nu} + \sum_{\nu\in C(\mu)} I_2(\nu)I_{g-1,\nu}\\
&+ \sum_{g_1+g_2=g}\ \sum_{\nu^1\cup\nu^2\in C(\mu)} I_3(\nu^1,\nu^2)I_{g_1,\nu_1}I_{g_2,\nu_2}\Bigg)
\end{aligned}$$

where

$$\tilde{H}_{g,\mu} = \int_{[\overline{\mathcal{M}}_{g,0}(\mathbf{P}^1,\mu)]^{\mathrm{vir}}} \mathrm{Br}^* H^r$$

is some relative Gromov-Witten invariant of $(\mathbf{P}^1,\infty)$, and $C(\mu), J(\mu), I_1, I_2, I_3$ are defined as in [**20**]. In fact, as proved in [**27**], this is double Hurwitz numbers. So we have

$$\begin{aligned}
(2g-2+|\mu|+l(\mu))I_{g,\mu} = &\sum_{\nu\in J(\mu)} I_{g,\nu} + \sum_{\nu\in C(\mu)} I_2(\nu)I_{g-1,\nu}\\
&+ \sum_{g_1+g_2=g}\ \sum_{\nu^1\cup\nu^2\in C(\mu)} I_3(\nu^1,\nu^2)I_{g_1,\nu_1}I_{g_2,\nu_2},
\end{aligned}$$

which is equivalent to the statement that the generating function $\Psi(\lambda;p)$ of $I_{g,\mu}$ also satisfies the cut-and-join equation.

Any solution $\Theta(\lambda;p)$ to the cut-and-join equation (14) is uniquely determined by its initial value $\Theta(0;p)$, so it remains to show that $\Psi(0;p) = \Phi(0;p)$. Note that $2g-2+|\mu|+l(\mu)=0$ if and only if $g=0$ and $\mu=(1)$, so

$$\Psi(0;p) = H_{0,(1)}p_1, \quad \Phi(0;p) = I_{0,(1)}p_1.$$

It is easy to see that $H_{0,(1)} = I_{0,(1)} = 1$, so

$$\Psi(0;p) = \Phi(0;p).$$

One can see geometrically that the relative Gromov-Witten invariant $\tilde{H}_{g,\mu}$ is equal to the Hurwitz number $H_{g,\mu}$. This together with (14) gives a proof of the ELSV formula presented in [**27**, Section 7] in the spirit of [**10**]. Note that $\tilde{H}_{g,\mu} = H_{g,\mu}$ is not used in the proof described above.

On the other hand we can deduce the ELSV formula as the limit of the Mariño-Vafa formula. By the Burnside formula, one easily gets the following expression (see e.g., [**29**]):

$$\begin{aligned}\Phi(\lambda;p) &= \log\left(\sum_{\mu}\left(\sum_{|\nu|=|\mu|}\frac{\chi_\nu(\mu)}{z_\mu}e^{\kappa_\nu\lambda/2}\frac{\dim R_\nu}{|\nu|!}\right)p_\mu.\right)\\ &= \sum_{n\geq 1}\frac{(-1)^{n-1}}{n}\sum_{\mu}\sum_{\cup_{i=1}^n \mu_i=\mu}\prod_{i=1}^{n}\sum_{|\nu_i|=|\mu_i|}\frac{\chi_{\nu_i}(\mu_i)}{z_{\mu_i}}e^{\kappa_{\nu_i}\lambda/2}\frac{\dim R_{\nu_i}}{|\nu_i|!}p_\mu.\end{aligned}$$

The ELSV formula reads

$$\Psi(\lambda;p) = \Phi(\lambda;p)$$

where the left hand side is a generating function of Hodge integrals $I_{g,\mu}$, and the right hand side is a generating function of representations of symmetric groups. So the ELSV formula and the MV formula are of the same type.

Actually, the ELSV formula can be obtained by taking a particular limit of the MV formula $G(\lambda;\tau;p) = R(\lambda;\tau;p)$. More precisely, it is straightforward to check that

$$\begin{aligned}&\lim_{\tau\to 0} G(\lambda\tau;\frac{1}{\tau};(\lambda\tau)p_1,(\lambda\tau)^2p_2,\ldots)\\ &= \sum_{|\mu|\neq 0}\sum_{g=0}^{\infty}\sqrt{-1}^{2g-2+|\mu|+l(\mu)}I_{g,\mu}\lambda^{2g-2+|\mu|+l(\mu)}p_\mu\\ &= \Psi(\sqrt{-1}\lambda;p)\end{aligned}$$

and

$$
\begin{aligned}
&\lim_{\tau\to 0} R(\lambda\tau; \frac{1}{\tau}; (\lambda\tau)p_1, (\lambda\tau)^2 p_2, \ldots) \\
&= \log\left(\sum_{\mu}\left(\sum_{|\nu|=|\mu|} \frac{\chi_\nu(C(\mu))}{z_\mu} e^{\sqrt{-1}\kappa_\nu\lambda/2} \lim_{t\to 0}(t^{|\nu|}V_\nu(t))\right) p_\mu\right) \\
&= \log\left(\sum_{\mu}\left(\sum_{|\nu|=|\mu|} \frac{\chi_\nu(C(\mu))}{z_\mu} e^{\sqrt{-1}\kappa_\nu\lambda/2} \frac{1}{\prod_{x\in\nu} h(x)}\right) p_\mu\right) \\
&= \Phi(\sqrt{-1}\lambda; p)
\end{aligned}
$$

where we have used

$$
\frac{1}{\prod_{x\in\nu} h(x)} = \frac{\dim R_\nu}{|\nu|!}.
$$

See [**30**] for more details. In this limit, the cut-and-join equation of $G(\lambda;\tau;p)$ and $R(\lambda;\tau;p)$ reduces to the cut-and-join equation of $\Psi(\lambda;p)$ and $\Phi(\lambda;p)$, respectively.

15. A Localization Proof of the Witten Conjecture

The Witten conjecture for moduli spaces states that the generating series F of the integrals of the ψ classes for all genera and any number of marked points satisfies the KdV equations and the Virasoro constraint. For example, the Virasoro constraint states that F satisfies

$$
L_n \cdot F = 0, \;\; n \geq -1
$$

where L_n denote certain Virasoro operators as given below.

Witten conjecture was first proved by Kontsevich [**16**] using a combinatorial model of the moduli space and matrix model, with later approaches by Okounkov-Pandhripande [**43**] using ELSV formula and combinatorics, and by Mirzakhani [**42**] using Weil-Petersson volumes on moduli spaces of bordered Riemann surfaces.

I will present a much simpler proof by using functorial localization and asymptotics. This was done [**14**] jointly with Y.-S. Kim. This is also motivated by methods in proving conjectures from string duality. It should have more applications.

The basic idea of our proof is to directly prove the following recursion formula which, as derived in physics by Dijkgraaf, Verlinde and Verlinde by using quantum field theory, implies the Virasoro and the KdV equation for the generating series F of the integrals of the ψ classes:

THEOREM 15.1. *We have identity*

$$\left\langle \tilde{\sigma}_n \prod_{k\in S} \tilde{\sigma}_k \right\rangle_g = \sum_{k\in S}(2k+1)\left\langle \tilde{\sigma}_{n+k-1} \prod_{l\neq k} \tilde{\sigma}_l \right\rangle_g$$
$$+\frac{1}{2}\sum_{a+b=n-2}\left\langle \tilde{\sigma}_a\tilde{\sigma}_b \prod_{l\neq a,b} \tilde{\sigma}_l \right\rangle_{g-1}$$
$$+\frac{1}{2}\sum_{\substack{S=X\cup Y,\\ a+b=n-2,\\ g_1+g_2=g}}\left\langle \tilde{\sigma}_a \prod_{k\in X} \tilde{\sigma}_k \right\rangle_{g_1}\left\langle \tilde{\sigma}_b \prod_{l\in Y} \tilde{\sigma}_l \right\rangle_{g_2}.$$

Here $\tilde{\sigma}_n = (2n+1)!!\psi^n$ and

$$\left\langle \prod_{j=1}^{n} \tilde{\sigma}_{k_j} \right\rangle_g = \int_{\overline{\mathcal{M}}_{g,n}} \prod_{j=1}^{n} \tilde{\sigma}_{k_j}.$$

The notation $S = \{k_1, \ldots, k_n\} = X \cup Y$.

To prove the above recursion relation, we first apply the functorial localization to the natural branch map from moduli space of relative stable maps $\overline{\mathcal{M}}_g(\mathbf{P}^1, \mu)$ to projective space $\mathbf{P}^r$ where $r = 2g-2+|\mu|+l(\mu)$ is the dimension of the moduli. Since the push-forward of 1 is a constant in this case, we easily get the cut-and-join equation for one Hodge integral

$$I_{g,\mu} = \frac{1}{|\mathrm{Aut}\ \mu|}\prod_{i=1}^{n}\frac{\mu_i^{\mu_i}}{\mu_i!}\int_{\overline{\mathcal{M}}_{g,n}}\frac{\Lambda_g^{\vee}(1)}{\prod(1-\mu_i\psi_i)}.$$

As given in the previous section, we have

$$(2g-2+|\mu|+l(\mu))I_{g,\mu}$$
$$= \sum_{\nu\in J(\mu)} I_{g,\nu} + \sum_{\nu\in C(\mu)} I_2(\nu)I_{g-1,\nu}$$
$$+ \sum_{g_1+g_2=g}\ \sum_{\nu^1\cup\nu^2\in C(\mu)} I_3(\nu^1,\nu^2)I_{g_1,\nu^1}I_{g_2,\nu^2}.$$

Write $\mu_i = Nx_i$. Let N goes to infinity and expand in x_i, and we get:

$$\sum_{i=1}^{n} \Big[\frac{(2k_i+1)!!}{2^{k_i+1}k_i!} x_i^{k_i} \prod_{j\neq i} \frac{x_j^{k_j-\frac{1}{2}}}{\sqrt{2\pi}} \int_{\overline{\mathcal{M}}_{g,n}} \prod \psi_j^{k_j}$$

$$- \sum_{j\neq i} \frac{(x_i+x_j)^{k_i+k_j-\frac{1}{2}}}{\sqrt{2\pi}} \prod_{l\neq i,j} \frac{x_l^{k_l-\frac{1}{2}}}{\sqrt{2\pi}} \int_{\overline{\mathcal{M}}_{g,n-1}} \psi^{k_i+k_j-1} \prod \psi_l^{k_l}$$

$$- \frac{1}{2} \sum_{k+l=k_i-2} \frac{(2k+1)!!(2l+1)!!}{2^{k_i}\, k_i!} x_i^{k_i} \prod_{j\neq i} \frac{x_j^{k_j-\frac{1}{2}}}{\sqrt{2\pi}} \Big[\int_{\overline{\mathcal{M}}_{g-1,n+1}} \psi_1^k \psi_2^l \prod \psi_j^{k_j}$$

$$+ \sum_{\substack{g_1+g_2=g,\\ \nu_1\cup\nu_2=\nu}} \int_{\overline{\mathcal{M}}_{g_1,n_1}} \psi_1^k \prod \psi_j^{k_j} \int_{\overline{\mathcal{M}}_{g_2,n_2}} \psi_1^l \prod \psi_j^{k_j} \Big] \Big] = 0.$$

Performing Laplace transforms on the x_i's, we get the recursion formula which implies both the KdV equations and the Virasoro constraints. For example, the Virasoro constraints state that the generating series

$$\tau(\tilde{t}) = \exp \sum_{g=0}^{\infty} \left\langle \exp \sum_n \tilde{t}_n \tilde{\sigma}_n \right\rangle_g$$

satisfies the equations:

$$L_n \cdot \tau = 0, \qquad (n \geq -1)$$

where L_n denote the Virasoro differential operators

$$L_{-1} = -\frac{1}{2}\frac{\partial}{\partial \tilde{t}_0} + \sum_{k=1}^{\infty} \left(k+\frac{1}{2}\right) \tilde{t}_k \frac{\partial}{\partial \tilde{t}_{k-1}} + \frac{1}{4}\tilde{t}_0^2$$

$$L_0 = -\frac{1}{2}\frac{\partial}{\partial \tilde{t}_1} + \sum_{k=0}^{\infty} \left(k+\frac{1}{2}\right) \tilde{t}_k \frac{\partial}{\partial \tilde{t}_k} + \frac{1}{16}$$

$$L_n = -\frac{1}{2}\frac{\partial}{\partial \tilde{t}_{n-1}} + \sum_{k=0}^{\infty} \left(k+\frac{1}{2}\right) \tilde{t}_k \frac{\partial}{\partial \tilde{t}_{k+n}} + \frac{1}{4}\sum_{i=1}^{n} \frac{\partial^2}{\partial \tilde{t}_{i-1}\partial \tilde{t}_{n-i}}.$$

We remark that the same method can be used to derive very general recursion formulas in Hodge integrals and general Gromov-Witten invariants. We hope to report these results on a later occasion.

16. Final Remarks

We have briefly reviewed our recent results on both the geometric and the topological aspect of the moduli spaces of Riemann surfaces. Although significant progress has been made in understanding the geometry and topology of the moduli spaces of Riemann surfaces, there are still many problems that remain to be solved in both aspects.

For the geometric aspect, it will be interesting to understand the convergence of the Ricci flow starting from the Ricci metric to the Kähler-Einstein metric, the representations of the mapping class group on the middle dimensional L^2-cohomology of these metrics, and the index theory associated to these complete Kähler metrics. Recently, we showed that the metrics on the logarithm cotangent bundle induced by the Weil-Petersson metric, the Ricci metric and the perturbed Ricci metric are good in the sense of Mumford [**36**]. Also the perturbed Ricci metric is the first complete Kähler metric on the moduli spaces with bounded negative Ricci and holomorphic sectional curvature and bounded geometry, and we believe this metric must have more interesting applications. Another question is which of these metrics are actually identical. We hope to report on the progress of the study of these problems on a later occasion.

For the topological aspect it will be interesting to have closed formulas to compute Hodge integrals involving more Hodge classes, and to use our complete understanding of the Gromov-Witten theory in the open formal toric Calabi-Yau manifolds to understand the compact Calabi-Yau case. We strongly believe that there is a more interesting and grand duality picture between Chern-Simons invariants for three dimensional manifolds and the Gromov-Witten invariants for open toric Calabi-Yau manifolds. Our proofs of the Mariño-Vafa formula, and the setup of the mathematical foundation for topological vertex theory and the results of Peng and Kim all together have just opened a small window for a more splendid picture.

Finally, although we have worked on two quite different aspects of the moduli spaces, we strongly believe that the methods and results we have developed and obtained in these seemingly unrelated aspects will eventually merge together to give us a completely clear understanding of the moduli spaces of Riemann surfaces.

References

[1] M. Aganagic, A. Klemm, M. Marino, and C. Vafa, *The topological vertex*, preprint, hep-th/0305132.

[2] M. Aganagic, M. Marino, and C. Vafa, *All loop topological string amplitudes from Chern-Simons theory*, preprint, hep-th/0206164.

[3] L. Ahlfors and L. Bers, *Riemann's mapping theorem for variable metrics*, Ann. of Math. (2) **72** (1960), 385–404.

[4] S.Y. Cheng and S.-T. Yau, *Differential equations on Riemannian manifolds and their geometric applications*, Comm. Pure Appl. Math. **28(3)** (1975), 333–354.

[5] S.Y. Cheng and S.-T. Yau. *On the existence of a complete Khler metric on noncompact complex manifolds and the regularity of Fefferman's equation*, Comm. Pure Appl. Math. **33(4)** (1980), 507–544.

[6] P. Deligne and D. Mumford, *The irreducibility of the space of curves of given genus*, Inst. Hautes Études Sci. Publ. Math. **36** (1969), 75–109.

[7] I.P. Goulden and D.M. Jackson, *Combinatorial enumeration*, John Wiley & Sons, 1983.

[8] T. Ekedahl, S. Lando, M. Shapiro, and A. Vainshtein, *Hurwitz numbers and intersections on moduli spaces of curves*, Invent. Math. **146**(**2**) (2001), 297–327.

[9] I.P. Goulden, D.M. Jackson, and A. Vainshtein, *The number of ramified coverings of the sphere by the torus and surfaces of higher genera*, Ann. of Comb. **4** (2000), 27–46.

[10] T. Graber and R. Vakil, *Hodge integrals and Hurwitz numbers via virtual localization*, Compositio Math. **135**(**1**) (2003), 25–36.

[11] A. Iqbal, *All genus topological amplitudes and 5-brane webs as Feynman diagrams*, preprint, hep-th/0207114.

[12] S. Katz and C.-C. Liu, *Enumerative geometry of stable maps with Lagrangian boundary conditions and multiple covers of the disc*, Adv. Theor. Math. Phys. **5** (2001), 1–49.

[13] Y. Kim, *Computing Hodge integrals with one lambda-class*, preprint, math-ph/0501018.

[14] Y. Kim and K. Liu, *A simple proof of Witten conjecture through localization*, preprint, math.AG/0508384.

[15] S. Kobayashi, *Hyperbolic complex spaces*, Grundlehren der Mathematischen Wissenschaften [Fundamental Principles of Mathematical Sciences], **318**, Springer-Verlag, Berlin, 1998.

[16] M. Kontsevich, *Intersection theory on the moduli space of curves and the matrix Airy function*, Comm. Math. Phys. **147**(**1**) (1992), 1–23.

[17] J. Li, *Stable Morphisms to singular schemes and relative stable morphisms*, J. Differential Geom. **57** (2001), 509–578.

[18] J. Li, *Relative Gromov-Witten invariants and a degeneration formula of Gromov-Witten invariants*, J. Differential Geom. **60** (2002), 199–293.

[19] J. Li, *Lecture notes on relative GW-invariants*, preprint.

[20] A.M. Li, G. Zhao, and Q. Zheng, *The number of ramified coverings of a Riemann surface by Riemann surface*, Comm. Math. Phys. **213**(**3**) (2000), 685–696.

[21] J. Li, C.-C. Liu, K. Liu, and J. Zhou, *A mathematical theory of the topological vertex*, preprint, math.AG/0411247.

[22] J. Li, K. Liu, and J. Zhou, *Topological string partition functions as equivariant indices*, preprint, math.AG/0412089.

[23] B. Lian, C.-H. Liu, K. Liu, and S.-T. Yau, *The S^1-fixed points in quot-schemes and mirror principle computations*, Contemp. Math. **322** (2003), 165–194.

[24] B. Lian, K. Liu, and S.-T. Yau, *Mirror Principle*, I, Asian J. Math. **1** (1997), 729–763.

[25] B. Lian, K. Liu, and S.-T. Yau, *Mirror Principle*, III, Asian J. Math. **3** (1999), 771–800.

[26] C.-C. Liu, *Formulae of one-partition and two-partition Hodge integrals*, preprint, math.AG/0502430.

[27] C.-C. Liu, K. Liu, and J. Zhou, *On a proof of a conjecture of Mariño-Vafa on Hodge Integrals*, Math. Res. Letters **11** (2004), 259–272.

[28] C.-C. Liu, K. Liu, and J. Zhou, *A proof of a conjecture of Mariño-Vafa on Hodge Integrals*, J. Differential Geometry **65** (2003), 289–340.

[29] C.-C, Liu, K. Liu, and J. Zhou, *A formula of two-partition Hodge integrals*, preprint, math.AG/0310272.

[30] C.-C, Liu, K. Liu, and J. Zhou, *Mariño-Vafa formula and Hodge integral identities*, preprint, math.AG/0308015.

[31] K. Liu, *Mathematical results inspired by physics*, Proc. ICM 2002, Vol. III, 457–466.

[32] K. Liu, *Geometric height inequalities*, Math. Research Letter **3** (1996), 693–702.

[33] K. Liu, *Localization and string duality*, to appear in Proceedings of the International Conference in Complex Geometry, ECNU 2005.

[34] K. Liu, X. Sun, and S.-T. Yau, *Canonical metrics on the moduli space of Riemann surface*, I, Journal of Differential Geometry **68** (2004), 571–637.

[35] K. Liu, X. Sun, and S.-T. Yau, *Canonical metrics on the moduli space of Riemann surface*, II, Journal of Differential Geometry **69** (2005), 163–216.
[36] K. Liu, X. Sun, and S.-T. Yau, *Good metrics on the moduli space of Riemann surface* preprint, 2005.
[37] K. Liu, X. Sun, and S.-T. Yau, *Geometric aspects of the moduli space of Riemann surface*, Sciences in China, 2005.
[38] I.G. MacDonald, *Symmetric functions and Hall polynomials*, 2nd edition, Claredon Press, 1995.
[39] M. Mariño and C. Vafa, *Framed knots at large N*, Orbifolds in mathematics and physics (Madison, WI, 2001), 185–204, Contemp. Math., **310**, Amer. Math. Soc., Providence, RI, 2002.
[40] H. Masur, *Extension of the Weil-Petersson metric to the boundary of Teichmuller space*, Duke Math. J. **43(3)** (1976), 623–635.
[41] C.T. McMullen, *The moduli space of Riemann surfaces is Kähler hyperbolic*, Ann. of Math. (2) **151(1)** (2000), 327–357.
[42] M. Mirzakhani, *Simple geodesics and Weil-Petersson volumes of moduli spaces of bordered Riemann surfaces*, preprint, 2003.
[43] A. Okounkov and R. Pandharipande, *Gromov-Witten theory, Hurwitz numbers, and Matrix models*, I, preprint, math.AG/0101147
[44] A. Okounkov and R. Pandharipande, *Hodge integrals and invariants of the unknots*, preprint, math.AG/0307209.
[45] P. Pan, *A simple proof of Gopakumar-Vafa conjecture for local toric Calabi-Yau manifolds*, preprint, math.AG/0410540.
[46] R. Schoen and S.-T. Yau, *Lectures on differential geometry*, Conference Proceedings and Lecture Notes in Geometry and Topology, I. International Press, Cambridge, MA, 1994.
Lecture notes prepared by Wei Yue Ding, Kung Ching Chang [Gong Qing Zhang], Jia Qing Zhong and Yi Chao Xu, Translated from the Chinese by Ding and S.Y. Cheng, Preface translated from the Chinese by Kaising Tso.
[47] G. Schumacher, *The curvature of the Petersson-Weil metric on the moduli space of Kähler-Einstein manifolds*, in 'Complex analysis and geometry', Univ. Ser. Math., 339–354, Plenum, New York, 1993.
[48] Y.T. Siu, *Curvature of the Weil-Petersson metric in the moduli space of compact Kähler-Einstein manifolds of negative first Chern class*, in 'Contributions to several complex variables', Aspects Math., E9, 261–298, Vieweg, Braunschweig, 1986.
[49] S. Trapani, *On the determinant of the bundle of meromorphic quadratic differentials on the Deligne-Mumford compactification of the moduli space of Riemann surfaces*, Math. Ann. **293(4)** (1992), 681–705.
[50] E. Witten, *Quantum field theory and the Jones polynomial*, Commun. Math. Phys. **121** (1989), 351–399.
[51] E. Witten, *Two-dimensional gravity and intersection theory on moduli space*, Surveys in differential geometry (Cambridge, MA, 1990), 243–310, Lehigh Univ., Bethlehem, PA, 1991.
[52] S.A. Wolpert, *Chern forms and the Riemann tensor for the moduli space of curves*, Invent. Math. **85(1)** (1986), 119–145.
[53] S.A. Wolpert, *Asymptotics of the spectrum and the Selberg zeta function on the space of Riemann surfaces*, Comm. Math. Phys. **112(2)** (1987), 283–315.
[54] S.A. Wolpert, *The hyperbolic metric and the geometry of the universal curve*, J. Differential Geom. **31(2)** (1990), 417–472.
[55] S.-T. Yau, *A general Schwarz lemma for Kähler manifolds*, Amer. J. Math. **100(1)** (1978), 197–203.
[56] S.-T. Yau, *On the Ricci curvature of a compact Kähler manifold and the complex Monge-Ampère equation*, I, Comm. Pure Appl. Math. **31(3)** (1978), 339–411.

[57] S.-T. Yau, *Nonlinear analysis in geometry*, Monographies de L'Enseignement Mathématique [Monographs of L'Enseignement Mathématique], **33**, L'Enseignement Mathématique, Geneva, 1986; Série des Conférences de l'Union Mathématique Internationale [Lecture Series of the International Mathematics Union], **8**.

[58] J. Zhou, *Hodge integrals, Hurwitz numbers, and symmetric groups*, preprint, math.AG/0308024.

[59] J. Zhou, *A conjecture on Hodge integrals*, preprint.

[60] J. Zhou, *Localizations on moduli spaces and free field realizations of Feynman rules*, preprint.

Center of Mathematical Sciences, Zhejiang University, Hangzhou, China
E-mail address: `liu@cms.zju.edu.cn`

and

Department of Mathematics, Univ. of California at Los Angeles
Los Angeles, CA 90095-1555
E-mail address: `liu@math.ucla.edu`

Applications of minimal surfaces to the topology of three-manifolds

William H. Meeks, III

1. Introduction

In this paper, I will mention some applications of minimal surfaces to the geometry and topology of three-manifolds that I discussed in my lecture at the Current Developments in Mathematics Conference for 2004.

The first important application of minimal surfaces to the geometry of three-manifolds was given by Schoen and Yau [**22**] in their study of Riemannian three-manifolds of positive scalar curvature and their related proof of the positive mass conjecture in general relativity. The techniques that they developed in their proof of this conjecture continue to be useful in studying relationships between stable minimal surfaces and the topology of Riemannian manifolds.

Around 1978, Meeks and Yau gave geometric versions of three classical theorems in three-dimensional topology. These classical theorems concern the existence of certain embedded surfaces. In the geometric versions of these theorems, Meeks and Yau proved the existence of essentially cononical solutions, which are given by area minimizing surfaces. They referred to these theorems as the Geometric Dehn's Lemma, Geometric Loop Theorem and the Geometric Sphere Theorem.

As an application of these special geometric minimal surface solutions to these classical topological theorems, Meeks and Yau gave new equivariant versions of these theorems in the presence of a differential finite group action. Their Equivariant Loop Theorem turned out to be the final missing step in the solution of the Smith Conjecture concerning the standardness of the smooth action of finite cyclic groups on the three-sphere S^3. These and related results will be discussed in Section 2.

This material is based upon work by the NSF under Award No. DMS - 0405836. Any opinions, findings, and conclusions or recommendations expressed in this publication are those of the authors and do not necessarily reflect the views of the NSF.

Recently, Colding and Minicozzi [**1**] gave an application of minimal surfaces to study the Ricci flow on a closed Riemannian three-manifold that is a homotopy sphere. They proved that on such a manifold the Ricci flow has finite extinction time, which means that a one-parameter family $g(t)$ of metrics evolving by Ricci flow becomes singular in finite time. A sketch of their proof of this result appears in Section 3.

2. Embedded least-area surfaces in three-manifolds.

In this section, I will review some of the results and background material for the minimal surface analogues of the classical Dehn's Lemma, Loop Theorem and the Sphere Theorem in three-manifold topology. We begin this discussion with the statement of the minimal surface version of Dehn's Lemma by Meeks and Yau. We recall that the least-energy map referred to in the statement of the theorem below has least-area and is conformal on the interior of D, where D is the unit disk in $\mathbb{C}$.

THEOREM 2.1 (Geometric Dehn's Lemma [**13**]). *Let M be a compact three-manifold with convex boundary. If Γ is a simple closed curve in ∂M which is homotopically trivial in M, then:*

1. *There exists a map $f\colon D \to M$ of least-energy such that $f|_{\partial D}$ is a parametrization of Γ.*
2. *Any map $f\colon D \to M$ given in (1) is injective and a smooth immersion of the interior of D.*
3. *Such an f is as regular as Γ along ∂D and if Γ is of class C^2, then f is an immersion.*
4. *If f_1 and f_2 are two such solutions and $f_1(\mathrm{Int}(D)) \bigcap f_2(\mathrm{Int}(D)) \neq \emptyset$, then $f_2 = f_1 \circ \varphi$, where φ is a conformal diffeomorphism of D.*

The proof of the existence of a least-energy $f\colon D \to M$ follows from Morrey's solution of Plateau's problem in a homogeneously regular n-manifold (without boundary). We now sketch how this result follows. In the case ∂M is smooth and strictly convex, we proved that M embeds as a subdomain of a homogeneously regular three-manifold $\widetilde{M}$ such that any compact minimal surface Σ in $\widetilde{M}$, with $\partial\Sigma \subset M \subset \widetilde{M}$, is contained in M. Hence, the Morrey least-energy solution to Plateau's problem for a $\Gamma \subset M \subset \widetilde{M}$ in $\widetilde{M}$, actually is contained in M. For the general case where ∂M is geodesically convex and perhaps just continuous, then one uses an approximation procedure to obtain a Morrey solution to the classical Plateau problem for $\Gamma \subset \partial M$, which has finite least-energy.

By Osserman [**19**] and Gulliver [**4**], one obtains that $f|_{\mathrm{Int}(D)}$ is an immersion. Statement 3 that the immersion f is as regular as ∂D follows from results of Lewy [**9**] in the case Γ is analytic, and when Γ is of class C^2 from results of Hildebrandt [**7**]. The nonexistence of boundary branch points for

$f\colon D \to M$, when Γ is of class C^2, easily follows from the C^2-regularity of f and the convexity of ∂M.

The proof of statement 2 in the case Γ and M are analytic is given by a topological argument, called the tower construction, used to prove the classical Dehn's Lemma in three-manifold topology. We give the proof of this analytic case at the end of this section. The proof of injectivity in the case of a general Γ and a general M with convex boundary is accomplished by approximation arguments.

The proof of statement 4 in the analytic case is a straightforward modification of an argument similar to the one used in the proof of statement 2 in the analytic case. This argument is based on a cut and paste argument, which we now explain. Suppose that D_1, D_2 are two least-area embedded disks in M with $\partial D_1 = \partial D_2 = \Gamma$, which intersect transversely at some interior point of the disks. Then, there is a simple closed curve γ in the intersection which bounds subdisks $\widetilde{D}_1 \subset D_1$ and $\widetilde{D}_2 \subset D_2$. Without loss of generality, we may assume that Area$(\widetilde{D}_1) \leq$ Area$(\widetilde{D}_2)$. Then cut $\widetilde{D}_2$ out of D_2 and replace it by $\widetilde{D_1}$, to obtain a piecewise smooth disk $\widehat{D_2} = (D_2 - \widetilde{D}_2) \cup \widetilde{D}_1$ with $\partial \widehat{D} = \Gamma$, which is not smooth, but has the same least-area as D_2. But, the area of $\widehat{D}_2$ can be decreased along γ, and so, a small perturbation of $\widehat{D}_2$ has less area than D_2, contradicting our least-area assumption for D_2.

Since every compact three-manifold has a smooth metric that is a product metric in a small ε-product neighborhood of its boundary, every compact three-manifold admits a metric with convex boundary. Thus, statement 2 yields the classical topological result.

COROLLARY 2.2 (Dehn's Lemma). *A smooth simple closed curve on the boundary of a three-manifold, which is homotopically trivial in the three-manifold, is the boundary of a smooth embedded disk.*

One of our original motivations for proving our Geometric Dehn's Lemma was to prove the following now classical result.

COROLLARY 2.3 ([**13**]). *Let Γ be a simple closed curve in $\mathbb{R}^3$ that is extremal (it lies on the boundary of its convex hull). Then, Γ bounds a disk of finite area and any classical Douglas solution to Plateau's problem for Γ is an embedded minimal disk.*

We now discuss the free boundary value problem that arises in our proof of the Geometric Loop Theorem. For this, we first consider a special easier to visualize case. Consider a three-manifold M which is a smooth solid torus (possibly knotted) in $\mathbb{R}^3$ whose boundary has nonnegative mean curvature. Courant [**3**] considered the classical free boundary valued problem for M and proved that there exists a branched minimal disk $f\colon D \to M$ of least-area such that $f(\partial D)$ represents a homotopically nontrivial curve in the boundary torus ∂M. The following theorem shows that such an f is a smooth embedding with $f(D)$ orthogonal to ∂M along ∂D.

Theorem 2.4 ([**14**] and [**12**]). *Let M be a compact three-manifold, whose boundary has nonnegative mean curvature. Let S be the disjoint union of some components of ∂M. Let K be the kernel of the map $i_*\colon \pi_1(S) \to \pi_1(M)$, where i is the inclusion map. Then:*

(1) *There are a finite number of smooth conformal maps $f_1, \ldots, f_k$ from the unit disk D into M, so that,*
   (a) *f_1 has least-area among all maps from D into M whose boundary σ_1 represents a nontrivial element in K.*
   (b) *For each i, f_i has minimal area among all maps from D into M whose boundary σ_1 does not belong to the smallest normal subgroup of $\pi_1(S)$ containing $[\sigma_1], \ldots, [\sigma_{i-1}]$.*
   (c) *The disks $f_i(D)$ are orthogonal to ∂M along their boundary σ_i.*
   (d) *K is the smallest normal subgroup of $\pi_1(S)$ containing $[\sigma_1], \ldots, [\sigma_k]$.*

(2) *Any set of conformal mappings $f_1, \ldots, f_k$ satisfying properties* (a) *and* (b) *are embeddings and have mutually disjoint images.*

(3) *If $g_1, \ldots, g_l$ is another set of conformal mappings satisfying* (a) *and* (b), *then any two mappings in the set $\{f_1, \ldots, f_k, g_1, ..., g_l\}$ either are equal up to conformal reparametrization or have disjoint images.*

We just remark that the strategy in the proof of the Geometric Loop Theorem is similar to the proof of the Geometric Dehn's Lemma in most respects. However, in the proof of the Geometric Loop Theorem, we needed to prove the existence of a least-energy solution $f\colon D \to M$ to the free boundary value problem; in the previous case, we could refer more directly to Morrey's solution to the classical Plateau problem.

The above theorem has the following topological corollary.

Corollary 2.5 (Loop Theorem). *If M is a three-manifold and there exists a homotopically nontrivial curve in ∂M, which is homotopically trivial in M. Then, there exists an embedded disk $(D, \partial D) \subset (M, \partial M)$ with ∂D homotopically nontrivial in ∂M.*

Since minimal surfaces are rather cononical, our geometric methods have potential applications beyond those obtained by the classical topological solutions. Indeed, a moment's thought shows that applications of geometric solutions to study smooth compact group actions will be most fruitful because these groups can be considered as groups of isometries of some Riemannian metric and minimal surfaces must behave well under such actions. In this way, we were able to prove Dehn's Lemma, Loop Theorem and the Sphere Theorem (to be discussed) in equivariant form. Combining an observation of Gordan and Literland, the following equivariant loop theorem and a theorem of W. Thurston (which also depends on a theorem of H. Bass), one settled in the affirmative the conjecture of P.A. Smith on the unknottedness of the fixed point set of a finite cyclic group action on S^3 (see **[15]** for details).

COROLLARY 2.6 (Equivariant Loop Theorem ([**12**])). *If G is a smooth finite group action on a compact three-manifold M with compressible boundary (some homotopy nontrivial curve in ∂M is homotopically trivial in M), then there exists an embedded disk $(D, \partial D) \subset (M, \partial M)$ with ∂D homotopically nontrivial in ∂M such that the G orbit of D is an embedded two-manifold.*

We now state the Geometric Sphere Theorem and its corresponding Equivariant Sphere Theorem as a corollary.

THEOREM 2.7 (Geometric Sphere Theorem [**12**]). *Let M be a three-manifold with convex boundary. Then, there exist conformal maps $f_1, \dots, f_k$ from S^2 into M such that:*

(1) *$f_1 \colon S^2 \to M$ is homotopically nontrivial and minimizes area among all homotopically nontrivial maps from S^2 into M. For each i, f_i does not belong to the $\pi_1(M)$ submodule of $\pi_2(M)$ generated by $\{f_1, \dots, f_{i-1}\}$ and f_i minimizes area among all such maps.*
(2) *$\{f_1, \dots, f_k\}$ generates $\pi_2(M)$ as a $\pi_1(M)$ module.*
(3) *For any set of maps $\{g_1, \dots, g_l\}$ from S^2 into M that satisfy property (1), then g_i is either a conformal embedding or a two-to-one covering map whose image is an embedded real projective plane RP^2. Furthermore, if $\{f_1, \dots, f_k\}$ and $\{g_1, \dots, g_l\}$ are two sets of mappings satisfying property* (1)*, then, for all i and j, either the images of f_i and g_j are disjoint or f_i and g_j are equal up to conformal reparametrization.*

THEOREM 2.8. *Suppose M is a compact orientable three-manifold and* $M = \left(\begin{array}{c} k_1 \\ \#P_1 \\ i=1 \end{array}\right) \# \left(\begin{array}{c} k_2 \\ \#P_2 \\ i=1 \end{array}\right) \# \dots \# \left(\begin{array}{c} k_n \\ \#P_n \\ i=1 \end{array}\right)$, *where $\#$ denotes connected sum and $P_1, P_2, \dots, P_n$ are distinct prime orientable three-manifolds such that P_i is not homotopically equivalent to S^3 or $S^2 \times S^1$. Suppose there is a finite group G of diffeomorphisms acting effectively on M. Then:*

(1) *There is a natural homomorphism $\sigma \colon G \to \Pi_{i=1}^n S(k_i)$, where $S(k_i)$ is the permutation group on k_i letters.*
(2) *There is a natural injective homomorphism $\tau : \mathrm{Ker}\ (\sigma) \to \Pi_{i=1}^n \mathrm{Diff}\ (P_i)$.*
(3) *Let $\pi_j \colon \Pi_{i=1}^n S(k_i) \to S(k_j)$ be the projection on the jth coordinate. Then, there is a natural injective homomorphism $\alpha \colon \mathrm{Ker}(\pi_j \circ \sigma) \to \mathrm{Diff}(P_j)$.*
(4) *If $k_1 = k_2 = \dots k_n = 1$, then G act effectively on S^2 and effectively on each P_j as a finite group of diffeomorphisms with some fixed point. In particular, in this case G is isomorphic to a finite subgroup of the orthogonal group $O(3)$.*

Our equivariant sphere theorem also shows that for a finite group acting smoothly on the connected sum of compact nonsimply connected prime orientable manifolds with fundamental group nonisomorphic to the integers, the action must split equivariantly up to the permutations of the factors. Hence, basically when we study finite group actions on a three-manifold, we can assume the manifold is prime. The sphere theorem also enables us to deal with finite groups acting on noncompact manifolds. For example, combining the above mentioned affirmative answer to the Smith conjecture and the sphere theorem, we prove in [**15**] that finite cyclic groups acting smoothly on $\mathbb{R}^3$ must be conjugate to the linear action and that every finite subgroup of $\mathrm{Diff}(\mathbb{R}^3)$ is isomorphic to a subgroup of $O(3)$. In fact, these results and further work by Thurston show that every finite subgroup of $\mathrm{Diff}(\mathbb{R}^3)$ is conjugate to a subgroup of $O(3) \subset \mathrm{Diff}(\mathbb{R}^3)$. In [**11**], we generalized Theorem 3 to the case where P_i is not S^3 but may be a homotopy three-sphere.

2.1. The proof of the embedding of the analytic case of the Geometric Dehn's Lemma. In this section, we give a simplified proof of the basic topological construction used in the proof of the geometric Dehn's Lemma in [**16**]. We give the proof only in the analytic setting.

THEOREM 2.9. *Suppose M is a compact analytic Riemannian three-manifold. Suppose that D is the closed unit disk in the plane and γ is an analytic curve on ∂M and that $f\colon D \to M$ is a least-area (energy) map with $f(\partial D) = f(D) \cap \partial N = \gamma$. Then f is injective.*

The proof of the theorem will depend on the following sequence of lemmas.

LEMMA 2.10. *$f\colon D \to M$ is an analytic immersion.*

PROOF. By the regularity theorems of Gulliver [**4**] and Osserman [**19**], f is an immersion on the interior of the D. The function f is analytic on $\mathrm{Int}(D)$ by Morrey's interior regularity theorem [**18**]. The map f is analytic on D by the boundary regularity theorems by Lewy [**9**] and by Hildebrandt [**7**]. By a theorem of Gulliver-Lesley [**5**], f is an immersion on D. □

LEMMA 2.11. *$f\colon D \to M$ is simplicial with respect to fixed triangulations of D and M.*

PROOF. By Lemma 2.10, f is analytic and it follows that $f(D)$ is a semi-analytic subset of M. Also, it follows from the triangulation theorems in [**10**] that the semi-analytic subset $f(D)$ of M is a two-dimensional subcomplex of some triangulation of M. Since f is an immersion, the triangulation of $f(D)$ induces a triangulation of D such that $f\colon D \to M$ is simplicial. □

LEMMA 2.12. *Suppose D_1 and D_2 are distinct analytic embedded disks in an open Riemannian three-manifold N and that D_1 and D_2 have least-area with respect to their boundary curves. If $D_1 \cap D_2 \subset \mathrm{Int}(D_1) \cap \mathrm{Int}(D_2)$, then $D_1 \cap D_2 = \emptyset$.*

PROOF. Suppose first that D_1 and D_2 are in general position which is the generic case. If $D_1 \cap D_2$ is nonempty, then $D_1 \cap D_2$ is a compact one-dimensional submanifold of $\mathrm{Int}(D_1)$ and of $\mathrm{Int}(D_2)$. By the classification of one-dimensional submanifolds, $D_1 \cap D_2$ is a finite collection of simple closed curves. Let γ be a component in $D_1 \cap D_2$. Then the Jordan curve theorem implies that γ is the boundary of a subdisk $\widetilde{D}_1$ of D_1 and a subdisk $\widetilde{D}_2$ of D_2.

Suppose that the area of $\widetilde{D}_1$ is less than or equal to the area of $\widetilde{D}_2$. Then consider the new piecewise smooth disk:

$$D_3 = (D_2 - \widetilde{D_2}) \cup \widetilde{D_1}.$$

The area of D_3 is less than or equal to the area D_2. The area of D_3 can now be decreased along γ, which contradicts the hypothesis that D_2 has least-area with respect to its boundary curve.

If D_1 and D_2 are not in general position, then there are two ways to reduce to the general position case. The first way is by approximation. The second is by way of the following assertion:

ASSERTION 2.13. If $D_1 \cap D_2 \subset \mathrm{Int}(D_1) \cap \mathrm{Int}(D_2)$ is nonempty, then $D_1 \cap D_2$ contains a simple closed curve.

Proof of Assertion 2.13. Since D_1 and D_2 are analytic, $\Gamma = D_1 \cap D_2$ is a compact triangulable analytic subset of $\mathrm{Int}(D_1)$. We first note that Γ has no isolated vertices. If Γ had an isolated vertex p, then p would correspond to a point on D_1 where D_1 is locally on one side of D_2. By the maximum principal for minimal surfaces, D_1 and D_2 intersect in an open set near p, and so, the vertex p is not isolated. Also, Γ cannot contain a 2-simplex, because by the uniqueness of analytic continuation, D_1 and D_2 must agree on an open set that goes to the boundary of D_1 or D_2. However, this is impossible, since the intersection of D_1 and D_2 does not, by hypothesis, include points on the boundaries.

The argument used above shows that Γ is a one-dimensional subcomplex of some triangulation of D_1 and Γ contains no isolated vertices. Analytic one-dimensional subsets of a disk have an even number of edges at every vertex. This implies that Γ represents a one-cycle in the simplicial one-chains of D_1 using $\mathbb{Z}_2$-coefficients. Since the first homology group with $\mathbb{Z}_2$ coefficients of D_1 is zero, geometric intersection theory implies that Γ must disconnect D_1. A boundary curve of an inner-most component of $D_1 - \Gamma$ is the required simple closed curve in the assertion, and so, the assertion is proved.

We now return to the proof of Lemma 2.12. The existence of a simple closed curve in $D_1 \cap D_2$ together with the disk replacement argument used in the general position case gives a contradiction. Hence, $D_1 \cap D_2$ must be empty which proves the lemma. □

LEMMA 2.14. *Suppose N is a triangulated three- dimensional manifold and $f\colon D \to N$ is a simplicial immersion of a disk with respect to some* *triangulation T of D. Then there exists a subdivision of the triangulation of N, so that $f\colon D \to N$ is still simplicial with respect to T and such that the simplicial neighborhood of $f(D)$ is a simplicial regular neighborhood of $f(D)$ The simplicial neighborhood of $f(D)$ is the union of the simplices which intersect $f(D)$.*

PROOF. This elementary result follows after subdividing two times the triangulation of N. Each time the subdivision includes the baricenters of the simplices which are not contained in $f(D)$. This proves Lemma 2.14. □

We now carry out the construction of a tower for $f\colon D \to M$ in order to simplify the self-intersection or singular set for $f\colon D \to M$, which by Lemma 2.11 is simplicial. First, let N_1 be a simplicial regular neighborhood of $f(D)$ given in Lemma 2.14. After restricting the range space of f to N_1, there is a new map $f_1\colon D \to N_1$. If N_1 is not simply connected, then let $P_1\colon \widetilde{N}_1 \to N_1$ be the universal covering space of N_1 and let $\widetilde{f}_1\colon D \to \widetilde{N}_1$ be a lift of f_1 to this covering space. Then restricting the range space of $\widetilde{f}_1$ to a regular neighborhood N_2 of $\widetilde{f}_1(D)$, we get another map $f_2\colon D \to N_2$.

If N_2 is not simply connected, then we can repeat the construction in the previous paragraph to get a lift $\widetilde{f}_2\colon D \to \widetilde{N}_2$ to the universal covering space $P_2\colon \widetilde{N}_2 \to N_2$ of N_2. After restricting the lift $\widetilde{f}_2$ to the regular neighborhood N_3 of $\widetilde{f}(D)$, we get $f_3\colon D \to N_3$.

Repeating k-times, the construction outlined above yields a tower

$$P_k \circ P_{k-1} \circ \ldots \circ P_1\colon N_k \to N_1,$$

where $P_i\colon N_{i+1} \to N_i$ is the restriction of $P_i\colon \widetilde{N}_i \to N_i$ to N_{i+1}.

Each N_i in the above tower is a Riemannian manifold with respect to the pulled back metric. Each of the lifts $f_i\colon D \to N_i$ is a solution to Plateau's problem for the simple closed curve $f_i(\partial D)$ with respect to this metric. Otherwise, there is an immersion $g\colon D \to N_i$ with $g(\partial D) = f_i(\partial D)$ and with respect to the pulled back metric on D, Area$(g) <$ Area $(f_i) =$ Area (f), which is impossible.

By Lemmas 2.11 and 2.14, we may assume that each map $f_i\colon D \to N_i$ in the tower is simplicial with respect to a fixed triangulation T for which $f_i\colon D \to N_1$ is simplicial. Note that the triangulation on N_i is induced from the triangulation on N_{i-1} pulled back to $\widetilde{N}_i$ by $P_i\colon \widetilde{N}_i \to N_i$. We now use this fact to prove that the tower construction terminates, after some finite number n of steps, with N_k being simply connected, where n is at most equal to the number of simplices in $T \times T$. We will consider T to be a collection of open simplices and vertices.

LEMMA 2.15. *If $S(f_i) = \{(\sigma, \tau) \in T \times T \mid \sigma \neq \tau$ and $f(\sigma) = f(\tau)\}$, then $S(f_{i+1})$ is a proper subset of $S(f_i)$. Hence, the tower construction terminates at some k with N_k simply connected.*

PROOF. Since $f_i = P_i \circ f_{i+1}$ where P_{i+1} is a simplicial map, then $S(f_{i+1}) \subset S(f_i)$. If $S(f_{i+1}) = S(f_i)$, then $h = P_i|_{f_{i+1}(D)}$ induces a homeomorphism between $f_{i+1}(D)$ and $f_i(D)$. Using h we can define a lift of the inclusion map $i\colon f_i(D) \to N_i$ to $\widetilde{N}_i$ by $\widetilde{i}\colon f_i(D) \to \widetilde{N}_i$, where $\widetilde{i} = h^{-1} \circ i$. Since N_i is a regular neighborhood of $f_i(D)$, then $i_*\colon \pi_1(f_i(D)) \to \pi_1(N_i)$ is an isomorphism. Since $\widetilde{N}_i$ is simply connected, the lifting criterion for maps in covering space theory implies that N_i is simply connected. Thus, we may assume that $S(f_{i+1}) < S(f_i)$, which proves the lemma. □

LEMMA 2.16. *The lift $f_k\colon D \to N_k$ is one-to-one.*

We first show:

ASSERTION 2.17. The boundary of N_k consists of spheres.

PROOF. Since N_k is simply connected, $H_1(N_k, \mathbb{Z}_2) = 0$. Since the pairing between homology and cohomology with coefficients in a field is non-degenerate, $H^1(N_k, \mathbb{Z}_2) = 0$. Poincare duality then shows that $H_2(N_k, \partial N_k, \mathbb{Z}_2) = 0$. From the following part of the long exact sequence in homology for the pair $(N_k, \partial N_k)$,

$$\to H_2(N_k, \partial N_k, \mathbb{Z}_2) \to H_1(\partial N_k, \mathbb{Z}_2) \to H_1(N_k, \mathbb{Z}_2) \to,$$

one computes that $H_1(\partial N_k, \mathbb{Z}_2) = 0$. This shows that the first homology group with $\mathbb{Z}_2$ coefficients is zero for each boundary component of N_k. By the classification theorem for compact surfaces, each component of the boundary of N_k is a sphere which proves the assertion. □

PROOF. We now prove Lemma 2.16. We shall now use the fact that the boundary of N_k consists entirely of spheres to show that $f_k\colon D \to N_k$ is an embedding. First note that since N_k is a simplicial regular neighborhood, there is, after a subdivision, a simplicial retraction $S\colon N_k \to f_k(D)$ whose restriction $R = S|_{\partial N_k} \to f_k(D)$ has the following property: R covers each open two simplex of $f_k(D)$ exactly two times and R restricted to $\partial N_k - f_k(\partial D)$ is locally one-to-one. The existence of such a retraction follows directly from the definition of a simplicial regular neighborhood and the collapsing properties of such a neighborhood onto an immersed codimension-one simplicial submanifold whose boundary is the intersection of the submanifold with the boundary of the ambient manifold. For a proof of the existence, we refer the reader to [**16**].

By Assertion 2.17, the curve $\gamma_k = f_k(\partial D)$ is contained in a sphere S in ∂N_k. The Jordan curve theorem implies that the simple closed curve γ_k disconnects the sphere S into two disks D_1 and D_2. Now consider the following inequalities:

$$\text{Area}(R|_{D_1}) + \text{Area}(R|_{D_2}) \leq \text{Area}(R|_{\partial N_k}) \leq 2 \text{ Area } (f_k).$$

The last inequality follows from the fact that area is carried by two-simplices and $F|_{\partial N_k}$ covers each two-simplex of $f_k(D)$ twice.

Since f_k is a solution to Plateau's problem for γ_k, the above area inequality implies that $R|_{D_1}$ and $R|_{D_2}$ are also disks of least-area with γ_k for boundary. However, if f_k is not an embedding, then the area of $R|_{D_1}$ and $R|_{D_2}$ can be decreased along a self-intersection curve of $f_k(D)$. Since this contradicts the least-area property of f_k, the map f_k must be an embedding which proves the lemma. □

We now complete the proof of Theorem 2.9. If $f\colon D \to M$ is not an embedding, then we may assume by the previous lemma that k is greater than one and $f_{k-1}\colon D \to N_{k-1}$ is not one-to-one. Let E be the embedded disk $i \circ f_k(D) \subset \widetilde{N}_{k-1}$, where $i\colon N_k \to \widetilde{N}_{k-1}$ is the inclusion map. Since f_{k-1} is not one-to-one and $N_{k-1} = \widetilde{N}_{k-1}/G$ where G is the group of covering transformations, then there exists a nontrivial covering transformation $\tau\colon \widetilde{N}_{k-1} \to \widetilde{N}_{k-1}$ such that $\tau(E) \cap E$ is nonempty. Since the covering transformation τ is an isometry of $\widetilde{N}_{k-1}$, the disk $\tau(E)$ has least-area with respect to its boundary curve. The hypothesis in the theorem that $f(\partial D) = f(D) \cap \partial M = \gamma$ implies that $E \cap \tau(E) \subset \mathrm{Int}(E) \cap \tau(\mathrm{Int}(E))$. Lemma 2.12 shows this containment is impossible, which implies that $f\colon D \to M$ must in fact be an embedding. This completes the proof of the Theorem 2.9.

3. Application of minimal surfaces to the problem of finite extinction time for the Ricci flow.

In this section, we review some results on minimal surfaces by Colding and Minicozzi that have an application to the question of finite extinction time for the Ricci flow on certain Riemannian three-manifolds. These results and related discussion are taken from the papers in [**1**] and [**2**].

Let M be a smooth closed orientable three-manifold and let $g(t)$ be a one-parameter family of metrics on M evolving by the Ricci flow, so

$$\partial_t g = -2\mathrm{Ric}_{M_t}.$$

For the remainder of this section, we will assume that M is a prime three-manifold, which is nonaspherical which just means that some homotopy group $\pi_k(M)$ is nonzero for some $k > 1$. Recall that a closed orientable three-manifold is irreducible if every embedded two-sphere in the manifold is the boundary of a ball. Note that $S^2 \times S^1$ is the only compact orientable three-manifold which is prime but not irreducible. If M is irreducible, then the sphere theorem in the previous section, implies $\pi_2(M) = 0$, and the Hurewicz isomorphism theorem implies in this case that $\pi_3(M) \neq 0$. Since $\pi_3(S^2 \times S^1) = \pi_3(S^2) = \mathbb{Z}$, we see that for the manifold in the case we are considering, $\pi_3(M) \neq 0$.

Consider the space of continuous maps from S^2 to M. This space is naturally a fiber bundle over M. Using this fact and suspension on the long exact sequence of related homotopy groups, Micallif and Moore, in Lemma 3 in [**17**], proved that this space is not simply connected.

Fix a continuous map

$$\beta\colon [0,1] \to C^0 \cap L_1^2(S^2, M),$$

where $\beta(0)$ and $\beta(1)$ are constant maps and so that β is in the nontrivial homotopy class $[\beta]$. We define the width $W = W(g,[\beta])$ by

$$W(g) = \min_{\gamma\in[\beta]} \max_{s\in[0,1]} \text{Energy}(\gamma(s)).$$

The next theorem, Theorem 0.3 in [**1**], gives an upper bound for the derivative of $W(g(t))$ under the Ricci flow, which forces the solution $g(t)$ to become extinct in finite time. We remark that Perelmann [**20**] has also found a proof that $g(t)$ becomes extinct in finite time in this situation.

THEOREM 3.1 ([**1**] and [**20**]). *Let M be a closed orientable prime nonaspherical three-manifold equipped with a Riemannian metric $g = g(0)$. Under the Ricci flow, the width $W(g(t))$ satisfies*

$$\frac{d}{dt} W(g(t)) \leq -4\pi + \frac{3}{4(t+C)} W(g(t)),$$

in the sense of the limsup of forward difference quotients. Hence, $g(t)$ must become extinct in finite time.

Suppose that $\Sigma \subset M$ is a closed immersed surface (not necessarily minimal), then results of Hamilton [**6**] give

$$\frac{d}{dt}\,|_{t=0}\, \text{Area}_{g(t)}(\Sigma) = -\int_\Sigma [R - \text{Ric}_M(\mathbf{n},\mathbf{n})].$$

If Σ is also minimal, then

$$\begin{aligned}\frac{d}{dt}\,|_{t=0}\, \text{Area}_{g(t)}(\Sigma) &= -2\int_\Sigma K_\Sigma - \int_\Sigma [|A|^2 + \text{Ric}_M(\mathbf{n},\mathbf{n})] \\ &= -\int_\Sigma K_\Sigma - \frac{1}{2}\int_\Sigma [|A|^2 + R].\end{aligned}$$

Here, K_Σ is the Gaussian curvature of Σ, and $\mathbf{n}$ is a unit normal for Σ. A is the second fundamental form of Σ, so that $|A|^2$ is the sum of the squares of the principal curvatures, Ric_M is the Ricci curvature of M, and R is the scalar curvature of M. (The curvature is normalized so that on the unit S^3 the Ricci curvature is 2 and the scalar curvature is 6.) To get the above equation, one uses that by the Gauss equations and minimality of Σ

$$\text{K}_\Sigma = \text{K}_M - \frac{1}{2}|A|^2,$$

where K_M is the sectional curvature of M on the two-plane tangent to Σ.

The first lemma in [**1**] gives an upper bound for the rate of change of area of minimal two-spheres, and we give their proof of it.

LEMMA 3.2. *If $\Sigma \subset M$ is a branched minimal immersion of the two-sphere, then*

$$\frac{d}{dt}\,|_{t=0}\, \text{Area}_{g(t)}(\Sigma) \leq -4\pi - \frac{\text{Area}_{g(0)}(\Sigma)}{2} \min_M R(0).$$

PROOF. Let $\{p_i\}$ be the set of branch points of Σ and $b_i > 0$ the order of branching at p_i. From above, we have

$$\frac{d}{dt}\,|_{t=0}\ \mathrm{Area}_{g(t)}(\Sigma) \leq -\int_\Sigma K_\Sigma - \frac{1}{2}\int_\Sigma R = -4\pi - 2\pi\sum b_i - \frac{1}{2}\int_\Sigma R,$$

where the equality used the Gauss-Bonnet theorem with branch points. □

The evolution equation for the scalar curvature $R = R(t)$ of M_t under Ricci flow (see [**6**]) is given by the following equation and gives rise to a related inequality:

$$\partial_t R = \Delta R + 2|\mathrm{Ric}|^2 \geq \Delta R + \frac{2}{3}R^2.$$

A maximum principle argument, then gives for some constant C,

$$R(t) \geq -\frac{3}{2(t+C)}.$$

Plugging this estimate for $R(t)$ into Lemma 3.2, then yields:

$$\frac{d}{dt}|_{t=0}\mathrm{Area}_{g(0)}(\Sigma) \leq -4\pi + \frac{3\mathrm{Area}(\Sigma)}{4(t+C)}.$$

What Colding and Minicozzi do next is to derive a related forward difference quotient for $W(g(t))$. Namely, they show that there is a constant $\widetilde{C}$, so that, given $\varepsilon > 0$, there exists an $\overline{h} > 0$ such that for $0 < h < \overline{h}$, then

$$\frac{W(g(\tau+h)) - W(g(\tau))}{h} \leq -4\pi + \widetilde{C}\varepsilon + \frac{3}{4(\tau+C)}W(g(\tau)) + \widetilde{C}h.$$

Taking $\varepsilon \to 0$ gives the differential inequality in Theorem 3.1.

Colding and Minicozzi derive the above related forward difference quotient inequality by applying the previous estimate for the derivative of the areas of minimal two-spheres Σ which arise in the next proposition. This proposition asserts the existence of a special sequence γ^i of sweep-outs, where for some $s_j, s_j \in [0,1]$, the spheres $\gamma^j_{s_j}$ converge to a collection of branched minimal spheres with total energy $W(g)$. This proposition is Theorem 4.2.1 by Jost in [**8**]. This result depends on the theory of minimal spheres using the concept of α-energy and the fact that α-energy functional is a Morse function on the appropriate spaces. This theory was first developed by Sachs and Uhlenbeck [**21**] with improvements by Meeks and Yau [**12**] and Siu and Yau (see Chapter VIII in [**23**] to prove that there was no loss of energy in the limit as $\alpha \to 0$). The index-one bound for the minimal spheres described below is not stated explicitly in [**8**] but follow by the arguments in [**17**].

PROPOSITION 3.3. *Given a metric g on M and a nontrivial $[\beta] \in \pi_1(C^0 \cap L^2_1(S^2, M))$, there exists a sequence of sweep-outs $\gamma^j \colon [0,1] \to C^0 \cap L^2_1(S^2, M)$ with $\gamma^j \in [\beta]$ so that*

$$W(g) = \lim_{j\to\infty}\max_{s\in[0,1]}\mathrm{Energy}(\gamma^j_s).$$

Furthermore, there exist $s_j \in [0,1]$ *and branched conformal minimal immersions* $u_0, \dots, u_m \colon S^2 \to M$ *with index at most one so that, as* $j \to \infty$*, the maps* $\gamma^j_{s_j}$*, converge to* u_0 *weakly in* S^2_1 *and uniformly on compact subsets of* $S^2/\{x_1, \dots, x_k\}$*, and*

$$W(g) = \sum_{i=0}^{m} \text{Energy}(u_i) = \lim_{j\to\infty} \text{Energy}(\gamma^j_{s_j}).$$

Finally, for each $e > 0$*, there exists a point* x_{k_i} *and a sequence of conformal dilations* $D_{i,j} \colon S^2 \to S^2$ *about* x_{k_i}*, so that the maps* $\gamma^j_{s_j} \circ D_{i,j}$ *converge to* u_i*.*

Finally, we show that the differential inequality for $W(g(t))$ given in Theorem 3.1 implies finite extinction time for the Ricci flow. Namely, rewriting this inequality as $\frac{d}{dt}(W(g(t))(t+C)^{3/4}) \leq -4\pi(t+C)^{-3/4}$ and then integrating gives

$$(T+C)^{-3/4}W(g(T)) \leq C^{-3/4}W(g(0)) - 16\pi[(T+C)^{1/4} - C^{1/4}].$$

Since $W \geq 0$ by definition and the right hand side of the equation would become negative for T sufficiently large, the theorem follows.

References

[1] T.H. Colding and W.P. Minicozzi II, *Estimates for the extinction time for the Ricci flow on certain 3-manifolds and a question of Perelman*, Journal of the AMS **18**, 347–559.

[2] T.H. Colding and W.P. Minicozzi II, *An excursion into geometric analysis*, in 'Surveys of Differential Geometry IX - Eigenvalues of Laplacian and other geometric operators', 83–146, International Press, edited by Alexander Grigor'yan and Shing Tung Yau, 2004, MR2195407, Zbl 1076.53001.

[3] R. Courant, *Dirichlet's Principle, Conformal Mapping and Minimal Surfaces*, Interscience Publishers, Inc., New York, 1950.

[4] R. Gulliver, *Regularity of minimizing surfaces of prescribed mean curvature*, Ann. of Math. **97** (1973), 275–305, MR0317188, Zbl 0246.53053.

[5] R. Gulliver and F. Lesley, *On boundary branch points of minimizing surfaces*, Arch. Rational Mech. Anal. **52** (1973), 20–25, MR0346641, Zbl 0263.53009.

[6] R.S. Hamilton, *The formation of singularities in the Ricci flow*, in 'Surveys in differential geometry', Vol. II (Cambridge, MA 1993), 1–119, International Press, Cambridge, MA, 1995.

[7] S. Hildebrandt, *Boundary behavior of minimal surfaces*, Archive Rational Mech. Anal. **35** (1969), 47–81.

[8] J. Jost, *Two-dimensional geometric variational problems*, J. Wiley and Sons, Chichester, NY, 1991.

[9] H. Lewy, *On the boundary behavior of minimal surfaces*, Proceedings of the National Academy **37** (1951), 103–110.

[10] S. Lojasiewicz, *Triangulation of semianalytic sets*, Ann. Scuola Norm. Sup. Pisa **18** (1964), 449–474.

[11] W.H. Meeks III, L. Simon, and S.-T. Yau, *The existence of embedded minimal surfaces, exotic spheres and positive Ricci curvature*, Annals of Math. **116** (1982), 221–259, MR0678484, Zbl 0521.53007.

[12] W.H. Meeks III and S.-T. Yau, *Topology of three dimensional manifolds and the embedding problems in minimal surface theory*, Annals of Math. **112** (1980), 441–484.

[13] W.H. Meeks III and S.-T. Yau, *The classical Plateau problem and the topology of three-dimensional manifolds*, Topology **21**(**4**) (1982), 409–442, MR0670745, Zbl 0489.57002.

[14] W.H. Meeks III and S.-T. Yau, *The existence of embedded minimal surfaces and the problem of uniqueness*, Math. Z. **179** (1982), 151–168, MR0645492, Zbl 0479.49026.

[15] W.H. Meeks III and S.-T. Yau, *Compact group actions on* $\mathbb{R}^3$, in 'Conference on the Smith Conjecture', Academic Press, 1984.

[16] W.H. Meeks III and S.-T. Yau, *The topological uniqueness of complete minimal surfaces of finite topological type*, Topology **31**(**2**) (1992), 305–316, MR1167172, Zbl 0761.53006.

[17] M. Micallef and J.D. Moore, *Minimal two-spheres and the topology of manifolds with positive curvature on totally isotropic two-planes*, Annals of Math. **127** (1988), 199–227.

[18] C.B. Morrey, *The problem of Plateau on a Riemannian manifold*, Annals of Math. **49** (1948), 807–851, MR0027137, Zbl 0033.39601.

[19] R. Osserman. *A proof of the regularity everywhere to Plateau's problem*, Annals of Math. **91**(**2**) (1970), 550–569, MR0266070, Zbl 0194.22302.

[20] G. Perelman, *Finite extinction time for the solutions to the Ricci flow on certain three-manifolds*, math.DG/0307245.

[21] J. Sacks and K. Uhlenbeck, *Minimal immersions of closed Riemannian surfaces*, Transactions of the AMS **271** (1982), 639–652.

[22] R. Schoen and S.-T. Yau, *Existence of incompressible minimal surfaces and the topology of three dimensional manifolds with non-negative scalar curvature*, Annals of Math. **110** (1979), 127–142.

[23] R. Schoen and S.-T. Yau, *Lectures on harmonic maps*, International Press, 1997.

Mathematics Department, University of Massachusetts
Amherst, MA 01003
E-mail address: bill@gang.umass.edu

An integral equation for spacetime curvature in general relativity

Vincent Moncrief

ABSTRACT. A key step in the proof of global existence for Yang-Mills fields, propagating in curved, 4-dimensional, globally hyperbolic, background spacetimes, was the derivation and reduction of an integral equation satisfied by the curvature of an arbitrary solution to the Yang-Mills field equations. This article presents the corresponding derivation of an integral equation satisfied by the curvature of a vacuum solution to the Einstein field equations of general relativity. The resultant formula expresses the curvature at a point in terms of a 'direct' integral over the past light cone from that point, a so-called 'tail' integral over the interior of that cone and two additional integrals over a ball in the initial data hypersurface and over its boundary. The tail contribution and the integral over the ball in the initial data surface result from the breakdown of Huygens' principle for waves propagating in a general curved, 4-dimensional spacetime.

By an application of Stokes' theorem and some integration by parts lemmas, however, one can re-express these 'Huygens-violating' contributions purely in terms of integrals over the cone itself and over the 2-dimensional intersection of that cone with the initial data surface. Furthermore, by exploiting a generalization of the parallel propagation, or Cronström, gauge condition used in the Yang-Mills arguments, one can explicitly express the frame fields and connection one-forms in terms of curvature. While global existence is certainly false for general relativity one anticipates that the resulting integral equation may prove useful in analyzing the propagation, focusing and (sometimes) blow up of curvature during the course of Einsteinian evolution and thereby shed light on the natural alternative conjecture to global existence, namely Penrose's *cosmic censorship conjecture.*

1. Introduction

Global existence fails to hold for many, otherwise reasonable solutions to the Einstein field equations. Examples of finite-time blowup include solutions developing black holes and solutions evolving to form cosmological big bang or big crunch singularities. The singularities that arise in such examples often, but not always, involve the blowup of certain spacetime curvature invariants. More subtle types of singular behavior include the formation of Cauchy horizons, at which the curvature can remain bounded, but across which global hyperbolicity, and hence classical determinism, is lost.

Examples of this latter phenomenon are provided by the Kerr and Kerr-Newman rotating black hole spacetimes and by non-isotropic, cosmological models of Taub-NUT-type wherein violations of strong causality (as signaled by the occurrence of closed timelike curves or the appearance of naked curvature singularities) develop beyond the Cauchy horizons arising in these solutions. On the other hand a variety of arguments and calculations strongly suggest that such Cauchy horizons, when they occur, are highly unstable-giving way, under generic perturbations, to the formation of strong curvature singularities that block the extension of such perturbed solutions beyond their maximal Cauchy developments.

Considerations such as these led Roger Penrose to propose the so-called (strong) *cosmic censorship conjecture* [**12**] according to which (in a here deliberately loosely stated form):

globally hyperbolic solutions to the Einstein field equations evolving from non-singular Cauchy data are generically inextendible beyond their maximal Cauchy developments.

For the non-vacuum cases of this conjecture it is natural to consider only those matter sources which exhibit, in the absence of gravitational coupling, the global existence property at least in Minkowski space but perhaps also (being somewhat more cautious) in generic globally hyperbolic 'background' spacetimes. Otherwise, rather straightforward counterexamples can be presented involving, for instance, self-gravitating perfect fluids that evolve to blow up in a nakedly singular but stable fashion [**15, 16**]. But Penrose's conjecture was never intended to suggest that Einsteinian gravity should miraculously hide the defects of inadequate models of matter inside black holes or cause their singularities to harmlessly merge with big bang or big crunch cosmological singularities.

There are a number of known types of relativistic matter sources that do exhibit the desired global existence property, but one is currently so far from a proof of cosmic censorship that their inclusion into the picture only presents an unwanted distraction from the more essential issues. Thus it seems natural to set these complications aside until genuine progress can be made in the vacuum special case.

On the other hand there is one particular class of matter fields whose study seems to be directly relevant to the analysis of the vacuum gravitational equations-namely the class of Yang-Mills fields propagating in a given, 4-dimensional, globally hyperbolic, background spacetime. First of all, these are examples of sources for which global existence results (for the case of compact Yang-Mills gauge groups) have already been established in both flat [**4, 5**] and curved [**2**] background spacetimes. Secondly however, the vacuum Einstein equations, when expressed in the Cartan formalism and combined with the Bianchi identities, imply that the spacetime curvature tensor, written as a matrix of two-forms, satisfies a propagation equation of precisely (curved-space) Yang-Mills type.

But in contrast to the case of 'pure' Yang-Mills fields this Einsteinian curvature propagation equation is coupled to another equation (the vanishing torsion condition) which links the connection one-form field to the (orthonormal) frame field and thus reinstates that frame (or metric) as the fundamental dynamical variable of general relativity. An additional, related distinction from conventional Yang-Mills theory is that the effective Yang-Mills gauge group for Einsteinian gravity, when formulated in this way, is the *non-compact* group of Lorentz transformations which acts (locally) to generate automorphisms of the bundle of orthonormal frames while leaving the metric invariant.

An initially disconcerting consequence of this non-compactness of the effective gauge group is that the associated, canonical Yang-Mills stress-energy tensor (a symmetric, second rank tensor quadratic in the curvature) need no longer have a positive definite energy density (as it always does in conventional, compact gauge group, Yang-Mills theory) and indeed this tensor vanishes identically in the gravitational case. Fortunately however the Bel-Robinson tensor (a fourth rank, totally symmetric tensor quadratic in curvature and having positive definite 'energy' density) is available to take over its fundamental role [**3**].

The proofs of flat and curved space global existence for conventional (compact gauge group) Yang-Mills fields given, respectively, in References [**5**] and [**2**] use a combination of light cone estimates and energy arguments that exploit, on the one hand, an integral equation satisfied by the curvature of the Yang-Mills connection and, on the other, the properties of the associated, canonical stress-energy tensor mentioned above. For the case of curved, globally hyperbolic, background spacetimes the proof guarantees only that the Yang-Mills connection, expressed in a suitable gauge, cannot blow up until the background spacetime itself blows up, for example by evolving to form a black hole or cosmological singularity or by developing a Cauchy horizon. But even linear Maxwell fields typically blow up at such singular boundaries or Cauchy horizons, so one could hardly expect better regularity in the nonlinear case.

Of course in general relativity there is no given, 'background' geometry at all and global existence is much too strong a conjecture for the gravitational field as the aforementioned examples and arguments show. Spacetime curvature does indeed blow up in many otherwise reasonable instances of Einsteinian evolution and this blowup is anticipated to be a stable feature of such solutions and not merely the artifact of, say, some special symmetry or other 'accidental' property of the spacetime under study. Cosmological solutions may only persist for a finite (proper) time in one or both temporal directions whereas timelike geodesics falling into a black hole may encounter divergent curvature, representing unbounded tidal 'forces', in a finite proper time.

But if Penrose's conjecture is true then global hyperbolicity is at least a generic feature of maximally extended Einstein spacetimes that evolve from non-singular Cauchy data and general relativity is thereby effectively rescued from an otherwise seemingly fatal breakdown of classical determinism. If, on the other hand, cosmic censorship is false then the implied breakdown of determinism may well render Einstein's equations inadequate as a classical theory of the gravitational field.

There is currently no clear-cut strategy for trying to prove the cosmic censorship conjecture but it nevertheless seems evident that a better understanding of how spacetime curvature propagates, focuses and (in some circumstances) blows up in the course of Einsteinian evolution will be essential for progress on this fundamental problem. For that reason one might hope that a further development of the "Yang-Mills analogy", wherein the parallel issues of curvature propagation, focusing and blowup for 'pure' Yang-Mills fields have already been somewhat successfully analyzed, could yield significant insights for understanding the still-wide-open gravitational problem.

One of the key steps in the 'pure' Yang-Mills analysis was the derivation of an integral equation satisfied by the curvature of an arbitrary solution to the field equations. This integral equation resulted from combining the Yang-Mills equations and their Bianchi identities in a well-known way to derive a wave equation satisfied by curvature and by then applying the fundamental solution of the associated wave operator to derive an integral expression for the curvature at an arbitrary point (within the domain of local existence for the solution in question) in terms of integrals over the past light cone of that point to the initial, Cauchy hypersurface. An additional key step was the transformation of this integral formula through the use of the parallel propagation, or Cronström, gauge condition [**5, 2, 1**] to eliminate the connection one-form explicitly in favor of the curvature itself. Certain resulting integrals over the light cone, from its vertex back to the initial data surface, could be bounded in terms of the Yang-Mills energy flux, defined via the aforementioned, canonical stress-energy tensor, and thence in terms of the actual energy on the initial hypersurface.

In the simplest, flat space setting of Ref. [**5**] a Gronwall lemma argument was employed to prove that the natural (gauge-invariantly-defined) L^∞ -norm of curvature is always bounded in terms of the (equally gauge-invariant) conserved total energy, with all reference to the artifice of the Cronström or parallel propagation gauge, used in the intermediate steps, effectively eliminated. Thus equipped with an a priori pointwise bound on curvature one completed the proof of global existence by showing that an appropriate Sobolev norm of the connection one-form, when evolved in the so-called 'temporal gauge', cannot blow up in finite time by a straightforward, higher order energy argument. A more elaborate argument was needed for the case of the curved backgrounds treated in Ref. [**2**] but the essential role played by the corresponding integral equation for curvature remained unaltered.

In the flat space argument one avoided certain complications, resulting from the breakdown of Huygens' principle for the complete gauge-covariant wave operator appearing in the curvature propagation equation, by splitting that operator into a pure flat-space wave operator (which does of course obey Huygens' principle in four-dimensional Minkowski space) and a collection of lower order, Huygens-violating, connection terms which were moved over and included with the 'source' terms in the full, inhomogeneous wave equation for curvature. One then derived the integral formula for curvature by applying the well-known fundamental solution for the flat space wave operator to the redefined source terms and then eliminating the connection terms in the redefined source, in favor of curvature, through an application of the Cronström gauge argument mentioned above.

This same operator splitting technique was also employed for curved backgrounds in Ref. [**2**] but there, since the ordinary tensor wave operator itself violates Huygens' principle (in a generic background), new terms in the resulting 'representation formula' for Yang-Mills curvature arose which had no direct analogue in the operator-split, flat space argument. These new, so-called *tail* terms appeared as integrals over the *interior* of the past light cone from an arbitrary point to the initial hypersurface and over the interior of the three ball in the initial hypersurface bounded by the intersection of the past light cone with this initial surface. Fortunately, however, these tail terms produced only a slight complication in the argument for the curved-space 'pure' Yang-Mills problem because all of the Huygen's-violating, tail contributions to the fundamental solution for the residual tensor wave operator (remaining after the aforementioned operator splitting is carried out) are functionals only of the given, *background* metric and thus are independent of the Yang-Mills field under study. Their contributions can therefore always be bounded by constants dependent only upon the background geometry but independent of the solution in question.

In this article we derive an integral equation satisfied by the curvature tensor of a vacuum solution to Einstein's equations by applying the fundamental solution of the associated, curved-space tensor wave operator to the source terms in the curvature propagation equation defined after an analogous operator splitting, within the Cartan formulation for the field equations, has been carried out. For this purpose we exploit the general theory of such wave operators developed over the years by Hadamard, Sobolev, Reisz, Choquet-Bruhat, Friedlander and others [**8**]. We then transform the resulting expression, by an application of Stokes' theorem and some integration-by-parts arguments, to rewrite the Huygen's-violating tail contribution integrals in terms of other integrals over the past light cone itself. A generalization of Cronström's argument is given which shows that not only the connection but also the frame field can be explicitly expressed in terms of curvature by exploiting a natural parallel propagation gauge condition in conjunction with the standard Hadamard/Friedlander constructions.

While the aforementioned calculations exploit an operator split version of the curvature propagation equation (written as an evolution equation for a matrix of two-forms), we also show how the same result can be derived, without using the Cartan formalism or associated operator splitting, by applying the Hadamard/Friedlander fundamental solution for the wave operator acting on a fourth rank tensor to the purely (fourth rank) tensorial form of the curvature propagation equation. At the other extreme one could presumably arrive at the same result in still another way by converting all the indices on the curvature tensor to frame indices, carrying out a maximal operator splitting to include the connection terms with the source and then applying the fundamental solution for the purely scalar wave operator to the wave equation for each component. We have not performed this latter derivation but strongly suspect that it leads to the same, 'canonical' result obtained in the other two ways.

In view of the foregoing remarks it may seem that we have gained little in emphasizing the use of the Cartan formalism and its associated 'Yang-Mills analogy' in analyzing the field equations but one should keep in mind that the derivation of this integral equation for curvature is only the first step in a proposed sequence of arguments wherein one hopes to exploit the Cronström-type formulas to re-express all the fundamental variables in terms of the curvature (written in Cartan fashion as a matrix of two-forms) and derive estimates for curvature by analogy with those obtained in Refs. [**5**] and [**2**]. Until such arguments are carried out it will not be evident whether the Cartan formulation is actually essential for the analysis or only a convenience for those familiar with the 'pure' Yang-Mills derivations.

Of course one cannot simply expect to copy the pattern of the 'pure' Yang-Mills arguments and thereby derive a global existence result for the Einstein equations. First of all we know that any such conclusion must be false but it is worth recalling here that the Yang-Mills arguments did not imply unqualified regularity of the Yang-Mills field but only implied that the field could not blow up until the background spacetime itself blew up. In general relativity though there is of course no *background* spacetime and the vanishing torsion condition, which links the metric to the connection, has no analogue in pure Yang-Mills theory.

One rather explicit obstruction to simply copying the 'pure', curved-space Yang-Mills argument is that one cannot simply bound the Bel-Robinson energy fluxes (which fortunately do bound certain relevant light cone integrals) in terms of the Bel-Robinson energy defined on the initial data hypersurface. While the Bel-Robinson tensor does in fact obey the vanishing divergence condition whose analogue, in the case of the canonical stress energy tensor, permitted the derivation of such a bound in the pure Yang-Mills problem, the Christoffel symbols occurring as coefficients in this equation are no longer background quantities and thus no longer a priori under control as they were in the arguments of Ref. [**2**].

However the full definition of a Bel-Robinson energy expression (and its associated fluxes) depends upon the additional choice of a timelike vector field on spacetime. If one had the luxury of choosing a timelike Killing or even conformal Killing field in defining these quantities then the corresponding Bel-Robinson energy would be a strictly conserved quantity and a significant portion of the needed arguments would revert to the simple form available in the flat space (or conformally stationary curved space) 'pure' Yang-Mills problem wherein the canonical (positive definite, gauge invariant) energy is strictly conserved. But such an assumption is absurdly restrictive in the case of Einstein's equations for which the small set of vacuum solutions admitting a globally defined timelike conformal Killing field is essentially known explicitly [**6**].

But whereas the presence of a conformal Killing field is out of the question for generic Einstein spacetimes there is nevertheless a potential utility in identifying what we might call *quasi-local, approximate Killing and conformal Killing fields* and trying to exploit these in a 'quasi-local, approximate' variant of the arguments that assume a strict Killing or conformal Killing field. The idea we have in mind is spelled out more explicitly in the concluding technical section of this article wherein we show that the parallel propagated frame fields (determined by parallel propagation of a frame chosen at the vertex of each light cone) satisfy Killing's equations approximately with an error term that is explicitly computable in terms of curvature and that tends to zero at a well-defined rate as one approaches the vertex of the given cone. The flux of the corresponding quasi-local energy (built from the chosen vector field and the Bel-Robinson tensor) will of course not be strictly equal (as it would for a truly conserved energy) to the energy contained on an initial data slice but the error will be estimable in terms of an integral involving the (undifferentiated) curvature tensor. The question of how best to use this observation to obtain optimal estimates from the integral equation for curvature is one we hope to address in future work.

The idea of exploiting the 'Yang-Mills analogy' to analyze Einstein's equations is certainly nothing new and has been proposed previously by Eardley and van Putten, for example, with a view towards numerical applications [**14**]. Furthermore the global existence of Yang-Mills fields propagating in Minkowski space has been proven by a completely independent argument, which avoids light cone estimates, in a paper by Klainerman and Machedon [**10**]. During a visit to the Erwin Schrödinger Institute in the summer of 2004 the author described the preliminary results for this paper with

Sergiu Klainerman who then, together with Igor Rodnianski, independently succeeded to derive an integral equation for curvature using a significantly different approach from that described herein [**11**]. Since the two formulations are quite dissimilar (in that, for example they do not use the frame formalism, the Hadamard/Friedlander analysis or the parallel propagation gauge condition) it is not yet clear whether the resultant integral equations are ultimately equivalent or perhaps genuinely different. Klainerman and Rodnianski trace the origins of their approach back through some fundamental papers by Choquet-Bruhat [**7**] and Sobolev [**13**] whereas the sources for our approach, as we have indicated, trace more directly back through the work of Friedlander [**8**] and Hadamard [**9**]

2. Propagation Equations for Spacetime Curvature

In this section we rederive the familiar wave equation satisfied by the curvature tensor of a vacuum spacetime and then reexpress that equation in a form which parallels the one satisfied by the Yang-Mills curvature in a vacuum background. One could generalize both forms by allowing the spacetime to be non-vacuum but since we shall not deal with sources for Einstein's equations in this paper, we simplify the presentation by setting

$$R^{\alpha}{}_{\mu\alpha\nu} := R_{\mu\nu} = 0. \tag{2.1}$$

The Bianchi identities give

$$R^{\alpha}{}_{\beta\gamma\delta;\mu} + R^{\alpha}{}_{\beta\delta\mu;\gamma} + R^{\alpha}{}_{\beta\mu\gamma;\delta} = 0 \tag{2.2}$$

so that, upon contracting and exploiting the algebraic symmetries of the curvature tensor, one gets

$$R_{\gamma\delta\beta\alpha}{}^{;\alpha} = R_{\beta\gamma;\delta} - R_{\beta\delta;\gamma}. \tag{2.3}$$

Imposing the vacuum field equations this yields

$$D_{\alpha} R_{\gamma\delta\beta}{}^{\alpha} := R_{\gamma\delta\beta\alpha}{}^{;\alpha} = 0 \tag{2.4}$$

where we have introduced D_α as an alternative to $;\alpha$ to symbolize covariant differentiation.

Taking a divergence of the Bianchi identity (2.2) yields

$$R^{\alpha}{}_{\beta\gamma\delta;\mu}{}^{;\mu} = R^{\alpha}{}_{\beta\gamma\mu;\delta}{}^{;\mu} - R^{\alpha}{}_{\beta\delta\mu;\gamma}{}^{;\mu}. \tag{2.5}$$

Commuting covariant derivatives on the right hand side and exploiting the field equations (2.1) together with Eq. (2.4), which follows from them, and using the algebraic Bianchi identity

$$R^{\alpha}{}_{[\beta\gamma\delta]} = 0 \tag{2.6}$$

to simplify the resulting expression finally gives

$$(2.7)\qquad \begin{aligned} D^{\mu}D_{\mu}R^{\alpha}{}_{\beta\gamma\delta} :&= R^{\alpha}{}_{\beta\gamma\delta;\mu}{}^{;\mu} \\ &= -R_{\gamma\delta}{}^{\rho\sigma}R^{\alpha}{}_{\beta\rho\sigma} \\ &\quad + 2R^{\alpha}{}_{\rho\delta\sigma}R_{\beta}{}^{\rho}{}_{\gamma}{}^{\sigma} - 2R^{\alpha}{}_{\rho\gamma\sigma}R_{\beta}{}^{\rho}{}_{\delta}{}^{\sigma}. \end{aligned}$$

This is the fundamental wave equation satisfied by the curvature tensor of a vacuum spacetime.

Now, following the notation of the appendix we set

$$(2.8)\qquad R^{\hat{a}}{}_{\hat{b}\mu\nu;\alpha} = \theta^{\hat{a}}_{\lambda}h^{\sigma}_{\hat{b}}R^{\lambda}{}_{\sigma\mu\nu;\alpha}$$

and expand out the right hand side of this expression to get

$$(2.9)\qquad \begin{aligned} D_{\alpha}R^{\hat{a}}{}_{\hat{b}\mu\nu} :&= R^{\hat{a}}{}_{\hat{b}\mu\nu;\alpha} \\ &= \theta^{\hat{a}}_{\lambda}h^{\sigma}_{\hat{b}}R^{\lambda}{}_{\sigma\mu\nu;\alpha} \\ &= \nabla_{\alpha}R^{\hat{a}}{}_{\hat{b}\mu\nu} + \omega^{\hat{a}}{}_{\hat{c}\alpha}R^{\hat{c}}{}_{\hat{b}\mu\nu} \\ &\quad - \omega^{\hat{c}}{}_{\hat{b}\alpha}R^{\hat{a}}{}_{\hat{c}\mu\nu} \end{aligned}$$

where we have defined

$$(2.10)\qquad \begin{aligned} \nabla_{\alpha}R^{\hat{a}}{}_{\hat{b}\mu\nu} :&= (R^{\hat{a}}{}_{\hat{b}\mu\nu})_{,\alpha} \\ &\quad - \Gamma^{\delta}_{\mu\alpha}R^{\hat{a}}{}_{\hat{b}\delta\nu} - \Gamma^{\delta}_{\nu\alpha}R^{\hat{a}}{}_{\hat{b}\mu\delta}. \end{aligned}$$

The operator ∇_{α} captures only that part of the full spacetime covariant derivative operator D_{α} that acts on the coordinate basis indices μ and ν of $R^{\hat{a}}{}_{\hat{b}\mu\nu}$ and ignores the contributions arising from the frame indices $\hat{a}$ and $\hat{b}$. These latter contributions are explicitly added back in Eq. (2.9) for the full spacetime covariant derivative of $R^{\hat{a}}{}_{\hat{b}\mu\nu}$ where they appear as the terms containing the Lorentz connection $\omega^{\hat{a}}{}_{\hat{b}\nu}$. We extend the definitions of D_{α} and ∇_{α} to operators on tensors of arbitrary type in the obvious way; D_{α} is the full spacetime covariant derivative operator while ∇_{α} ignores frame indices and acts only on spacetime coordinate indices.

This splitting of the full covariant derivative into a spacetime coordinate contribution and a frame or "internal space" contribution is parallel to what one has in Yang-Mills theory wherein the Yang-Mills connection $A^{\hat{a}}{}_{\hat{b}\nu}$ plays the role of the Lorentz connection $\omega^{\hat{a}}{}_{\hat{b}\nu}$ but in which the internal space Lie algebra indices refer to the chosen gauge group and not to the Lorentz group. In Yang-Mills theory of course the spacetime metric and its Christoffel connection are prescribed a priori and have no relation to the internal space connection $A^{\hat{a}}{}_{\hat{b}\nu}$.

Rewriting the Bianchi identity (2.2) in this notation one gets

$$(2.11)\qquad D_{\mu}R^{\hat{a}}{}_{\hat{b}\gamma\delta} + D_{\gamma}R^{\hat{a}}{}_{\hat{b}\delta\mu} + D_{\delta}R^{\hat{a}}{}_{\hat{b}\mu\gamma} = 0$$

or more explicitly, using the aforementioned splitting of D_α

$$\begin{aligned}(2.12)\qquad &\nabla_\mu R^{\hat{a}}{}_{\hat{b}\gamma\delta} + \omega^{\hat{a}}{}_{\hat{c}\mu} R^{\hat{c}}{}_{\hat{b}\gamma\delta} - R^{\hat{a}}{}_{\hat{c}\gamma\delta}\omega^{\hat{c}}{}_{\hat{b}\mu} \\ &+ \nabla_\gamma R^{\hat{a}}{}_{\hat{b}\delta\mu} + \omega^{\hat{a}}{}_{\hat{c}\gamma} R^{\hat{c}}{}_{\hat{b}\delta\mu} - R^{\hat{a}}{}_{\hat{c}\delta\mu}\omega^{\hat{c}}{}_{\hat{b}\gamma} \\ &+ \nabla_\delta R^{\hat{a}}{}_{\hat{b}\mu\gamma} + \omega^{\hat{a}}{}_{\hat{c}\delta} R^{\hat{c}}{}_{\hat{b}\mu\gamma} - R^{\hat{a}}{}_{\hat{c}\mu\gamma}\omega^{\hat{c}}{}_{\hat{b}\delta} \\ &= 0\end{aligned}$$

wherein one sees the internal space (frame) contributions arising as a set of matrix commutators of the Lorentz connection and curvature. This has exactly the structure of the corresponding Bianchi identity for Yang-Mills theory and reproduces that formula if one makes the substitutions of $F^{\hat{a}}{}_{\hat{b}\mu\nu}$ for $R^{\hat{a}}{}_{\hat{b}\mu\nu}$ and $A^{\hat{a}}{}_{\hat{b}\mu}$ for $\omega^{\hat{a}}{}_{\hat{b}\mu}$ with the "spacetime" covariant derivative ∇_μ playing the same role in each equation. The full spacetime/gauge covariant derivative bears the same relation to the pure "spacetime" covariant derivative as D_α does to ∇_α in Eq. (2.9) when the same substitutions are made.

On the other hand, a Yang-Mills curvature does not have the full algebraic symmetries of the Riemann curvature and, for closely related reasons, one cannot form the analogue of the Ricci tensor from $F^{\hat{a}}{}_{\hat{b}\mu\nu}$. Thus equation (2.1) has no analogue in Yang-Mills theory. If Eq. (2.4) however is first reexpressed as

$$(2.13)\qquad D^\alpha R^{\hat{a}}{}_{\hat{b}\beta\alpha} := g^{\alpha\gamma} D_\gamma R^{\hat{a}}{}_{\hat{b}\beta\alpha} = 0$$

then it corresponds precisely to the (source-free) Yang-Mills equation which, by definition, is

$$\begin{aligned}(2.14)\qquad D^\alpha F^{\hat{a}}{}_{\hat{b}\beta\alpha} &:= g^{\alpha\gamma} D_\gamma F^{\hat{a}}{}_{\hat{b}\beta\alpha} \\ &:= g^{\alpha\gamma}\{\nabla_\gamma F^{\hat{a}}{}_{\hat{b}\beta\alpha} + A^{\hat{a}}{}_{\hat{c}\gamma} F^{\hat{c}}{}_{\hat{b}\beta\alpha} \\ &\quad - F^{\hat{a}}{}_{\hat{c}\beta\alpha} A^{\hat{c}}{}_{\hat{b}\gamma}\} \\ &= 0.\end{aligned}$$

In addition, $F^{\hat{a}}{}_{\hat{b}\mu\nu}$ is defined in terms of $A^{\hat{a}}{}_{\hat{b}\mu}$ by the precise analogue of the equation (A.17) which expresses $R^{\hat{a}}{}_{\hat{b}\mu\nu}$ in terms of $\omega^{\hat{a}}{}_{\hat{b}\mu}$, namely

$$\begin{aligned}(2.15)\qquad F^{\hat{a}}{}_{\hat{b}\mu\nu} &= \partial_\mu A^{\hat{a}}{}_{\hat{b}\nu} - \partial_\nu A^{\hat{a}}{}_{\hat{b}\mu} \\ &\quad + A^{\hat{a}}{}_{\hat{d}\mu} A^{\hat{d}}{}_{\hat{b}\nu} - A^{\hat{a}}{}_{\hat{d}\nu} A^{\hat{d}}{}_{\hat{b}\mu}.\end{aligned}$$

Note that this formula does not involve the spacetime metric or its Christoffel symbols. In fact, the Christoffel symbols entering into the definition of ∇_α also cancel in Eq. (2.12) which entails only the exterior derivatives of the two-forms $F^{\hat{a}}{}_{\hat{b}\mu\nu} dx^\mu \wedge dx^\nu$ when the aforementioned substitutions are made there. On the other hand, Eq. (2.14) involves the metric and its Christoffel symbols explicitly and these quantities enter thereby into the wave equation for Yang-Mills curvature which played a central role in the Chruściel-Shatah analysis [**2**] of Yang-Mills fields on a curved background spacetime.

Returning to the wave equation for space time curvature (2.7), we now write it in the Cartan formalism which is, for us, motivated by the rather close analogy with Yang-Mills theory. Setting

$$R^{\hat{a}}{}_{\hat{b}\mu\nu;\lambda}{}^{;\lambda} = \theta^{\hat{a}}_{\rho}\, h^{\sigma}_{\hat{b}}\, R^{\rho}{}_{\sigma\mu\nu;\alpha\beta}\, g^{\alpha\beta} \tag{2.16}$$

and expanding out the right hand side using the notation introduced above one now gets

$$\begin{aligned} g^{\alpha\beta}\{&\nabla_\beta[\nabla_\alpha R^{\hat{a}}{}_{\hat{b}\mu\nu} + \omega^{\hat{a}}{}_{\hat{c}\alpha}R^{\hat{c}}{}_{\hat{b}\mu\nu} \\ &- R^{\hat{a}}{}_{\hat{c}\mu\nu}\omega^{\hat{c}}{}_{\hat{b}\alpha}] \\ &+ \omega^{\hat{a}}{}_{\hat{c}\beta}[\nabla_\alpha R^{\hat{c}}{}_{\hat{b}\mu\nu} + \omega^{\hat{c}}{}_{\hat{d}\alpha}R^{\hat{d}}{}_{\hat{b}\mu\nu} \\ &- R^{\hat{c}}{}_{\hat{d}\mu\nu}\omega^{\hat{d}}{}_{\hat{b}\alpha}] \\ &- [\nabla_\alpha R^{\hat{a}}{}_{\hat{c}\mu\nu} + \omega^{\hat{a}}{}_{\hat{d}\alpha}R^{\hat{d}}{}_{\hat{c}\mu\nu} - R^{\hat{a}}{}_{\hat{d}\mu\nu}\omega^{\hat{d}}{}_{\hat{c}\alpha}]\omega^{\hat{c}}{}_{\hat{b}\beta}\} \\ &= -R_{\mu\nu}{}^{\rho\sigma}R^{\hat{a}}{}_{\hat{b}\rho\sigma} \\ &+ 2R^{\hat{a}}{}_{\hat{c}\mu\sigma}R^{\hat{c}}{}_{\hat{b}\nu}{}^{\sigma} - 2R^{\hat{a}}{}_{\hat{c}\nu\sigma}R^{\hat{c}}{}_{\hat{b}\mu}{}^{\sigma}. \end{aligned} \tag{2.17}$$

Rearranging this slightly, one can write it in the form

$$\begin{aligned} \nabla^\alpha\nabla_\alpha R^{\hat{a}}{}_{\hat{b}\mu\nu} &+ R_{\mu\nu}{}^{\rho\sigma}R^{\hat{a}}{}_{\hat{b}\rho\sigma} \\ &= 2R^{\hat{a}}{}_{\hat{c}\mu\sigma}R^{\hat{c}}{}_{\hat{b}\nu}{}^{\sigma} - 2R^{\hat{a}}{}_{\hat{c}\nu\sigma}R^{\hat{c}}{}_{\hat{b}\mu}{}^{\sigma} \\ &- g^{\alpha\beta}\{\nabla_\beta[\omega^{\hat{a}}{}_{\hat{c}\alpha}R^{\hat{c}}{}_{\hat{b}\mu\nu} - R^{\hat{a}}{}_{\hat{c}\mu\nu}\omega^{\hat{c}}{}_{\hat{b}\alpha}] \\ &\quad + \omega^{\hat{a}}{}_{\hat{c}\beta}[\nabla_\alpha R^{\hat{c}}{}_{\hat{b}\mu\nu} + \omega^{\hat{c}}{}_{\hat{d}\alpha}R^{\hat{d}}{}_{\hat{b}\mu\nu} \\ &\quad - R^{\hat{c}}{}_{\hat{d}\mu\nu}\omega^{\hat{d}}{}_{\hat{b}\alpha}] \\ &\quad - [\nabla_\alpha R^{\hat{a}}{}_{\hat{c}\mu\nu} + \omega^{\hat{a}}{}_{\hat{d}\alpha}R^{\hat{d}}{}_{\hat{c}\mu\nu} - R^{\hat{a}}{}_{\hat{d}\mu\nu}\omega^{\hat{d}}{}_{\hat{c}\alpha}]\omega^{\hat{c}}{}_{\hat{b}\beta}\} \end{aligned} \tag{2.18}$$

where we have put $\nabla^\alpha = g^{\alpha\beta}\nabla_\beta$. The operator acting on $R^{\hat{a}}{}_{\hat{b}\mu\nu}$ on the left hand side of this equation ignores the frame indices and has exactly the same form as the wave operator that acts on the Faraday tensor $F_{\mu\nu}$ of a solution to Maxwell's equation on a vacuum background spacetime.

3. Normal Charts and Parallel Propagated Frames

In any Riemannian or pseudo-Riemannian (e.g., Lorentzian) manifold (V, g) one can construct, using the exponential map, a normal coordinate chart on some neighborhood of an arbitrary point in that manifold. Within our framework let $q \,\epsilon\, V$ be an arbitrary point of V and choose an orthonormal frame $\{\tilde{e}_\mu\}$ at the point q. Tangent vectors $\tilde{v} \,\epsilon\, T_qV$ can then be expressed as $\tilde{v} = x^\mu \tilde{e}_\mu$ and, for each such $\tilde{v}$, one can construct the affinely parameterized geodesic of (V, g) which begins (with parameter value zero) at the point q with initial tangent vector $\tilde{v}$. If the components $\{x^\mu\}$ are constrained to a sufficiently small neighborhood of the origin in the relevant real number space each such geodesic will extend (at parameter value unity) to a uniquely defined point $p \in V$ in some (normal) neighborhood of the point q. More precisely one proves that this (exponential) mapping determines a diffeomorphism between a neighborhood of the origin in the relevant real number space and a corresponding neighborhood of the point q in the manifold V. As usual, such neighborhoods are called normal neighborhoods and the corresponding coordinates $\{x^\mu\}$ normal coordinates. This construction breaks down only when distinct geodesics emerging from q begin to intersect away from q.

Note that by construction one has $\tilde{e}_\mu = \frac{\partial}{\partial x^\mu} \mid_q$ though of course away from q the (normal) coordinate basis fields $\{\frac{\partial}{\partial x^\mu}\}$ will no longer be orthonormal. It is not difficult to show that when the metric and Christoffel connection are expressed in normal coordinates about q (with $x^\mu(q) = 0$) they obey

$$g_{\mu\nu}(0) = \eta_{\mu\nu}, \quad \Gamma^\alpha_{\mu\nu}(0) = 0 \tag{3.1}$$

at the point q. More remarkable are the formulas

$$g_{\mu\nu}(x)x^\nu = g_{\mu\nu}(0)x^\nu = \eta_{\mu\nu}x^\nu \tag{3.2}$$

and

$$\Gamma^\alpha_{\mu\nu}(x)x^\mu x^\nu = 0 \tag{3.3}$$

satisfied throughout an arbitrary normal coordinate chart [**17**]. We shall give an alternative proof of these equations later in this section.

An important feature of normal coordinates based at q is that the geodesics through q are expressed simply as straight lines in such coordinates. In other words the curves defined by

$$x^\mu(\lambda) = x^\mu \cdot \lambda \ , \ \lambda \in [0, 1] \tag{3.4}$$

are all geodesics beginning at q for any $\{x^\mu(p)\}$ lying in the range of the chosen chart.

The geodesic with $x^\mu = x^\mu(p)$ connects q (at $\lambda = 0$) to p (at $\lambda = 1$) and is the unique geodesic, lying entirely within the chart domain, to have this property. Note that the tangent vector to this geodesic at the point p is given by $\tilde{v}_p = x^\mu(p)\frac{\partial}{\partial x^\mu} \mid_p$. Thus the vector field $\tilde{v} = x^\mu \frac{\partial}{\partial x^\mu}$ is, away from q, everywhere tangent to the geodesic from q which determines that arbitrary point p via the exponential map.

On any such normal coordinate chart domain we now introduce a preferred orthonormal frame field $\{h_{\hat{a}}\}$ as follows. Choose $h_{\hat{a}} \mid_q = \delta^\mu_{\hat{a}} \tilde{e}_\mu$ at the point q and extend each such frame field to a normal neighborhood of q by parallel propagation along the geodesics emerging from q in the construction of the normal chart. Such parallel propagation automatically preserves orthonormality and thus yields an orthonormal frame field $\{h_{\hat{a}}\}$ defined throughout the chart domain. The dual, co-frame field $\{\theta^{\hat{a}}\}$ can either be obtained algebraically by computing $\theta^{\hat{a}}_{\mu} = \eta^{\hat{a}\hat{b}} g_{\mu\nu} h^{\nu}_{\hat{b}}$ in the normal coordinate system or, equivalently, from parallel propagation of the co-frame field $\{\theta^{\hat{a}}\}\,|_q$ defined at q along the geodesics emerging from q. This works naturally since parallel propagation of both $\{\theta^{\hat{a}}\}$ and $\{h_{\hat{a}}\}$ along these geodesics automatically preserves the duality relations

$$\langle \theta^{\hat{a}}, h_{\hat{b}} \rangle := \theta^{\hat{a}}_{\mu} h^{\mu}_{\hat{b}} = \delta^{\hat{a}}_{\hat{b}}. \tag{3.5}$$

Here and below we let $\langle , \rangle$ signify the natural pairing of a one-form and a vector.

From the foregoing construction it follows that $\nabla_{\tilde{v}} h_{\hat{a}} = 0$ where $\tilde{v} = x^{\mu} \frac{\partial}{\partial x^{\mu}}$ is the geodesic tangent field previously defined and $\nabla_{\tilde{v}}$ is the directional covariant derivative operator. More explicitly this yields

$$(\nabla_{\tilde{v}} h_{\hat{a}})^{\mu} = v^{\nu}(h^{\mu}_{\hat{a},\nu} + \Gamma^{\mu}_{\gamma\nu} h^{\gamma}_{\hat{a}}) = x^{\nu}(h^{\mu}_{\hat{a},\nu} + \Gamma^{\mu}_{\gamma\nu} h^{\gamma}_{\hat{a}}) = 0. \tag{3.6}$$

Contracting with $\theta^{\hat{c}}_{\mu}$ one gets the equivalent equation

$$\begin{aligned} \theta^{\hat{c}}_{\mu}(\nabla_{\tilde{v}} h_{\hat{a}})^{\mu} &= v^{\nu}(\theta^{\hat{c}}_{\mu} h^{\mu}_{\hat{a},\nu} + \theta^{\hat{c}}_{\mu} h^{\gamma}_{\hat{a}} \Gamma^{\mu}_{\gamma\nu}) \\ &= x^{\nu}(\theta^{\hat{c}}_{\mu} h^{\mu}_{\hat{a},\nu} + \theta^{\hat{c}}_{\gamma} h^{\mu}_{\hat{a}} \Gamma^{\gamma}_{\mu\nu}) \\ &= x^{\nu} \omega^{\hat{c}}{}_{\hat{a}\nu} = 0. \end{aligned} \tag{3.7}$$

In other words parallel propagation of the orthonormal frame $\{h_{\hat{a}}\}$ along $\tilde{v}$ corresponds to the equation

$$\langle \omega^{\hat{c}}{}_{\hat{a}}, \tilde{v} \rangle = \omega^{\hat{c}}{}_{\hat{a}\nu} x^{\nu} = 0 \tag{3.8}$$

holding throughout the normal coordinates chart where, as before $\omega^{\hat{c}}{}_{\hat{a}} = \omega^{\hat{c}}{}_{\hat{a}\nu} dx^{\nu}$ is the connection one-form defined by this choice of chart and frame.

Equation (3.8) is completely analogous to the Cronström gauge condition for a Yang-Mills connection $A^{\hat{a}}{}_{\hat{b}} = A^{\hat{a}}{}_{\hat{b}\nu} dx^{\nu}$ introduced in [**1**] and exploited in [**5**] and [**2**] to establish global existence for solutions to the Yang-Mills equation in flat and curved spacetimes respectively. In Yang-Mills theory the gauge condition,

$$A^{\hat{a}}{}_{\hat{b}\nu} x^{\nu} = 0 \tag{3.9}$$

(again imposed throughout a normal coordinate chart on spacetime) results from parallel propagation in the internal space whereas here it results from parallel propagation in the space of orthonormal frames tangent to spacetime. As in Yang-Mills theory one can exploit this choice of gauge to compute the connection one-forms $\omega^{\hat{c}}{}_{\hat{a}}$ directly from the curvature two-forms $R^{\hat{c}}{}_{\hat{a}}$, reversing the order of the usual calculation. In the chosen gauge Eq. (A17) gives immediately

$$x^{\nu} R^{\hat{c}}{}_{\hat{a}\mu\nu} = -x^{\nu} \frac{\partial}{\partial x^{\nu}} \omega^{\hat{c}}{}_{\hat{a}\mu} - \omega^{\hat{c}}{}_{\hat{a}\mu} \tag{3.10}$$

or, equivalently, along the geodesic curve $x^{\mu}(\lambda) = x^{\mu} \cdot \lambda$, that

$$-\frac{d}{d\lambda}[\lambda \omega^{\hat{c}}{}_{\hat{a}\mu}(x(\lambda)] = \lambda x^{\nu} R^{\hat{c}}{}_{\hat{a}\mu\nu}(x(\lambda)). \tag{3.11}$$

Integrating this from $\lambda = 0$ to $\lambda = 1$ gives

$$\omega^{\hat{c}}{}_{\hat{a}\mu}(x) = -\int_0^1 d\lambda\, \lambda x^\nu R^{\hat{c}}{}_{\hat{a}\mu\nu}(x \cdot \lambda) \tag{3.12}$$

in exact parallel to Cronström's formula for $A^{\hat{c}}{}_{\hat{a}\mu}$ in terms of $F^{\hat{c}}{}_{\hat{a}\mu\nu}$.

In general relativity however, one can go further and compute the (co-) frame field $\{\theta^a\}$ (which has no analogue in Yang-Mills theory) directly in terms of the connection and hence in terms of curvature. To see this first note that the tangent vector to any of the (normal) geodesics through q is given by

$$\frac{dx^\mu(\lambda)}{d\lambda} = \frac{d}{d\lambda}(x^\mu \lambda) = x^\mu \tag{3.13}$$

and thus is independent of λ. Since this tangent vector is (by the definition of geodesics) parallel propagated along the geodesic its natural pairing with a parallel propagated one-form such as $\theta^{\hat{a}}$ is necesarily independent of the curve parameter λ. Equating these pairings at $\lambda = 0$ and $\lambda = 1$ gives

$$\theta^{\hat{a}}_\nu(0)x^\nu = \theta^{\hat{a}}_\nu(x)x^\nu \tag{3.14}$$

$\forall\{x^\nu\}$ within the normal neighborhood. Squaring this formula gives immediately

$$\begin{aligned} \eta_{\hat{a}\hat{b}}\theta^{\hat{a}}_\nu(0)\theta^{\hat{b}}_\mu(0)x^\nu x^\mu &= g_{\mu\nu}(0)x^\mu x^\nu \\ &= \eta_{\hat{a}\hat{b}}\theta^{\hat{a}}_\nu(x)\theta^{\hat{b}}_\mu(x)x^\nu x^\mu = g_{\mu\nu}(x)x^\mu x^\nu \end{aligned} \tag{3.15}$$

which is related to, but weaker than, Equation (3.2). We shall reproduce the strong form momentarily.

The zero torsion condition is given by

$$\begin{aligned} &\partial_\nu\theta^{\hat{c}}_\mu(x) - \partial_\mu\theta^{\hat{c}}_\nu(x) \\ &+ \omega^{\hat{c}}_{\hat{a}\nu}(x)\theta^{\hat{a}}_\mu(x) - \omega^{\hat{c}}{}_{\hat{a}\mu}(x)\theta^{\hat{a}}_\nu(x) \\ &= 0. \end{aligned} \tag{3.16}$$

Contracting this with x^ν and using Eq. (3.8) one obtains

$$\begin{aligned} &x^\nu\partial_\nu(\theta^{\hat{c}}_\mu(x)) - \partial_\mu[x^\nu\theta^{\hat{c}}_\nu(x)] \\ &+ \theta^{\hat{c}}_\mu(x) - \omega^{\hat{c}}{}_{\hat{a}\mu}(x)(x^\nu\theta^{\hat{a}}_\nu(x)) \\ &= 0. \end{aligned} \tag{3.17}$$

But making use of the result in Eq. (3.14) we can reexpress this as

$$\begin{aligned} &x^\nu\partial_\nu(\theta^{\hat{c}}_\mu(x)) - \partial_\mu[x^\nu\theta^{\hat{c}}_\nu(0)] \\ &+ \theta^{\hat{c}}_\mu(x) - \omega^{\hat{c}}{}_{\hat{a}\mu}(x)[x^\nu\theta^{\hat{a}}_\nu(0)] \\ &= x^\nu\partial_\nu(\theta^{\hat{c}}_\mu(x)) + \theta^{\hat{c}}_\mu(x) - \theta^{\hat{c}}_\mu(0) \\ &- \omega^{\hat{c}}{}_{\hat{a}\mu}(x)[x^\nu\theta^{\hat{a}}_\nu(0)] \\ &= 0 \end{aligned} \tag{3.18}$$

which can be written as

$$(3.19)\qquad \begin{aligned} & x^\nu \partial_\nu [\theta^{\hat{c}}_\mu(x) - \theta^{\hat{c}}_\mu(0)] \\ & + [\theta^{\hat{c}}_\mu(x) - \theta^{\hat{c}}_\mu(0)] \\ & = \omega^{\hat{c}}{}_{\hat{a}\mu}(x)[\theta^{\hat{a}}_\nu(0)x^\nu], \end{aligned}$$

a transport equation for the quantity $\theta^{\hat{c}}_\mu(x) - \theta^{\hat{c}}_\mu(0)$. Along a geodesic $x^\mu(\lambda) = x^\mu \cdot \lambda$ through q one thus has

$$(3.20)\qquad \begin{aligned} & \frac{d}{d\lambda}[\lambda(\theta^{\hat{c}}_\mu(x(\lambda)) - \theta^{\hat{c}}_\mu(0))] \\ & = \omega^{\hat{c}}{}_{\hat{a}\mu}(x(\lambda))[x^\nu \cdot \lambda\ \theta^{\hat{a}}_\nu(0)]. \end{aligned}$$

Integrating this form $\lambda = 0$ to $\lambda = 1$ one gets

$$(3.21)\qquad \begin{aligned} \theta^{\hat{c}}_\mu(x) &= \theta^{\hat{c}}_\mu(0) \\ & + \int_0^1 d\lambda [\omega^{\hat{c}}{}_{\hat{a}\mu}(\lambda x)(\lambda x^\nu \theta^{\hat{a}}_\nu(0))] \end{aligned}$$

which is the desired expression for $\theta^{\hat{c}}_\mu(x)$.

Combined with Eq. (3.12) this allows us to express both the connection and the frame one-forms directly in terms of curvature by explicit integral formulas. Given the (co-) frame $\{\theta^{\hat{a}}\}$ one can of course compute the frame fields $\{h_{\hat{a}}\}$ and the metric algebraically.

To show how Eq. (3.21) implies Eq. (3.2) we use the former to evaluate

$$(3.22)\qquad \begin{aligned} g_{\mu\nu}(x)x^\nu &= \eta_{\hat{c}\hat{d}}\theta^{\hat{c}}_\mu(x)\theta^{\hat{d}}_\nu(x)x^\nu \\ &= \eta_{\hat{c}\hat{d}}\left(\theta^{\hat{c}}_\mu(0) + \int_0^1 d\lambda[\omega^{\hat{c}}{}_{\hat{a}\mu}(\lambda x)(\lambda x^\gamma \theta^{\hat{a}}_\gamma(0))]\right) \\ &\quad \cdot \left(\theta^{\hat{d}}_\nu(0)x^\nu + \int_0^1 d\sigma[\omega^{\hat{d}}{}_{\hat{b}\nu}(\sigma x)x^\nu(\sigma x^\delta \theta^{\hat{b}}_\delta(0))]\right) \\ &= \left(\eta_{\hat{c}\hat{d}}\theta^{\hat{c}}_\mu(0) + \int_0^1 d\lambda[\omega_{\hat{d}\hat{a}\mu}(\lambda x)(\lambda x^\gamma \theta^{\hat{a}}_\gamma(0))]\right)\theta^{\hat{d}}_\nu(0)x^\nu \\ &= \eta_{\hat{c}\hat{d}}\theta^{\hat{c}}_\mu(0)\theta^{\hat{d}}_\nu(0)x^\nu + \int_0^1 d\lambda[\omega_{\hat{d}\hat{a}\mu}(\lambda x)\lambda x^\gamma \theta^{\hat{a}}_\gamma(0) \cdot x^\nu \theta^{\hat{d}}_\nu(0)] \\ &= g_{\mu\nu}(0)x^\nu \end{aligned}$$

where we have used the parallel propagation condition, $\omega^{\hat{a}}{}_{\hat{b}\nu}(x)x^\nu = 0$, and the metric compatibility condition, $\omega_{\hat{a}\hat{b}\nu}(x) = -\omega_{\hat{b}\hat{a}\nu}(x)$, to simplify the intermediate expressions.

Equation (3.3) is normally proven directly from the geodesic equation specialized to normal coordinates. Using the duality relations $\theta^{\hat{a}}_\mu\ h^\mu_{\hat{c}} = \delta^{\hat{a}}{}_{\hat{c}}$ and $\theta^{\hat{a}}_\mu\ h^\nu_{\hat{a}} = \delta_\mu{}^\nu$ however, we can reexpress Eq. (A.13) in the equivalent form

$$(3.23)\qquad \Gamma^\lambda_{\delta\nu} = h^\lambda_{\hat{a}}\ \theta^{\hat{c}}_\delta\ \omega^{\hat{a}}{}_{\hat{c}\nu} + h^\lambda_{\hat{a}}\ \theta^{\hat{a}}_{\delta,\nu}.$$

Thus since $\omega^{\hat{a}}{}_{\hat{c}\nu}(x)x^\nu = 0$ we get

$$\Gamma^\lambda_{\delta\nu}(x)x^\delta x^\nu = h^\lambda_{\hat{a}}(x)\theta^{\hat{a}}_{\delta,\nu}(x)x^\delta x^\nu. \tag{3.24}$$

But using Eq. (3.14), one gets

$$\begin{aligned} x^\nu x^\delta(\theta^{\hat{a}}_{\delta,\nu}(x)) &= x^\nu\{\partial_\nu[\theta^{\hat{a}}_\delta(x)x^\delta] - \theta^{\hat{a}}_\nu(x)\} \\ &= x^\nu\{\partial_\nu[\theta^{\hat{a}}_\delta(0)x^\delta] - \theta^{\hat{a}}_\nu(x)\} \\ &= x^\nu\{\theta^{\hat{a}}_\nu(0) - \theta^{\hat{a}}_\nu(x)\} = 0 \end{aligned} \tag{3.25}$$

where the last step follows from Eq. (3.21) and the parallel propagation condition $\omega^{\hat{c}}{}_{\hat{a}\nu}(x)x^\nu = 0$. Thus $\Gamma^\lambda_{\delta\nu}(x)x^\delta x^\nu = 0$ in normal coordinates.

4. An Integral Equation for the Curvature Tensor

In Section 2 we rederived the fundamental wave equation satisfied by the curvature tensor of a vacuum spacetime and expressed this, via the Cartan formalism, as a curved space Yang-Mills equation coupled to the vanishing torsion condition. The latter equation, which relates the frame field determining the spacetime metric to the connection, has no analogue in a "pure" Yang-Mills problem but here of course provides the fundamental link between the metric and its curvature.

In the Cartan formalism wherein one regards the curvature tensor as a matrix of two-forms, $R^{\hat{a}}{}_{\hat{b}\mu\nu}dx^\mu \wedge dx^\nu$, or equivalently as a two-form with values in the matrix Lie algebra for the Lorentz group $SO(3,1)$, the wave operator (defined by the left-hand side of Eq. (2.18)) takes the form (for each separate matrix element) of the same wave operator that acts on the Faraday tensor $F_{\mu\nu}dx^\mu \wedge dx^\nu$ of a solution to Maxwell's equations. In particular, the frame indices play completely inert roles on the left-hand side of Eq. (2.18) which leaves the different matrix elements uncoupled.

We want to derive an integral equation satisfied by curvature by applying the fundamental solution for this wave operator to the "source" term defined by the right hand side of Eq. (2.18), using Eqs. (3.12) and (3.21) to eliminate the connection and frame in favor of curvature in much the same way that one previously used Cronström's formula to eliminate the Yang-Mills connection in favor of its curvature in studies of the flat and curved space pure Yang-Mills fields. The theory developed in Friedlander's book [**8**] (which builds on the fundamental work of Hadamard, Riesz, Sobolev, Choquet-Bruhat and others) applies to this wave operator (as well as to others we shall consider later) and allows one to write an integral formula for the solution of the corresponding Cauchy problem on so-called causal domains of the spacetime (i.e., on geodesically convex domains which are also globally hyperbolic in a suitable sense [**18**]). For Friedlander, who treats only linear problems, the integral formula in question is a genuine representation formula for the solution of the associated wave equation whereas for us it only yields an integral equation satisfied by the relevant solution to the Cauchy problem.

Of course not every solution to Eq. (2.18) corresponds to a solution of Einstein's field equations. It is necessary, in order to avoid introducing spurious solutions, to restrict the Cauchy data appearing in the Friedlander formula by imposing those first order equations upon the curvature which results from the Bianchi identities when the Ricci tensor vanishes (the vacuum condition). The Friedlander formalism applies to all solutions of the relevant wave equation and hence in particular to the solutions of physical interest.

To simplify the notation, let us write $F_{\mu\nu}$ for any particular matrix element $R^{\hat{a}}{}_{\hat{b}\mu\nu}$ of curvature (surpressing the inert frame indices $\hat{a}, \hat{b}$) and $f_{\mu\nu}$ for the corresponding source term so that Eq. (2.18) now takes the form

$$F_{\mu\nu;\gamma}{}^{;\gamma} + R_{\mu\nu}{}^{\alpha\beta} F_{\alpha\beta} = f_{\mu\nu}. \tag{4.1}$$

With reference to Fig. 5.3.1 of Friedlander's book, let p be a point in some causal domain of $({}^{(4)}V, g)$ and S be a spacelike hypersurface within this domain such that every past-directed causal geodesic from p meets S. Further, let C_p be the mantle of the (truncated) past light cone from p to S, σ_p be the (two-dimensional) intersection of C_p with S and let D_p be the interior of this truncated cone and designate by S_p the (three-dimensional) intersection of D_p with S. Finally, let T_p designate the expanding lightlike hypersurface which intersects S in σ_p.

Friedlander's representation formula for the field at point p is given in local coordinates by [**19**]:

$$\begin{aligned} F_{\alpha\beta}(x) = {} & \frac{1}{2\pi} \int\limits_{C_p} U^{\mu'\nu'}_{\alpha\beta}(x,x') f_{\mu'\nu'}(x') \mu_\Gamma(x') \\ & + \frac{1}{2\pi} \int\limits_{D_p} (V^+)^{\mu'\nu'}_{\alpha\beta}(x,x') f_{\mu'\nu'}(x') \mu(x') \\ & + \frac{1}{2\pi} \int\limits_{S_p} *[(V^+)^{\mu'\nu'}_{\alpha\beta}(x,x') \nabla^{\gamma'} F_{\mu'\nu'}(x') \\ & \qquad - F_{\mu'\nu'}(x') \nabla^{\gamma'} (V^+)^{\mu'\nu'}_{\alpha\beta}(x,x')] \\ & + \frac{1}{2\pi} \int\limits_{\sigma_p} \{ U^{\mu'\nu'}_{\alpha\beta}(x,x') [2(\nabla^{\gamma'} t(x'))(\nabla_{\gamma'} F_{\mu'\nu'}(x')) \\ & \qquad + F_{\mu'\nu'}(x') \,\Box\, t(x')] \\ & - \langle \nabla t(x'), \nabla' \Gamma(x,x') \rangle (V^+)^{\mu'\nu'}_{\alpha\beta}(x,x') F_{\mu'\nu'}(x') \} \mu_{t,\Gamma}(x'). \end{aligned} \tag{4.2}$$

Here $U^{\mu'\nu'}_{\alpha\beta}(x,x') = \kappa(x,x') \tau^{\mu'\nu'}_{\alpha\beta}(x,x')$ where κ is the transport biscalar defined by Eq. (4.2.17) of Ref. [**8**] and given in local coordinates by Eq.

(4.2.18) or (4.2.19) of that reference and $\tau^{\mu'\nu'}_{\alpha\beta}(x,x')$ is the transport bitensor (or propagator) defined in Section (5.5) of Friedlander. The latter is expressible explicitly in terms of an orthonormal frame parallel propagated from p along the geodesic issuing from that point.

The measure $\mu(x')$ is the standard spacetime volume measure given in local coordinates by $\sqrt{-\det g_{\mu\nu}(x')}d^4x'$ whereas the measure on the light cone $\mu_\Gamma(x')$ is a Leray form defined such that

$$d_{x'}\Gamma(x,x') \wedge \mu_\Gamma(x') = \mu(x') \tag{4.3}$$

where $\Gamma(x,x')$ is the optical function (squared geodesic distance within a causal domain) introduced in Sect. (1.2) of Friedlander (c.f., Theorem 1.2.3). Leray forms are introduced in Sect. (2.9) and developed further in Sect. (4.5) of this same reference and the coordinate expression for the dual $*v$ of a vector v is given there by Eq. (2.9.3). This is needed in the boundary integral over S_p whereas μ_Γ arises in that over C_p. The two-dimensional Leray form $\mu_{t,\Gamma}(x')$ needed for the integral over σ_p, is defined such that (c.f., Lemma 5.3.3. of Ref. [**8**])

$$dt(x') \wedge d_{x'}\Gamma(x,x') \wedge \mu_{t,\Gamma}(x') = \mu(x') \tag{4.4}$$

where $t(x')$ is the null field defined by Lemma 5.3.2 of Friedlander. Note also in this reference the needed expressions for $(\Box\, t)\mu_{t,\Gamma}$ and $\langle \nabla t, \nabla\Gamma\rangle$ given respectively by Eqs. (5.3.20) and (5.3.19) of this same section.

The tail field $(V^+)^{\mu'\nu'}_{\alpha\beta}(x,x')$ is the solution of a characteristic initial value problem for the homogeneous wave equation. By virtue of the self-adjoincy of our Eq. (4.1) and the reciprocity relations derived by Friedlander in Sect. (5.2) (which apply as well to the tensor case as discussed in Sect. (5.5)) the tail bitensor V^+ satisfies the wave equation

$$\begin{aligned}(V^+)^{\mu'\nu';\gamma'}_{\alpha\beta;\gamma'}(x,x') &+ R^{\mu'\nu'}{}_{\delta'\gamma'}(x')(V^+)^{\delta'\gamma'}_{\alpha\beta}(x,x') \\ &+ R^{\mu'}_{\delta'}(x')(V^+)^{\nu'\delta'}_{\alpha\beta}(x,x') - R^{\nu'}_{\delta'}(x')(V^+)^{\mu'\delta'}_{\alpha\beta}(x,x') \\ &= 0\end{aligned} \tag{4.5}$$

wherein the indices $\alpha\beta$ and coordinates x^μ play inert roles. In the foregoing formulas, as well as below, the notations ∇_γ and $;\gamma$ are used interchangeably. The initial data for V^+ is computable on the light cone C_p where it reduces to the bitensor field that Friedlander expresses as V_0. The transport equation determining V_0 is provided by Friedlander's Eq. (5.5.23) and its explicit solution is given in his Eq. (5.5.25).

5. Transformations of the Tail Field Integrals

Define the tail field contributions to $F_{\alpha\beta}(x)$ by

(5.1)
$$\begin{aligned} F^{\text{tail}}_{\alpha\beta}(x) := \frac{1}{2\pi}\int\limits_{D_p} (V^+)^{\mu'\nu'}_{\alpha\beta}(x,x')f_{\mu'\nu'}(x')\mu(x') \\ + \frac{1}{2\pi}\int\limits_{S_p} *[(V^+)^{\mu'\nu'}_{\alpha\beta}(x,x')\nabla^{\gamma'}F_{\mu'\nu'}(x') \\ - F_{\mu'\nu'}(x')\nabla^{\gamma'}(V^+)^{\mu'\nu'}_{\alpha\beta}(x,x')] \\ - \frac{1}{2\pi}\int\limits_{\sigma_p} \langle\nabla t(x'),\nabla'\Gamma(x,x')\rangle(V^+)^{\mu'\nu'}_{\alpha\beta}(x,x')F_{\mu'\nu'}(x')\mu_{t,\Gamma}(x'). \end{aligned}$$

This consists of all the terms that would vanish if Huygen's principle were valid since in that case $V^+ = 0$ but, in a curved spacetime, these terms are generally non-zero.

Let us reexpress the source f through the use of the wave equation for F as

$$f_{\mu'\nu'}(x') = (PF)_{\mu'\nu'}(x') \tag{5.2}$$

where P is the second order linear, self-adjoint operator defined by the left hand side of Eq. (4.1). Recalling Eq. (4.5) which can be written as

$$(PV^+)^{\mu'\nu'}_{\alpha\beta}(x,x') = 0 \tag{5.3}$$

where P acts at x' and the indices α, β and x are inert, one finds that the integrand $(V^+)^{\mu'\nu'}_{\alpha\beta}(x,x')f_{\mu'\nu'}(x')$ can be expressed as

(5.4)
$$\begin{aligned} &(V^+)^{\mu'\nu'}_{\alpha\beta}(x,x')f_{\mu'\nu'}(x') \\ &= (V^+)^{\mu'\nu'}_{\alpha\beta}(x,x')(PF)_{\mu'\nu'}(x') - (PV^+)^{\mu'\nu'}_{\alpha\beta}(x,x')F_{\mu'\nu'}(x') \\ &= \nabla_{\gamma'}\{(V^+)^{\mu'\nu'}_{\alpha\beta}(x,x')(\nabla^{\gamma'}F_{\mu'\nu'}(x')) \\ &\quad - (\nabla^{\gamma'}(V^+)^{\mu'\nu'}_{\alpha\beta}(x,x'))F_{\mu'\nu'}(x')\} \end{aligned}$$

where the curvature terms have canceled from the final expression by virtue of the self-adjoint structure of the wave operator P. Thus the integrand in the volume integral over D_p can be reexpressed as a total divergence. It is worth noting that the scalar field analogue to the above observation is given at the end of p.187 in Friedlander's book.

Using Eq. (5.4) to reexpress the integral over D_p in the equation for $F^{\text{tail}}_{\alpha\beta}(x)$ and using Stokes' theorem to rewrite this volume integral as a boundary integral over $\partial D_p = C_p \cup S_p$, one arrives at the result that

$$(5.5)\quad F^{\text{tail}}_{\alpha\beta}(x) = \frac{1}{2\pi}\int_{C_p} *[(V^+)^{\mu'\nu'}_{\alpha\beta}(x,x')\nabla^{\gamma'}F_{\mu'\nu'}(x') - F_{\mu'\nu'}(x')\nabla^{\gamma'}(V^+)^{\mu'\nu'}_{\alpha\beta}(x,x')] - \frac{1}{2\pi}\int_{\sigma_p}\langle\nabla t(x'), \nabla'\Gamma(x,x')\rangle(V^+)^{\mu'\nu'}_{\alpha\beta}(x,x')F_{\mu'\nu'}(x')\mu_{t,\Gamma}(x')$$

where the orientation chosen for the integral over the null cone C_p corresponds to a normal field directed towards the vertex p. The cancelation of the two boundary integrals over S_p parallels that shown by Friedlander for the scalar case in his Eq. (5.3.14) (wherein however it was assumed that the support of the scalar field did not meet C_p). One can also think of deriving Eq. (5.5) from Eq. (5.1) by pushing the surface S_p forward, holding its boundary σ_p fixed, until it merges in the limit with C_p. Friedlander remarks in his Section (5.4) that the representation formula for the characteristic initial value problem can be derived in a similar manner wherein, however, one pushes $S_p$towards the past rather than towards the future.

Though we have succeeded to reexpress the tail contributions in terms of integrals only over C_p and σ_p the resulting formula is still not in a satisfactory state from the point of view of the ultimate applications we have in mind. This is so, in large measure, because Eq. (5.5) contains derivatives of the unknown curvature and it would be hopeless to try to derive estimates for the undifferentiated curvature from an integral equation involving the derivatives of this same quantity.

Fortunately, however, in the integral over C_p in Eq. (5.5) for $F^{\text{tail}}_{\alpha\beta}(x)$ only derivatives of $(V^+)^{\mu'\nu'}_{\alpha\beta}(x,x')$ and $F_{\mu'\nu'}(x')$ tangential to the null generators of the light cone are involved. The point is that since C_p is a null surface its normal ($\nabla^{\gamma'}\Gamma(x,x')$ in Friedlander's notation) is in fact tangential to the cone and hence the dual operator ($*$ in Eq. (5.5)) produces only these tangential derivatives in the integrand. Thus one is at liberty to integrate by parts and throw the directional derivative onto V^+ for example and thereby remove it from F. In effect, Friedlander exploited this freedom (though in the opposite way) in recasting the integral over $C_{p'}$ in his representation formula for the characteristic initial value problem into a form in which only tangential derivatives of F were involved. For our purposes, though it is essential to avoid the necessity of computing tangential derivatives to F and to recall that the tangential derivative of V^+ is given rather explicitly by Friedlander's Eq. (5.5.23) for this latter quantity (which coincides with V_0 on C_p). On the other hand, this integration by parts produces an additional contribution to the integral over C_p (since $\nabla^{\gamma'}\Gamma(x,x')$ gets differentiated) and a boundary contribution which modifies the integral over σ_p. We shall carry out these further reductions in the following section and thereby arrive at our final integral equation for curvature within the framework of the Cartan formalism.

The reader may be wondering though why it should be possible, as we have argued, to transform the tail contributions, which result from the failure of Huygens' principle to hold in a general spacetime, into a form (involving only integrals over C_p and σ_p) which seems to have miraculously restored Huygens' principle. The resolution of this seeming paradox results from noting that even for a truly linear problem (where the meaning of Huygens' principle is clearly defined) the transformed "representation" formula requires knowledge of the unknown field $F_{\mu\nu}$, on the light cone C_p and not merely on σ_p, the intersection of the cone with the initial hypersurface. Thus the transformed equation is not really a representation formula at all, even in the linear case, whereas initially (in Eq. (4.2)) it was. For the non-linear problems that we are interested in however, a genuine representation formula (for the solution of the Cauchy problem) is out of the question and it is far more convenient to have the tail contributions transformed, as we have done, to integrals over C_p and σ_p alone.

6. Reduction of the Tail Contributions

To simplify the notation slightly let us write Eq. (5.5) in the form

$$F^{\text{tail}}_{\alpha\beta}(x) = {}^{I}F^{\text{tail}}_{\alpha\beta}(x) + {}^{II}F^{\text{tail}}_{\alpha\beta}(x) \tag{6.1}$$

where ${}^{I}F^{\text{tail}}_{\alpha\beta}(x)$ is the integral over C_p and ${}^{II}F^{\text{tail}}_{\alpha\beta}(x)$ that over σ_p. Reexpressing the dual $*v$ to a vector v via Eq. (2.9.3) of Ref. [**8**] (see also p. 194 of this reference)

$$*v(x') = \langle v(x'), \text{grad}'\Gamma(x,x')\rangle \mu_\Gamma(x') \tag{6.2}$$

one gets the more explicit formula for ${}^{I}F^{\text{tail}}_{\alpha\beta}(x)$

$$\begin{aligned} {}^{I}F^{\text{tail}}_{\alpha\beta}(x) = \frac{1}{2\pi} \int_{C_p} \mu_\Gamma(x')\{\nabla^{\gamma'}\Gamma(x,x')[(V^+)^{\mu'\nu'}_{\alpha\beta}(x,x')\nabla_{\gamma'}F_{\mu'\nu'}(x') \\ - F_{\mu'\nu'}(x')\nabla_{\gamma'}(V^+)^{\mu'\nu'}_{\alpha\beta}(x,x')]\}. \end{aligned} \tag{6.3}$$

The key point here is that only derivatives tangential to the null generators of the cone C_p appear in the integrand. This allows one to integrate by parts to eliminate derivatives of $F_{\mu'\nu'}$ in favor of (tangential) derivatives of $(V^+)^{\mu'\nu'}_{\alpha\beta}$ which, in turn, may be evaluated from the transport equation (cf. Eq. (5.5.23) of Ref. [**8**]) which determines this quantity along C_p. Carrying out these operations and writing $(V_0)^{\mu'\nu'}_{\alpha\beta}(x,x')$ for the restriction of $(V^+)^{\mu'\nu'}_{\alpha\beta}(x,x')$ to C_p one arrives at

$$(6.4)\quad {}^I F^{\text{tail}}_{\alpha\beta}(x) = \frac{1}{2\pi}\int\limits_{C_p} \mu_\Gamma(x')\{(\nabla^{\gamma'}\Gamma(x,x'))\nabla_{\gamma'}((V_0)^{\mu'\nu'}_{\alpha\beta}(x,x')F_{\mu'\nu'}(x'))$$
$$+ F_{\mu'\nu'}(x')[PU^{\mu'\nu'}_{\alpha\beta}(x,x') + (\Box'\Gamma(x,x') - 4)(V_0)^{\mu'\nu'}_{\alpha\beta}(x,x')]\}$$

where P is the wave operator defined in Eq. (5.2) above and where, as mentioned above, we can write

$$(6.5)\qquad U^{\mu'\nu'}_{\alpha\beta}(x,x') = \kappa(x,x')\tau^{\mu'\nu'}_{\alpha\beta}(x,x')$$

with the parallel transport "propagator" $\tau^{\mu'\nu'}_{\alpha\beta}$ expressible in terms of our orthonormal frame as

$$(6.6)\qquad \tau^{\mu'\nu'}_{\alpha\beta}(x,x') = h^{\mu'}_{\hat{e}}(x')\theta^{\hat{e}}_{\alpha}(x)h^{\nu'}_{\hat{f}}(x')\theta^{\hat{f}}_{\beta}(x).$$

One can evaluate the first integral in the above expression for ${}^I F^{\text{tail}}_{\alpha\beta}(x)$ by first transforming from normal coordinates $\{x^{\mu'}\}$ to spherical null coordinates defined by

$$(6.7)\qquad \begin{aligned} x^{1'} &= r'\sin\,\theta\,\cos\varphi \\ x^{2'} &= r'\,\sin\,\theta\sin\varphi \\ x^{3'} &= r'\,\cos\theta \\ t' = x^{0'} &= \frac{u+v}{2},\ r' = \frac{v-u}{2} \\ r' &= \sqrt{\Sigma(x^{i'})^2} \end{aligned}$$

so that

$$(6.8)\qquad u = t' - r',\ v = t' + r'$$

with $\Gamma = -uv$ everywhere and $v = 0$ on C_p. In terms of these coordinates it is straightforward to show that

$$(6.9)\qquad \Gamma^{;\alpha}\frac{\partial}{\partial x^\alpha} = 2v\frac{\partial}{\partial v} + 2u\frac{\partial}{\partial u}$$

and that the Leray form

$$(6.10)\qquad \mu_\Gamma = \frac{\sqrt{-\det{(g_{\mu\nu})}}}{u}du\wedge d\theta\wedge d\varphi$$

satisfies

$$(6.11)\qquad \mu = d\Gamma\wedge\mu_\Gamma = \sqrt{-\det{(g_{\mu\nu})}du\wedge dv\wedge d\theta\wedge d\varphi}$$

as required by its definition (where $\det{(g_{\mu\nu})}$ is the determinant of g in the spherical null coordinates). Substituting these expressions into the integral in question one easily arrives at

$$(6.12)\quad \frac{1}{2\pi}\int\limits_{C_p} \mu_\Gamma(x')(\nabla^{\gamma'}\Gamma(x,x'))\nabla_{\gamma'}[(V_0)^{\mu'\nu'}_{\alpha\beta}(x,x')F_{\mu'\nu'}(x')]$$

$$= \frac{1}{2\pi}\int\limits_{C_p} du\wedge d\theta\wedge d\varphi\left[\frac{\partial}{\partial u}[2\sqrt{-\det(g_{\gamma\delta})}(V_0)^{\mu'\nu'}_{\alpha\beta}(x,x')F_{\mu'\nu'}(x')]\right]$$

$$+\frac{1}{2\pi}\int\limits_{C_p} \mu_\Gamma(x')[(4-\nabla_{\gamma'}\nabla^{\gamma'}\Gamma(x,x'))(V_0)^{\mu'\nu'}_{\alpha\beta}(x,x')F_{\mu'\nu'}(x')]$$

$$= -\frac{1}{2\pi}\int\limits_{\sigma_p} d\theta\wedge d\varphi\left\{2\sqrt{-\det(g_{\gamma\delta})}[(V_0)^{\mu'\nu'}_{\alpha\beta}(x,x')F_{\mu'\nu'}(x')]\right\}$$

$$+\frac{1}{2\pi}\int\limits_{C_p} \mu_\Gamma(x')[(4-\Box'\Gamma(x,x'))(V_0)^{\mu'\nu'}_{\alpha\beta}(x,x')F_{\mu'\nu'}(x')].$$

Evaluating the metric form restricted to C_p one gets

(6.13)

$$ds^2\,\big|_{C_p} = -dudv + {}^{(2)}V_\theta dv d\theta + {}^{(2)}V_\varphi dv d\varphi$$

$$+\ {}^{(2)}g_{AB}dx^A dx^B + \left(-\frac{1}{4}\,{}^{(4)}g^{uu} + \frac{1}{4}\,{}^{(2)}g_{AB}\;{}^{(2)}V^A\;{}^{(2)}V^B\right)dv^2$$

where $\{x^A; A = 1,2\} = \{\theta,\varphi\}$ and where ${}^{(2)}g_{AB}dx^Adx^B$ and ${}^{(2)}V_A dx^A = {}^{(2)}g_{AB}\;{}^{(2)}V^B dx^A$ are (at each fixed u on the hypersurface C_p defined by $v = 0$) a 2-dimensional Riemannian metric and one-form respectively. Thus, on C_p

$$(6.14)\qquad 2\sqrt{-\det g_{\gamma\delta}}\,\big|_{C_p} = \sqrt{\det {}^{(2)}g_{AB}}\,\big|_{C_p}$$

so that

$$(6.15)\quad \frac{1}{2\pi}\int\limits_{C_p} \mu_\Gamma(x')\{(\nabla^{\gamma'}\Gamma(x,x'))\nabla_{\gamma'}[(V_0)^{\mu'\nu'}_{\alpha\beta}(x,x')F_{\mu'\nu'}(x')]\}$$

$$= -\frac{1}{2\pi}\int\limits_{\sigma_p} \sqrt{\det {}^{(2)}g_{AB}}d\theta\wedge d\varphi[(V_0)^{\mu'\nu'}_{\alpha\beta}(x,x')F_{\mu'\nu'}(x')]$$

$$+\frac{1}{2\pi}\int\limits_{C_p} \mu_\Gamma(x')[(4-\Box'\Gamma(x,x'))(V_0)^{\mu'\nu'}_{\alpha\beta}(x,x')F_{\mu'\nu'}(x')].$$

It is easy to see from the metric form (6.13) that $\sqrt{\det {}^{(2)}g_{AB}}d\theta\wedge d\varphi$ is just the invariant 2-surface area element induced on σ_p (defined in coordinates by $v = 0$, $u = u(\theta,\varphi)$) by the spacetime metric. Writing this as $d\sigma_p$

and combining Eqs. (6.4) and (6.15) we get

$$(6.16)\qquad {}^{I}F^{\text{tail}}_{\alpha\beta}(x) = -\frac{1}{2\pi}\int_{\sigma_p} d\sigma_p[(V_0)^{\mu'\nu'}_{\alpha\beta}(x,x')F_{\mu'\nu'}(x')] + \frac{1}{2\pi}\int_{C_p} \mu_\Gamma(x')F_{\mu'\nu'}(x')(PU^{\mu'\nu'}_{\alpha\beta}(x,x'))$$

where the terms involving $(\Box'\Gamma(x,x')-4)$ have cancelled. Adding this result to the expression for ${}^{II}F^{\text{tail}}_{\alpha\beta}(x)$ and recalling Friedlander's formula for the measure $\mu_{t,\Gamma}(x')$ given by his Eq. (5.3.19),

$$(6.17)\qquad \langle\nabla t, \nabla\Gamma\rangle\mu_{t,\Gamma} = -d\sigma_p$$

one finds that the two remaining integrals in $F^{\text{tail}}_{\alpha\beta}(x)$ involving the non-local quantity $(V_0)^{\mu'\nu'}_{\alpha\beta}(x,x')$ also cancel leaving

$$(6.18)\qquad F^{\text{tail}}_{\alpha\beta}(x) = \frac{1}{2\pi}\int_{C_p} \mu_\Gamma(x')(F_{\mu'\nu'}(x')PU^{\mu'\nu'}_{\alpha\beta}(x,x'))$$

so that our expression for $F_{\alpha\beta}(x)$ (c.f. Eq. (5.1)) now becomes

$$(6.19)\qquad \begin{aligned} F_{\alpha\beta}(x) = {} & \frac{1}{2\pi}\int_{C_p} U^{\mu'\nu'}_{\alpha\beta}(x,x')f_{\mu'\nu'}(x')\mu_\Gamma(x') \\ & \frac{1}{2\pi}\int_{\sigma_p} \{U^{\mu'\nu'}_{\alpha\beta}(x,x')[2(\nabla^{\gamma'}t(x'))(\nabla_{\gamma'}F_{\mu'\nu'}(x')) \\ & + F_{\mu'\nu'}(x')\Box't(x')]\}\mu_{t,\Gamma}(x') \\ & + \frac{1}{2\pi}\int_{C_p} \mu_\Gamma(x')(F_{\mu'\nu'}(x')PU^{\mu'\nu'}_{\alpha\beta}(x,x')). \end{aligned}$$

The integral over σ_p in the above formula involves first derivatives of the unknown field $F_{\alpha\beta}$ but only on the initial, Cauchy hypersurface where these quantities must be given.

Upon substituting the explicit form for the source terms $f_{\mu'\nu'}(x')$ into Eq. (6.19) we shall encounter integrals of the type

$$(6.20)\qquad I = \frac{1}{2\pi}\int_{C_p} \mu_\Gamma(x')(\nabla^{\gamma'}\Omega_{\gamma'})$$

where $\Omega_{\gamma'}$ is a one-form which (thanks to its explicit dependence upon $\omega^{\hat{a}}{}_{\hat{b}\gamma'}$ which satisfies the Cronström gauge condition) obeys $\Gamma^{;\gamma'}\Omega_{\gamma'} = 0$ everywhere throughout the causal domain containing C_p. This special fact allows us to successfully integrate the 4-divergence over the 3-manifold C_p and obtain a boundary integral over σ_p. In deriving this result, we must compute derivatives of the equation $\Gamma^{;\gamma'}\Omega_{\gamma'} = 0$ in directions transversal to the cone C_p

so it is essential that this equation hold not just on C_p but (at least to first order) off the cone as well.

By introducing coordinates $\{\overline{x^\mu}\} = \{\bar{u}, \bar{v}, \bar{\theta}, \bar{\varphi}\}$ of the form

$$\bar{u} = \bar{u}(u, \theta, \varphi),\ \bar{v} = v,\ \bar{\theta} = \theta,\ \bar{\varphi} = \varphi \tag{6.21}$$

adapted to the domain of integration so that σ_p coincides with a surface $\bar{u} =$ constant lying in C_p one can carry out the integration explicitly to find that

$$I = \frac{1}{2\pi} \int\limits_{\sigma_p} d\sigma_p (\xi^\mu \Omega_\mu) \tag{6.22}$$

where, as before, $d\sigma_p$ is the invariant surface area element induced upon σ_p by the spacetime metric and in which $\xi^\mu \partial_\mu$ is a future pointing null vector, orthogonal to σ_p and normalized such that

$$\xi^\mu \Gamma_{;\mu} = 1. \tag{6.23}$$

In Friedlander's terminology, this vector is tangent to the null generators of the null surface T_p which contains σ_p. As we shall see, the boundary term arising in this way will combine naturally with the integral over σ_p in Eq. (6.19).

We now reinstate the heretofore inert indices on the curvature and its source by letting $F_{\mu\nu} \to R^{\hat{a}}{}_{\hat{b}\mu\nu}$ and $f_{\mu\nu} \to f^{\hat{a}}{}_{\hat{b}\mu\nu}$ so that Eq. (6.19) becomes

$$\begin{aligned} R^{\hat{a}}{}_{\hat{b}\alpha\beta}(x) = &-\frac{1}{2\pi} \int\limits_{C_p} (U^{\mu'\nu'}_{\alpha\beta}(x,x') f^{\hat{a}}{}_{\hat{b}\mu'\nu'}(x')) \mu_\Gamma(x') \\ &+ \frac{1}{2\pi} \int\limits_{\sigma_p} \{U^{\mu'\nu'}_{\alpha\beta}(x,x')[2(\nabla^{\gamma'} t(x'))(\nabla_{\gamma'} R^{\hat{a}}{}_{\hat{b}\mu'\nu'}(x')) \\ &+ R^{\hat{a}}{}_{\hat{b}\mu'\nu'}(x') \Box' t(x')]\} \mu_{t,\Gamma}(x') \\ &+ \frac{1}{2\pi} \int_{C_p} \mu_\Gamma(x') (R^{\hat{a}}{}_{\hat{b}\mu'\nu'}(x') (P U^{\mu'\nu'}_{\alpha\beta}(x,x'))). \end{aligned} \tag{6.24}$$

Upon inserting the explicit formula for $f^{\hat{a}}{}_{\hat{b}\mu'\nu'}$ from Eq. (2.18) and rewriting it slightly one finds that it contains the divergence integral

$$\begin{aligned} \mathcal{D}^{\hat{a}}{}_{\hat{b}\alpha\beta}(x) := \frac{1}{2\pi} \int\limits_{C_p} \mu_\Gamma(x') \{ \nabla^{\sigma'} [2\omega^{\hat{a}}{}_{\hat{c}\sigma'}(x') U^{\mu'\nu'}_{\alpha\beta}(x,x') R^{\hat{c}}{}_{\hat{b}\mu'\nu'}(x') \\ - 2\omega^{\hat{c}}{}_{\hat{b}\sigma'} U^{\mu'\nu'}_{\alpha\beta}(x,x') R^{\hat{a}}{}_{\hat{c}\mu'\nu'}(x')] \} \end{aligned} \tag{6.25}$$

which includes the only terms in the integrals over C_p which contain derivatives of curvature. Exploiting the argument above to reduce this expression to an integral over σ_p one finds that

$$\mathcal{D}^{\hat{a}}{}_{\hat{b}\alpha\beta}(x) = \frac{1}{2\pi}\int_{\sigma_p} d\sigma_p\{\xi^{\sigma'}[2\omega^{\hat{a}}{}_{\hat{c}\sigma'}(x')U^{\mu'\nu'}_{\alpha\beta}(x,x')R^{\hat{c}}{}_{\hat{b}\mu'\nu'}(x') - 2\omega^{\hat{c}}{}_{\hat{b}\sigma'}(x')U^{\mu'\nu'}_{\alpha\beta}(x,x')R^{\hat{a}}{}_{\hat{c}\mu'\nu'}(x')]\}. \tag{6.26}$$

The remaining integral over σ_p in Eq. (6.24) can be reexpressed, thanks to Eq. (6.17) as

$$\begin{aligned}\mathcal{S}^{\hat{a}}{}_{\hat{b}\alpha\beta}(x) &:= \frac{1}{2\pi}\int_{\sigma_p}\{U^{\mu'\nu'}_{\alpha\beta}(x,x')[2(\nabla^{\gamma'}t(x'))(\nabla_{\gamma'}R^{\hat{a}}{}_{\hat{b}\mu'\nu'}(x')) \\ &\quad + R^{\hat{a}}{}_{\hat{b}\mu'\nu'}(x')\Box't(x')]\}\mu_{t,\Gamma}(x') \\ &= \frac{1}{2\pi}\int_{\sigma_p} d\sigma_p\Big\{2U^{\mu'\nu'}_{\alpha\beta}(x,x')(\xi^{\sigma'}\nabla_{\sigma'}R^{\hat{a}}{}_{\hat{b}\mu'\nu'}(x')) \\ &\quad + U^{\mu'\nu'}_{\alpha\beta}(x,x')R^{\hat{a}}{}_{\hat{b}\mu'\nu'}(x,x')\left(\frac{\Box't(x')}{(\langle\nabla't,\nabla'\Gamma\rangle(x'))}\right)\Big\}.\end{aligned} \tag{6.27}$$

Defining (via Friedlander's Eqs. (5.3.7) and (5.3.20)) the dilation θ of $d\sigma_p$ along the bicharacteristics of T_p by

$$\theta(x') = \frac{\Box't(x')}{\langle\nabla't,\nabla'\Gamma\rangle(x')} \tag{6.28}$$

and combining the integrals $\mathcal{D}^{\hat{a}}{}_{\hat{b}\alpha\beta}(x)$ and $\mathcal{S}^{\hat{a}}{}_{\hat{b}\alpha\beta}(x)$ one gets

$$\begin{aligned}&\mathcal{D}^{\hat{a}}{}_{\hat{b}\alpha\beta}(x) + \mathcal{S}^{\hat{a}}{}_{\hat{b}\alpha\beta}(x) \\ &\quad = \frac{1}{2\pi}\int_{\sigma_p} d\sigma_p\{2U^{\mu'\nu'}_{\alpha\beta}(x,x')(\xi^{\sigma'}D_{\sigma'}R^{\hat{a}}{}_{\hat{b}\mu'\nu'}(x')) \\ &\qquad + U^{\mu'\nu'}_{\alpha\beta}(x,x')R^{\hat{a}}{}_{\hat{b}\mu'\nu'}(x')\theta(x')\}\end{aligned} \tag{6.29}$$

where now $D_{\sigma'}$ is the *total* spacetime covariant derivative defined in Section 2. The addition of $\mathcal{D}^{\hat{a}}{}_{\hat{b}\alpha\beta}$ to $\mathcal{S}^{\hat{a}}{}_{\hat{b}\alpha\beta}$ has contributed precisely the terms needed to convert $\nabla_{\sigma'}$ to $D_{\sigma'}$ in the formula above.

Writing out the factor $PU^{\mu'\nu'}_{\alpha\beta}(x,x')$ more explicitly as

$$
\begin{aligned}
(6.30)\qquad & PU^{\mu'\nu'}_{\alpha\beta}(x,x') \\
&= \nabla^{\gamma'}\nabla_{\gamma'}U^{\mu'\nu'}_{\alpha\beta}(x,x') \\
&\quad + R^{\mu'\nu'}{}_{\delta'\gamma'}(x')U^{\delta'\gamma'}_{\alpha\beta}(x,x') \\
&= \left(\frac{\nabla^{\gamma'}\nabla_{\gamma'}\kappa(x,x')}{\kappa(x,x')}\right)U^{\mu'\nu'}_{\alpha\beta}(x,x') + R^{\mu'\nu'}{}_{\delta'\gamma'}(x')U^{\delta'\gamma'}_{\alpha\beta}(x,x') \\
&\quad + 2(\nabla^{\gamma'}\kappa(x,x'))(\nabla_{\gamma'}\tau^{\mu'\nu'}_{\alpha\beta}(x,x')) \\
&\quad + \kappa(x,x')(\nabla^{\gamma'}\nabla_{\gamma'}(\tau^{\mu'\nu'}_{\alpha\beta}(x,x'))),
\end{aligned}
$$

where $\tau^{\mu'\nu'}_{\alpha\beta}(x,x')$ is defined via Eq. (6.6), one can evaluate the derivatives of $\tau^{\mu'\nu'}_{\alpha\beta}(x,x')$ using Eqs. (A.10) - (A.13) which yield

$$
(6.31)\qquad h^{\gamma}{}_{\hat{a};\nu} = h^{\gamma}{}_{\hat{c}}\omega^{\hat{c}}{}_{\hat{a}\nu}
$$

so that

$$
\begin{aligned}
(6.32)\qquad \nabla_{\gamma'}\tau^{\mu'\nu'}_{\alpha\beta}(x,x') &= \omega^{\hat{d}}{}_{\hat{e}\gamma'}h^{\mu'}_{\hat{d}}(x')\theta^{\hat{e}}_{\alpha}(x)h^{\nu'}_{\hat{f}}(x')\theta^{\hat{f}}_{\beta}(x) \\
&\quad + \omega^{\hat{d}}{}_{\hat{f}\gamma'}h^{\mu'}_{\hat{e}}(x')\theta^{\hat{e}}_{\alpha}(x)h^{\nu'}_{\hat{d}}(x')\theta^{\hat{f}}_{\beta}(x)
\end{aligned}
$$

with a similar expanded formula for $\nabla^{\gamma'}(\nabla_{\gamma'}\tau^{\mu'\nu'}_{\alpha\beta}(x,x'))$. The latter will clearly entail factors of the type $(\nabla^{\gamma'}\omega^{\hat{d}}{}_{\hat{e}\gamma'})$ as well as factors quadratic in the connection coefficients $\omega^{\hat{d}}{}_{\hat{e}\gamma'}$. Written out explicitly it becomes:

$$
\begin{aligned}
(6.33)\qquad & \nabla^{\gamma'}\nabla_{\gamma'}\tau^{\mu'\nu'}_{\alpha\beta}(x,x') \\
&= (\nabla^{\gamma'}\omega^{\hat{d}}{}_{\hat{e}\gamma'})h^{\mu'}_{\hat{d}}(x')\theta^{\hat{e}}_{\alpha}(x)h^{\nu'}_{\hat{f}}(x')\theta^{\hat{f}}_{\beta}(x) \\
&\quad + (\nabla^{\gamma'}\omega^{\hat{d}}{}_{\hat{f}\gamma'})h^{\mu'}_{\hat{e}}(x')\theta^{\hat{e}}_{\alpha}(x)h^{\nu'}_{\hat{d}}(x')\theta^{\hat{f}}_{\beta}(x) \\
&\quad \omega^{\hat{d}}{}_{\hat{e}\gamma'}g^{\gamma'\sigma'}(x')[h^{\mu'}_{\hat{c}}(x')\omega^{\hat{c}}{}_{\hat{d}\sigma'}(x')h^{\nu'}_{\hat{f}}(x') \\
&\quad + h^{\mu'}_{\hat{d}}(x')\omega^{\hat{c}}{}_{\hat{f}\sigma'}(x')h^{\nu'}_{\hat{c}}(x')]\theta^{\hat{e}}_{\alpha}(x)\theta^{\hat{f}}_{\beta}(x) \\
&\quad + \omega^{\hat{d}}{}_{\hat{f}\gamma'}g^{\gamma'\sigma'}(x')[h^{\mu'}_{\hat{c}}(x')\omega^{\hat{c}}{}_{\hat{e}\sigma'}(x')h^{\nu'}_{\hat{d}}(x') \\
&\quad + h^{\mu'}_{\hat{e}}(x')\omega^{\hat{c}}{}_{\hat{d}\sigma'}(x')h^{\nu'}_{\hat{c}}(x')]\theta^{\hat{e}}_{\alpha}(x)\theta^{\hat{f}}_{\beta}(x).
\end{aligned}
$$

Assembling the various pieces of the formula for $R^{\hat{a}}{}_{\hat{b}\alpha\beta}(x)$ we thus get:

$$
\begin{aligned}
(6.34)\qquad & R^{\hat{a}}{}_{\hat{b}\alpha\beta}(x) \\
&= \frac{1}{2\pi}\int\limits_{C_p}\mu_{\Gamma}(x')\{U^{\delta'\gamma'}_{\alpha\beta}(x,x')[-2R^{\hat{a}}{}_{\hat{c}\delta'\sigma'}(x')R^{\hat{c}}{}_{\hat{b}\gamma'}{}^{\sigma'}(x')
\end{aligned}
$$

$$
\begin{aligned}
&+ 2R^{\hat{a}}{}_{\hat{c}\gamma'\sigma'}(x')R^{\hat{c}}{}_{\hat{b}\delta'}{}^{\sigma'}(x') + R^{\hat{a}}{}_{\hat{b}\mu'\nu'}(x')R_{\delta'\gamma'}{}^{\mu'\nu'}(x')]\} \\
&+ \frac{1}{2\pi}\int\limits_{C_p} \mu_\Gamma(x')\{U^{\mu'\nu'}_{\alpha\beta}(x,x') \\
&\quad \cdot [g^{\lambda'\sigma'}(x')\{\omega^{\hat{a}}{}_{\hat{c}\sigma'}[\omega^{\hat{c}}{}_{\hat{d}\lambda'}R^{\hat{d}}{}_{\hat{b}\mu'\nu'}(x') - R^{\hat{c}}{}_{\hat{d}\mu'\nu'}\omega^{\hat{d}}{}_{\hat{b}\lambda'}(x')] \\
&\quad - [\omega^{\hat{a}}{}_{\hat{d}\lambda'}R^{\hat{d}}{}_{\hat{c}\mu'\nu'}(x') - R^{\hat{a}}{}_{\hat{d}\mu'\nu'}\omega^{\hat{d}}{}_{\hat{c}\lambda'}(x')]\omega^{\hat{c}}{}_{\hat{b}\sigma'}(x')\} \\
&\quad - [(\nabla^{\sigma'}\omega^{\hat{a}}{}_{\hat{c}\sigma'})R^{\hat{c}}{}_{\hat{b}\mu'\nu'}(x') - R^{\hat{a}}{}_{\hat{c}\mu'\nu'}(\nabla^{\sigma'}\omega^{\hat{c}}{}_{\hat{b}\sigma'})]]\} \\
&+ \frac{1}{2\pi}\int\limits_{C_p} \mu_\Gamma(x')\{-2(\nabla^{\sigma'}U^{\mu'\nu'}_{\alpha\beta}(x,x'))\omega^{\hat{a}}{}_{\hat{c}\sigma'}(x')R^{\hat{c}}{}_{\hat{b}\mu'\nu'}(x') \\
&\quad + 2(\nabla^{\sigma'}U^{\mu'\nu'}_{\alpha\beta}(x,x'))R^{\hat{a}}{}_{\hat{c}\mu'\nu'}(x')\omega^{\hat{c}}{}_{\hat{b}\sigma'}(x')\} \\
&+ \frac{1}{2\pi}\int\limits_{\sigma_p} d\sigma_p\{2U^{\mu'\nu'}_{\alpha\beta}(x,x')(\xi^{\sigma'}D_{\sigma'}R^{\hat{a}}{}_{\hat{b}\mu'\nu'}(x')) \\
&\quad + U^{\mu'\nu'}_{\alpha\beta}(x,x')R^{\hat{a}}{}_{\hat{b}\mu'\nu'}(x')\theta(x')\} \\
&+ \frac{1}{2\pi}\int\limits_{C_p} \mu_\Gamma(x')\{R^{\hat{a}}{}_{\hat{b}\mu'\nu'}[(\nabla^{\gamma'}\nabla_{\gamma'}\kappa(x,x'))\tau^{\mu'\nu'}_{\alpha\beta}(x,x') \\
&\quad + 2(\nabla^{\gamma'}\kappa(x,x'))(\nabla_{\gamma'}\tau^{\mu'\nu'}_{\alpha\beta}(x,x')) \\
&\quad + \kappa(x,x')(\nabla^{\gamma'}\nabla_{\gamma'}\tau^{\mu'\nu'}_{\alpha\beta}(x,x'))]\}
\end{aligned}
$$

where, of course, the factors involving

$$
\nabla^{\sigma'}U^{\mu'\nu'}_{\alpha\beta}(x,x') = (\nabla^{\sigma'}\kappa(x,x'))\tau^{\mu'\nu'}_{\alpha\beta}(x,x') + \kappa(x,x')\nabla^{\sigma'}\tau^{\mu'\nu'}_{\alpha\beta}(x,x') \tag{6.35}
$$

can be expanded out as in the foregoing paragraph.

In this explicit form the result seems quite complicated but it is straightforward to reexpress it as

$$
\begin{aligned}
&R^{\alpha}{}_{\beta\gamma\delta}(x) \\
&= \frac{1}{2\pi}\int\limits_{C_p} \mu_\Gamma(x')\{\nabla^{\kappa'}\nabla_{\kappa'}(U^{\alpha\nu'\rho'\sigma'}_{\mu'\beta\gamma\delta}(x,x'))R^{\mu'}{}_{\nu'\rho'\sigma'}(x') \\
&\quad + U^{\alpha\nu'\rho'\sigma'}_{\mu'\beta\gamma\delta}(x,x')[R^{\lambda'\xi'}{}_{\rho'\sigma'}(x')R_{\lambda'\xi'}{}^{\mu'}{}_{\nu'}(x') \\
&\quad - 2R^{\mu'}{}_{\lambda'\sigma'\xi'}(x')R_{\nu'}{}^{\lambda'}{}_{\rho'}{}^{\xi'}(x') \\
&\quad + 2R^{\mu'}{}_{\lambda'\rho'\xi'}(x')R_{\nu'}{}^{\lambda'}{}_{\sigma'}{}^{\xi'}(x')]\} \\
&\quad + \frac{1}{2\pi}\int\limits_{\sigma_p} d\mu_\sigma\{U^{\alpha\,\nu'\rho'\sigma'}_{\mu'\,\beta\gamma\delta}(x,x')[2\xi^{\lambda'}(\nabla_{\lambda'}R^{\mu'}{}_{\nu'\rho'\sigma'}(x')) \\
&\qquad + R^{\mu'}{}_{\nu'\rho'\sigma'}(x')\theta(x')]\}
\end{aligned}
\tag{6.36}
$$

where

$$(6.37)\qquad U^{\alpha\,\nu'\rho'\sigma'}_{\mu'\,\beta\gamma\delta}(x,x') = \kappa(x,x')\theta^{\hat{a}}_{\mu'}(x')h^{\alpha}_{\hat{a}}(x)h^{\nu'}_{\hat{b}}(x')\theta^{\hat{b}}_{\beta}(x)h^{\rho'}_{\hat{c}}(x')\theta^{\hat{c}}_{\gamma}(x)h^{\sigma'}_{\hat{d}}(x')\theta^{\hat{d}}_{\delta}(x)$$

the parallel propagator for tensors of type $\binom{1}{3}$. Equation (6.36) can be derived much more directly by simply applying the Friedlander formalism to the wave equation (2.7) for curvature treated as a 4-th rank tensor and then proceeding as above to recast the tail terms in the representation formula in terms of integrals over C_p which can in turn be simplified by the methods of the present section.

However, we have already emphasized the potential usefulness of the Cartan formulation in carrying out the sought-after light cone estimates for curvature because of its close resemblance to the integral equation for curvature arising in Yang-Mills theory. In references [**5**] and [**2**] it was necessary to express the integral equation for (Yang-Mills) curvature in the form analogous to Eq. (6.34) above in order to exploit the Cronström gauge conditions and derive bounds on the curvature tensor. Thus we anticipate that the expanded form of the integral expression for gravitational curvature, given by Eq. (6.34), will play an important role in subsequent work to derive estimates for the spacetime curvature of a solution to Einstein's equations.

7. Approximate Quasi-Local Killing and Conformal Killing Fields

As is well-known the Bel-Robinson tensor for a vacuum spacetime can be used to construct a conserved positive definite "energy" (essentially an L^2-norm of spacetime curvature) for any timelike Killing or conformal Killing field admitted by the metric. This follows from exploiting its total symmetry as a 4-th rank tensor and the vanishing of its divergence and trace in much the same way that one can use the (trace-free) stress energy tensor of a matter field to construct the conserved energy associated to a Killing or conformal Killing field of the "background". Except for "test" matter fields propagating on a stationary or self-similar background however this observation is of little value in practice since the imposition of a Killing or conformal Killing symmetry is far too restrictive a condition to enforce on physically interesting gravitational fields.

On the other hand it may not be necessary to have a strictly conserved energy in order to get adequate analytical control of some mathematically relevant energy norm. For example, in their treatment of Yang-Mills fields propagating in a background spacetime, Chruściel and Shatah exploited the observation that the (gauge invariant, positive definite) L^2-norm of Yang-Mills curvature cannot blow up until the spacetime itself blows up (through becoming singular or developing a Cauchy horizon at its boundary) [**2**]. This fact, which follows from the vanishing of the divergence of the Yang-Mills stress energy tensor and the fact that its components are pointwise bounded by the energy density, was essentially as useful in practice as a fully conserved energy would have been had it existed. When the spacetime itself though is the object of dynamical study this argument (applied to the Bel-Robinson tensor) is of less interest since it requires pointwise control of the connection to yield a mere L^2 bound on the curvature and there is no a priori reason for the Christoffel components to be so bounded.

For this reason it seems potentially useful, especially in the gravitational case, to look for approximate Killing or conformal Killing fields, in a general spacetime, that could in turn be employed to construct corresponding approximately conserved energies. With this in mind we show below that the orthonormal frame fields $\{h^{\mu}_{\hat{a}} \frac{\partial}{\partial x^{\mu}}\}$ defined, as in Section 3, by parallel propagation of a fixed frame at a point p along the radial geodesics issuing from that point, satisfy Killing's equations in an approximate sense that becomes more and more exact (at a well-defined rate) as one approaches the point p along an arbitrary radial geodesic. The error term, or so-called deformation tensor, which measures precisely the failure of Killing's equations to be satisfied, will be shown to be explicitly expressible in terms of radial integrals of spacetime curvature which vanish linearly (in normal coordinates centered at point p) as one approaches this vertex radially.

In a similar way we shall show that the gradient, $\nabla\Gamma$, of the "optical function" Γ (representing squared geodesic distance from the vertex p) satisfies an approximate form of the conformal (in fact homothetic) Killing equations with an error term that vanishes quadratically (in terms of normal coordinates) as one approaches p radially. Both $\nabla\Gamma$ and any timelike linear combination of the $\{h^{\mu}_{\hat{a}} \frac{\partial}{\partial x^{\mu}}\}$ provide timelike vector fields inside the past lightcone from point p (and restricted to a causal domain of p) and thus allow the definition of corresponding positive definite and approximately conserved energy expressions for curvature inside this past lightcone. The timelike character of a frame field such as $\{h^{\mu}_{\hat{0}} \frac{\partial}{\partial x^{\mu}}\}$ is of course not confined to the interior of the cone and its associated energy is therefore positive definite throughout the causal domain in which it remains well-defined.

These approximate Killing and conformal Killing fields should perhaps (for lack of a better term) be called *quasi-local* since they only approach satisfaction of the relevant Killing equations as one approaches the preferred vertex that was used in their construction. The potential (quasi-local) application that we have in mind for such objects can be described loosely as follows. Suppose that some future directed timelike geodesic γ approaches a singular boundary point for the spacetime under study and that we wish to derive bounds on the rate at which curvature can blow up as γ nears its (singular) endpoint. For each point p lying on γ we can construct the past lightcone from p and parallel propagate the (unit, timelike) tangent to γ at p throughout a causal domain for p to get a timelike, approximate Killing field of the type described above (which will however vary with the choice of the "moving" point p). By exploiting the associated approximately conserved energy we might reasonably hope to estimate (with some controllable error) the energy flux through the past light cone from p, back to some "initial" hypersurface, in terms on the energy defined (by an integral over the ball bounded by the intersection of the light cone with this surface) on this initial "slice". Since control of these (Bel-Robinson) energy fluxes is sure to play a vital role in carrying out the light cone estimates we propose to derive later, the possibility of bounding them in terms of initial data is sure to provide a key step in the hoped-for argument to bound curvature pointwise in terms of its L^2-norms.

If, as in Section 3, $\{h^{\nu}_{\hat{b}} \frac{\partial}{\partial x^{\nu}}\}$ is an orthonormal frame field constructed by parallel propagation of a fixed frame at p along the radial geodesics spraying out from p to fill out a causal domain of this point, then the corresponding co-frame field $\{\theta^{\hat{a}}_{\mu} dx^{\mu}\}$ is given by $\theta^{\hat{a}}_{\mu} = \eta^{\hat{a}\hat{b}} g_{\mu\nu} h^{\nu}_{\hat{b}}$. Using the defining formula for the connection coefficients $\omega^{\hat{a}}{}_{\hat{b}\nu}$,

$$\theta^{\hat{a}}_{\mu,\nu} - \Gamma^{\lambda}_{\mu\nu}\theta^{\hat{a}}_{\lambda} = \theta^{\hat{a}}_{\mu;\nu} = -\omega^{\hat{a}}{}_{\hat{b}\nu}\theta^{\hat{b}}_{\mu} \tag{7.1}$$

one computes the Killing form of $\{\theta^{\hat{a}}_{\mu}\}$ to find

$$\theta^{\hat{a}}_{\mu;\nu} + \theta^{\hat{a}}_{\nu;\mu} = -\omega^{\hat{a}}{}_{\hat{b}\nu}\theta^{\hat{b}}_{\mu} - \omega^{\hat{a}}{}_{\hat{b}\mu}\theta^{\hat{b}}_{\nu} \tag{7.2}$$

with the right hand side representing the error for Killing's equation. The frame fields approach a fixed orthonormal (co-) frame at the vertex point p but the connection components satisfy the "Cronström" formula given (taking $x^{\mu}(p) = 0$) by

$$\omega^{\hat{c}}{}_{\hat{a}\mu}(x) = -\int_0^1 d\lambda \; \lambda x^{\nu} R^{\hat{c}}{}_{\hat{a}\mu\nu}(\lambda x) \tag{7.3}$$

and thus vanish to order $0(x)$, for any metric with pointwise bounded curvature, as one approaches the vertex along a radial geodesic. A key observation, from our point of view, is that only undifferentiated curvature enters into this equation for the error. By contrast one can show that the coordinate basis fields $\{\frac{\partial}{\partial x^{\mu}}\}$ (of a normal coordinate system based at p, with $x^{\mu}(p) = 0$) also satisfy Killing's equations approximately, with an error that vanishes linearly with the $\{x^{\mu}\}$, but, in this case, we do not have a formula for the error that depends only upon undifferentiated curvature (though it is conceivable that one exists). Thus we are inclined to strongly prefer the parallel propagated frame fields as natural candidates for our quasi-local approximate Killing fields. Though not commuting in general (as the coordinate basis fields would of course do) these fields nevertheless satisfy an approximate Lie algebra relation, with linearly vanishing error terms, since their commutator is given by

$$h^{\mu}_{\hat{a}} h^{\nu}_{\hat{b};\mu} - h^{\mu}_{\hat{b}} h^{\nu}_{\hat{a};\mu} = [h_{\hat{a}}, h_{\hat{b}}]^{\nu} = h^{\nu}_{\hat{f}}[h^{\mu}_{\hat{a}}\, \omega^{\hat{f}}{}_{\hat{b}\mu} - h^{\mu}_{\hat{b}}\, \omega^{\hat{f}}{}_{\hat{a}\mu}]. \tag{7.4}$$

Now, consider the "optical" function Γ, introduced in Section 4, and its gradient $\nabla\Gamma$ which, in normal coordinates, satisfies

$$(\nabla\Gamma)^\beta = \Gamma^{;\beta} = g^{\alpha\beta}\Gamma_{,\alpha} = 2g^{\alpha\beta}(x)g_{\alpha\nu}(x)x^\nu = 2x^\beta. \tag{7.5}$$

One expects that $\nabla\Gamma = 2x^\beta\frac{\partial}{\partial x^\beta}$ should generalize the well-known, corresponding homothetic Killing field of Minkowski space and indeed, by construction, this vector field is timelike inside the lightcone from p, null on the cone itself and spacelike outside since, in general we have

$$\langle\nabla\Gamma, \nabla\Gamma\rangle_g = g^{\alpha\beta}\Gamma_{,\alpha}\Gamma_{,\beta} = 4\Gamma \tag{7.6}$$

and Γ represents the squared geodesic distance from the cone vertex p.

Computing the Killing and conformal Killing forms for $\nabla\Gamma$ one gets

$$\begin{aligned}
&\Gamma_{;\alpha\beta} + \Gamma_{;\beta\alpha} = 4g_{\alpha\beta} + 2x^\nu g_{\alpha\beta,\nu},\\
&\Gamma_{;\alpha\beta} + \Gamma_{;\alpha\beta} - \frac{1}{2}g_{\alpha\beta}g^{\mu\nu}\Gamma_{;\mu\nu}\\
&\quad = 2x^\nu g_{\alpha\beta,\nu} - \frac{1}{2}g_{\alpha\beta}g^{\gamma\delta}(x^\nu g_{\gamma\delta,\nu})
\end{aligned} \tag{7.7}$$

where the error term on the right hand side of the last equation is simply the trace free part of $2x^\nu g_{\alpha\beta,\nu}$ (evaluated in normal coordinates). This latter quantity can be calculated using the same transport formula (derived from the zero torsion equation) that we used in Section 3 to express the frame field in terms of the connection. The result is

$$\begin{aligned}
&x^\beta g_{\mu\nu,\beta}(x)\\
&= \eta_{\hat{a}\hat{b}}\Bigg\{\theta^{\hat{b}}_\nu(x)\left[\omega^{\hat{a}}{}_{\hat{f}\mu}(x)(x^\gamma\theta^{\hat{f}}_\gamma(0)) - \int_0^1 d\lambda[\omega^{\hat{a}}{}_{\hat{f}\mu}(\lambda x)(\lambda x^\gamma\theta^{\hat{f}}_\gamma(0))]\right]\\
&\quad + \theta^{\hat{a}}_\mu(x)\left[\omega^{\hat{b}}{}_{\hat{f}\nu}(x)(x^\gamma\theta^{\hat{f}}_\gamma(0)) - \int_0^1 d\lambda[\omega^{\hat{b}}{}_{\hat{f}\nu}(\lambda x)(\lambda x^\gamma\theta^{\hat{f}}_\gamma(0))]\right]\Bigg\}
\end{aligned} \tag{7.8}$$

wherein $\theta^{\hat{b}}_\nu(x)$ and $\omega^{\hat{b}}{}_{\hat{f}\nu}(x)$ are given explicitly in terms of integrals of curvature by Eqs. (3.21) and (3.12). Thus in this case the error term vanishes quadratically with the normal coordinates as one approaches the vertex at $x^\mu(p) = 0$ though here of course the vector field itself, $\nabla\Gamma = 2x^\beta\frac{\partial}{\partial x^\beta}$ vanishes linearly. The divergence of this approximate conformal Killing field is given, through the trace of the first of Eqs. (7.7), by

$$\begin{aligned}
\Gamma^{;\alpha}{}_{;\alpha} &= 8 + x^\nu(g^{\alpha\beta}g_{\alpha\beta,\nu})\\
&= 8 + 2x^\nu\frac{(\sqrt{-\det g})_{,\nu}}{\sqrt{-\det g}}
\end{aligned} \tag{7.9}$$

which coincides with a well-known equation for the d'Alembertian of Γ given by Friedlander [**20**]. Thus the divergence is constant up to a quadratically vanishing error which suggests that we regard $\nabla\Gamma$ as approximately homothetic.

In Minkowski space, the vector fields $\{h^{\mu}_{\hat{a}}\frac{\partial}{\partial x^{\mu}}, \Gamma^{;\mu}\frac{\partial}{\partial x^{\mu}}\}$ form a Lie subalgebra of the algebra of generators of the conformal group. Here of course this algebra can at most be approximate but, for completeness, we compute the remaining commutators of $\nabla\Gamma$ with the frame fields $\{h^{\mu}_{\hat{a}}\frac{\partial}{\partial x^{\mu}}\}$. The Lie brackets are given initially by

$$[h_{\hat{a}}, \nabla\Gamma]^{\nu} = h^{\mu}_{\hat{a}}\Gamma^{;\nu}{}_{;\mu} - \Gamma^{;\mu}h^{\nu}_{\hat{a};\mu} \tag{7.10}$$

but we can simplify this by noting that the equations

$$\begin{aligned} h^{\nu}_{\hat{a};\mu} &= \omega^{\hat{f}}{}_{\hat{a}\mu}h^{\nu}_{\hat{f}}, \\ \Gamma^{;\beta} &= 2x^{\beta} \end{aligned} \tag{7.11}$$

together with the parallel propagation gauge condition, $x^{\mu}\omega^{\hat{f}}{}_{\hat{a}\mu} = 0$, imply that,

$$\Gamma^{;\mu}h^{\nu}_{\hat{a};\mu} = 0 \tag{7.12}$$

and thus (using Eq. (7.9)) that

$$\begin{aligned} [h_{\hat{a}}, \nabla\Gamma]^{\nu} &= h^{\mu}_{\hat{a}}g^{\lambda\nu}\Gamma_{;\lambda\mu} \\ &= h^{\mu}_{\hat{a}}g^{\lambda\nu}(2g_{\lambda\mu} + x^{\beta}g_{\lambda\mu,\beta}) \\ &= 2h^{\nu}_{\hat{a}} + h^{\mu}_{\hat{a}}g^{\lambda\nu}(x^{\beta}g_{\lambda\mu,\beta}). \end{aligned} \tag{7.13}$$

Hence we recover the flat space result up to a quadratically vanishing "error" in the would-be Lie algebra. Though we did not need it to derive the foregoing results, it is useful to note that

$$x^{\nu}\Gamma^{\beta}_{\mu\nu} = \frac{1}{2}g^{\alpha\beta}(x^{\nu}g_{\mu\alpha,\nu}) \tag{7.14}$$

which follows from the normal coordinate identity $g_{\mu\nu}(x)x^{\nu} = g_{\mu\nu}(0)x^{\nu}$ by differentiating to get

$$x^{\nu}g_{\mu\nu,\alpha}(x) = g_{\mu\alpha}(0) - g_{\mu\alpha}(x) \tag{7.15}$$

and then antisymmetrizing in μ and α to arrive at

$$x^{\nu}(g_{\mu\nu,\alpha}(x) - g_{\alpha\nu,\mu}(x)) = 0. \tag{7.16}$$

Without this result, the direct calculation of $\Gamma_{;\alpha\beta}$, beginning with $\Gamma_{;\alpha} = 2g_{\alpha\nu}(x)x^{\nu}$, would not yield a symmetric formula in α and β as it must.

While one could continue along the above lines and define approximate Killing and conformal Killing analogues for Lorentz rotation, boost and inversion generators with formulas like $x^1h^{\nu}_{\hat{2}}\frac{\partial}{\partial x^{\nu}} - x^2h^{\nu}_{\hat{1}}\frac{\partial}{\partial x^{\nu}}$, $x^0h^{\nu}_{\hat{1}}\frac{\partial}{\partial x^{\nu}} + x^1h^{\nu}_{\hat{0}}\frac{\partial}{\partial x^{\nu}}$, etc., these would not be timelike throughout the regions (interiors of past light cones from vertices with $x^{\mu}(p) = 0$) of interest and so would not yield positive energy expressions. While their approximate conformal Lie algebra relations might be of interest to develop, we shall not pursue that issue here.

Appendix A. Notation, conventions and basic definitions

Much of our analysis will be carried out in rather specially chosen charts and associated frames. For the moment however, to introduce the notation that we shall use throughout, we consider an arbitrary chart and an arbitrary (orthonormal) frame. In coordinates $\{x^\mu\}$ defined on some domain of our spacetime manifold V we write the Lorentzian metric g in the standard form

$$g = g_{\mu\nu} dx^\mu \otimes dx^\nu \tag{A.1}$$

and introduce an orthonormal frame $\{h_{\hat{a}}\} = \{h^\mu_{\hat{a}} \frac{\partial}{\partial x^\mu}\}$ and dual, coframe $\{\theta^{\hat{a}}\} = \{\theta^{\hat{a}}_\mu dx^\mu\}$ for this (locally expressed) metric. The orthonormality and duality relations satisfied by these fields are summarized as follows:

$$\begin{aligned}
h_{\hat{a}} = h^\mu_{\hat{a}} \frac{\partial}{\partial x^\mu},\ \ \theta^{\hat{a}} = \theta^{\hat{a}}_\mu dx^\mu \ \ &\text{coordinate basis expression}\\
g_{\mu\nu} h^\mu_{\hat{a}} h^\nu_{\hat{b}} = \eta_{\hat{a}\hat{b}},\ \ g^{\mu\nu}\theta^{\hat{a}}_\mu \theta^{\hat{b}}_\nu = \eta^{\hat{a}\hat{b}} \ \ &\text{orthonormality relations}\\
\theta^{\hat{a}}_\mu h^\mu_{\hat{b}} = \delta^{\hat{a}}_{\hat{b}},\ \ \theta^{\hat{a}}_\mu h^\nu_{\hat{a}} = \delta^\nu_\mu \ \ &\text{duality relations}\\
g_{\mu\nu} = \eta_{\hat{a}\hat{b}} \theta^{\hat{a}}_\mu \theta^{\hat{b}}_\nu,\ \ g^{\mu\nu} = \eta^{\hat{a}\hat{b}} h^\mu_{\hat{a}} h^\nu_{\hat{b}} \ \ &\text{metric formulas}\\
\frac{\partial}{\partial x^\nu} = \theta^{\hat{a}}_\nu h_{\hat{a}},\ \ dx^\nu = h^\nu_{\hat{a}} \theta^{\hat{a}} \ \ &\text{coordinate basis vectors and forms}\\
\theta^{\hat{a}}_\mu = \eta^{\hat{a}\hat{b}} g_{\mu\nu} h^\nu_{\hat{b}},\ \ h^\mu_{\hat{a}} = \eta_{\hat{a}\hat{b}} g^{\mu\nu} \theta^{\hat{b}}_\nu \ \ &\text{component relations.}
\end{aligned} \tag{A.2}$$

Here $(\eta_{\hat{a}\hat{b}})$ is the standard Minkowski metric

$$\eta = (\eta_{\hat{a}\hat{b}}) = \begin{pmatrix} -1 & & & 0 \\ & 1 & & \\ & & 1 & \\ 0 & & & 1 \end{pmatrix} \tag{A.3}$$

and $\eta^{-1} = (\eta^{\hat{a}\hat{b}})$ is its inverse. Many of the formulas we shall derive in this section hold true for arbitrary spacetime dimensions and also for Riemannian metrics instead of Lorentzian ones if η is replaced by a Euclidean metric. Tensors are expressed in coordinate and orthonormal bases as

$$\begin{aligned}
\mathcal{S} &= \mathcal{S}^{\mu\nu\ldots}{}_{\gamma\delta\ldots} \frac{\partial}{\partial x^\mu} \otimes \frac{\partial}{\partial x^\nu} \otimes \ldots \otimes dx^\gamma \otimes dx^\delta \otimes \ldots \\
&= \mathcal{S}^{\hat{a}\hat{b}\ldots}{}_{\hat{e}\hat{f}\ldots} h_{\hat{a}} \otimes h_{\hat{b}} \otimes \ldots \otimes \theta^{\hat{e}} \otimes \theta^{\hat{f}} \otimes \ldots
\end{aligned} \tag{A.4}$$

with components related by

$$\mathcal{S}^{\hat{a}\hat{b}\ldots}{}_{\hat{e}\hat{f}}\ \cdots = \mathcal{S}^{\mu\nu\ldots}{}_{\gamma\delta\ldots} \theta^{\hat{a}}_\mu \theta^{\hat{b}}_\nu \ldots h^\gamma_{\hat{e}} h^\delta_{\hat{f}} \ldots. \tag{A.5}$$

In particular, the metric g and its inverse g^{-1} take the forms

$$g = \eta_{\hat{a}\hat{b}} \theta^{\hat{a}} \otimes \theta^{\hat{b}},\ \ g^{-1} = \eta^{\hat{a}\hat{b}} h_{\hat{a}} \otimes h_{\hat{b}}. \tag{A.6}$$

For all differentiable tensor fields, we have the conventional (coordinates basis) expressions for the covariant derivatives of scalar, vector and one-form (or co-vector) fields respectively given by

$$\text{(A.7)} \qquad \begin{aligned} \varphi_{;\alpha} &= \varphi_{,\alpha} = \frac{\partial \varphi}{\partial x^{\alpha}} \quad \text{scalar} \\ v^{\mu}_{;\nu} &= v^{\mu}_{,\nu} + \Gamma^{\mu}_{\gamma\nu} v^{\gamma} \quad \text{vector} \\ \lambda_{\mu;\nu} &= \lambda_{\mu,\nu} - \Gamma^{\gamma}_{\mu\nu}\lambda_{\gamma} \quad \text{co-vector} \end{aligned}$$

where $\{\Gamma^{\mu}_{\alpha\beta}\}$ are the Christoffel symbols of g given by

$$\text{(A.8)} \qquad \Gamma^{\mu}_{\alpha\beta} = \frac{1}{2}\, g^{\mu\nu}(g_{\alpha\nu,\beta} + g_{\beta\nu,\alpha} - g_{\alpha\beta,\nu}).$$

The frame components of these formulas take the form

$$\text{(A.9)} \qquad \begin{aligned} \varphi_{;\hat{a}} &= h^{\mu}_{\hat{a}}\varphi_{;\mu} = h^{\mu}_{\hat{a}}\frac{\partial \varphi}{\partial x^{\mu}} \\ v^{\hat{a}}_{;\hat{b}} &= \theta^{\hat{a}}_{\mu} h^{\nu}_{\hat{b}} v^{\mu}_{;\nu} = v^{\hat{a}}_{,\hat{b}} + \Gamma^{\hat{a}}_{\hat{c}\hat{b}} v^{\hat{c}} = h^{\nu}_{\hat{b}}\left(\frac{\partial v^{\hat{a}}}{\partial x^{\nu}}\right) + \Gamma^{\hat{a}}_{\hat{c}\hat{b}} v^{\hat{c}} \\ \lambda_{\hat{a};\hat{b}} &= h^{\mu}_{\hat{a}} h^{\nu}_{\hat{b}} \lambda_{\mu;\nu} = \lambda_{\hat{a},\hat{b}} - \Gamma^{\hat{c}}_{\hat{a}\hat{b}}\lambda_{\hat{c}} = h^{\nu}_{\hat{b}}\left(\frac{\partial \lambda_{\hat{a}}}{\partial x^{\nu}}\right) - \Gamma^{\hat{c}}_{\hat{a}\hat{b}}\lambda_{\hat{c}} \end{aligned}$$

where

$$\text{(A.10)} \qquad \Gamma^{\hat{a}}_{\hat{c}\hat{b}} = h^{\nu}_{\hat{b}}\ \theta^{\hat{a}}_{\mu}\ \{h^{\gamma}_{\hat{c}}\Gamma^{\mu}_{\gamma\nu} + h^{\mu}_{\hat{c},\nu}\}.$$

We shall also write

$$\text{(A.11)} \qquad \Gamma^{\hat{a}}_{\hat{c}\hat{b}} = h^{\nu}_{\hat{b}}\omega^{\hat{a}}{}_{\hat{c}\nu}$$

and express the connection one-forms, $\omega^{\hat{a}}{}_{\hat{c}}$ as

$$\text{(A.12)} \qquad \omega^{\hat{a}}{}_{\hat{c}} := \omega^{\hat{a}}{}_{\hat{c}\nu} dx^{\nu} = \omega^{\hat{a}}{}_{\hat{c}\nu} h^{\nu}_{\hat{b}}\theta^{\hat{b}} = \Gamma^{\hat{a}}_{\hat{c}\hat{b}}\theta^{\hat{b}}$$

which is equivalent to setting

$$\text{(A.13)} \qquad \omega^{\hat{a}}{}_{\hat{c}\nu} = \theta^{\hat{a}}_{\mu} h^{\gamma}_{\hat{c}}\Gamma^{\mu}_{\gamma\nu} + \theta^{\hat{a}}_{\mu}(h^{\mu}_{\hat{c},\nu}).$$

Defining

$$\text{(A.14)} \qquad \omega_{\hat{b}\hat{c}} = \eta_{\hat{a}\hat{b}}\omega^{\hat{a}}{}_{\hat{c}} = \eta_{\hat{a}\hat{b}}\omega^{\hat{a}}{}_{\hat{c}\nu} dx^{\nu}$$

one easily verifies that

$$\text{(A.15)} \qquad \omega_{\hat{b}\hat{c}\nu} = -\omega_{\hat{c}\hat{b}\nu}$$

which captures the metric compatibility of the chosen connection (i.e., the fact that $g_{\mu\nu;\alpha} = 0$). The vanishing of torsion for the Christoffel connection (i.e., the fact that $\Gamma^{\mu}_{\alpha\beta} = \Gamma^{\mu}_{\beta\alpha}$) takes the form

$$\text{(A.16)} \qquad \partial_{\nu}\theta^{\hat{c}}_{\mu} - \partial_{\mu}\theta^{\hat{c}}_{\nu} + \omega^{\hat{c}}{}_{\hat{a}\nu}\theta^{\hat{a}}_{\mu} - \omega^{\hat{c}}{}_{\hat{a}\mu}\theta^{\hat{a}}_{\nu} = 0$$

which can also be regarded as an equation determining the connection one forms, $\omega^{\hat{c}}{}_{\hat{a}\nu}\, dx$ in terms of the (co-) frame fields $\theta^{\hat{c}} = \theta^{\hat{c}}_{\mu} dx^{\mu}$.

In this same notation the Riemann curvature tensor takes the form

$$R^{\hat{c}}{}_{\hat{a}\mu\nu} = \theta^{\hat{c}}_{\gamma} h^{\lambda}_{\hat{a}} R^{\gamma}{}_{\lambda\mu\nu} \tag{A.17}$$
$$= \partial_\mu \omega^{\hat{c}}{}_{\hat{a}\nu} - \partial_\nu \omega^{\hat{c}}{}_{\hat{a}\mu} + \omega^{\hat{c}}{}_{\hat{d}\mu} \omega^{\hat{d}}{}_{\hat{a}\nu} - \omega^{\hat{c}}{}_{\hat{d}\nu} \omega^{\hat{d}}{}_{\hat{a}\mu}$$

which, since

$$R^{\hat{c}}{}_{\hat{a}\mu\nu} = -R^{\hat{c}}{}_{\hat{a}\nu\mu} \tag{A.18}$$

and

$$R_{\hat{b}\hat{a}\mu\nu} = -R_{\hat{a}\hat{b}\mu\nu} \tag{A.19}$$

where

$$R_{\hat{b}\hat{a}\mu\nu} := \eta_{\hat{b}\hat{c}} R^{\hat{c}}{}_{\hat{a}\mu\nu} \tag{A.20}$$

may be regarded as a two-form which takes values in the space of anti-symmetric Lorentz matrices. In view of Eq. (A.15) the connection one-form can be thought of as taking values in this same space which in turn represents the Lie algebra of (local) Lorentz transformations that can act on the frame fields while leaving the spacetime metric invariant.

Regarding connection and curvature as one and two-forms which take their values in the Lie algebra of some "internal" gauge group (in this case the Lorentz group of frame transformations) is parallel to what one does in Yang-Mills theory. There the principle bundle connection one-form $A_\mu dx^\mu$ and its curvature two-form $F_{\mu\nu} dx^\mu \wedge dx^\nu$ take their values in a matrix representation of the Lie algebra $\mathfrak{g}$ of some gauge "internal" Lie group G. By attaching (in a slightly unconventional way) row and column indices to label the matrix elements of these geometric objects, one could express their components as $A^{\hat{a}}{}_{\hat{b}\mu}$ and $F^{\hat{a}}{}_{\hat{b}\mu\nu}$ respectively, in parallel to the notation we have used above. The expression for $F^{\hat{a}}{}_{\hat{b}\mu\nu}$ in terms of $A^{\hat{a}}{}_{\hat{b}\mu}$ is identical in form to that for $R^{\hat{a}}{}_{\hat{b}\mu\nu}$ in terms of $\omega^{\hat{a}}{}_{\hat{b}\mu}$ given in Eq. (A.17) above. There are numerous other precise correspondences between Yang-Mills theory and Cartan's formulation of general relativity but there are also significant differences. For example in Yang-Mills theory, even if formulated on a curved background spacetime, there is no relationship between the connection one-form $A_\mu dx^\mu$ and the spacetime connection as expressed through the Christoffel symbols $\{\Gamma^{\mu}_{\alpha\beta}\}$ since the former does not derive from a metric or frame field whereas the latter does. Furthermore, the gauge groups for physically interesting Yang-Mills theories are normally required to be compact whereas the corresponding "gauge" group for general relativity is the non-compact (local) Lorentz group of orthonormal frame transformations. The compactness normally assumed for a gauge group G allows one to define an energy momentum tensor, quadratic in the Yang-Mills curvature, which has positive definite energy density. The corresponding second rank symmetric tensor, quadratic in the spacetime curvature, vanishes identically in Einstein's theory. Fortunately, the fourth-rank, totally symmetric

Bel-Robinson tensor and its associated positive definite "energy" density supply the needed replacements for these important objects.

Acknowledgements. This project is a natural continuation of the early work with Douglas Eardley on the Yang-Mills-Higgs equations. I am grateful for Eardley's numerous vital contributions to that collaboration and for his recognition of the relevance of the Yang-Mills problem to the gravitational one. I have also benefitted from numerous conversations with Piotr Chruściel, Yvonne Choquet-Bruhat, Lars Andersson, Sergiu Klainerman, Igor Rodnianski and Hans Lindblad. In addition I am grateful for the hospitality and support of the Albert Einstein Institute (Golm, Germany), the Erwin Schrödinger Institute (Vienna, Austria), the Institut des Hautes Études Scientifiques (Bures-sur-Yvette, France), The Isaac Newton Institute (Cambridge, UK), the Kavli Institute for Theoretical Physics (Santa Barbara, California), Caltech University (Pasadena, CA) and Stanford University and the American Institute of Mathematics (Palo Alto, California) where portions of this research were carried out. This research was supported by the NSF grant PHY-0354391 to Yale University.

References

[1] C. Cronström, *A Simple and Complete Lorentz Covariant Gauge Condition*, Phys. Lett. B **90** (1980), 267–269.

[2] P.T. Chruściel and J. Shatah, *Global Existence of Solutions of the Yang-Mills Equations on Globally Hyperbolic Four Dimensional Lorentzian Manifolds*, Asian J. Math. **1** (1997), 530–548.

[3] P.T.Chruściel and H. Friedrich (editors), *The Einstein Equations and the Large Scale Behavior of Gravitational Fields*, See for example the discussion in Section 4 of "Future Complete Vacuum Spacetimes" by L. Andersson and V. Moncrief, Birkhäuser, 2004.

[4] D.M. Eardley and V. Moncrief, *The Global Existence of Yang-Mills-Higgs Fields in 4-Dimensional Minkowski Space* I. *Local Existence and Smoothness Properties*, Commun. Math. Phys. **83** (1982), 171–191.

[5] D.M. Eardley and V. Moncrief, *The Global Existence of Yang-Mills-Higgs Fields in 4-Dimensional Minkowski Space* II. *Completion of Proof*, Commun. Math. Phys. **83** (1982), 193–212.

[6] D.M. Eardley, J. Isenberg, J.E. Marsden and V. Moncrief, *Homothetic and Conformal Symmetries of Solutions to Einstein's Equations*, Comm. Math. Phys. **106** (1986), 137–158.

[7] Y. Fourès-Bruhat, *Théorèmes d'existence pour certains systèmes d'équations aux dérivées partielles non linéaires*, Acta Math. **88** (1952), 141–225.

[8] F.G. Friedlander, *The Wave Equation on a Curved Space-Time*, Cambridge University Press, 1975.

[9] J. Hadamard, *Lectures on Cauchy's Problem in Linear Partial Differential Equations*, Silliman Lecture Series, Yale University Press, 1921.

[10] S. Klainerman and M. Machedon, *Finite energy solutions of the Yang-Mills equation in* $\mathbf{R}^{3+1}$, Annals of Math. **142** (1995), 39–119.

[11] S. Klainerman and I. Rodnianski, *A First Order Covariant Hadamard Parametrix for Curved Space-Time*, Preliminary Verison, preprint, Princeton University, 2005.

[12] R. Penrose, *Singularities and Time Asymmetry*, in 'General Relativity, an Einsetin Centenary Survey' (S.W. Hawking and W. Israel, eds.), Cambridge University Press, 1979.

[13] S. Sobolev, *Méthode nouvelle à resoudre le problème de Cauchy pour les équations linéaires hyperboliques normales*, Math. Sb. (N.S.) **1** (1936), 39–71.

[14] H.P.M. van Putten and D.M. Eardley, *Nonlinear Wave Equations for Relativity*, Phys. Rev. D **53** (1996), 3056–3063.

[15] P. Yodzis, H.-J. Seifert, and H. Müller zum Hagen, *On the Occurrence of Naked Singularities in General Relativity*, Commun. Math. Phys. **34** (1973), 135–148.

[16] P. Yodzis, H.-J. Seifert, and H. Müller zum Hagen, *On the Occurrence of Naked Singularities in General Relativity*, II, Commun. Math. Phys. **37** (1974), 29–40.

[17] See the discussion of normal coordinate systems and their properties in, for example, Ref. [**8**], Sect. 1.2.

[18] See Sect. 4.4 of Ref. [**8**] for a definition and discussion of *causal domains*.

[19] See Sect. 5.5 of Ref. [**8**], especially Theorem 5.5.2 for the representation formula in the case of tensor wave equations.

[20] See Sect. 4.2 of Ref. [**8**], in particular the discussion on p. 132.

Department of Physics and Department of Mathematics,
Yale University, New Haven, Connecticut

Topological strings and their physical applications

Andrew Neitzke and Cumrun Vafa

ABSTRACT. We give an introductory review of topological strings and their application to various aspects of superstrings and supersymmetric gauge theories. This review includes developing the necessary mathematical background for topological strings, such as the notions of Calabi-Yau manifold and toric geometry, as well as physical methods developed for solving them, such as mirror symmetry, large N dualities, the topological vertex and quantum foam. In addition, we discuss applications of topological strings to $\mathcal{N} = 1, 2$ supersymmetric gauge theories in 4 dimensions as well as to BPS black hole entropy in 4 and 5 dimensions. (These are notes from lectures given by the second author at the 2004 Simons Workshop in Mathematics and Physics.)

CONTENTS

1. Introduction

The topological string grew out of attempts to extend computations which occurred in the physical string theory. Since then it has developed in many interesting directions in its own right. Furthermore, the study of the topological string yielded an unanticipated but very exciting bonus: it has turned out that the topological string has many physical applications far beyond those that motivated its original construction!

In a sense, the topological string is a natural locus where mathematics and physics meet. Unfortunately, though, the topological string is not very well-known among physicists; and conversely, although mathematicians are able to understand what the topological string is mathematically, they are generally less aware of its physical content. These lectures are intended as a short overview of the topological string, hopefully accessible to both groups, as a place to begin. When we have the choice, we mostly focus on specific examples rather than the general theory. In general, we make no pretense at being complete; for more details on any of the subjects we treat, one should consult the references.

These lectures are organized as follows; for a more detailed overview of the individual sections, see the beginning of each section. We begin by introducing Calabi-Yau spaces, which are the geometric setting within which the topological string lives. In Section 2, we define these spaces, give some examples, and briefly explain why they are relevant for the physical string. Next, in Section 3, we discuss a particularly important class of Calabi-Yaus which can be described by "toric geometry"; as we explain, toric geometry is convenient mathematically and also admits an enlightening physical realization, which has been particularly important for making progress in the topological string.

With this background out of the way, we can then move on to the topological string itself, which we introduce in Section 4. There we give the definition of the topological string, and discuss its geometric meaning, with particular emphasis on the "simple" case of genus zero. Having defined the topological string the next question is how to compute its amplitudes, and in Section 5 we describe a variety of methods for computing topological string amplitudes at all genera, including mirror symmetry, large N dualities and direct target space analysis.

Having computed all these amplitudes one would like to use them for something; in Section 6, we consider the physical applications of the topological string. We consider applications to $\mathcal{N} = 1, 2$ supersymmetric gauge theories as well as to BPS black hole counting in four and five dimensions.

Finally, in Section 7 we briefly describe some speculations on a "topological M-theory" which could give a nonperturbative definition and unification of the two topological string theories.

2. Calabi-Yau spaces

Before defining the topological string, we need some basic geometric background. In this section we introduce the notion of "Calabi-Yau space." We begin with the mathematical definition and a short discussion of the reason why Calabi-Yau spaces are relevant for physics. Next we give some representative examples of Calabi-Yau spaces in dimensions 1, 2 and 3, both compact and non-compact. We end the section with a short overview of a particularly important non-compact Calabi-Yau threefold, namely the conifold, and the topology changing transition between its "deformed" and "resolved" versions.

2.1. Definition of Calabi-Yau space. We begin with a review of the notion of "Calabi-Yau space." There are many definitions of Calabi-Yau spaces, which are not quite equivalent to one another; but here we will not be too concerned about such subtleties, and all the spaces we will consider are Calabi-Yau under any reasonable definition. For us a Calabi-Yau space is a manifold X with a Riemannian metric g, satisfying three conditions:

- **I. X is a complex manifold.** This means X looks locally like $\mathbb{C}^n$ for some n, in the sense that it can be covered by patches admitting local complex coordinates

$$z_1, \ldots, z_n, \tag{2.1}$$

  and the transition functions between patches are holomorphic. In particular, the real dimension of X is $2n$, so it is always even. Furthermore the metric g should be Hermitian with respect to the complex structure, which means

$$g_{ij} = g_{\overline{i}\overline{j}} = 0, \tag{2.2}$$

  so the only nonzero components are $g_{i\overline{j}}$.

- **II. X is Kähler.** This means that locally on X there is a real function K such that

$$g_{i\overline{j}} = \partial_i \partial_{\overline{j}} K. \tag{2.3}$$

  Given a Hermitian metric g one can define its associated Kähler form, which is of type $(1,1)$,

$$k = g_{i\overline{j}} \mathrm{d}z_i \wedge \mathrm{d}\overline{z_j}. \tag{2.4}$$

  Then the Kähler condition is $\mathrm{d}k = 0$.

- **III. X admits a global nonvanishing holomorphic n-form.** In each local coordinate patch of X one can write many such forms,

$$\Omega = f(z_1, \ldots, z_n) \mathrm{d}z_1 \wedge \cdots \wedge \mathrm{d}z_n, \tag{2.5}$$

  for an arbitrary holomorphic function f. The condition is that such an Ω exists globally on X. For compact X there is always at most one such Ω up to an *overall* scalar rescaling; its existence is equivalent to the topological condition

$$c_1(TX) = 0, \tag{2.6}$$

  where TX is the tangent bundle of X.

If conditions I, II, and III are satisfied there is an important consequence. Namely, according to Yau's Theorem [**1**], X admits a metric g for which the Ricci curvature vanishes:

$$R_{i\overline{j}} = 0. \tag{2.7}$$

Except in the simplest examples, it is difficult to determine the Ricci-flat Kähler metrics on Calabi-Yau spaces. Nevertheless it is important and useful to know that such a metric *exists*, even if we cannot construct it explicitly. One thing we can construct explicitly is the *volume form* of the Ricci-flat metric; it is (up to a scalar multiple)

$$\text{vol} = \Omega \wedge \overline{\Omega}. \tag{2.8}$$

Strictly speaking Yau's Theorem as stated above applies to compact X, and has to be supplemented by suitable boundary conditions at infinity for the holomorphic n-form Ω when X is non-compact. For physical applications we do not require that X be compact; in fact, as we will see, many topological string computations simplify in the non-compact case, and this is also the case which is directly relevant for the connections to gauge theory.

2.2. Why Calabi-Yau? Before turning to examples, let us briefly explain the role that the Calabi-Yau conditions play in superstring theory. First, why are we interested in Riemannian manifolds at all? The reason is that they provide a class of candidate backgrounds on which the strings could propagate. The requirement that the background X be complex and Kähler turns out to have a rather direct consequence for the physics of observers living in the target space: namely, it implies that these observers will see supersymmetric physics. Since supersymmetry is interesting phenomenologically, this is a natural condition to impose. Finally, the requirement that X be Ricci-flat is even more fundamental: string theory would not even make sense without it, as we will sketch in Section 4.

In addition to these motivations from the physical superstring, once one specializes to the topological string, one finds other reasons to be interested in Calabi-Yau spaces and particularly Calabi-Yau threefolds; so we will revisit the question "why Calabi-Yau?" in Section 4.4. Although the Calabi-Yau conditions can be relaxed to give "generalized Calabi-Yau spaces," with correspondingly more general notions of topological string, the examples which have played the biggest role in the development of the theory so far are honest Calabi-Yaus. Therefore, in this review we focus on the honest Calabi-Yau case.

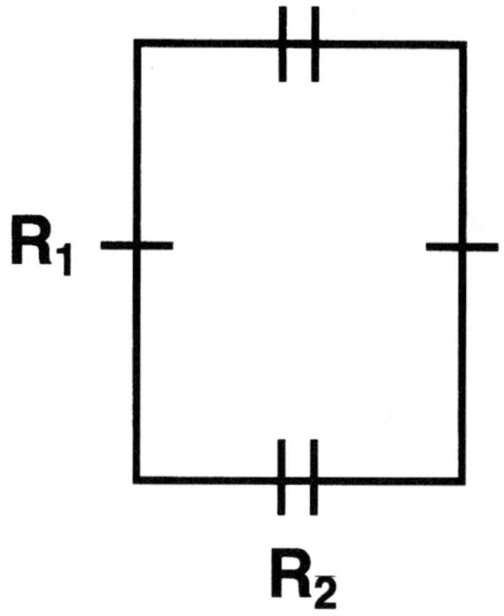

FIGURE 1. A rectangular torus; the top and bottom sides are identified, as are the left and right sides.

2.3. Examples of Calabi-Yau spaces.

2.3.1. *Dimension* 1. We begin with the case of complex dimension $n = 1$. In this case one can easily list all the Calabi-Yau spaces.

Example 2.1 (The complex plane). The simplest example is just the complex plane $\mathbb{C}$, with a single complex coordinate z, and the usual flat metric

$$g_{z\bar{z}} = -2\mathrm{i}\,. \tag{2.9}$$

In this case the holomorphic 1-form is simply

$$\Omega = \mathrm{d}z. \tag{2.10}$$

Example 2.2 (The punctured complex plane, aka the cylinder). The next simplest example is $\mathbb{C}^\times = \mathbb{C} \setminus \{0\}$, with its cylinder metric

$$g_{z\bar{z}} = -2\mathrm{i}\,/|z|^2, \tag{2.11}$$

and holomorphic 1-form

$$\Omega = \mathrm{d}z/z. \tag{2.12}$$

Example 2.3 (The 2-torus). Finally there is one compact example, namely the torus $T^2 = S^1 \times S^1$. We can picture it as a rectangle which we have glued together at the boundaries, as shown in Figure 1.

This torus has an obvious flat metric, namely the metric of the page; this metric depends on two parameters R_1, R_2 which are the lengths of the sides, so we say we have a two-dimensional "moduli space" of Calabi-Yau metrics on T^2, parameterized by the pair (R_1, R_2). It is convenient to repackage the moduli of T^2 into

$$A = \mathrm{i}\,R_1 R_2, \tag{2.13}$$

$$\tau = \mathrm{i}\,R_2/R_1. \tag{2.14}$$

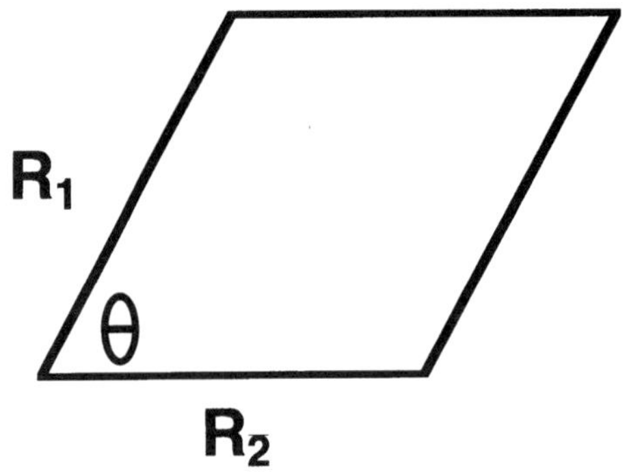

FIGURE 2. A 2-torus with a more general metric; again, opposite sides of the figure are identified.

Then A describes the overall area of the torus, or its "size," while τ describes its complex structure, or its "shape." A remarkable fact about string theory is that it is in fact invariant under the exchange of size and shape,

$$A \leftrightarrow \tau. \tag{2.15}$$

This is the simplest example of "mirror symmetry," which we will discuss further in Section 5.1. Here we just note that the symmetry (2.15) is quite unexpected from the viewpoint of classical geometry; for example, when combined with the obvious geometric symmetry $R_1 \leftrightarrow R_2$, it implies that string theory is invariant under $A \leftrightarrow 1/A$!

We could also consider a more general 2-torus, as shown in Figure 2, again with the flat metric inherited from the plane. This is still a Calabi-Yau space. It is natural to include such tori in our moduli space by letting the parameter τ have a real part as well as an imaginary part: namely, one can define the torus to be the quotient $\mathbb{C}/(\mathbb{Z} \oplus \tau\mathbb{Z})$, equipped with the Kähler metric inherited from $\mathbb{C}$. But then in order for the symmetry (2.15) to make sense, A should also be allowed to have a real part; in string theory this real part is naturally provided by an extra field, known as the "B field." For general X this B field is a class in $H^2(X, \mathbb{R})$, which should be considered as part of the moduli of the Calabi-Yau space along with the metric; it naturally combines with k to give the complex 2-form $k + \mathrm{i}\,B$. In our case $X = T^2$, $H^2(X, \mathbb{R})$ is 1-dimensional, and it exactly provides the missing real part of A.

Finally, let us introduce some terminology which will recur repeatedly throughout this review. We call τ a "complex modulus" of T^2 because changing τ changes the complex structure of the torus. In contrast, we can change A just by changing the (complexified) Kähler metric without changing the complex structure, so we call A a "Kähler modulus."

2.3.2. *Dimension* 2. Now let us move to Calabi-Yau spaces of complex dimension 2. Here the supply of examples is somewhat richer. First there is a trivial example:

Example 2.4 (Cartesian products). One can obtain Calabi-Yau spaces of dimension 2 by taking Cartesian products of the ones we had in dimension 1, e.g. $\mathbb{C}^2, \mathbb{C} \times \mathbb{C}^\times, \mathbb{C} \times T^2$.

Next we move on to the nontrivial compact examples. Up to diffeomorphism there are only two, namely the four-torus T^4 and the "K3 surface." We focus here on K3.

Example 2.5 (K3). The fastest way to construct a K3 surface is to obtain it as a quotient $T^4/\mathbb{Z}_2$, using the $\mathbb{Z}_2$ identification

$$(x_1, x_2, x_3, x_4) \sim (-x_1, -x_2, -x_3, -x_4), \tag{2.16}$$

where the x_i are coordinates on T^4 (so they are periodically identified.) Strictly speaking, this quotient gives a singular K3 surface, with 16 singular points which are the fixed points of (2.16). The singular points can be "blown up" (this roughly means replacing them by embedded 2-spheres, see e.g. [**2**]) to obtain a smooth K3 surface. In string theory both singular K3 surfaces and smooth K3 surfaces are allowed; the singular ones correspond to a particular sublocus of the moduli space of K3 surfaces.

One can also define the K3 surface directly by means of algebraic equations. To begin with we introduce an auxiliary space $\mathbb{CP}^n$, which is also important in its own right:

Example 2.6 (Complex projective space). $\mathbb{CP}^n$ consists of all $(n+1)$-tuples $(z_1, \ldots, z_{n+1}) \in \mathbb{C}^{n+1}$, excluding the point $(0, 0, \ldots, 0)$, modulo the identification

$$(z_1, \ldots, z_{n+1}) \sim (\lambda z_1, \ldots, \lambda z_{n+1}), \tag{2.17}$$

for all $\lambda \in \mathbb{C}^\times$. Then $\mathbb{CP}^n$ is an n-dimensional complex manifold, roughly because we can use the identification (2.17) to eliminate one coordinate. $\mathbb{CP}^n$ is not Ricci-flat, so it is not a Calabi-Yau space.

A useful special case to remember is $\mathbb{CP}^1$, which is simply the Riemann sphere S^2. The same is not true in higher dimensions, though — e.g. $\mathbb{CP}^2$ is not topologically the same as S^4 (the latter is not even a complex manifold.)

Having introduced complex projective space, now we return to the job of constructing K3. We consider the equation

$$P_4(z_1, \ldots, z_4) = 0, \tag{2.18}$$

where P_4 is some homogeneous polynomial of degree 4. Then we define K3 to be the set of solutions to (2.18) inside the complex projective space $\mathbb{CP}^3$. Since $\mathbb{CP}^3$ is 3-dimensional and (2.18) is 1 complex equation, K3 so defined will be 2-dimensional. (Note that in order for this definition to make sense it is important that P_4 is a *homogeneous* polynomial — otherwise the condition (2.18) would not be well-defined after the identification (2.17).)

Different choices for the polynomial P_4 give rise to different K3 surfaces, in the sense that they have different complex structures, although they are all diffeomorphic. P_4 has 20 complex coefficients, but the equation (2.18) is obviously independent of the overall scaling of P_4, so this rescaling does not affect the complex structure of the resulting K3; all the other coefficients do affect the complex structure, so one gets a 19-parameter family of K3 surfaces from this construction. These 19 parameters are the analog of the single parameter τ in Example 2.3.[1]

So far we have only discussed K3 as a complex manifold, but it is indeed a Calabi-Yau space, as we now explain. It is easy to see that it is Kähler since it inherits a Kähler metric from $\mathbb{CP}^4$. To see that it has a Ricci-flat Kähler metric one can invoke Yau's Theorem, as we mentioned in Section 2.1; that reduces the task to showing that K3 satisfies the topological condition $c_1 = 0$. By using the "adjunction formula" from algebraic geometry **[2]** one finds that given a polynomial equation of degree d inside $\mathbb{CP}^{k-1}$, the resulting hypersurface X has

$$c_1(X) \sim (d-k)c_1(\mathbb{CP}^{k-1}). \tag{2.19}$$

In this case we took $d = k = 4$, so $c_1(X) = 0$ as desired. This shows the existence of the desired Calabi-Yau metric. However, the explicit form of the metric is not known, except at special points in the moduli space.

Example 2.7 (ALE spaces)**.** The "asymptotically locally Euclidean," or "ALE," spaces form an important class of non-compact Calabi-Yaus of complex dimension 2. Roughly speaking, these spaces are are obtained as $\mathbb{C}^2/G$, where G is a finite subgroup of $SU(2)$ acting linearly on $\mathbb{C}^2$. (The condition that $G \subset SU(2)$ implies that it preserves the holomorphic 2-form on $\mathbb{C}^2$, so that it descends to a holomorphic 2-form on $\mathbb{C}^2/G$, which is therefore a Calabi-Yau.) More precisely, the ALE space is not quite $\mathbb{C}^2/G$; that quotient has a singularity at the origin, because that point is fixed by the linear action of G. One obtains the ALE space by a local modification near the origin known as "resolving" the singularity. This resolution replaces the singularity by a number of $\mathbb{CP}^1$'s localized near the origin. The number of $\mathbb{CP}^1$'s which one gets and their intersection numbers with one another are determined by the group G; for example, if $G = \mathbb{Z}_n$ one gets $n-1$ such $\mathbb{CP}^1$'s C_j, $j = 1, \dots, n-1$, with intersection numbers

$$C_i \cap C_i = -2, \tag{2.20}$$

$$C_i \cap C_j = 1 \quad \text{if } |i-j| = 1, \tag{2.21}$$

$$C_i \cap C_j = 0 \quad \text{if } |i-j| > 1. \tag{2.22}$$

These intersection numbers are exactly the Cartan matrix of the Lie algebra $A_{n-1} = su(n)$. So the curves C_i are playing the role of the simple roots of A_{n-1}. This "coincidence" also extends to other choices for $G \subset SU(2)$. One possibility is that G can be a double cover of the dihedral group on n elements; in this case resolving the singularity gives the simple roots of

[1]These are not quite all the complex moduli of K3 — there is one more complex deformation possible, for a total of 20, but after making this deformation one gets a surface which cannot be realized by algebraic equations inside $\mathbb{CP}^3$.

$D_{n-1} = so(2n-2)$. The other possibilities for G are the "exceptional subgroups" of $SU(2)$, namely double covers of the tetrahedral, octahedral and dodecahedral groups, and these give the simple roots of E_6, E_7, E_8 respectively. This relation between singularities $\mathbb{C}^2/G$ and simply-laced Lie algebras is known as an "ADE classification." The meaning of the Lie algebras which appear here will become more clear in Section 6.1 where they will be related to gauge symmetries.

After resolving the singularity of $\mathbb{C}^2/G$, one obtains the ALE space, which admits a Calabi-Yau metric. In fact, as with our other examples, it has a whole moduli space of such metrics: in particular, for each of the curves C_i obtained by resolving the singularity, there is a Kähler modulus $t_i = \int_{C_i} k + \mathrm{i}B$ determining its size. In the limit $t_i \to 0$ the metric reduces to that of the singular space $\mathbb{C}^2/G$. In this sense one can think of the singularity of $\mathbb{C}^2/G$ as containing a number of "zero size $\mathbb{CP}^1$'s."

2.3.3. *Dimension* 3. Now we move to the case which is most interesting for topological string theory. In $d=3$ the problem of classifying Calabi-Yau spaces is far more complicated, even if we restrict to compact Calabi-Yaus; while in $d=1$ we had just T^2, and in $d=2$ just T^4 and K3, in $d=3$ it is not even known whether the number of compact Calabi-Yau spaces up to diffeomorphism is finite. So we content ourselves with a few examples.

Example 2.8 (The quintic threefold). The quintic threefold is defined similarly to our algebraic construction of K3 in Example 2.5; namely we consider the equation

$$P_5(z_1, \dots, z_5) = 0, \tag{2.23}$$

where P_5 is homogeneous of degree 5. The solutions of (2.23) inside $\mathbb{CP}^4$ give a 3-dimensional space which we call the "quintic threefold." It is a Calabi-Yau space again using (2.19) just as we did for K3.

The quintic threefold has 101 complex moduli, and is in some sense the simplest compact Calabi-Yau threefold. As such it has been extensively studied, e.g., as the first example of full-fledged mirror symmetry.

Example 2.9 (Local $\mathbb{CP}^2$). One non-compact Calabi-Yau can be obtained by starting with four complex coordinates (x, z_1, z_2, z_3), subject to the condition $(z_1, z_2, z_3) \neq (0,0,0)$, and making the identification

$$(x, z_1, z_2, z_3) \sim (\lambda^{-3}x, \lambda z_1, \lambda z_2, \lambda z_3) \tag{2.24}$$

for all $\lambda \in \mathbb{C}^\times$. Mathematically, this space is known as the total space of the line bundle $\mathcal{O}(-3) \to \mathbb{CP}^2$; we can think of it as obtained by starting with the $\mathbb{CP}^2$ spanned by z_1, z_2, z_3 and adjoining the extra coordinate x. See Figure 3. Locally on $\mathbb{CP}^2$, our space has the structure of $\mathbb{CP}^2 \times \mathbb{C}$. In this sense it has "4 compact directions" and "2 non-compact directions."

The rule (2.24) characterizes the behavior of x under rescalings of the homogeneous coordinates on $\mathbb{CP}^2$, or equivalently, it determines how x transforms as one moves between different coordinate patches on $\mathbb{CP}^2$.

FIGURE 3. A crude representation of the local $\mathbb{CP}^2$ geometry, $\mathcal{O}(-3) \to \mathbb{CP}^2$.

Although the local $\mathbb{CP}^2$ geometry is non-compact, it can arise naturally even if we start with a compact Calabi-Yau — namely, it describes the geometry of a Calabi-Yau space containing a $\mathbb{CP}^2$, in the limit where we focus on the immediate neighborhood of the $\mathbb{CP}^2$.

Example 2.10 (Local $\mathbb{CP}^1$). Similar to the last example, we can start with four complex coordinates (x_1, x_2, z_1, z_2), subject to the condition $(z_1, z_2) \neq (0,0)$, and make the identification

$$(x_1, x_2, z_1, z_2) \sim (\lambda^{-1} x_1, \lambda^{-1} x_2, \lambda z_1, \lambda z_2) \tag{2.25}$$

for all $\lambda \in \mathbb{C}^\times$. This gives the total space of the line bundle $\mathcal{O}(-1) \oplus \mathcal{O}(-1) \to \mathbb{CP}^1$. Similarly to the previous example, it is obtained by starting with $\mathbb{CP}^1$, which has "2 compact directions," and then adjoining the coordinates x_1, x_2, which contribute "4 non-compact directions." See Figure 4.

This example is also known as the "resolved conifold," a name to which we will return in Section 2.4.

Example 2.11 (Local $\mathbb{CP}^1 \times \mathbb{CP}^1$). Another standard example comes by starting with five complex coordinates (x, y_1, y_2, z_1, z_2), with $(y_1, y_2) \neq (0,0)$ and $(z_1, z_2) \neq (0,0)$, and making the identification

$$(x, y_1, y_2, z_1, z_2) \sim (\lambda^{-2} \mu^{-2} x, \lambda y_1, \lambda y_2, \mu z_1, \mu z_2) \tag{2.26}$$

for all $\lambda, \mu \in \mathbb{C}^\times$. This gives the total space of the line bundle $\mathcal{O}(-2,-2) \to \mathbb{CP}^1 \times \mathbb{CP}^1$. It has four compact directions and two non-compact directions.

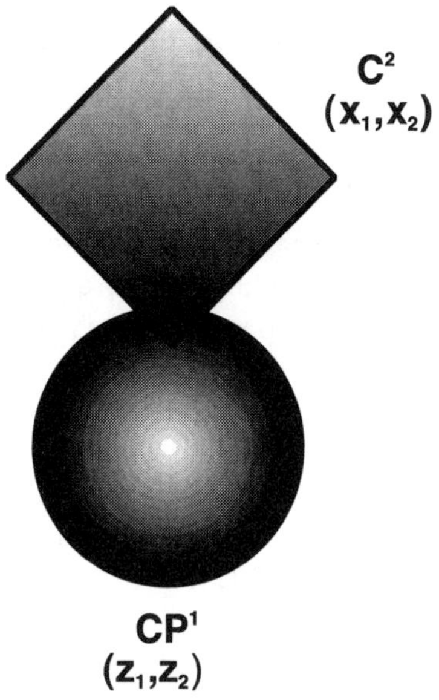

FIGURE 4. A crude representation of the local $\mathbb{CP}^1$ geometry, $\mathcal{O}(-1) \oplus \mathcal{O}(-1) \to \mathbb{CP}^1$.

Example 2.12 (Deformed conifold)**.** All the local examples we discussed so far were "rigid," in other words, they had no deformations of their complex structure.[2] Now let us consider an example which is not rigid. Starting with the complex coordinates $(x, y, z, t) \in \mathbb{C}^4$, this time without any projective identification, we look at the space of solutions to

$$xy - zt = \mu. \tag{2.27}$$

This gives a Calabi-Yau 3-fold for any value $\mu \in \mathbb{C}$, so μ spans the 1-dimensional moduli space of complex structures. If $\mu = 0$ then the Calabi-Yau has a singularity at $(x, y, z, t) = (0, 0, 0, 0)$, known as the "conifold singularity." For finite μ it is smooth. Since we obtain the smooth Calabi-Yau from the singular one just by varying the parameter μ, which deforms the complex structure, we call the smooth $\mu \neq 0$ version the "deformed conifold." We will discuss it in more detail in Section 2.4.

2.4. Conifolds. In the last section we introduced the singular conifold

$$xy - zt = 0, \tag{2.28}$$

and the deformed conifold

$$xy - zt = \mu. \tag{2.29}$$

[2]Strictly speaking, this is a delicate statement in the non-compact case since we should specify what kind of boundary conditions we are imposing at infinity. When we say that these local examples are rigid we essentially mean that the compact part, $\mathbb{CP}^1$ or $\mathbb{CP}^2$, has no complex deformations.

Since the deformed conifold is such an important example it will be useful to describe it in another way. Namely, by a change of variables we can rewrite (2.29) as

$$x_1^2 + x_2^2 + x_3^2 + x_4^2 = r. \tag{2.30}$$

Describing it this way it is easy to see that there is an S^3 in the geometry, namely, just look at the locus where all $x_i \in \mathbb{R}$. The full geometry where we include also the imaginary parts of x_i is in fact diffeomorphic to the cotangent bundle, T^*S^3.

This space is familiar to physicists as the phase space of a particle which moves on S^3; it has three "position" variables labeling a point $x \in S^3$ and three "momenta" spanning the cotangent space at x. Now we want to describe its geometry "near infinity," i.e., at large distances, similar to how we might describe the infinity of Euclidean $\mathbb{R}^3$ as looking like a large S^2. In the case of T^*S^3 the position coordinates are bounded, so looking near infinity means choosing large values for the momenta, which gives a large S^2 in the cotangent space $\mathbb{R}^3$. Therefore the infinity of T^*S^3 should look like some S^2 bundle over the position space S^3, i.e., locally on S^3 it should look like $S^2 \times S^3$. It turns out that this is enough to imply that it is even *globally* $S^2 \times S^3$.

So at infinity the deformed conifold has the geometry of $S^2 \times S^3$. As we move from infinity toward the origin both S^2 and S^3 shrink, until the S^2 disappears altogether, leaving just an S^3 with radius r, which is the core of the T^*S^3 geometry (the zero section of the cotangent bundle.) This is depicted on the left side of Figure 5.

Now let us describe another way of smoothing the conifold singularity. First rewrite (2.28) as

$$\det \begin{pmatrix} x & z \\ t & y \end{pmatrix} = 0. \tag{2.31}$$

This equation is equivalent to the existence of nontrivial solutions to

$$\begin{pmatrix} x & z \\ t & y \end{pmatrix} \begin{pmatrix} \xi_1 \\ \xi_2 \end{pmatrix} = 0. \tag{2.32}$$

Indeed, away from $(x, y, z, t) = (0, 0, 0, 0)$, (2.31) just states that the matrix has rank 1, so (ξ_1, ξ_2) solving (2.32) are unique up to an overall rescaling. So away from $(x, y, z, t) = (0, 0, 0, 0)$ one could describe the singular conifold as the space of solutions to (2.32), with $(\xi_1, \xi_2) \neq (0, 0)$, and with the identification

$$(\xi_1, \xi_2) \sim (\lambda\xi_1, \lambda\xi_2) \tag{2.33}$$

where $\lambda \in \mathbb{C}^\times$. But at $(x, y, z, t) = (0, 0, 0, 0)$ something new happens: any pair (ξ_1, ξ_2) now solves (2.32). Taking into account (2.33), (ξ_1, ξ_2) parameterize a $\mathbb{CP}^1$ of solutions. In summary, (2.28) and (2.32),(2.33) are equivalent, except that $(x, y, z, t) = (0, 0, 0, 0)$ describes a single point in (2.28), but a whole $\mathbb{CP}^1$ in (2.32),(2.33). We refer to the space described

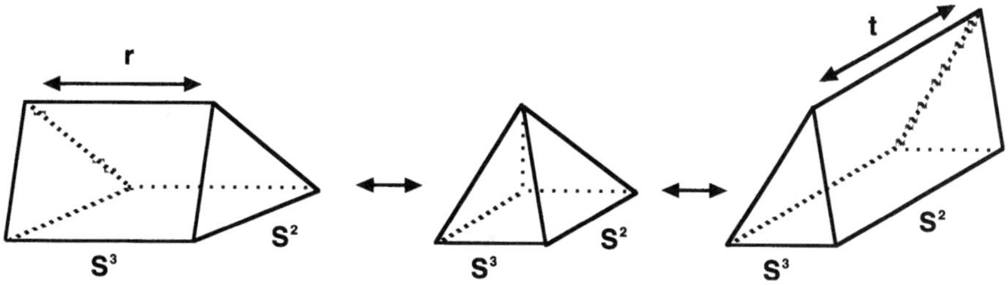

FIGURE 5. The three conifold geometries: from left to right, deformed, singular and resolved. Both geometries look like $S^2 \times S^3$ near infinity (the bottom of the figure); they are distinguished by whether the S^2 or the S^3 shrinks to zero size in the interior (the top of the figure).

by (2.32),(2.33) as the "resolved conifold." (In fact, it is isomorphic to the local $\mathbb{CP}^1$ geometry of Example 2.10.)

Mathematically this discussion would be summarized by saying that the resolved conifold is obtained by making a "small resolution" of the conifold singularity. We emphasize, however, that physically it is natural to consider this as a continuous process, contrary to the usual mathematical description in which it seems to be a discrete jump. This is because physically we consider the full Calabi-Yau metric rather than just the complex structure. Namely, the resolved conifold has a single Kähler modulus for its Calabi-Yau metric,[3] naturally parameterized by

$$t = \mathrm{vol}\,(\mathbb{CP}^1) = \int_{\mathbb{CP}^1} k + \mathrm{i}\,B. \tag{2.34}$$

In the limit $t \to 0$, the $\mathbb{CP}^1$ shrinks to a point and the Calabi-Yau metric on the resolved conifold approaches the Calabi-Yau metric on the singular conifold. So the resolved conifold is obtained by a Kähler deformation of the metric without changing the complex structure.[4]

In summary, we have two different non-compact Calabi-Yau geometries, as depicted in Figure 5: the deformed conifold, which has one complex modulus r and no Kähler moduli, and the resolved conifold, which has no complex moduli but one Kähler modulus t; we can interpolate from one space to the other by passing through the singular conifold geometry. The deformed conifold has a single S^3 at its heart, whose size is determined by r, while the resolved conifold has a single S^2, whose size is determined by t.

Note that from the perspective of Figure 5, the S^2 and S^3 which appear when we resolve the singular conifold seem very natural; in some sense they

[3]Once again, we are here considering only variations of the metric which preserve suitable boundary conditions at infinity.

[4]Mathematically, the resolved conifold and the singular conifold are not the same as complex manifolds, but they are birationally equivalent. Physically we want to consider birationally equivalent spaces as really having the same complex structure.

were both in the game even before resolving, as we see from the $S^2 \times S^3$ at infinity. All three cases — deformed, singular, and resolved — look the same at infinity; they differ only near the tip of the cone. This is exactly what we expect since we were trying to study only localized deformations.

We will return to the conifold repeatedly in later sections. For more information about its geometry, including the explicit Calabi-Yau metrics, see [**3**].

3. Toric geometry

Now we want to introduce a particularly convenient representation of a special class of algebraic manifolds, which includes and generalizes some of the examples we considered above. Mathematically this representation is called "toric geometry"; for a more detailed review than we present here, see e.g. [**4**]. As we will see, toric manifolds have two closely related virtues: first, they are easily described in terms of a finite amount of combinatorial data; second, they can be concretely realized via two-dimensional field theories of a particularly simple type.

We begin with the simplest of all toric manifolds.

Example 3.1 ($\mathbb{C}^n$)**.** Consider the n-complex-dimensional manifold $\mathbb{C}^n$, with complex coordinates $(z_1, \ldots, z_n)$ and the standard flat metric, and parameterize it in an idiosyncratic way: writing

$$z_i = |z_i| e^{\mathrm{i}\,\theta_i}, \tag{3.1}$$

choose the coordinates $((|z_1|^2, \theta_1), \ldots, (|z_n|^2, \theta_n))$. This coordinate system emphasiz es the symmetry $U(1)^n$ which acts on $\mathbb{C}^n$ by shifts of the θ_i. It is also well suited to describing the symplectic structure given by the Kähler form k:

$$k = \sum_i \mathrm{d}z_i \wedge \mathrm{d}\overline{z_i} = \sum_i \mathrm{d}|z_i|^2 \wedge \mathrm{d}\theta_i. \tag{3.2}$$

Roughly, splitting the coordinates into $|z_i|^2$ and θ_i gives a factorization

$$\mathbb{C}^n \approx \mathcal{O}^{n+} \times T^n, \tag{3.3}$$

where $\mathcal{O}^{n+}$ denotes the "positive orthant" $\{|z_i|^2 \geq 0\}$, represented (for $n = 3$) in Figure 6.

Namely, at each point of $\mathcal{O}^{n+}$ we have the product of n circles obtained by fixing $|z_i|$ and letting θ_i vary. However, when $|z_i|^2 = 0$ the circle $|z_i|e^{\mathrm{i}\,\theta_i}$ degenerates to a single point. Therefore (3.3) is not quite precise, because the "fiber" T^n degenerates at each boundary of the "base" $\mathcal{O}^{n+}$; which circle of T^n degenerates is determined by which $|z_i|^2$ vanishes, or more geometrically, by the direction of the unit normal to the boundary. When $m > 1$ of the $|z_i|^2$ vanish, which occurs at the intersection locus of m faces of the orthant, the corresponding m circles of T^n degenerate. At the origin all n cycles have degenerated and T^n shrinks to a single point.

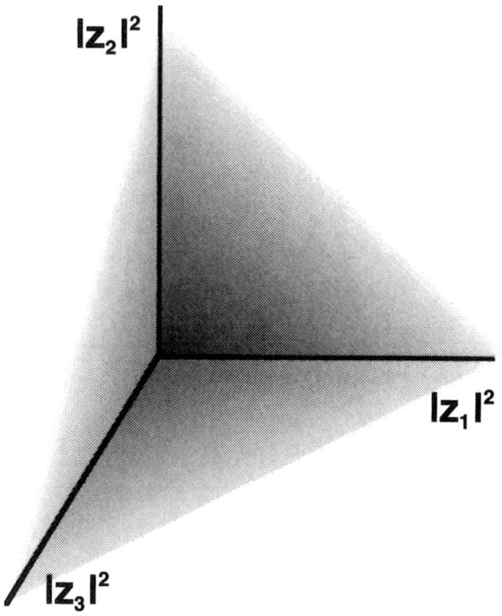

FIGURE 6. The positive octant $\mathcal{O}^{3+}$, which is the toric base of $\mathbb{C}^3$.

In this sense all the information about the symplectic manifold $\mathbb{C}^3$ is contained in Figure 6, which is called the "toric diagram" for $\mathbb{C}^3$. When looking at this diagram one always has to remember that there is a T^3 over the generic point, and that this T^3 degenerates at the boundaries, in a way determined by the unit normal. Despite the fact that the T^3 becomes singular at the boundaries, the full geometry of $\mathbb{C}^3$ is of course smooth. (Of course, all this holds for general n as well as $n = 3$, but the analogue of Figure 6 would be hard to draw in the general case.)

Example 3.2 (Complex projective space). Next we want to give a toric representation for $\mathbb{CP}^n$. We first give a slightly different quotient presentation of this space than the one we used in (2.17): namely, for any $r > 0$, we start with the $2n+1$-sphere

$$|z_1|^2 + \cdots + |z_{n+1}|^2 = r, \tag{3.4}$$

and then make the identification

$$(z_1, \ldots, z_{n+1}) \sim (e^{\mathrm{i}\theta} z_1, \ldots, e^{\mathrm{i}\theta} z_{n+1}) \tag{3.5}$$

for all real θ. This is equivalent to our original "holomorphic quotient" definition, where we did not impose (3.4) but worked modulo arbitrary rescalings of the z_i instead of just phase rescalings; indeed, starting from that definition one can make a rescaling to impose (3.4), and afterward one still has the freedom to rescale by a phase as in (3.5). The presentation we are using now is more closely rooted in symplectic geometry.

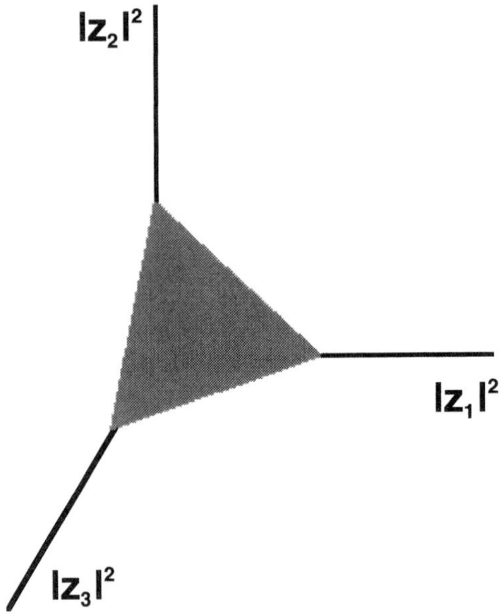

FIGURE 7. The toric base of $\mathbb{CP}^2$; geometrically it is just the two-dimensional interior of a triangle, but here we show it naturally embedded in $\mathbb{R}^3$ and cut out by the condition (3.4).

This toric presentation is also natural from the physical point of view, as we now briefly discuss. The physical theory which describes the worldsheet of the superstring propagating on $\mathbb{CP}^n$ is a two-dimensional quantum field theory known as the "supersymmetric nonlinear sigma model into $\mathbb{CP}^n$." We will not discuss this sigma model in detail, but the crucial point is that in this case it can be obtained as the IR limit of an $\mathcal{N} = (2,2)$ supersymmetric *linear* sigma model with $U(1)$ gauge symmetry [**5**]. Specifically, the coordinates z_i appear as the scalar components of 4 chiral superfields, all with $U(1)$ charge 1. Then the physics of the vacua of the linear sigma model exactly mirrors our toric construction of $\mathbb{CP}^n$; namely, the constraint (3.4) is imposed by the D-terms, and the quotient (3.5) is the identification of gauge equivalent field configurations. This construction, which we will generalize below when we discuss other toric varieties, turns out to be extremely useful for the study of the topological string on such spaces; we will see some examples of its utility in later sections.

Note that in our toric presentation of $\mathbb{CP}^n$ we have the parameter $r > 0$, which did not appear in the holomorphic quotient. This parameter appears naturally in the gauged linear sigma model (as a Fayet-Iliopoulos parameter), where one sees directly that it corresponds to the size of $\mathbb{CP}^n$.

Now we want to use this presentation to draw the toric diagram. As for $\mathbb{C}^n$, the toric base lies in the space coordinatized by the $|z_i|^2$. In the present case we have to impose (3.4), so the base turns out to be an n-dimensional simplex; for example, in the case of $\mathbb{CP}^2$ it is just a triangle, as shown in

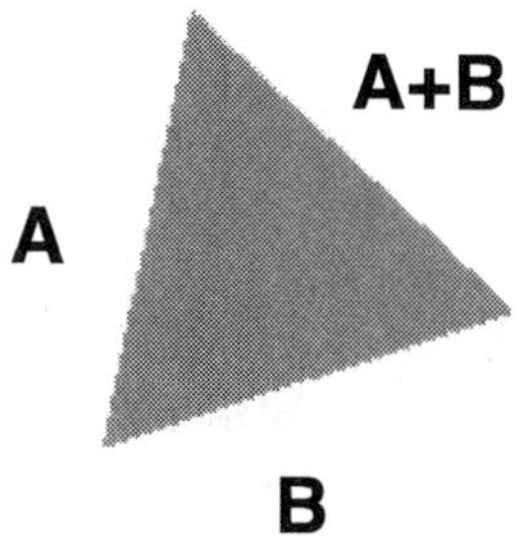

FIGURE 8. The toric base of $\mathbb{CP}^2$. Over each boundary a cycle of the fiber T^2 collapses; if we label the basis cycles as A and B, then the collapsing cycle over each boundary is as indicated.

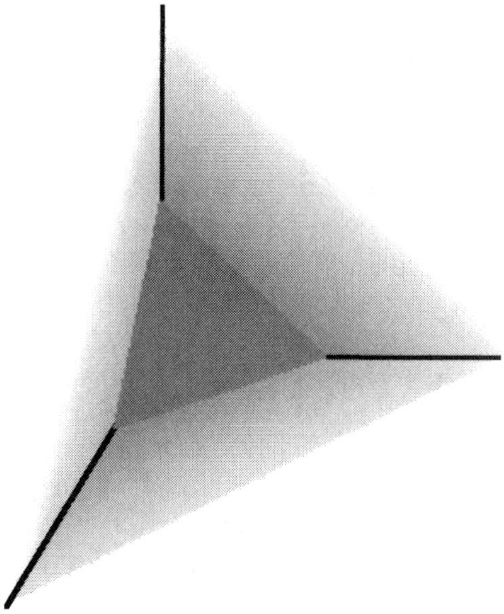

FIGURE 9. The toric base of the local $\mathbb{CP}^2$ geometry.

Figure 7. Over each point of the base we have a T^2 fiber generated by shifts of θ_i (naively this would give a T^3 for $\theta_1, \theta_2, \theta_3$, but the identification (3.5) reduces this to T^2.) A cycle of T^2 collapses over each boundary of the triangle, as indicated in Figure 8.

Example 3.3 (Local $\mathbb{CP}^2$). To get a toric presentation of a Calabi-Yau manifold we have to choose a non-compact example. The construction is closely analogous to what we did above to construct $\mathbb{CP}^n$; namely, for $r > 0$, we start with

$$-3|z_0|^2 + |z_1|^2 + |z_2|^2 + |z_3|^2 = r, \tag{3.6}$$

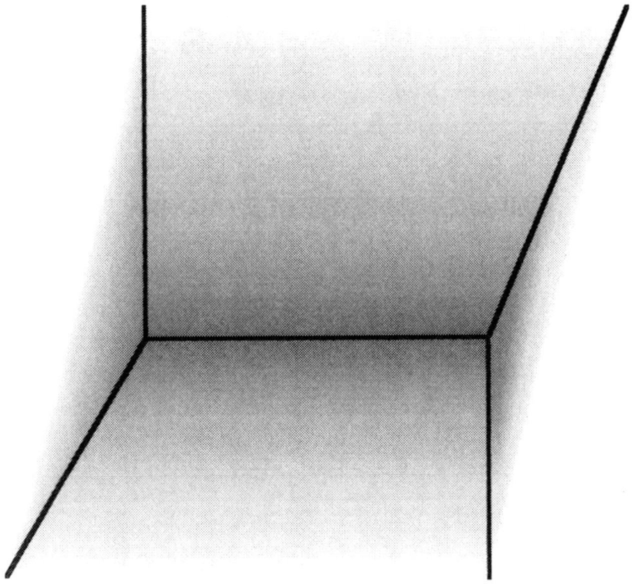

FIGURE 10. The toric base of the local $\mathbb{CP}^1$ geometry.

and then make the additional identification

$$(z_0, z_1, z_2, z_3) \sim (e^{-3\mathrm{i}\,\theta} z_0, e^{\mathrm{i}\,\theta} z_1, e^{\mathrm{i}\,\theta} z_2, e^{\mathrm{i}\,\theta} z_3), \tag{3.7}$$

for any real θ. In the gauged linear sigma model of [**5**] this is realized by taking four chiral superfields with $U(1)$ charges $(-3, 1, 1, 1)$. Actually, the fact that the local $\mathbb{CP}^2$ geometry is Calabi-Yau can also be understood naturally in the gauged linear sigma model: the condition $c_1 = 0$ turns out to be equivalent to the statement that the sum of the $U(1)$ charges vanishes, which in turn implies vanishing of the 1-loop beta function.

We can also draw the toric diagram for this case. Introducing the notation $p_i = |z_i|^2$, the base is spanned by the four real coordinates p_0, p_1, p_2, p_3, subject to the condition (3.6), which can be solved to eliminate p_0,

$$p_0 = \frac{1}{3}(p_1 + p_2 + p_3 - r). \tag{3.8}$$

The condition that all $p_i > 0$ then becomes

$$p_1 + p_2 + p_3 > r, \tag{3.9}$$

$$p_1 > 0, \tag{3.10}$$

$$p_2 > 0, \tag{3.11}$$

$$p_3 > 0. \tag{3.12}$$

So the toric base is the positive octant in $\mathbb{R}^3$ with a corner chopped off, as shown in Figure 9. The triangle at the corner represents the $\mathbb{CP}^2$ at the core of the geometry, just as in the previous example.

Example 3.4 (Local $\mathbb{CP}^1$). A similar construction gives the toric diagram for the local $\mathbb{CP}^1$ geometry, $\mathcal{O}(-1) \oplus \mathcal{O}(-1) \to \mathbb{CP}^1$, from Example 2.10. One obtains in this case Figure 10. One feature of interest is the $\mathbb{CP}^1$ at the core of the geometry, which can be easily seen as the line segment in the middle. (To see that the line segment indeed represents the topology of $\mathbb{CP}^1$, recall that along this segment two of the three circles of the fiber T^3 are degenerate, so that one just has an S^1 in the fiber; moving along the segment, this S^1 then sweeps out a $\mathbb{CP}^1$; indeed, the S^1 degenerates at the two ends of the segment, which are identified with the north and south poles of $\mathbb{CP}^1$.) Furthermore it is easy to read off the volume of this $\mathbb{CP}^1$ from the toric diagram: the Kähler form in this geometry is $k = \mathrm{d}p_i \wedge \mathrm{d}\theta_i$, and integrating it just gives $2\pi\Delta p$, i.e., the length of the line segment![5]

This example illustrates a general feature: finite segments (or more generally finite simplices) of the toric diagram correspond to compact cycles in the geometry, and the sizes of the simplices correspond to the volumes of the cycles.

Example 3.5 (Local $\mathbb{CP}^1 \times \mathbb{CP}^1$). We can give a toric construction for this case as well, again parallel to the holomorphic construction we gave above; in gauged linear sigma model terms it would correspond to having 5 chiral superfields and two $U(1)$ gauge groups, with the charges $(-2, 1, 1, 0, 0)$ and $(-2, 0, 0, 1, 1)$. (Note that the charges under both $U(1)$ groups sum to zero as required for one-loop conformality.) The corresponding toric diagram is the "oubliette" shown in Figure 11.

Our list of toric Calabi-Yaus has included only non-compact examples, but we should note that it is also possible to construct compact Calabi-Yaus using the techniques of toric geometry. Indeed, we have already done so in Examples 2.5 and 2.8, where we started with the toric manifolds $\mathbb{CP}^3$ and $\mathbb{CP}^4$ respectively and then imposed some extra algebraic relations on the coordinates to obtain a Calabi-Yau. A similar construction can be performed starting with a more general toric manifold, and this gives a large class of interesting examples of compact Calabi-Yau spaces. This construction is also natural from the physical point of view: in the gauged linear sigma model, imposing an algebraic relation on the coordinates corresponds to introducing a superpotential.

4. The topological string

With the geometrical preliminaries behind us, we are now ready to move on to physics. In this section we will sketch the definition of the topological string. First we describe the two-dimensional field theories which are underlying the physical string theory. Next we discuss the "twisting" procedure which converts the ordinary field theory into its topological cousin, and how

[5]We are using a fact about Kähler geometry, namely, the volume of a holomorphic cycle is just obtained by integrating k over the cycle.

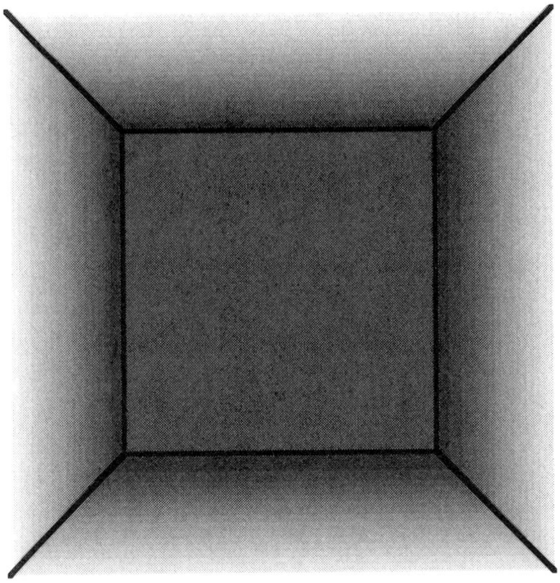

FIGURE 11. The toric base of the local $\mathbb{CP}^1 \times \mathbb{CP}^1$ geometry.

to extend this field theory to the full-fledged string theory. After this discussion we will be in a position to appreciate why Calabi-Yau threefolds are particularly relevant spaces for the topological string. We then plunge into a discussion of the two different variations of the topological string (A and B models) and their observables, with a brief intermezzo on their holomorphic properties, and finish with a description of exactly what is computed by the topological string at genus zero.

4.1. Sigma models and $\mathcal{N} = (2,2)$ supersymmetry. The string theories in which we will be interested (both the ordinary physical version and the topological version) have to do with maps from a surface Σ to a target space X. Roughly, in string theory one integrates over all such maps $\phi : \Sigma \to X$ as well as over metrics on Σ, weighing each map by the Polyakov action:[6]

$$\int \mathrm{D}\phi \, \mathrm{D}g \; e^{-\int_\Sigma |\partial\phi|^2}. \tag{4.1}$$

The integral over ϕ alone defines a two-dimensional quantum field theory which is called a "sigma model into X"; its saddle points are locally area-minimizing surfaces in X. Because we are integrating both over ϕ and over metrics on Σ, one often describes the string theory as obtained by coupling the sigma model to two-dimensional quantum gravity.

[6]Actually, this is the Polyakov action for the bosonic string; we are really interested in the superstring, for which there are extra fermionic degrees of freedom, but we are suppressing those for simplicity.

Classically, the sigma model action depends only on the conformal class of the metric g, so that the integral over metrics can be reduced to an integral over conformal structures — or equivalently, to an integral over complex structures on Σ. For the string theory to be well defined we need this property to persist at the quantum level, but this turns out to be a nontrivial restriction on the allowed X; namely, requiring that the sigma model should be conformally invariant even after including one-loop quantum effects on Σ, one finds the condition that X should be Ricci flat.

For generic X one might expect even more conditions to appear when one considers higher-loop quantum effects; this does happen in the bosonic string, but mercifully not in the superstring provided that X is Kähler. The reason why the Kähler condition is so effective in suppressing quantum corrections is that it is related to $(2,2)$ supersymmetry of the 2-dimensional sigma model, and hence implies bose/fermi cancellations in loops on the worldsheet.[7] This $(2,2)$ supersymmetry is also crucial for the definition of the topological string, so we now discuss it in more detail.

The statement of $\mathcal{N}=2$ supersymmetry means that there are 4 worldsheet currents

$$J, G^+, G^-, T, \tag{4.2}$$

with spins $1, \frac{3}{2}, \frac{3}{2}, 2$ respectively, and with prescribed operator product relations. These operators get interpreted as follows: T is the usual energy-momentum tensor; $G^\pm$ are conserved supercurrents for two worldsheet supersymmetries; J is the conserved current for the $U(1)$ R-symmetry of the $\mathcal{N}=2$ algebra, under which $G^\pm$ have charges ± 1. The modes of these currents act on the Hilbert space of the worldsheet theory.

In the case of the sigma model on X, these currents are analogous (in the "B-model" case — see below) to the operators

$$\deg, \overline{\partial}, \overline{\partial}^\dagger, \Delta \tag{4.3}$$

acting on $\Omega^*(LX)$, the space of differential forms on the loop space of X. (This analogy arises because the loop space is roughly the configuration space of the sigma model on X.) This identification suggests that among the operator product relations of the $\mathcal{N}=2$ algebra should be

$$(G^+)^2 \sim 0, \tag{4.4}$$

$$(G^-)^2 \sim 0, \tag{4.5}$$

$$G^+G^- \sim T + J; \tag{4.6}$$

these relations indeed hold and they will play a particularly important role in what follows.

---

[7]Note that this "worldsheet" supersymmetry is different from the spacetime supersymmetry we discussed in the previous section, although the Kähler condition on X is ultimately responsible for both, and there are arguments which relate one to the other.

In the case where X is Calabi-Yau, so that the sigma model is conformal, we can make a further refinement: each of the currents (4.2) is a sum of two commuting copies, one "left-moving" (holomorphic) and one "right-moving" (antiholomorphic). We thus obtain two copies of the $\mathcal{N} = 2$ algebra, which we write $(J, G^{\pm}, T)$ and $(\overline{J}, \overline{G}^{\pm}, \overline{T})$; this split structure is referred to as $\mathcal{N} = (2,2)$ supersymmetry. This structure of $\mathcal{N} = (2,2)$ superconformal field theory — the operators listed above as well as the Hilbert space on which they act — should be regarded as an invariant associated to the Calabi-Yau manifold X; from it one can recover various more well-known invariants such as the Dolbeault cohomology groups of X, but the full superconformal field theory is a considerably more subtle object, as we will see.

4.2. Twisting the $\mathcal{N} = (2,2)$ supersymmetry. Given an $\mathcal{N} = (2,2)$ superconformal field theory as described in the previous section, there is an important construction which produces a "topological" version of the theory. One can think of this procedure as analogous to the passage from the de Rham complex $\Omega^*(X)$ to its cohomology $H^*(X)$: while the cohomology contains less information than the full de Rham complex, the information it does contain is far more easily organized and understood. So how do we construct this topological version of the SCFT? Guided by the relation $(G^+)^2 \sim 0$ and the above analogy, we might try to form the cohomology of the zero mode of G^+. In fact this is not quite possible, because G^+ has the wrong spin, namely $3/2$; in order to obtain a scalar zero mode we need to begin with an operator of spin 1.

This problem can be overcome, as explained in [**6**] (see also [**7**]), by "twisting" the sigma model. The twist can be understood in various ways, but one way to describe it is as a shift in the operator T:

$$T_{\text{new}} = T_{\text{old}} - \frac{1}{2}\partial J. \tag{4.7}$$

This shift has the effect of changing the spins of all operators by an amount proportional to their $U(1)$ charge q,

$$S_{\text{new}} = S_{\text{old}} - \frac{1}{2}q. \tag{4.8}$$

After this shift the operators (G^+, J) have spin 1 while (T, G^-) have spin 2.[8] Now we can define $Q = G_0^+$, which makes sense on arbitrary Σ and obeys $Q^2 = 0$, and restrict our attention to only observables which are annihilated by Q.

In this context one often calls Q a "BRST operator," since the restriction to observables annihilated by a nilpotent fermionic Q is precisely how one implements gauge invariance in the BRST formalism for quantization of gauge theories. Here we have not obtained Q from the BRST procedure. Nevertheless, the structure of the twisted $\mathcal{N} = 2$ algebra is isomorphic to one which *is* obtained from the usual BRST procedure, namely that of the

[8]Note that although $G^{\pm}$ now have integer spin, they still obey fermionic statistics!

bosonic string. In that case one has currents (Q, J_{ghost}) of spin 1 and (T, b) of spin 2, where (Q, b) are the BRST current and antighost corresponding to the diffeomorphism symmetry on the bosonic string worldsheet[9]; the isomorphism to the twisted $\mathcal{N} = (2,2)$ algebra is

$$(G^+, J, T, G^-) \leftrightarrow (Q, J_{\text{ghost}}, T, b). \tag{4.9}$$

4.3. Constructing the string correlation functions. In the last subsection we noted that the twisted $\mathcal{N} = 2$ algebra is isomorphic to a subalgebra of the symmetry algebra of the bosonic string. In particular, this subalgebra includes the b antighost, which is the crucial element needed for the computation of correlation functions in the bosonic string. Namely, the b antighost provides the link between CFT correlators, computed on a fixed worldsheet Σ, and string correlators, which involve integrating over all metrics on Σ; one sees this link by performing the Faddeev-Popov procedure, which reduces the integral over metrics on Σ to an integral over the moduli space $\mathcal{M}_g$ of genus g Riemann surfaces, with the b ghosts providing the measure. The genus g free energy of the bosonic string obtained in this way is[10]

$$\int_{\mathcal{M}_g} \left\langle \left| \prod_{i=1}^{3g-3} b(\mu_i) \right|^2 \right\rangle. \tag{4.10}$$

Here the symbol $\langle \cdots \rangle$ denotes a CFT correlation function. The $3g-3$ μ_i are "Beltrami differentials," anti-holomorphic 1-forms on Σ with values in the holomorphic tangent bundle; they span the space of infinitesimal deformations of the $\overline{\partial}$ operator on Σ, which is the tangent space to $\mathcal{M}_g$. Then $b(\mu_i)$ is an operator obtained by integrating the b-ghost against μ_i:

$$b(\mu) = \int_\Sigma b_{zz} \mu^z_{\bar{z}}. \tag{4.11}$$

More abstractly, b is an operator-valued 1-form on $\mathcal{M}_g$, so the expectation value of the product of $3g-3$ copies of b gives a holomorphic $(3g-3)$-form; taking both the holomorphic and antiholomorphic pieces we then get a $(6g-6)$-form, which can be integrated over $\mathcal{M}_g$.

Now comes the important point: since the twisted $\mathcal{N} = 2$ superconformal algebra is isomorphic to the algebra appearing in the bosonic string, we can now define a "topological string" from the correlation functions of the $\mathcal{N} = (2,2)$ SCFT on fixed Σ, by repeating (4.10) with b replaced by G^-:

$$F_g = \int_{\mathcal{M}_g} \left\langle \left| \prod_{i=1}^{3g-3} G^-(\mu_i) \right|^2 \right\rangle. \tag{4.12}$$

---

[9]We are using the notation Q both for the current in the bosonic string and for its zero mode.

[10]Strictly speaking this is the answer for $g > 1$; the expression (4.10) has to be slightly modified for $g = 0, 1$ because the sphere and torus admit nonzero holomorphic vector fields.

The formula (4.12) should also be understood as coming from coupling the twisted $\mathcal{N} = (2, 2)$ theory to topological gravity — see [**6**] for some discussion.

One then defines the full topological string free energy to be

$$\mathcal{F} = \sum_{g=0}^{\infty} \lambda^{2-2g} F_g, \tag{4.13}$$

where λ is the "string coupling constant" weighing the contributions at different genera.[11] Finally, the topological string partition function is defined as

$$Z = \exp \mathcal{F}. \tag{4.14}$$

4.4. Why Calabi-Yau threefolds? From our present point of view, the construction of the topological string would have made sense starting from any $\mathcal{N} = (2, 2)$ SCFT, and in particular, the sigma model on any Calabi-Yau space X would suffice. On the other hand, for the physical string, there is a good reason to focus on Calabi-Yau threefolds. Namely, if we look for backgrounds which could resemble the real world, we find an obvious constraint: to a first approximation, the real world looks like 4-dimensional Minkowski space M. On the other hand, conformal invariance of the SCFT coupled to worldsheet supergravity requires the total dimension of spacetime to be 10. To reconcile these two statements one is naturally led to consider backgrounds $M \times X$, where X is some compact 6-dimensional space, small enough that it cannot be seen directly, either by the naked eye or by any experiment we have so far been able to do. Studying string theory on $M \times X$, one finds that the *internal* properties of X lead to physical consequences for the observers living in M. Conversely, the four-dimensional perspective on the string theory computations sheds a great deal of light on the geometry of X, as we will see.

Remarkably, it turns out that the case of Calabi-Yau threefolds is special for the topological string as well. Namely, although one can define F_g for any Calabi-Yau d-fold, this F_g actually vanishes for all $g \neq 1$ unless $d = 3$! This follows from considerations of charge conservation: namely, the topological twisting turns out to introduce a background $U(1)$ charge $d(g-1)$. In order for the correlator appearing in (4.12) to be nonvanishing, the insertions which appear must exactly compensate this background charge; but the insertions consist of $3g-3$ G^- operators, so they have total charge $-3(g-1)$, hence the correlator vanishes unless $d = 3$.

4.5. A and B twists. We are almost ready to discuss the geometric meaning of the topological string, but there is one subtlety to take care of first. In Section 4.2 we glossed over an important point: although we chose the operator G^+ for our BRST supercharge Q, we could equally well have

[11] This expression is only perturbative; it should be understood in the sense of an asymptotic series in λ.

chosen G^-. The latter possibility corresponds to an opposite twist where we replace (4.7) by

$$T_{\text{new}} = T_{\text{old}} + \frac{1}{2}\partial J. \tag{4.15}$$

With this twist it is G^- rather than G^+ which will have spin 1. We have a similar freedom in the antiholomorphic sector, so altogether there are four possible choices of twist, corresponding to choosing for the BRST operators

$$(G^+, \overline{G}^+) : \text{A model} \tag{4.16}$$

$$(G^-, \overline{G}^-) : \overline{\text{A}} \text{ model} \tag{4.17}$$

$$(G^+, \overline{G}^-) : \text{B model} \tag{4.18}$$

$$(G^-, \overline{G}^+) : \overline{\text{B}} \text{ model} \tag{4.19}$$

We have listed each choice together with the name usually given to the corresponding topological string. The $\overline{\text{A}}$ ($\overline{\text{B}}$) model is related to the A (B) model in a trivial way, namely, all correlators are just related by an overall complex conjugation; so essentially we have two distinct ways to make a topological string theory from a given Calabi-Yau X, namely the A and B models.

4.6. Observables and correlation functions. So far we have described how to start with the Calabi-Yau space X and construct two topological string theories called the A and B models. Now let us begin to discuss the observables of these models and the meaning of the correlation functions.

In the A model case, the combined BRST operator $Q + \overline{Q}$ turns out to be the d operator on X, and its cohomology is the de Rham cohomology $H^*_{\text{dR}}(X)$. It is natural to impose an additional "physical state" constraint which leads to considering only the degree $(1,1)$ part of this cohomology. A $(1,1)$ form corresponds to a deformation of the Kähler form on X, so finally, the observables of the A model which we are considering are deformations of the Kähler moduli of X. Furthermore, one can show directly that correlation functions computed in the A model are *independent* of the chosen complex structure on X; namely, one shows that the operators which deform the complex structure are Q-exact, so that they decouple from the computation of the string amplitudes.

In the B model case the space of physical states in the BRST cohomology again consists of objects of bidegree $(1,1)$, but this time the complex in question is the $\overline{\partial}$ cohomology with values in $\wedge^* TX$, so the observables are $(0,1)$-forms with values in TX, i.e. Beltrami differentials on X. As we discussed before, these Beltrami differentials correspond to deformations of the complex structure of X; so the observables of the B model are deformations of complex structure. Similarly to the A model case, one shows that the B model correlation functions are independent of the Kähler structure.

In sum,

$$\text{A model on } X \leftrightarrow \text{Kähler moduli of } X, \tag{4.20}$$

$$\text{B model on } X \leftrightarrow \text{complex moduli of } X. \tag{4.21}$$

Now, what do the correlation functions in the A and B models actually mean mathematically? Usually the correlation functions in a quantum field theory are hard to define because of the complexity inherent in the path integral over an infinite-dimensional field space. In the present case we are indeed computing a path integral $\int e^{-S}$, but this path integral is significantly simplified by the fermionic Q symmetry [**7**]: it reduces to a sum of local contributions from the fixed points of Q! The rest of the field space contributes zero, because one can introduce coordinates in which Q acts by an infinitesimal shift of a Grassmann coordinate θ, and then note that the integral over that one coordinate gives

$$\int \mathrm{d}\theta \, e^{-S} = 0. \tag{4.22}$$

This follows from the standard rules for Grassmann integration, and the fact that Q is a symmetry of the path integral, so that S is independent of θ.

So the path integral is localized on Q-invariant configurations. In the B model these turn out to be simply the constant maps $\phi : \Sigma \to X$, obeying $\mathrm{d}\phi = 0$. In this sense the string worldsheet reduces to a point on X, so the B model is "local," and its correlation functions are those of a field theory on X. It turns out that these correlation functions compute quantities determined by the *periods* of the holomorphic 3-form Ω, which are sensitive to changes in the complex structure.

In the A model, on the other hand, one finds the condition $\overline{\partial}\phi = 0$, which requires only that the map $\phi : \Sigma \to X$ be holomorphic; such a map is called a *worldsheet instanton*. In nontrivial instanton sectors the string worldsheet does not reduce to a point. From this point of view the fact that the A model depends on Kähler moduli is easy to understand; it arises simply because each worldsheet instanton is weighted by the factor

$$e^{-\int_C k} \tag{4.23}$$

i.e., the area of the curve $C \subset X$ which is the image of the string worldsheet in X. The sum over instanton sectors is a complicated structure, non-local from the point of view of X, and therefore the A model does not reduce straightforwardly to a field theory on X.

Note that the B model moduli (the periods) are naturally complex numbers themselves, while the A model moduli (volumes of 2-cycles) are real numbers, so we seem to have a serious asymmetry between the two moduli spaces and hence between the A and B models. As we mentioned earlier, the symmetry between the two moduli spaces is restored by including an extra class $B \in H^2(X, \mathbb{R})$. When B is included, the weighting factor for a

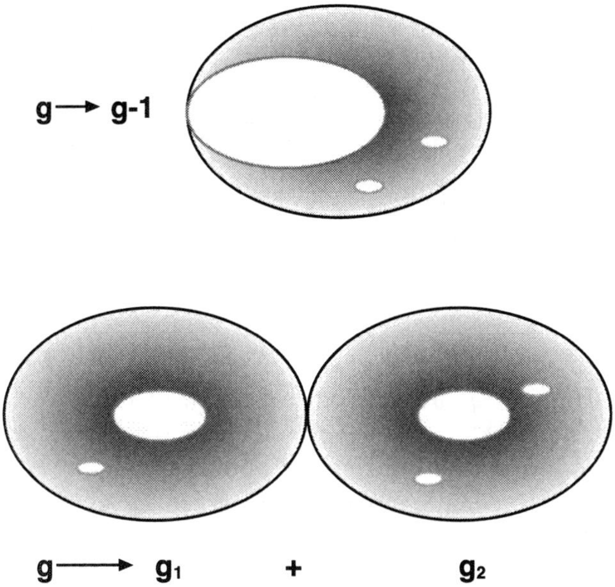

FIGURE 12. Degenerations of a Riemann surface of genus g, corresponding to boundary components of the moduli space $\mathcal{M}_g$.

worldsheet instanton becomes

$$e^{-\int_C k+iB}. \tag{4.24}$$

We will combine k and B into a single modulus $t = k + iB \in H^2(X, \mathbb{C})$.

4.7. Holomorphic anomaly. As we have discussed above, the A and B models each depend on only "half" the moduli of X, namely the Kähler and complex moduli respectively. In fact even more is true: in each case the partition function *formally* depends only holomorphically on its moduli. One sees this by computing the antiholomorphic derivative of a correlator, which amounts to inserting the operator corresponding to the antiholomorphic deformation into the correlation function. This operator is BRST-exact, so one might expect that it is decoupled from correlation functions of BRST-invariant operators. However, the G^- insertions in the definition (4.12) of the correlation function are not BRST-exact; taking this into account one finds that the antiholomorphic derivative of the correlator is the integral of a total derivative over the moduli space $\mathcal{M}_g$. Such an integral would vanish if the moduli space were compact, but since it is not compact one has to worry about contributions from the boundary; indeed there are such contributions, so the partition function is not quite holomorphic as a function of the moduli. Nevertheless its antiholomorphic dependence can be determined precisely; it is expressed in terms of a "holomorphic anomaly equation" derived in [**8, 9**]. Through the anomaly equation $\overline{\partial} F_g$ gets related to the $F_{g'}$

with $g' < g$, corresponding to boundaries of moduli space where some cycle of the genus g surface shrinks — see Figure 12.

In the case of the B model in genus 1, the holomorphic anomaly is familiar to mathematicians; it is related to the curvature of the determinant line bundle which obstructs the construction of a holomorphic $\det \overline{\partial}$ [**10**]. The full holomorphic anomaly in the B model, including all genera, can be interpreted as the statement that the partition function transforms as a *wavefunction* obtained by quantizing the symplectic space $H^3(X, \mathbb{R})$ [**11, 12**].

4.8. Genus zero. After all these preliminaries, we can begin to discuss the geometric content of the topological string. It is natural to begin with the simplest case, namely genus zero; it turns out that this case already contains a lot of interesting information about X.

4.8.1. *A model.* In the A model one finds for the genus zero free energy

$$F_0 = \int_X k \wedge k \wedge k + \sum_{n \in H_2(X,\mathbb{Z})} \sum_{m=1}^{\infty} d_n \frac{e^{-\langle n,t\rangle m}}{m^3}. \tag{4.25}$$

The first term is the classical contribution in the sense of worldsheet perturbation theory; it corresponds to the zero-instanton sector, where the string reduces to a point, and just gives the volume of X. The second term is more interesting since it contains information about worldsheet instantons. Its form is intuitive, at least if we focus on the $m = 1$ term: we sum over all $n \in H_2(X, \mathbb{Z})$, the homology classes of the image of the worldsheet, and weigh each instanton by the factor $e^{-\langle n,t\rangle}$ giving the complexified area. The interesting information is then contained in the number d_n which counts the number of holomorphic maps in the homology class n.[12]

The sum over m reflects the subtlety that there are contributions from "multi-wrappings," maps $\Sigma \to X$ which are m-to-one; these lead to a universal correction, determined by the geometry of maps $S^2 \to S^2$, captured by the factor $1/m^3$.

4.8.2. *B model.* To write the B model partition function we introduce a convenient coordinate system for the complex moduli space. To describe it we first discuss the space $H^3(X, \mathbb{C})$, which has the Hodge decomposition

$$\begin{array}{ccccccccc} H^3 & = & H^{3,0} & \oplus & H^{2,1} & \oplus & H^{1,2} & \oplus & H^{0,3}, \\ h^3 & = & 1 & + & h^{2,1} & + & h^{2,1} & + & 1. \end{array} \tag{4.26}$$

[12]Sometimes this number needs some extra interpreting from the mathematical point of view: it could be that the holomorphic maps are not isolated, so that there is a whole moduli space of such maps. Nevertheless, the virtual or "expected" dimension of this moduli space is always zero (when X is a Calabi-Yau threefold); roughly this means that one can define a sensible "number of maps" even when the actual dimension happens to be nonzero. The index computation showing that the virtual dimension vanishes when $d = 3$ is in fact isomorphic to the charge-conservation computation which showed that general F_g are only nontrivial when $d = 3$.

(The fact that $h^{3,0} = h^{0,3} = 1$ reflects the fact that a Calabi-Yau space has a unique nonvanishing holomorphic 3-form up to scalar multiple.) Therefore $H_3(X, \mathbb{R})$ has real dimension $2h^{2,1} + 2$. Now we choose a symplectic basis of $H_3(X, \mathbb{Z})$; this amounts to choosing 3-cycles A^i, B_j, for $i = 1, \dots, h^{2,1} + 1$ and $j = 1, \dots, h^{2,1} + 1$, with intersection numbers

$$A^i \cap A^j = 0, \quad B_i \cap B_j = 0, \quad A^i \cap B_j = \delta^i_j. \tag{4.27}$$

Note that $h^{2,1}(X)$ is the complex dimension of the moduli space of complex structures (this identification is obtained by using the holomorphic 3-form to convert Beltrami differentials to $(2,1)$-forms.) This suggests that we could try to get coordinates on the moduli space by defining

$$X^i = \int_{A^i} \Omega. \tag{4.28}$$

Actually this gives $h^{2,1}+1$ complex coordinates corresponding to the $h^{2,1}+1$ A cycles, one more than the $h^{2,1}$ needed to cover the moduli space. The reason for this overcounting is that Ω is not quite unique for a given complex structure — it is unique only up to an overall complex rescaling, so from (4.28), the X^i are also ambiguous up to an overall rescaling. Thus we have the right number of coordinates after accounting for this rescaling; and indeed the periods over the A cycles do determine the complex structure. Thus we say that the X^i give "homogeneous coordinates" on the moduli space.

What about the periods over the B cycles? Writing[13]

$$F_i = \int_{B_i} \Omega \tag{4.29}$$

it follows from the above that they must be expressible in terms of the A periods,

$$F_i = F_i(X^j). \tag{4.30}$$

(Of course, since our choice of symplectic basis was arbitrary, and in particular we could have interchanged the A and B cycles, one could equally well write $X^i = X^i(F_j)$.)

We are almost ready to write the B model genus zero free energy, but we need one more fact, namely the statement of "Griffiths transversality." Recall that $\Omega \in H^{3,0}(X, \mathbb{C})$. Now work in a local complex coordinate system in which $\Omega = f(z) dz_1 \wedge dz_2 \wedge dz_3$, and consider a variation of complex structure given by a Beltrami differential μ, which changes the local complex coordinates by $dz_i \mapsto dz_i + \mu_i^{\bar{j}} d\bar{z}_j$. Then expanding in dz and $d\bar{z}$ one sees that to first order in μ, the variation of Ω satisfies $\delta\Omega \in H^{3,0} \oplus H^{2,1}$, and

[13]There is an unfortunate clash of notation here; the F_i we define here are *not* the genus i free energy, although below we will consider the genus 0 free energy, which we will write simply as F!

the second-order variations similarly have $\delta\delta\Omega \in H^{3,0} \oplus H^{2,1} \oplus H^{1,2}$. This implies

$$\int_X \delta\Omega \wedge \Omega = 0, \tag{4.31}$$

$$\int_X \delta\delta\Omega \wedge \Omega = 0. \tag{4.32}$$

Using this fact and the "Riemann bilinear identity," which states that for closed 3-forms α, β one has

$$\int_X \alpha \wedge \beta = \int_{A^i} \alpha \int_{B_i} \beta - \int_{A^i} \beta \int_{B_i} \alpha, \tag{4.33}$$

one can prove that

$$\frac{\partial}{\partial X^i} F_j = \frac{\partial}{\partial X^j} F_i. \tag{4.34}$$

This is the integrability condition which allows one to define a new function F:

$$F_i = \frac{\partial}{\partial X^i} F. \tag{4.35}$$

The F so defined is the genus zero free energy of the B model. Strictly speaking, F is not quite a function on the complex moduli space, because it depends on the choice of the overall scaling of Ω; under $\Omega \mapsto \xi\Omega$ one has $F \mapsto \xi^2 F$. So F is homogeneous of degree 2 in the homogeneous coordinates X^i on the moduli space; geometrically speaking, it is a section of a line bundle over the moduli space rather than an honest function.[14] It is given by a simple formula

$$F = \frac{1}{2} X_i F^i. \tag{4.36}$$

4.8.3. *Comparing the A and B models.* We have just described the content of the A and B models at genus zero. Note that in contrast to the A model, which involved an infinite sum over worldsheet instantons, weighed by the integral coefficients d_n, the B model free energy was determined purely by "classical" geometry (the periods) and has no obvious underlying integral structure. These properties also persist to higher genera. In this sense one could say that the B model is easy to compute, and contains relatively boring information, while the A model is hard to compute but contains more interesting information. On the other hand, it is the A model free energy which is easier to define; at least formally it just counts holomorphic maps, whereas even to define the B model we had to introduce the notion of special coordinates!

[14] Even this more refined description is still a little misleading, because F also depends on the choice of A and B cycles, i.e. the choice of a special coordinate system. For a fixed such choice one obtains a homogeneous section F as we described; if one makes a symplectic transformation of the basis, F transforms by an appropriate Legendre transform.

5. Computing the topological amplitudes

Having defined the topological string theory and seen that it is related to some quantities of geometric interest, the next step is to learn how to compute the topological amplitudes at all genera. In principle they could be computed using their definition (4.12), i.e. by direct integration over the moduli space of Riemann surfaces. But this is too hard for all but the very simplest amplitudes; if this were the only method at our disposal, topological string theory would be just a mathematical curiosity. Instead it is a powerful tool, because a variety of techniques have been discovered which allow one to compute topological string amplitudes not only at tree level but to all genera! In this section we will summarize the various major techniques for computation of topological string amplitudes.

First we describe mirror symmetry, a technique which allows one to exploit the simplicity of the B model for computations in the A model. It was first applied in genus zero, since that is where the B model amplitudes are easiest to compute. The B model computation was subsequently extended to higher genera using the holomorphy of the amplitudes, thus effectively solving the mirror A model at higher genera; we briefly indicate how this extension goes. Next we discuss an alternative approach to computation of topological amplitudes which exploits a duality between the topological open and closed string; this approach yields results at all genera for a particular class of non-compact geometries. Along the way we sketch the meaning of branes in the topological string and their target space field theories. The results obtained from the open/closed duality suggest the existence of a more powerful method for computations of A model amplitudes in arbitrary toric geometries; the last three subsections are devoted to this method, known as the "topological vertex." First we describe what the vertex is; next we sketch a method of computing it using mirror symmetry and the symmetries of the B model; and finally we describe an interpretation directly in the A model, where the vertex is understood as a sum over fluctuations of Kähler geometry at the Planck scale, i.e., the quantum foam.

5.1. Mirror symmetry. In the last section we concluded that while the A model on a Calabi-Yau threefold M contains some very interesting geometric information about holomorphic curves in M, it is the B model which is easier to compute. Remarkably, it is possible to exploit the simplicity of the B model to make computations in the A model! Namely, the A model on a Calabi-Yau space M is often equivalent to a B model on a "mirror" Calabi-Yau space W. Therefore computations of the periods of W can be exploited to count holomorphic curves in M.

A good general reference for mirror symmetry is [**4**].

5.1.1. *T-duality.* To understand how such a surprising duality could be true, we consider an example which is in some sense underlying the whole phenomenon: bosonic string theory on a circle S^1 of radius R. The spectrum of physical states of this theory has one obvious quantum number, namely the number w of times the string is wound around S^1. It also has a second quantum number n corresponding to the momentum of the center of mass of the string going around the circle; this momentum is quantized in units of $1/R$, as is familiar from point particle quantum mechanics in compact spaces. The contribution to the worldsheet energy of a state from these two quantum numbers is (in units with $\alpha' = 1$)

$$E_{n,w} = (wR)^2 + \left(\frac{n}{R}\right)^2. \tag{5.1}$$

Note that the set of possible $E_{n,w}$ is invariant under the interchange $R \leftrightarrow 1/R$ — namely $E_{n,w}$ at radius R is the same as $E_{w,n}$ at radius $1/R$! This is the first clue that this interchange might be a symmetry of the full string theory; indeed, there is such a symmetry, called "T-duality," which can be rigorously understood from the worldsheet point of view, and has deep consequences for the target space physics. Indeed, all of the different approaches to understanding mirror symmetry involve T-duality in some essential way [**13, 14**].

Example 5.1 (Mirror symmetry for T^2)**.** The simplest example is one we already mentioned in Section 2.3.1. Namely, given a rectangular torus T^2 with radii R_1, R_2 and defining

$$A = \mathrm{i}\, R_1 R_2, \tag{5.2}$$

$$\tau = \mathrm{i}\, R_2/R_1, \tag{5.3}$$

exchanging $R_1 \leftrightarrow 1/R_1$ is equivalent to exchanging $A \leftrightarrow \tau$. This is an example of mirror symmetry for which M and its mirror W are both T^2, but with different metrics, i.e. different values of the moduli. Anyway, given that the physical string has this T-duality symmetry, one could ask how it gets implemented in the topological theory. Since T-duality exchanges complex and Kähler moduli it would be natural to conjecture that it exchanges the A and B models, and this is indeed the case; the A model on T^2 with Kähler modulus A computes exactly the same quantity as the B model on T^2 with complex modulus $\tau = A$.

Since T^2 has complex dimension $1 \neq 3$, most of the topological string is trivial as we explained in Section 4.4. However, one can still look at the one-loop free energy F_1, and mirror symmetry turns out to be an interesting statement already here. Namely, the B model at one loop computes the inverse of the determinant of the $\overline{\partial}$ operator acting on T^2, in keeping with the general principle that the B model has to do with *local* expressions on the target space. This determinant is the Dedekind η function,

$$\eta(q) = q^{1/24} \prod_{n=1}^{\infty} (1 - q^n), \tag{5.4}$$

where $q = e^{2\pi i \tau}$. On the other hand, the A model at one loop counts maps $T^2 \to T^2$, but according to mirror symmetry, it should also give the η function. This gives a natural interpretation of the integrality of the coefficients in the q-expansion of $1/\eta(q)$. Namely, q gets related to e^{-A} by the mirror map, and from the A model point of view the coefficient of e^{-nA} counts maps which wrap T^2 over itself n times. It can be checked directly that this counting is indeed correct.

5.1.2. *Mirror symmetry for threefolds.* Now what about the case of maximal interest, namely Calabi-Yau threefolds? Here also one might expect a mirror duality. Indeed, this duality was conjectured before a single non-trivial example was known, on the basis of lower-dimensional examples like the one discussed above, and also because from the point of view of the $\mathcal{N} = (2,2)$ algebra the difference between A and B models is purely a matter of convention — considered abstractly, the SCFT has no way of knowing whether it is an A model or a B model. This conjecture turned out to be spectacularly true, and by now many examples of mirror pairs are known, both compact and non-compact.

Here we sketch a physical proof given in [**14**] which encapsulates all known examples of mirror symmetry. Like the T-duality example we gave above, the proof is most naturally stated directly in the physical superstring rather than the topological string; but after twisting it reduces to an equivalence between a topological A model and a topological B model.

So we begin with a toric Calabi-Yau threefold M and realize it concretely via the gauged linear sigma model of [**5**], as we described in Example 3.2. Recall that this model is constructed from a set of chiral superfields Z_i representing the homogeneous coordinates of M, and that its space of vacua is M itself. Then to get the mirror theory to the sigma model on M one splits each Z_i into its modulus and phase as we did before when discussing the toric diagram,

$$Z_i \to (|Z_i|^2, \theta_i), \tag{5.5}$$

and then performs T-duality on the circle coordinatized by θ_i. The T-duality gives a new dual periodic coordinate ϕ_i, and we organize this coordinate together with $|Z_i|^2$ into a new "twisted chiral" superfield

$$Y_i = |Z_i|^2 + \mathrm{i}\,\phi_i. \tag{5.6}$$

Crucially, the dual description in terms of the Y_i has a superpotential:

$$W(Y) = \left(\sum_i Q_i Y_i - t\right)\Sigma + \sum_i e^{-Y_i}. \tag{5.7}$$

Here Σ is the twisted chiral superfield in the $U(1)$ vector multiplet, and Q_i are the $U(1)$ charges of the Z_i. The first term in (5.7) follows from a classical T-duality computation; the really interesting part is the second term. This term was derived in [**14**] from an instanton computation in the gauged linear sigma model. It can also be determined more indirectly (and more easily), as follows [**15**].[15] One compares masses of BPS particle states in

[15]For simplicity we just discuss one chiral superfield Z.

the original theory and in the mirror. In the original theory the field Z has momentum modes with BPS mass $|Q\Sigma|$. After T-duality these momentum modes become winding modes along the T-dual circle, i.e., they should correspond to classical BPS solitons where ϕ increases by $2\pi i$. For such a soliton interpolating between vacua to exist, $W(Y)$ must have critical points which are spaced by $2\pi i$. Moreover, since the BPS mass of a soliton interpolating from y_1 to y_2 is $|W(y_1) - W(y_2)|$, we see that this difference must be equal to $|Q\Sigma|$. Finally, because of the periodicity of ϕ_i, the instanton-generated superpotential can only involve Y_i through its exponential e^{-Y_i}. These conditions are enough to fix $W(Y)$ as in (5.7).

Integrating out Σ then yields the holomorphic constraints in the dual model,

$$\sum_i Q_i Y_i = t, \tag{5.8}$$

and the reduced superpotential,

$$W(Y) = \sum_i e^{-Y_i}. \tag{5.9}$$

The two equations (5.8), (5.9) contain all the information about the dual theory, as we now see in an example.

Example 5.2 (Mirror symmetry for local $\mathbb{CP}^2$)**.** Consider the local $\mathbb{CP}^2$ geometry $\mathcal{O}(-3) \to \mathbb{CP}^2$, for which we discussed the toric realization in Example 3.3. The gauged linear sigma model for this geometry involves four chiral superfields Z_i with charges Q_i,

$$\begin{aligned} Z = &\quad (Z_0, \quad Z_1, \quad Z_2, \quad Z_3), \\ Q = &\quad (-3, \quad 1, \quad 1, \quad 1). \end{aligned} \tag{5.10}$$

The holomorphic constraint in the dual model is

$$-3Y_0 + \sum_{i=1}^{3} Y_i = t, \tag{5.11}$$

and the superpotential is

$$W = \sum_{i=0}^{3} e^{-Y_i}. \tag{5.12}$$

It is convenient to make the change of variables

$$y_i = e^{-Y_i/3}. \tag{5.13}$$

Then, after eliminating Y_0 using (5.11), we are left with the superpotential

$$W = y_1^3 + y_2^3 + y_3^3 + e^{t/3} y_1 y_2 y_3. \tag{5.14}$$

So the mirror of the gauged linear sigma model is a gauged Landau-Ginzburg model, describing uncharged twisted chiral superfields Y_i which interact via the superpotential (5.9). More precisely, the mirror is an *orbifold* of the

Landau-Ginzburg model, because the change of variables (5.13) is not quite one-to-one; the y_i are ambiguous by cube roots of unity, and therefore we have to divide out by the group $\mathbb{Z}_3^2$ which multiplies the y_i by cube roots of unity while leaving W invariant. This is the generic situation: the mirror to an $\mathcal{N} = (2, 2)$ gauged linear sigma model is an orbifolded Landau-Ginzburg model. Note that the complexified Kähler modulus t of the original theory appears in the Landau-Ginzburg model as a modulus of the holomorphic superpotential.

From this Landau-Ginzburg realization one can directly compute the desired genus zero partition function. Nevertheless, one might ask: how is the Landau-Ginzburg theory related to our original claim that the sigma model on the Calabi-Yau geometry should have a mirror which is also a sigma model on a Calabi-Yau? The point is that the Landau-Ginzburg model with superpotential (5.14) is actually equivalent to a sigma model with Calabi-Yau target space: more precisely, one can interpolate from one to the other just by varying Kähler parameters, which are decoupled from the B model correlation functions. After so doing we obtain the mirror to the local $\mathbb{CP}^2$ geometry; it is simply given by the equation $W = 0$, modulo the orbifold action.

Let us look at this geometry a bit more closely. If the y_i are considered as homogeneous coordinates in projective space, then $W = 0$ describes an elliptic curve (torus) since it is a cubic equation in $\mathbb{CP}^2$. Passing to inhomogeneous coordinates we could rewrite it as an equation in two variables,

$$x^3 + z^3 + 1 + e^{t/3}xz = 0. \tag{5.15}$$

Indeed, the mirror geometry in this case is effectively an elliptic curve rather than a Calabi-Yau threefold, in the sense that the B model partition function can be computed solely from the geometry of the elliptic curve. This is a common phenomenon when computing mirrors of noncompact Calabi-Yaus. Nevertheless, the usual statement of mirror symmetry requires a threefold mirror to a threefold; to make contact with that formulation we should add two extra variables u, v which enter the geometry in a rather trivial way, replacing (5.15) by

$$x^3 + z^3 + 1 + e^{t/3}xz = uv. \tag{5.16}$$

These two variables u, v just contribute a quadratic term to the superpotential W in the Landau-Ginzburg realization, so they do not couple to the rest of the physics.

One can similarly derive mirror symmetry for compact Calabi-Yaus with linear sigma model realizations.

Example 5.3 (Mirror symmetry for the quintic threefold)**.** Recall the quintic threefold from Example 2.8. This space can be obtained by starting with the gauged linear sigma model for $\mathcal{O}(-5) \to \mathbb{CP}^4$ and then introducing a superpotential which reduces the space of vacua to the quintic hypersurface in $\mathbb{CP}^4$. Temporarily ignoring this superpotential and repeating the steps above, we get a Landau-Ginzburg model with

$$W = y_1^5 + y_2^5 + y_3^5 + y_4^5 + y_5^5 + e^{t/5} y_1 y_2 y_3 y_4 y_5, \tag{5.17}$$

modulo a $\mathbb{Z}_5^4$ symmetry multiplying the y_i by fifth roots of unity. Now what changes in the mirror if we include the superpotential in the original theory? Remarkably, it turns out that the only effect is to change the fundamental variables of the theory to the y_i instead of Y_i. (One might think that what is the "fundamental variable" is a matter of terminology, but concretely, it affects the measures of integration one uses when computing the B model periods.)

5.1.3. *Super mirror symmetry.* There is another point of view on mirror symmetry for compact Calabi-Yaus realized torically, which is in a sense more direct. Namely, it was observed in [**16**] that the A model on the quintic threefold is in fact equivalent to the A model on a weighted *super* projective space $\mathbb{CP}^{1,1,1,1,1|5}$, with five bosonic directions and one fermionic one. This space is compact but nevertheless can be constructed in a gauged linear sigma model without the need for a superpotential. Since it has $U(1)$ isometries, unlike the quintic threefold, one can T-dualize on phases directly to obtain the mirror. This requires a generalization of the mirror techniques of [**14**] to the case of a chiral superfield Θ whose lowest component is fermionic, which was worked out in [**15**].

The main difference from the bosonic case is that the number of fields is not conserved: namely, since the phase of Θ is bosonic, dualizing it gives a new bosonic chiral superfield X. But since Θ contributes central charge -1 instead of $+1$ in the sigma model, one also has to get two more fermionic fields η, χ on the mirror side, since the central charges on the two sides must be equal. As in the bosonic case the superpotential can be determined by comparing BPS masses, and it turns out to be

$$W(\Sigma, X, \eta, \chi) = -Q\Sigma(X - \eta\chi) + e^{-X}. \tag{5.18}$$

This superpotential defines the mirror Landau-Ginzburg model.

In addition to providing a streamlined derivation of the mirror periods for hypersurfaces in toric varieties, super mirror symmetry is important in its own right, particularly in light of a recent application of topological strings on supermanifolds to a twistorial reformulation of $\mathcal{N} = 4$ super Yang-Mills theory [**17**]. In that case the supermanifold in question is the super twistor space $\mathbb{CP}^{3|4}$, and computations in [**15**] showed that its mirror is (at least in the limit where $\mathbb{CP}^{3|4}$ has large volume) a quadric hypersurface in $\mathbb{CP}^{3|3} \times \mathbb{CP}^{3|3}$. This result may be relevant for gauge theory, since $\mathbb{CP}^{3|3} \times \mathbb{CP}^{3|3}$ can also be viewed as a twistor space, which is related at least classically to $\mathcal{N} = 4$ super Yang-Mills [**18**]; one might expect that a topological string on $\mathbb{CP}^{3|3} \times \mathbb{CP}^{3|3}$ could give an alternative twistorial version of $\mathcal{N} = 4$ super Yang-Mills [**15, 17, 19**].

5.2. Holomorphy and higher genera. So far we have discussed the topological amplitudes only at genus zero. More generally, one can compute all the F_g using the fact that they depend only holomorphically on moduli.[16] We think of F_g as a holomorphic section of a line bundle over the moduli space. Such objects are highly constrained — recall that a holomorphic line bundle over a compact space has only a finite-dimensional space of holomorphic sections. The Calabi-Yau moduli spaces under consideration are compact, or can be compactified by adding some points at infinity, where the singular behavior of the F_g can be constrained by geometrical considerations; hence the F_g are basically determined by holomorphy, up to a finite-dimensional ambiguity at each g [**9**]. Using some integrality properties of the F_g which we discuss in Section 6.3, this ambiguity can also be fixed; this leads to a practical method for computing the F_g, which has been applied to high degrees and genera [**20**].

5.3. Branes and large N dualities. Another approach to computing the F_g depends on the notion of "large N duality." Such dualities have played a starring role in the physical string theory over the last few years [**21, 22**]; as it turns out, they are equally important in the topological string [**23, 24**]. We now turn to an overview of how they are realized in this context.

5.3.1. *D-branes in the topological string.* Large N dualities relate open string theory in the presence of N D-branes to closed string theory in the gravitational background those D-branes produce; so in order to discuss their topological realization, we have to begin by explaining the notion of D-brane in the topological string.

From the worldsheet perspective, a D-brane simply corresponds to a boundary condition which can be consistently imposed on worldsheets with boundaries. In the topological case what we mean by "consistency" is that the boundary condition preserves the BRST symmetry. In the A model this condition implies that the boundary should be mapped into a Lagrangian submanifold L of the target Calabi-Yau X [**25**] ("Lagrangian" means that the dimension of L is half that of X and the Kähler form ω vanishes when restricted to L). Such an L should be thought of as a real section of X — a typical 1-dimensional model is the upper half-plane, which ends on the real axis L. If we allow open strings with boundaries on L, we say that we have a D-brane which is "wrapped" on L. We can also include a weighting factor N for each boundary, in which case we say we have N D-branes instead of one.

We will be interested in computing the partition function of the topological open string theory with branes. For this purpose it turns out that taking a target space viewpoint is very convenient: the dynamics of the open

[16] Actually, as we mentioned in Section 4.7, the F_g are not quite holomorphic; but the antiholomorphic dependence is completely determined by the anomaly equation of [**9**] and does not qualitatively affect the discussion to follow.

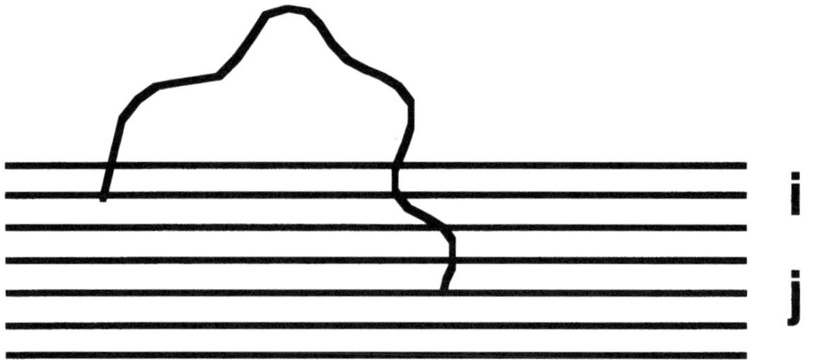

FIGURE 13. A stack of N branes carries a $U(N)$ gauge symmetry; the fundamental and antifundamental gauge indices arise from strings which can end on any of the N branes.

strings ending on branes can be completely described in terms of a string field theory on the branes. What field theory is it? Both in the physical and the topological string theory, the open strings produce a gauge theory on the branes in the low energy limit; for example, in the case of a stack of N coincident branes in oriented string theory in flat space, the fact that strings can end on any of the N branes leads to a $U(N)$ gauge theory. See Figure 13.

In the physical string the gauge theory of the open strings is rather complicated, although at low energies it reduces to Yang-Mills theory. But in the topological A model the situation is much simpler and one can work out the exact open string field theory describing a stack of N branes; it is again a gauge theory, but this time a topological gauge theory, namely $U(N)$ Chern-Simons theory. To see this we first note that our construction of the topological string (and specifically its coupling to worldsheet gravity) was modeled on the bosonic string, and therefore the open string field theory should also be the obtained by the same procedure one uses for the open bosonic string. In the open bosonic string it was shown in [**26**] that the string field theory is an abstract version of Chern-Simons, written

$$S = \int A * QA + \frac{2}{3} A * A * A. \tag{5.19}$$

Specializing to the case of the topological A model, using the dictionary $Q \leftrightarrow \mathrm{d}$, one can show [**25**] that this abstract Chern-Simons in this case boils down to the standard Chern-Simons action for a $U(N)$ connection A,

$$S = \int_L \mathrm{Tr} \left(A \wedge \mathrm{d}A + \frac{2}{3} A \wedge A \wedge A \right), \tag{5.20}$$

possibly corrected by terms involving holomorphic instantons ending on L.[17] In some interesting cases there are no holomorphic instantons and we just get pure Chern-Simons; this happens in particular in the case where L is the S^3 in the deformed conifold T^*S^3.

One can similarly consider the open string field theory on N B model branes. In the case where the branes wrap the full Calabi-Yau threefold X, one gets a holomorphic version of Chern-Simons, with action [**26**]

$$\int_X \Omega \wedge \mathrm{Tr}\left(A\overline{\partial}A + \frac{2}{3}A^3\right). \tag{5.21}$$

Here A is a $u(N)$-valued $(0,1)$ form on X, which we are combining with the $(3,0)$ form Ω so that the full action is a $(3,3)$ form as required. Starting from (5.21), one can also obtain the action for B model branes which wrap holomorphic 0,2,4-cycles inside X, by realizing such lower-dimensional branes as defects in the gauge field on a brane that fills X: the brane charges correspond respectively to the Chern classes c_3, c_2, c_1 of the gauge field.

As an aside, it is interesting that the branes which appear in the A model are wrapping Lagrangian cycles, which are 3-cycles for which the volume is naturally measured by the holomorphic 3-form Ω — the natural object in the B model! Similarly, in the B model the branes turn out to wrap holomorphic cycles, whose volume is measured by the A model field k. This crossover between the A and B models may be a hint of a deeper relation, possibly an S-duality, which is currently under investigation [**19, 27**].

5.3.2. *The geometric transition.* After these preliminaries on branes in the topological string, we are ready to use them to compute closed string amplitudes. The crucial point which makes such a computation possible is that the topological D-branes affect the closed string background; so we first explain how this works.

In the physical superstring D-branes are sources of Ramond-Ramond flux. In the A or B model topological string we expect something similar, but now the flux in question should be the Kähler 2-form or holomorphic 3-form respectively. More precisely, consider a Lagrangian subspace L, on which an A model brane could be wrapped. Since the total dimension of X is 6, we can consider a 2-cycle C which links the 3-dimensional L, similar to the way two curves can link one another inside a space of total dimension 3. The precise meaning of "link" is that $C = \partial S$ for some 3-cycle S which intersects L once; so C is homologically trivial as a cycle in X, although it becomes nontrivial if considered as a cycle in $X \setminus L$. Because C is homologically trivial

[17] In fact, one might ask how the appearance of the Chern-Simons action is consistent with the localization of the open string path integral on holomorphic configurations — one might have expected to get *only* the terms from holomorphic instantons. The resolution is that the localization has to be interpreted carefully because of the non-compactness of the field space; one has to include contributions from "degenerate instantons" in which the Riemann surface has collapsed to a Feynman diagram (with lines replaced by infinitesimal ribbons), and these diagrams precisely account for the Chern-Simons action.

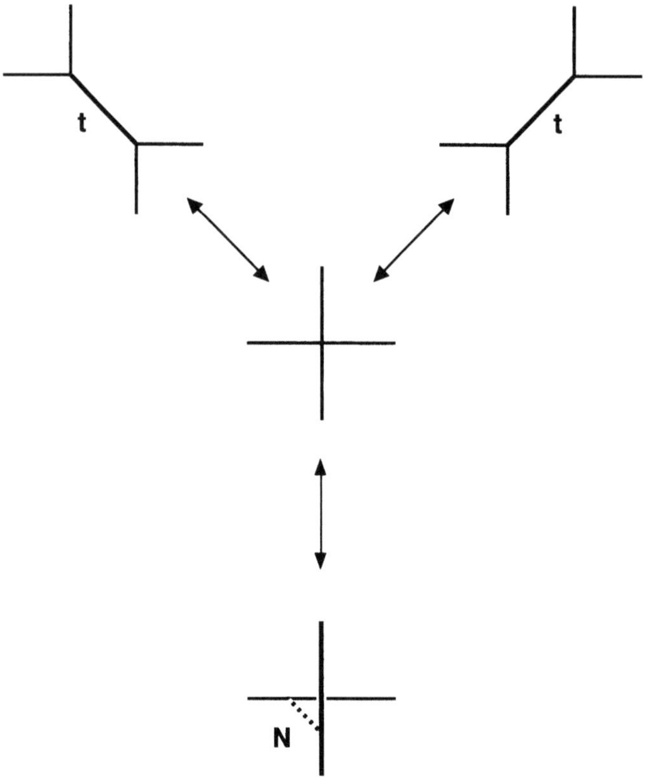

FIGURE 14. The geometric transition between the resolved conifold with Kähler parameter $t = Ng_s$ (above) and deformed conifold with N branes (below).

we must have $\int_C k = 0$ in X, since $\mathrm{d}k = 0$. Now the effect of wrapping N branes on L is to create a flux of the Kähler form through C, namely

$$\int_C k = Ng_s. \tag{5.22}$$

This can be understood by saying that the branes act as a δ-function source for k, i.e., the usual $\mathrm{d}k = 0$ is replaced by

$$\mathrm{d}k = Ng_s\delta(L). \tag{5.23}$$

Similarly, a B model brane on a 2-cycle Y induces a flux of Ω over a 3-cycle linking Y. Note that this phenomenon actually suggests a privileged role for 2-cycles; we could also have B model branes on 0, 4, or 6-cycles, but these branes do not induce gravitational backreaction since there is no candidate field for them to source.

Now let us describe an example in which the closed string backreaction from branes can be used to compute A model closed string amplitudes.

Example 5.4 (Large N duality for the conifold)**.** Consider the A model on the deformed conifold T^*S^3. This geometry is uninteresting from the point of view of the closed A model, since it has no 2-cycles and hence no Kähler moduli; but it contains the Lagrangian 3-cycle S^3 on which we can wrap A model branes. The effect of these branes on the closed string geometry is to create a flux Ng_s of the Kähler form k on the S^2 which links S^3. Now, á la AdS/CFT, let us try to describe the string theory on this geometry in terms of a background without branes. There is an obvious guess for the answer: as we discussed earlier, in addition to the deformed conifold which has a nontrivial S^3 at its core, there is also the resolved conifold which has a nontrivial S^2, and both geometries look the same at long distances. So it is natural to conjecture **[23]** that the dual geometry is the resolved conifold, where the nontrivial S^2 has volume $t = Ng_s$. In the resolved conifold there are no branes anymore, and indeed there is not even a nontrivial cycle where the branes could have been wrapped! The passage from one geometry to the other is referred to as a "geometric transition," and the key is that the A model partition function is the same both before and after the geometric transition.

The geometric transition is summarized in Figure 14. The pictures appearing in that figure require a bit of explanation, though: they are similar, but not identical, to the toric pictures which appeared in Section 3. The full geometry in this case is a $T^2 \times \mathbb{R}$ fibration over the whole of $\mathbb{R}^3$, rather than a T^3 fibration over some bounded region inside $\mathbb{R}^3$. The solid lines represent loci where one of the circles of the T^2 fiber degenerates. At the top of the figure we have the resolved conifold, with its Kähler modulus t (actually there are two different versions of the resolved conifold, related by a relatively mild topology changing transition called a "flop.") Here all the degeneration loci line in a common plane. As $t \to 0$ the resolved conifold approaches the singular conifold, shown in the middle of the figure; the degeneration locus then consists simply of two intersecting lines. Finally separating these two lines in space gives the deformed conifold T^*S^3. The Lagrangian submanifold S^3 can be seen in the resulting picture: namely, it is the T^2 fibration over the dotted line connecting the two degeneration loci.

On the deformed conifold the closed A model is trivial, because there are no Kähler moduli; so the partition function is just the partition function of $U(N)$ Chern-Simons theory on S^3, with level k determined by $g_s = 2\pi i/(k+N)$. On the resolved conifold there are no open strings, so one gets a prediction for the partition function of the closed A model there: namely, from the Chern-Simons side one expects

$$Z(g_s, t) = \prod_{n=1}^{\infty} (1 - q^n Q)^n, \tag{5.24}$$

where $q = e^{-g_s}$ and $Q = e^{-t}$. Note that this expansion has integral coefficients! This seems remarkable from the point of view of the closed string, and might make us wonder whether the closed string partition function has an interpretation as the answer to some counting problem. The answer is

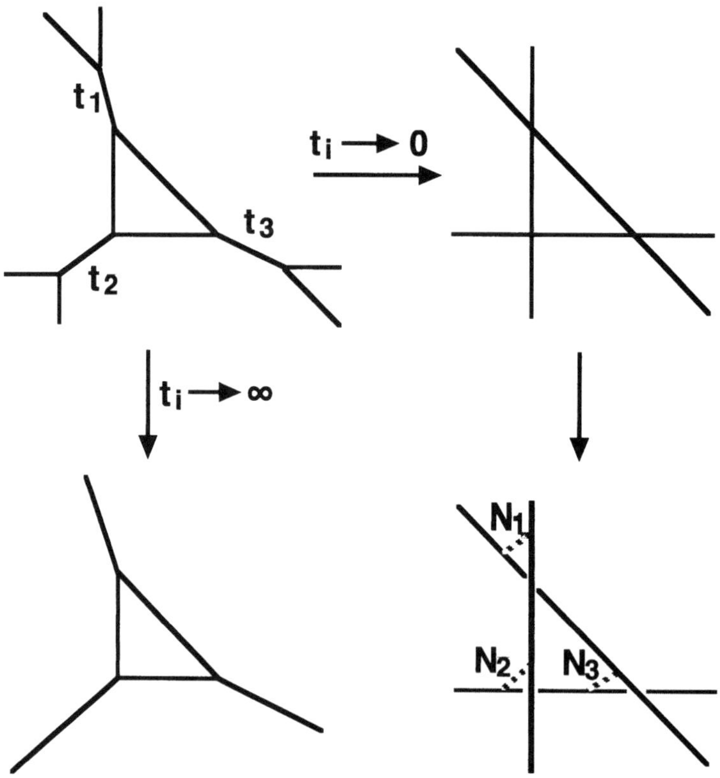

FIGURE 15. The geometric transition relating local $\mathbb{CP}^2$ (lower left) to a rigid geometry with $N_i \to \infty$ branes (lower right).

"yes," as we will see in Section 6.3 when we discuss the application of the topological string to counting BPS states in five dimensions.

One can also use open/closed duality to compute the closed string partition function in more complicated geometries [**28**], as we now discuss.

Example 5.5 (Large N duality for local $\mathbb{CP}^2$). For example, consider the local $\mathbb{CP}^2$ geometry. As shown in Figure 15, we can obtain this geometry as the $t_i = N_i g_s \to \infty$ limit of a geometry with three compact $\mathbb{CP}^1$'s. Namely, in the lower left corner we have local $\mathbb{CP}^2$, which we consider as the $t_i \to \infty$ limit of the more complicated geometry at upper left. This geometry is in turn related by three geometric transitions to the geometry at lower right, which has three Lagrangian S^3's represented by the dotted lines, each supporting N_i A model branes. In this way the closed A model partition function on local $\mathbb{CP}^2$ is identified with the open string partition function on these three stacks of branes; no Kähler moduli remain after the transitions, so the closed string does not contribute anything. Naively, this open string

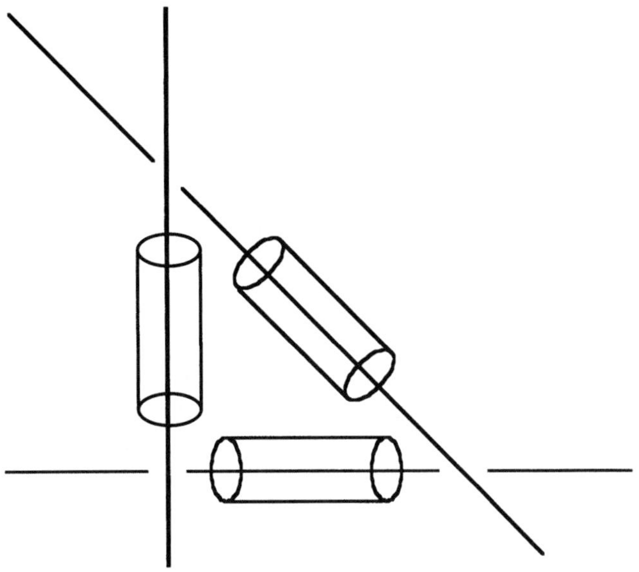

FIGURE 16. Worldsheet instantons, with each boundary on an S^3, which contribute to the A model amplitudes after the transition, or dually, to the A model amplitudes on local $\mathbb{CP}^2$.

partition function would be just the product of three copies of the Chern-Simons partition function, coming from the three S^3's. However, we have to remember that the open string field theory of the A model is not pure Chern-Simons theory; it includes corrections due to worldsheet instantons. In this toric case one can show that the only instantons which contribute are ones in which the worldsheets form tubes connecting two of the Lagrangian S^3's, as shown in Figure 16. Each such tube ends on an unknotted circle in S^3; so in a generic instanton sector each S^3 has two such circles on it, and a careful analysis shows that these circles are in fact linked, forming the "Hopf link." One therefore has to compute the Chern-Simons partition function including an operator associated to the link. This operator was determined in [**29**] and turns out to be given by a sum of Chern-Simons link invariants. Putting everything together [**28**], the full partition function at all genera is a sum over irreducible representations of $U(N)$:

$$Z = \sum_{R_1,R_2,R_3} e^{-t|R_1|} S_{R_1R_2} e^{-t|R_2|} S_{R_2R_3} e^{-t|R_3|} S_{R_3R_1}, \tag{5.25}$$

where $S_{RR'}$ is the Chern-Simons knot invariant of the Hopf link with representations R and R' on the two circles, as defined in [**30**], and $|R|$ is the number of boxes in a Young diagram representing R.

5.4. The topological vertex. Although the geometric transitions we described above lead to an all-genus formula for the A model partition function in the local $\mathbb{CP}^2$ geometry, the method of computation is somewhat

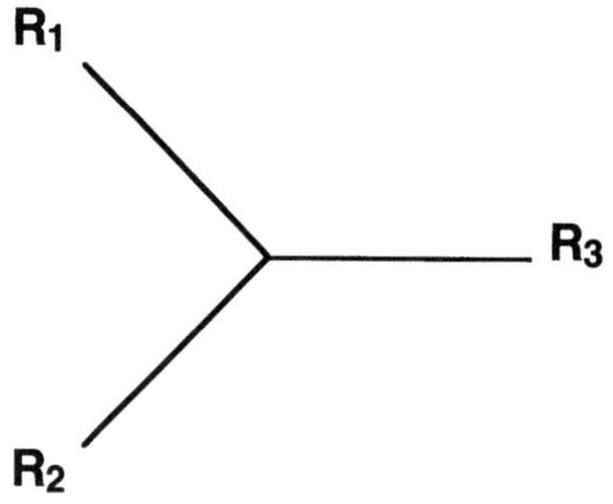

FIGURE 17. The topological vertex, which assigns a function of g_s to any three Young diagrams R_1, R_2, R_3.

unsatisfactory: one obtains local $\mathbb{CP}^2$ only after taking the $t_i \to \infty$ limit of a more complicated geometry. One might have hoped for a more intrinsic method of computation. Indeed there is such a method, and it generalizes to arbitrary toric diagrams, whether or not they come from geometric transitions! The method essentially involves treating the toric diagram (with fixed Kähler parameters) as if it were a Feynman diagram, with trivalent vertices and fixed Schwinger parameters. Namely, one can define a "topological vertex," $C_{R_1 R_2 R_3}(g_s)$, depending on three Young diagrams R_1, R_2, R_3 and on the string coupling g_s [**31**]. See Figure 17. Then one assigns a Young diagram R to each edge of the toric diagram, with a propagator $e^{-t|R|+mC_2(R)}$ for each internal edge, and a factor $C_{R_1 R_2 R_3}(g_s)$ for each vertex.[18] The assignment of representations to edges of the toric diagram is as follows: external edges always carry the trivial representation, while for internal edges one sums over all R.

Of course, the actual vertex $C_{R_1 R_2 R_3}(g_s)$ is rather complicated! It was originally determined in [**31**] using Chern-Simons theory along the lines discussed in Section 5.3.2. Since then two other methods of computing the vertex have appeared, which we will describe in the next two subsections.

5.5. Computing the vertex from W_∞ symmetries. First we briefly describe a target space approach to computing the topological string partition function [**32**]. Namely, suppose we study the A model on a non-compact threefold which has a toric realization as we discussed in Section 3. By mirror symmetry this is equivalent to the B model on a Calabi-Yau of the form

$$F(x, z) = uv, \tag{5.26}$$

with the corresponding holomorphic 3-form

$$\Omega = \frac{du \wedge dx \wedge dz}{u}. \tag{5.27}$$

[18] The integer m appearing in the propagator is related to the relative orientation of the 2-surfaces on which the propagator ends.

We view this geometry as a fibration over the (x, z) plane, with 1-complex-dimensional fibers. At points (x, z) with $F(x, z) = 0$ the fiber degenerates to $uv = 0$, which has two components $u = 0$ and $v = 0$; so $F(x, z) = 0$ characterizes the degeneration locus of the fibration. Contour integration around $u = 0$ on the fiber reduces Ω to

$$\omega = \mathrm{d}x \wedge \mathrm{d}z. \tag{5.28}$$

So the geometry of the Calabi-Yau threefold is captured by an algebraic curve $F(x, z) = 0$, embedded in the (x, z) space; this ambient space is furthermore equipped with the two-form ω. What are the symmetries of this structure? If F were identically zero, then we would just have the group of ω-preserving diffeomorphisms, which form the so-called "W_∞" symmetry. This infinite-dimensional symmetry is extremely powerful. Indeed, even when $F \neq 0$ and the W_∞ symmetry is spontaneously broken, it nevertheless gives constraints on the dynamics of the Goldstone modes which describe deformations of F. But these deformations exactly correspond to complex structure deformations of the Calabi-Yau geometry, which are the objects of study in the B model! Hence this W_∞ symmetry generates Ward identities which act on the closed string field theory of the B model (the "Kodaira-Spencer theory of gravity," described in [**9**].) In fact, these Ward identities are sufficient to completely determine the B model partition function at all genera (and hence the A model partition function on the original toric threefold) — see [**32**].

5.6. Quantum foam. In the last subsection we sketched a derivation of the topological vertex by applying mirror symmetry and then using the B model closed string target space theory. However, one can also obtain the vertex by a direct A model closed string target space computation [**33, 34, 35**]. The string field theory in question is a theory of "Kähler gravity" [**36**], which roughly sums over Kähler geometries with the weight $e^{-\int k^3/g_s^2}$. One can think of this summing over geometries as a kind of "quantum foam" — the spacetime itself is wildly fluctuating and "foamy" at small scales. This feature has long been expected for theories of quantum gravity, but in the case of the topological A model it turns out that one can describe this quantum foam very precisely; namely, there is a simple description of exactly which Kähler geometries should be summed over, and this description enables us to compute the topological vertex.

So let us begin with the problem of computing the A model partition function on the non-compact Calabi-Yau $\mathbb{C}^3$. The simplest geometry which contributes to the quantum foam in this case is simply $\mathbb{C}^3$ itself. The rest of the geometries that contribute may be obtained by making various blow-ups involving the origin $(0, 0, 0) \in \mathbb{C}^3$.[19] The simplest possibility is to just

[19]One might wonder what is special about the origin, since $\mathbb{C}^3$ has a translation symmetry. Actually, there is nothing special about the origin. We are using a toric realization of $\mathbb{C}^3$ to get a $U(1)^3$ action on the space of possible blow-ups, and the claim is that by

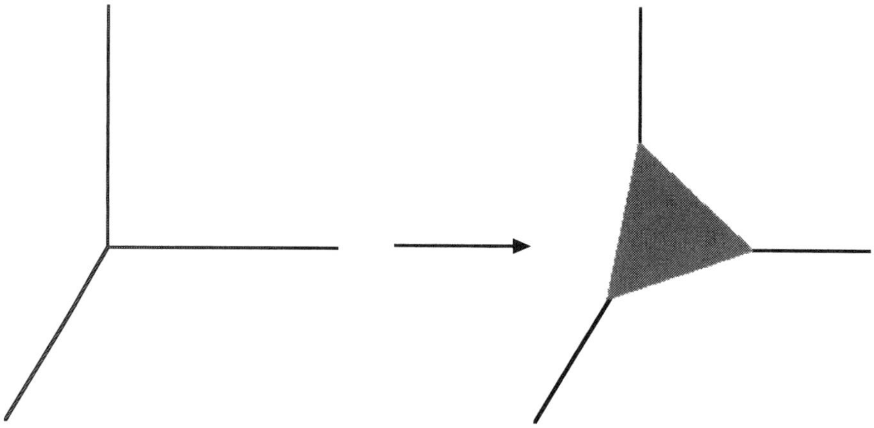

FIGURE 18. Blowing up the origin in $\mathbb{C}^3$ gives a new geometry which is not Calabi-Yau but still contributes to the target space sum in the A model.

blow up the origin once; this leads to the toric diagram shown in Figure 18, where the origin has been replaced by a single $\mathbb{CP}^2$. This new geometry is not Calabi-Yau; the only Calabi-Yau geometry which is asymptotically $\mathbb{C}^3$ is $\mathbb{C}^3$ itself. Nevertheless, it should be included in the target space A model sum; this is not unexpected, since a theory of quantum gravity should sum over off-shell configurations as well as on-shell ones. After blowing up the origin there is a new Kähler modulus t for the size of $\mathbb{CP}^2$; in the A model partition sum this modulus turns out to be quantized, $t = ng_s$, and we sum over all n. In the toric diagram the modulus t is reflected in the size of the triangle representing $\mathbb{CP}^2$, as we discussed in Section 3.

One can also do more complicated blow-ups. For example, after blowing up the origin of $\mathbb{C}^3$, one could then blow up a fixed point on the exceptional divisor $\mathbb{CP}^2$, as shown in Figure 19. We could then blow up another point on the resulting surface, then another, and so on. Continuing in this way one obtains a large class of toric manifolds which are asymptotically $\mathbb{C}^3$; a typical example is shown in Figure 20.

However, it turns out that these blow-ups are not the only configurations that contribute to the A model partition sum. Namely, for any toric manifold obtained by successive blow-ups of points, the interior of the toric diagram is always a convex set; but to reproduce the A model partition function one also has to include generalizations of Figure 20 in which the interior of the diagram is not required to be convex. These generalizations still have an algebro-geometric meaning, which can be roughly explained as follows [**34**].

standard localization techniques, the partition function can be computed considering only blow-ups which are torically invariant. Since the origin is the only point of $\mathbb{C}^3$ that is invariant in the toric representation we chose, this implies that we only consider blow-ups of the origin.

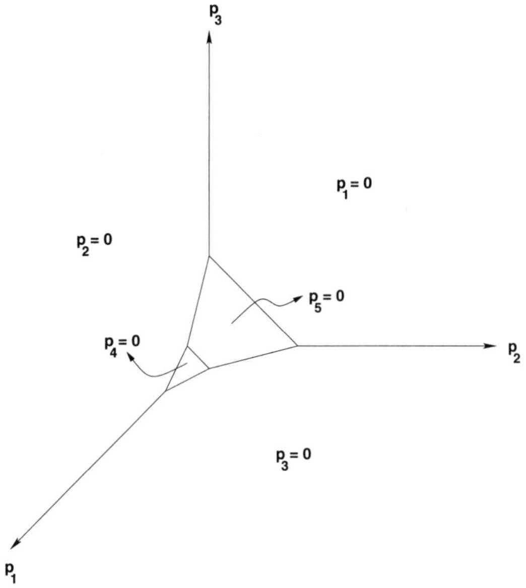

FIGURE 19. This toric diagram is the result of blowing up the origin of $\mathbb{C}^3$ and then blowing up a torically invariant point on the exceptional divisor $\mathbb{CP}^2$.

Consider the ring $R = \mathbb{C}[X, Y, Z]$ of algebraic functions on $\mathbb{C}^3$; these are just polynomials in the three complex coordinates. Given any ideal I in R, there is an construction known as "blowing up along I" [**37**], which yields a new algebraic variety, equipped with a line bundle $\mathcal{L}$. Holomorphic sections of this line bundle correspond precisely to elements of I. Note in particular that there are many ideals I which give the same algebraic variety but different bundles $\mathcal{L}$. We identify the first Chern class of $\mathcal{L}$ with the Kähler class k (so k is naturally quantized!)

In the partition sum we want to blow up not along arbitrary ideals but only over torically invariant ones; the coordinate ring R has a natural action of $U(1)^3$ which just multiplies X,Y and Z by phases, and we restrict to ideals I which are invariant under that action. These ideals are in 1-1 correspondence with 3-dimensional Young diagrams D (or equivalently to configurations of a "melting crystal," as described in [**33**].) The weight $e^{-\int k^3/g_s^2}$ for such a geometry obtained by blowing up an ideal is simply $q^{|D|}$, where $q = e^{-g_s}$ and $|D|$ is the number of boxes of the 3-dimensional Young diagram D, or equivalently the codimension of the corresponding ideal, or equivalently the relative number of sections of the line bundle $\mathcal{L}$.

Amazingly, the sum over all 3-dimensional Young diagrams with this simple weight gives the exact A model partition function on $\mathbb{C}^3$,

$$Z_A(\mathbb{C}^3) = \sum_D q^{|D|} = \prod_{i=1}^{n} (1 - q^n)^{-n}. \tag{5.29}$$

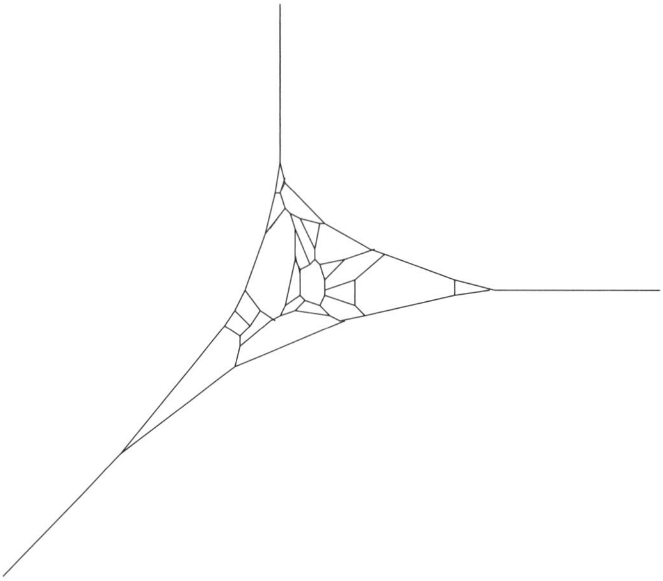

FIGURE 20. This toric diagram represents a typical result of blowing up the origin in $\mathbb{C}^3$, then blowing up a point on the exceptional divisor, then blowing up another torically invariant point on the exceptional set, and repeating many times.

This is the special case $C_{\cdots}$ of the topological vertex where the representations R_1, R_2, R_3 on the legs are trivial. More generally, one could consider infinite 3-d Young diagrams, which asymptote to fixed 2-d diagrams R_1, R_2, R_3 along the x, y, z directions; in this case the sum over diagrams gives the full topological vertex $C_{R_1 R_2 R_3}$!

6. Physical applications

So far we have mostly discussed the topological string in its own right. Now we turn to its physical applications. At first it might be a surprise that there are any physical applications at all. Remarkably, they do exist, and they are quite spectacular!

How is such a link possible? The topological string can be considered as a *localized* version of the physical string, i.e., it receives contributions only from special path-integral configurations, which can be identified with special configurations of the physical string. At the same time, there are some "BPS" observables of the physical string for which the physical string computation localizes on these same special configurations. In these cases the computations in the topological string and the physical string simply become isomorphic!

The main examples which have been explored so far are summarized in the table below:

physical theory	physical observable	topological theory
$\mathcal{N} = 2$, $d = 5, 4$ gauge theory	prepotential	A model
$\mathcal{N} = 1$, $d = 4$ gauge theory	superpotential	B model with branes/fluxes
spinning black holes in $d = 5$	BPS states	perturbative A model
charged black holes in $d = 4$	BPS states	nonperturbative A/B model?

Now we will discuss these applications in turn.

6.1. $\mathcal{N} = 2$ gauge theories. We begin with the application to $\mathcal{N} = 2$ gauge theories. First we describe the physical amplitudes of $\mathcal{N} = 2$ theories which are captured by the topological string; then we explain the particular geometries which give rise to interesting gauge theories; and finally we show how to use mirror symmetry to recover the Seiberg-Witten solution of $\mathcal{N} = 2$ theories.

6.1.1. *What the topological string computes.* To understand the connection between the topological string and $\mathcal{N} = 2$ gauge theories in $d = 4$, we begin by discussing the physical theory obtained by compactifying the Type II (A or B) superstring on a Calabi-Yau X. The holonomy of X breaks 3/4 of the supersymmetry, leaving 8 supercharges which make up the $\mathcal{N} = 2$ algebra in $d = 4$; the massless field content in $d = 4$ can then be organized into multiplets of $\mathcal{N} = 2$ supergravity as follows:

	vector	hyper	gravity
IIA on X	$h^{1,1}(X)$	$h^{2,1}(X) + 1$	1
IIB on X	$h^{2,1}(X)$	$h^{1,1}(X) + 1$	1

We will focus on the vector multiplets, for which the effective action is better understood. Each vector multiplet contains a single complex scalar, and these scalars correspond to the Kähler moduli of X in the Type IIA case, or the complex moduli in the Type IIB case.

The topological string computes particular F-terms in the effective action which involve the vector multiplets [**38, 9**]. These terms can be written conveniently in terms of the $\mathcal{N} = 2$ Weyl multiplet, which is a chiral superfield $\mathcal{W}_{\alpha\beta}$ with lowest component $F_{\alpha\beta}$.[20] Namely, forming the combination

$$\mathcal{W}^2 = \mathcal{W}_{\alpha\beta}\mathcal{W}_{\alpha'\beta'}\epsilon^{\alpha\alpha'}\epsilon^{\beta\beta'}, \tag{6.1}$$

[20]Here the "graviphoton" F is the field strength for the $U(1)$ vector in the supergravity multiplet, and α, β are spinor indices labeling the self-dual part of the full field strength $F_{\mu\nu}$, i.e., $F_{\mu\nu} = F_{\alpha\beta}(\gamma_\mu)^{\alpha\dot{\sigma}}(\gamma_\nu)^{\beta}_{\dot{\sigma}} + F_{\dot{\alpha}\dot{\beta}}(\gamma_\mu)^{\dot{\alpha}}_{\sigma}(\gamma_\nu)^{\dot{\beta}\sigma}$.

the terms in question can be written as

$$\int \mathrm{d}^4x \int \mathrm{d}^4\theta \, F_g(X^I)(\mathcal{W}^2)^g. \tag{6.2}$$

Now we can state the crucial link between physical and topological strings: the $F_g(X^I)$ which appears in (6.2) is precisely the genus g topological string free energy, written as a function of the vector multiplets X^I (so if we study Type IIB then the F_g appearing is the B model free energy, since the vector multiplets in that case parameterize the complex deformations, while for Type IIA F_g is the A model free energy.)

Note that each F_g contributes to a different term in the effective action and hence to a different physical process. To see this more clearly we can expand (6.2) in components; one term which appears is (for $g > 1$)

$$\int \mathrm{d}^4x \, F_g(X^I)(R_+^2 F_+^{2g-2}), \tag{6.3}$$

so $F_g(X^I)$ is the coefficient of a gravitational correction to the amplitude for scattering of $2g-2$ graviphotons. In the application to $\mathcal{N}=2$ gauge theory we will mostly be interested in F_0, which gets identified with the prepotential of the gauge theory, as one sees from (6.2).

6.1.2. *Compactifying on ALE fibrations.* Now let us focus on the specific geometries which will lead to interesting $\mathcal{N}=2$ gauge theories. In order to decouple gravity we should consider a non-compact Calabi-Yau space. The simplest example is an ALE singularity $\mathbb{C}^2/G$, as we discussed in Example 2.7. Recall from that example that one can think of the singularity of $\mathbb{C}^2/G$ as containing a number of zero size $\mathbb{CP}^1$'s, which naturally correspond to the simple roots of a Lie algebra $\mathfrak{g}$. Then considering Type IIA string theory on $\mathbb{C}^2/G$, one obtains massless states from D2-branes which wrap around these zero size $\mathbb{CP}^1$'s. These massless states get identified with gauge bosons in six dimensions, and it turns out that one gets a gauge theory with gauge symmetry $\mathfrak{g}$ (note in particular that the number of these gauge bosons agrees with the rank of $\mathfrak{g}$ as expected.)

But $\mathbb{C}^2/G$ is not quite the example we want; we want to get down to $d=4$ rather than $d=6$, and we also want to get down to 8 supercharges rather than 16. These goals can be simultaneously accomplished by fibering $\mathbb{C}^2/G$ over a genus g Riemann surface Σ_g; this can be done in a way so that the resulting six-dimensional space is a Calabi-Yau threefold X. Compactifying the Type IIA string on X gives an $\mathcal{N}=2$ theory with gauge group determined by G and with g adjoint hypermultiplets [**39**]. (The origin of these hypermultiplets can be roughly understood by starting with the gauge theory in $d=6$ and compactifying it on Σ_g; the electric and magnetic Wilson lines of the gauge theory give rise to the $4g$ scalar components of the g hypermultiplets.)

We first consider the special case $g=1$.

Example 6.1 ($\mathbb{C}^2/G \times T^2$). In this case the fibration of $\mathbb{C}^2/G$ over the Riemann surface T^2 is trivial, so the $\mathcal{N} = 2$ supersymmetry should be enhanced to $\mathcal{N} = 4$; this agrees with the fact that we get a single adjoint hypermultiplet, which is the required matter content for the $\mathcal{N} = 4$ theory. Furthermore, there is a relation

$$\text{vol}\,(T^2) = 1/g_{YM}^2. \tag{6.4}$$

T-dualizing on the two circles of T^2 then implies that the theory with coupling g_{YM} is equivalent to the theory with coupling $1/g_{YM}$ — so the existence of a string theory realization already implies the highly nontrivial Montonen-Olive duality of $\mathcal{N} = 4$ super Yang-Mills!

One could also consider the case $g > 1$, but in this case the gauge theory is not asymptotically free. We therefore focus on $g = 0$, and for simplicity we consider the case $G = \mathbb{Z}_2$.

Example 6.2 ($\mathbb{C}^2/\mathbb{Z}_2$ fibered over $\mathbb{CP}^1$). This geometry turns out to be just the local $\mathbb{CP}^1 \times \mathbb{CP}^1$ geometry we discussed in Example 2.11; one of the $\mathbb{CP}^1$'s is the base of the fibration, while the other is sitting in the fiber (obtained by resolving the singularity $\mathbb{C}^2/\mathbb{Z}_2$.) We call their sizes t_b and t_f respectively. Type II string theory on this geometry gives pure $\mathcal{N} = 2$ Yang-Mills in four dimensions, with gauge group $SU(2)$.

To "solve" this gauge theory á la Seiberg and Witten [**40**], one wants to compute its prepotential F_0, as a function of the Coulomb branch modulus. This modulus determines the mass of the W bosons, so in our geometric setup it gets identified with the Kähler parameter t_f (recall that the W bosons are obtained by wrapping branes over the fiber $\mathbb{CP}^1$.) The other Kähler parameter t_b is identified with the Yang-Mills coupling, through the relation

$$t_b = 1/g_{YM}^2. \tag{6.5}$$

Now, as we remarked above, the prepotential F_0 of the gauge theory should coincide with the F_0 computed by the genus zero A model topological string. We can obtain the exact solution for F_0 using mirror symmetry; namely, recalling that we have a toric realization for this geometry as discussed in Example 3.5, the techniques we illustrated in Section 5.1 can be straightforwardly applied. The mirror geometry is of the form $F(x, z) = uv$, where the Riemann surface $F(x, z) = 0$ turns out to be precisely the Seiberg-Witten curve encoding the solution of the model [**41**]! From this Seiberg-Witten curve one can read off all the desired information.

One frequently describes the Seiberg-Witten solution as counting gauge theory instantons in four dimensions, whereas in Section 4.8 we described the A model F_0 as counting genus zero worldsheet instantons in X. The connection between these two languages is clear: indeed, from (6.5) one sees that the worldsheet instantons which wrap n times around the base $\mathbb{CP}^1$ contribute with a factor e^{-n/g_{YM}^2} to F_0, and hence they correspond precisely to n-instanton effects in four dimensions.

One can similarly obtain any ADE gauge group just by making an appropriate choice of the finite group G. Conversely, anytime we have a toric geometry where the Kähler parameters arise by resolving some singularity, we expect that that toric geometry can be interpreted in terms of gauge theory. The zoo of $\mathcal{N} = 2$ theories one can "geometrically engineer" in this way includes cases with arbitrarily complicated product gauge groups and bifundamental matter content, as well as some exotic conformal fixed points in higher dimensions; see e.g., [**39, 41, 42, 43, 44, 45**]. To obtain the prepotentials for the geometrically-engineered theories is in principle straightforward via mirror symmetry, and it has been worked out in many cases, but it is not always easy — e.g. for the E_k singularities one would have a more difficult job, because to realize these geometries torically one has to include a superpotential, which makes the mirror procedure and computation of the mirror periods less straightforward.

Finally we should mention an important subtlety which we have so far glossed over: at generic values of g_{YM}^2 and the fiber moduli t_i, the string theory actually contains more information than just the four-dimensional gauge theory. This is to be expected since the F_0 of the gauge theory depends just on the Coulomb branch moduli t_i, while the F_0 of the A model has one more parameter: it also depends on the size of the base, which we identified with g_{YM}^2 at the string scale. To isolate the four-dimensional theory we have to take a decoupling limit in which g_{YM}^2 and t_i approach zero, which sends the string scale to infinity while keeping the masses of the W bosons on the Coulomb branch fixed [**42**]. If we do not take this decoupling limit, we get a theory which includes information about compactification on S^1 from five to four dimensions; from that point of view the four-dimensional instantons can be interpreted as particles of the five-dimensional theory which are running in loops, as was explained in [**46**].

6.2. $\mathcal{N} = 1$ gauge theories. So far we have seen that the IR dynamics in a large class of $\mathcal{N} = 2$ gauge theories can be completely solved using mirror symmetry. Now we want to move on to the $\mathcal{N} = 1$ case, where we will see that the topological string is similarly powerful.

How can we geometrically engineer an $\mathcal{N} = 1$ theory? Starting with compactification of Type II string theory on a Calabi-Yau space, we need to introduce an extra ingredient which reduces the supersymmetry by half. There are two natural possibilities: we can add either D-branes or fluxes. In both cases we want to preserve the four-dimensional Poincarè invariance; so if we use D-branes we have to choose them to fill the four uncompactified dimensions, and if we use fluxes we have to choose them entirely in the Calabi-Yau directions (i.e. the 0, 1, 2, 3 components of the flux should vanish.) In fact, the two possibilities are sometimes equivalent via a geometric transition in which branes are replaced by flux, as we discussed in Section 5.3.2.

In the next two subsections we describe the superpotentials which arise from these two ways of breaking from $\mathcal{N} = 2$ to $\mathcal{N} = 1$; these superpotentials can be computed by the topological string, and they are the basic objects we want to understand in the $\mathcal{N} = 1$ context, since they determine much of the IR physics. The form of the superpotentials obtained in the two cases is quite similar, and as we explain in the following section, this is not an accident; it follows from the equality of topological string partition functions before and after the geometric transition. This geometric transition is a practical tool for computation of the superpotentials, and we discuss some basic examples. Finally we discuss an alternative method of computing the superpotentials via holomorphic matrix models, which also gives an interesting new perspective on the geometric transition: the dual geometry emerges as a kind of effective theory of a density of eigenvalues in the large N limit!

6.2.1. *Breaking to $\mathcal{N} = 1$ with branes.* To engineer $\mathcal{N} = 1$ gauge theories, we begin with Type II string theory on a Calabi-Yau space X. This would give $\mathcal{N} = 2$ supersymmetry, but let us reduce it to $\mathcal{N} = 1$ by introducing N D-branes, which are wrapped on some cycle in the Calabi-Yau and also fill the four dimensions of spacetime. Then we obtain an $\mathcal{N} = 1$ theory in four dimensions, with $U(N)$ gauge symmetry, as we discussed in Section 5.3.1. (Note the difference from the geometric engineering we did in the $\mathcal{N} = 2$ case; there we obtained the gauge symmetry from a geometric singularity, but in the $\mathcal{N} = 1$ case it just comes from the N branes. As we will see, in this case the geometry is responsible for details of the gauge theory, specifically the form of the bare superpotential.)

We now want to expose a connection between this gauge theory and the topological string on X. In the $\mathcal{N} = 2$ case we saw that the genus zero topological string free energy F_0 computed the prepotential. After introducing D-branes in the topological string, we need not consider only closed worldsheets anymore; we can also consider open strings, i.e., Riemann surfaces with boundaries. Therefore we can define a free energy $F_{g,h}$, obtained by integrating over worldsheets with genus g and h holes, with each hole mapped to one of the D-branes; and we can ask whether this $F_{g,h}$ computes something relevant for the $\mathcal{N} = 1$ theory. The answer is of course "yes." (More precisely, as in the $\mathcal{N} = 2$ case, it turns out that $g = 0$ is the case relevant to the pure gauge theory; higher genera are related to gravitational corrections, which we will not discuss here.)

To write the terms which the topological string computes in the $\mathcal{N} = 1$ theory with branes, we need the "glueball" superfield S; this is a chiral superfield with lowest component $\operatorname{Tr} \psi_\alpha \psi^\alpha$, where ψ_α is the gluino. Organize the $F_{0,h}$ into a generating function:

$$F(S) = \sum_{h=0}^{\infty} F_{0,h} S^h. \tag{6.6}$$

The F-term the genus zero topological string computes in the $\mathcal{N}=1$ theory can then be written [**9**]

$$\int \mathrm{d}^4x \int \mathrm{d}^2\theta\, N \frac{\partial F}{\partial S}. \tag{6.7}$$

This term gives a superpotential for the glueball S, and it turns out that this superpotential captures a lot of the infrared dynamics of the gauge theory. More precisely, in addition to (6.7), one also has to include the term

$$\int \mathrm{d}^4x \int \mathrm{d}^2\theta\, \tau S, \tag{6.8}$$

which is simply the classical super Yang-Mills action in superfield notation, with

$$\tau = \frac{4\pi \mathrm{i}}{g_{YM}^2} + \frac{\theta}{2\pi}. \tag{6.9}$$

After including this extra term, one then has the glueball superpotential

$$W(S) = N\frac{\partial F}{\partial S} + \tau S. \tag{6.10}$$

In the IR one expects that the glueball field will condense to some value with $W'(S) = 0$, so one can determine the vacuum structure of the theory just by extremizing this $W(S)$, as we will see below in some examples.

6.2.2. *Breaking to $\mathcal{N}=1$ with fluxes.* Now what about the case where we introduce fluxes instead of branes? Consider the Type IIB superstring on a Calabi-Yau X. Recall from the last section that this theory has a prepotential term

$$\int \mathrm{d}^4x \int \mathrm{d}^4\theta\, F_0(X^I), \tag{6.11}$$

where F_0 is the B model topological string free energy at genus zero, and the X^I are the vector superfields, whose lowest components parameterize the complex structure moduli of X. How does this term change if we introduce N^I units of Ramond-Ramond three-form flux on the I-th A cycle?[21] In the $\mathcal{N}=2$ supergravity language, it turns out that this flux corresponds to the θ^2 component of the superfield X^I; turning on a vacuum expectation value for this component absorbs two θ integrals from (6.11), leaving behind an F-term in the $\mathcal{N}=1$ language [**47**],

$$\int \mathrm{d}^4x \int \mathrm{d}^2\theta\, N^I \frac{\partial F_0}{\partial X^I}. \tag{6.12}$$

As above, this term can be interpreted as a superpotential, this time for the moduli X^I. There is a natural extension to include a flux τ_I on the I-th B

[21] Recall that in writing the $\mathcal{N}=2$ supergravity Lagrangian we have chosen a splitting of $H_3(X)$ into A and B cycles, with the X^I representing the A cycle periods.

cycle:

$$W(X^I) = N^I \frac{\partial F_0}{\partial X^I} + \tau_I X^I. \tag{6.13}$$

This form of the superpotential was derived in [**48, 49**].

6.2.3. *The geometric transition, redux.* There is an obvious analogy between (6.10) and (6.13). Note though that the lowest component of the X^I which appears in (6.13) is a scalar field parameterizing a complex structure modulus, while the S which appears in (6.10) is a fermion bilinear, which naively cannot have a classical vacuum expectation value. Nevertheless, the analogy between the two sides seems to be suggesting that we should treat S also as an honest scalar, and we will do so in what follows.

So what do (6.10) and (6.13) have to do with one another? The crucial link is provided by the notion of "geometric transition," which we discussed in Section 5.3.2, but now in the context of the Type IIB superstring rather than the topological string:[22] start with a Calabi-Yau X which has a nontrivial 2-cycle. Then wrap N D5-branes on this 2-cycle, obtaining a $U(N)$ gauge theory. There is a dual geometry where the D5-branes disappear and are replaced by a 3-cycle A; in this dual geometry there are N units of Ramond-Ramond flux on the dual cycle B. The claim is that the physical string theories on these two geometries are equivalent in the IR, after we identify the glueball superfield S with the period of Ω over the A cycle in the dual geometry.[23] With this identification (6.10) and (6.13) are identical.

One can therefore use either the brane picture or the flux picture to compute the glueball superpotential. In this section we will discuss some examples of the use of the flux picture.

Example 6.3 (D5-branes on the resolved conifold). The simplest example of a geometric transition from branes to flux is provided by the resolved conifold, which just has a single 2-cycle $\mathbb{CP}^1$. So suppose we wrap M D5-branes on the $\mathbb{CP}^1$ of the resolved conifold. As one might expect, this simplest possible geometry leads to the simplest possible gauge theory in $d = 4$, namely $\mathcal{N} = 1$ super Yang-Mills. This theory has a well-known glueball superpotential, which we now derive from the flux picture and (6.12). The dual geometry after the transition is the deformed conifold, which has

[22]See [**47**] for a detailed discussion of the superstring version of the large N duality in the Type IIA case.

[23]On the face of it this claim might sound bizarre since the theory with branes should have $U(N)$ gauge symmetry in four dimensions; but since we are now talking about the effective theory in $d = 4$, what we should really compare is the IR dynamics, and we know that $\mathcal{N} = 1$ gauge theories confine, which reduces the $U(N)$ to $U(1)$ in the IR.

a compact S^3 and its dual B cycle, with corresponding periods

$$X = \int_A \Omega = \mu, \tag{6.14}$$

$$F = \int_B \Omega = \mu \log \mu. \tag{6.15}$$

(A simple way to check the formula for F is to note that it has the correct monodromy; as $\mu \to e^{2\pi i}\mu$ the B cycle gets transformed into a linear combination of the B cycle and the A cycle, corresponding to the fact that F gets shifted by the A period μ.) From the periods we immediately obtain the closed string F_0, via (4.36),

$$F_0 = \frac{1}{2} X F = \frac{1}{2}\mu^2 \log \mu. \tag{6.16}$$

Now to compare with the gauge theory we have to identify $\mu = S$ as we stated above. This leads to the superpotential

$$W(S) = N\frac{\partial F_0}{\partial S} - 2\pi \mathrm{i}\, \tau S = NS\log S - 2\pi \mathrm{i}\, \tau S. \tag{6.17}$$

This is the standard Veneziano-Yankielowicz glueball superpotential for $\mathcal{N} = 1$ super Yang-Mills [**50**].

By extremizing $W(S)$ one finds the expected N vacua of $\mathcal{N} = 1$ super Yang-Mills,[24]

$$S = \Lambda_0^3 \exp(2\pi \mathrm{i}\, j\tau/N) = \Lambda^3 \exp\left(2\pi \mathrm{i}\, j/N\right), \tag{6.18}$$

where $j = 1, \ldots, N$.

So far we have not used much of our topological-string machinery. But now we can consider a more elaborate example.

Example 6.4 (D5-branes on the multi-conifold)**.** Instead of the singular conifold geometry

$$u^2 + v^2 + y^2 + x^2 = 0, \tag{6.19}$$

which just has a single zero size $\mathbb{CP}^1$, consider

$$u^2 + v^2 + y^2 + W'(x)^2 = 0, \tag{6.20}$$

for some polynomial $W(x)$ of degree $n+1$. Writing

$$W'(x) = \prod_{i=1}^{n} (x - x_n), \tag{6.21}$$

the geometry has n conifold singularities located at the critical points $x_1, \ldots, x_n$ of W. The singularities can be resolved by blowing up to obtain n $\mathbb{CP}^1$'s at these n points (all these $\mathbb{CP}^1$'s are homologous, however, so in particular there is only one Kähler modulus describing the resolution.)

[24] We have not been careful to keep track of the cutoff Λ_0; if one does keep track of it, one finds that it combines with the bare coupling τ to give the QCD scale Λ which appears in (6.18).

We want to use this geometry to engineer an interesting $\mathcal{N} = 1$ gauge theory. To construct this gauge theory we consider M D5-branes. What are the possible supersymmetric configurations? We should expect that we can get a supersymmetric configuration by wrapping M_1 branes on the first $\mathbb{CP}^1$, M_2 on the second, and so on, and in this configuration we expect to realize a gauge symmetry $U(M_1) \times \cdots \times U(M_n)$. All these configurations can be naturally understood as different sectors of a single UV theory, which describes the dynamics of the M branes and includes a $U(M)$ adjoint chiral multiplet Φ, whose lowest component represents the x-coordinate of the branes.[25] The supersymmetric vacua described above then arise from configurations in which M_1 of the eigenvalues of Φ are equal to x_1, M_2 are equal to x_2 and so on.

A very natural way for this vacuum structure to arise is if the $U(M)$ gauge theory describing the branes has a bare superpotential $\operatorname{Tr} W(\Phi)$. This is indeed the case; one can derive this result from the holomorphic Chern-Simons action which, as we discussed earlier, is the topological open string field theory of the brane [**51**]. Namely, one shows from the holomorphic Chern-Simons action that, as one moves the 2-brane along a path, sweeping out a 3-cycle C, the classical action is shifted by $\int_C \Omega$; combined with the explicit form of Ω in the geometry (6.20) this gives the classical action for the brane at x as $W(x)$. This classical action in the topological string turns out to be the superpotential of the physical superstring. This superpotential computation can also be interpreted directly in the worldsheet language as coming from disc diagrams with boundary on the brane; to see this from the topological string one notes that $F_{0,1}$ contributes an S-independent term to (6.10), which gets interpreted as the desired bare superpotential.

Thus we have geometrically engineered an $\mathcal{N} = 1$ gauge theory, with $U(M_1) \times \cdots \times U(M_n)$ gauge group, one adjoint chiral multiplet Φ, and a superpotential $\operatorname{Tr} W(\Phi)$. To answer questions about the vacuum structure of this theory we now want to find the appropriate glueball superpotential, which is now a function of n different glueball fields S_i for the n gauge factors. As in the case of the conifold, one way to compute the superpotential is to consider the dual geometry in which the branes have disappeared and each of the n $\mathbb{CP}^1$'s has been replaced by an S^3. This geometry is written

$$u^2 + v^2 + y^2 + W'(x)^2 = f(x), \tag{6.22}$$

where $f(x)$ is a polynomial of degree $n - 1$ characterizing the deformation. This $f(x)$ depends on the M_i, and is completely fixed by the requirement that the period of Ω over the i-th S^3 is $M_i g_s$ (in keeping with the principle that the B model branes produce precisely this flux of Ω — this is precisely analogous to the fact that A model branes produce a flux of k, which we used in Section 5.3.2 to compute A model amplitudes.) This approach

[25] The adjoint scalar Φ is present even in the conifold case which we considered above, but there (as we will shortly see) it is accompanied by a quadratic superpotential $W(\Phi) = \Phi^2$, so Φ can be harmlessly integrated out to leave pure $\mathcal{N} = 1$ super Yang-Mills.

was followed in [**52**], and leads to a complete computation of the glueball superpotential.

6.2.4. *Holomorphic matrix models.* So far we have shown how to compute the glueball superpotential from a transition to a geometry where D5-branes are replaced by fluxes. Alternatively, one can avoid the geometric transitions altogether and compute directly in the gauge theory on the D5-branes. The idea is that since the glueball superpotential is computed by the topological string, one can avoid all the complexities of Yang-Mills theory, and use instead the topological open string field theory; as we explained above, in the case of the B model this is (the dimensional reduction of) holomorphic Chern-Simons. One finds that the whole computation of the topological string free energy is reduced to a computation in a holomorphic matrix model [**53, 54, 55**]. For example, in the case of the multi-conifold geometry of Example 6.4, one just has to integrate over a single $N \times N$ matrix Φ, with action $W(\Phi)/g_s$:

$$Z = \int \mathrm{d}^{N^2}\Phi\, e^{-W(\Phi)/g_s}. \tag{6.23}$$

The matrix model contains all the information that can be obtained from the open topological string in this background. For example, to compute the genus zero open topological string partition function — which determines the glueball superpotential — one just has to study the large N (planar) limit of the matrix model!

These models have turned out to be a quite powerful tool, which is applicable to geometries more general than the case we described here. They are also related in a beautiful way to the geometric transitions we described above: namely, the planar limit of the matrix model can be described as a saddle-point expansion around a particular distribution of the infinitely many eigenvalues, and this distribution turns out to capture the dual geometry in a precise way. In this sense the smooth geometry seems to be an emergent property, which only makes sense in the planar (classical) limit of the string theory.

6.3. BPS black holes in $d = 5$. So far we have discussed applications of the topological string to gauge theory, which involved only the genus zero free energy F_0. Now we want to discuss an application to black hole entropy, which is more sophisticated in the sense that it naturally involves all of the F_g. We ask the following question: given a compactification of M theory to five dimensions on a Calabi-Yau threefold X, how many BPS states are there with a particular spin and charge?

First, what do we mean by "charge"? M-theory compactified on X has a $U(1)$ gauge field for each 2-cycle of X, obtained by dimensional reduction of the M-theory 3-form C on the 2-cycle, i.e. via the ansatz $C_{\mu\alpha\beta} = A_\mu \omega_{\alpha\beta}$, where $\omega_{\alpha\beta}$ is the harmonic 2-form dual to the cycle in question. So we get

$U(1)^n$ gauge symmetry, where $n = b_2(X)$ is the number of independent 2-cycles. We also naturally get states which are charged under this $U(1)^n$; namely, an M2-brane wrapped on a 2-cycle gives a particle charged under the corresponding $U(1)$. Hence the charges in the theory are classified by the second homology of X, $Q \in H_2(X, \mathbb{Z})$.

So we could ask for the number of BPS states with given Q. But actually there is a finer question we can ask: namely, it turns out that in five dimensions it is possible for a state to have spin and still be BPS. The little group for a massive particle in this dimension is $SO(4) = SU(2)_L \times SU(2)_R$, giving rise to spins (j_L, j_R), and one can get BPS states so long as one requires either $j_L = 0$ or $j_R = 0$. So fixing, say, $j_R = 0$, we can ask for the number of BPS states with charge Q and spin j_L.

A convenient way of packaging this information is suggested by the notion of *elliptic genus*, which we now quickly recall in a related context.

Example 6.5 (The $\mathcal{N} = (1,1)$ elliptic genus)**.** Consider a theory with $\mathcal{N} = (1,1)$ supersymmetry in two dimensions. The partition function on a torus with modular parameter τ, with the natural boundary conditions, is

$$\mathrm{Tr}\,(-1)^F q^{L_0} \overline{q}^{\overline{L_0}}. \tag{6.24}$$

This partition function is relatively "boring" in the sense that it just computes the Witten index [**56**], which is independent of q and $\overline{q}$. But in an $\mathcal{N} = (1,1)$ theory one can define separate left and right-moving fermion number operators F_L, F_R, and we can use these to define a more interesting object, the *elliptic genus* [**57**],

$$\mathrm{Tr}\,(-1)^{F_R} q^{L_0} \overline{q}^{\overline{L_0}}. \tag{6.25}$$

The usual argument shows that (6.25) gets contributions only from states which have $\overline{L_0} = 0$, so it is independent of $\overline{q}$, but it is a nontrivial function of q, which has modular properties. Like the usual Witten index it has some rigidity properties, namely, it does not depend on small deformations of the theory; this follows from the fact that the coefficients in the q expansion are integral.

Now we turn to the case of interest for us.

Example 6.6 (The $d = 5$, $\mathcal{N} = 2$ elliptic genus)**.** Returning to the $d = 5$ BPS state counting, note that we have a splitting into left and right similar to the one for $\mathcal{N} = (1,1)$ theories, so instead of computing the ordinary index

$$\mathrm{Tr}\,(-1)^J e^{-\beta H} \tag{6.26}$$

we can consider an elliptic genus analogous to (6.25),

$$\mathrm{Tr}\,(-1)^{J_R} q^{J_L} e^{-\beta H}. \tag{6.27}$$

Like (6.25), this elliptic genus has a rigidity property: it is independent of the complex structure moduli of X, although it can and does depend continuously on the Kähler moduli t_i. This property is reminiscent of the A model topological string, and indeed it turns out that the A model partition function $Z_A(g_s, t_i)$ is precisely the elliptic genus (6.27), with the identification

$$q = e^{-g_s}, \tag{6.28}$$

as we will see below.[26] In this sense the spin-dependence of the BPS state counting gets related to the genus-dependence of the topological string.

Now, *why* is the A model partition function counting BPS states? Such a connection seems reasonable; after all, the A model counts holomorphic maps, and the image of a holomorphic map is a supersymmetric cycle on which a brane could be wrapped to give a BPS state. There is a more precise argument which explains the relation; it was worked out in [**58, 59**] and goes roughly as follows.

Consider the Type IIA string on X. As we mentioned earlier, there are certain F-terms in the effective four-dimensional action of this theory which are computed by the A model topological string, namely

$$\int \mathrm{d}^4x \int \mathrm{d}^4\theta \, F_g(t)(\mathcal{W}^2)^g + \text{c.c.}, \tag{6.29}$$

which when expanded in components give

$$\int \mathrm{d}^4x \, F_g(t)(R_+^2 F_+^{2g-2} + R_-^2 F_-^{2g-2}). \tag{6.30}$$

If we consider the Euclidean version of the theory, then in four dimensions we can turn on a self-dual graviphoton background $F_+ \neq 0$, $F_- = 0$, i.e. $\mathcal{W} \neq 0$, $\overline{\mathcal{W}} = 0$. Substituting this background into (6.30) we get a correction to the R_+^2 term,

$$\left(\sum_{g=0}^{\infty} F_g(t) F_+^{2g-2} \right) R_+^2. \tag{6.31}$$

Note that in (6.31) we have a sum over all genus topological string amplitudes, with the role of the topological string coupling played by the graviphoton field strength F_+.

To establish the relation between the topological string and the elliptic genus, we now want to show that one can compute the same R_+^2 correction in a graviphoton background in a different way which gives the elliptic genus. This second computation is based on Schwinger's computation of the correction to the vacuum energy from pair-production of charged particles in a background electric field. In the present context the relevant charged

[26]Strictly speaking, this is true once we rescale t_i by a factor β, which completely absorbs the β dependence.

particles are the quanta of charged $\mathcal{N}=2$ hypermultiplet fields obtained by quantization of the wrapped D2- and D0-branes; for a D2-brane wrapped on the cycle Q, bound to k D0-branes, the central charge is

$$Z = \langle Q, t\rangle + ik, \tag{6.32}$$

and the mass of the corresponding BPS state is $m = |Z|$. We compute the corrections to the effective action due to pair production of such states in the self-dual graviphoton background F. Since these states come in hypermultiplets, their contribution to the vacuum energy cancels, but it turns out that they make a nonzero contribution to the R_+^2 term: for example, a multiplet whose lowest component is scalar contributes to R_+^2 precisely as a scalar would have contributed to the vacuum energy.

Let us focus on the contribution to the R_+^2 correction from a BPS hypermultiplet with lowest component scalar, arising from a quantization of a D2-brane in homology class Q. Actually, since the D2-brane can be bound to D0-branes, these hypermultiplets will come in families: we will get one for each value of the D0 brane charge k. The Schwinger computation expresses the contribution from each of these hypermultiplets as a one-loop determinant; summing over k to treat the whole family at once gives

$$\sum_k \log\det(\Delta + m_k^2) = \sum_{k=-\infty}^{\infty} \int_\epsilon^\infty \frac{ds}{s} \frac{e^{-s(\langle Q,t\rangle + ik)}}{(2\sinh\frac{sF_+}{2})^2}. \tag{6.33}$$

(Here F_+ enters the determinant through the non-commutation of the covariant derivatives which appear in Δ.) The integral appearing in (6.33) looks formidable, but luckily we do not have to do it: the sum over k gives a δ-function which cancels the integral and also removes the awkward dependence on the cutoff ϵ. We get a simple result,

$$\sum_{n=1}^{\infty} \frac{1}{n} \frac{e^{-n\langle Q,t\rangle}}{(2\sinh\frac{nF_+}{2})^2}. \tag{6.34}$$

This is the contribution to the R_+^2 correction coming from a single family of BPS multiplets with lowest component scalar; alternatively, setting the topological string coupling $g_s = F_+$, we could interpret it as the contribution to the topological string free energy $\mathcal{F}(g_s, t_i)$ from this family,

$$\sum_{n=1}^{\infty} \frac{1}{n} \frac{e^{-n\langle Q,t\rangle}}{(2\sinh\frac{ng_s}{2})^2}. \tag{6.35}$$

Now what does all this have to do with the elliptic genus (6.27) in M-theory on X? We will argue that the exponential of (6.35) in fact agrees with the contribution to the elliptic genus from a single BPS hypermultiplet in five dimensions, obtained from quantization of an M2 brane in homology class Q. The first promising sign is that the exponential of (6.35) has a nice integer expansion: namely, it is

$$\prod_{n=1}^{\infty}(1-q^n e^{-\langle Q,t\rangle})^n. \tag{6.36}$$

To reproduce this, write the hypermultiplet field as ϕ in five dimensions. This ϕ can have excitations which are not Poincarè invariant but are still BPS. Namely, choosing complex coordinates z_1, z_2 for the Euclidean time-slice $\mathbb{R}^4$, we can write

$$\phi = \phi(z_1, z_2, \overline{z_1}, \overline{z_2}), \tag{6.37}$$

and the BPS excitations are the ones independent of $\overline{z_i}$. Expanding

$$\phi = \sum_{l,m\geq 0} \phi_{lm} z_1^l z_2^m, \tag{6.38}$$

we get a collection of creation operators ϕ_{lm}. The operator ϕ_{lm} creates $SU(2)_L$ spin $l+m+1$, so there are n of them that create spin n (and BPS mass $\langle Q,t\rangle$.) The second quantization of these operators then accounts for the factor (6.36).

This almost completes the identification between the topological string partition function and the elliptic genus, except that the hypermultiplets obtained from quantization of the wrapped M2-brane need not in general have lowest component scalar. From the discussion of the last paragraph, one easily sees how to modify the contribution to the elliptic genus if the lowest component has $SU(2)_L$ spin j: one just has to replace q^n by q^{n+j} in (6.36). We should also note that the creation operators ϕ_{lm} may be fermionic or bosonic depending on the net spin. Hence the most general form of the contribution to the elliptic genus is

$$\left[\prod_{n=1}^{\infty}(1-q^{n+j}e^{-\langle Q,t\rangle})^n\right]^{\pm 1}. \tag{6.39}$$

One can check that this also agrees with the result of the Schwinger computation of the R_+^2 correction in this more general case, and hence with the topological string.

What have we learned about the topological string? We can already obtain an interesting result by taking the $g_s \to 0$ limit in the contribution (6.35) to $\mathcal{F}$ from a single five-dimensional hypermultiplet: namely, we recover

$$\frac{1}{g_s^2}\sum_{n=1}^{\infty}\frac{e^{-n\langle Q,t\rangle}}{n^3}, \tag{6.40}$$

which is precisely the formula (4.25) for the contribution of an isolated genus zero curve to the A model F_0! So the counting of BPS states automatically reproduces the tricky $\sum_n 1/n^3$, which arose from multi-covering maps $S^2 \to S^2$ in the A model.

Indeed, from counting BPS states one obtains formulae for the multi-covering contributions at all genera, as well as "bubbling" terms which occur when part of the worldsheet degenerates to a surface of lower genus. All these terms are encapsulated in the general form of the topological A model free energy in terms of the five-dimensional BPS content, which we now write. It is convenient to choose a slightly exotic basis for the representation content: namely, we introduce the symbol $[j]$ for the $SU(2)_L$ representation $[2(0) \oplus (\frac{1}{2})]^{\otimes j}$. Any representation of $SU(2)_L$ can be written as a sum of the representations $[j]$ with integer coefficients (not necessarily positive). Then write $\mathcal{N}_{j,Q}$ for the number of times $[j]$ appears in the $SU(2)_L$ content of the BPS spectrum obtained by wrapping M2 branes on Q. Combining the results we catalogued above, one obtains

$$F(t, g_s) = \sum_{j \geq 0} \sum_{Q \in H_2(X,\mathbb{Z})} \mathcal{N}_{j,Q} \left(\sum_{n \geq 0} \left(2 \sinh \frac{n g_s}{2} \right)^{2j-2} e^{-n\langle Q,t \rangle} \right). \tag{6.41}$$

The formula (6.41) expresses all the complexity of the A model topological string at all genera in terms of the integer invariants $\mathcal{N}_{j,Q}$. Conversely, it gives an algorithm for computing the numbers $\mathcal{N}_{j,Q}$, which capture the degeneracy of BPS states, using the topological string.

The topological string thus completely captures the counting of BPS black hole states in compactifications of M-theory on Calabi-Yau threefolds. Nevertheless, despite the formidable computational techniques which are known for the topological string, it has not yet been possible to use it to verify one of the simplest predictions from black hole physics: namely, the asymptotic growth of the $\mathcal{N}_{j,Q}$ with Q should agree with the scaling of the BPS black hole entropy with the charge in $d = 5$,

$$S \sim \sqrt{Q^3 - j^2}. \tag{6.42}$$

6.4. BPS black holes in $d = 4$**.** In Section 6.3 we showed that the topological string counts BPS black hole states in $d = 5$. Remarkably, it turns out that the topological string is also relevant to black hole entropy in $d = 4$! This application is somewhat subtler than the $d = 5$ case, however. In the $d = 5$ case, using (6.41) one could recover the exact number of BPS states with fixed charge and spin j from the A model amplitudes up to genus j. In $d = 4$, the perturbative topological string will only give us coefficients of the asymptotic *growth* of the number of states as a function of the charge; to get the actual number of states with a given fixed charge would require some sort of nonperturbative completion of the topological string.

We are interested in computing the number of BPS states as a function of the charge — or more precisely an index, which counts the BPS states possibly weighed by $\pm$ signs. The charges in $d = 4$ are a little more subtle than in $d = 5$; namely, in $d = 4$, each $U(1)$ in the gauge group leads to both an electric and a magnetic charge. In Type IIA on X, there is a natural splitting of the charges into electric and magnetic; namely, D0- and D2-branes on X can be considered as electrically charged states, while D4- and D6-branes are magnetically charged states. In Type IIB, on the other hand, all of the charges are realized by D3-branes wrapping 3-cycles, so a general combination of electric and magnetic charges can be realized by a D3-brane wrapping a general 3-cycle, i.e. a choice of $C \in H_3(X, \mathbb{Z})$. In this case a splitting into electric and magnetic charges is obtained only after making a choice of symplectic basis (A and B cycles), as in (4.27). In this section we will use the IIB language.

How can the Calabi-Yau space X give us the number of BPS states, as a function of the charge C? The answer is very pretty. We first describe it to leading order in the limit of large C. It is convenient to express the answer in terms of S, the entropy: this turns out to be given by the "holomorphic volume" of the Calabi-Yau,

$$S(\Omega) = \frac{\mathrm{i}\pi}{4} \int_X \Omega \wedge \overline{\Omega}. \tag{6.43}$$

Here Ω is the holomorphic 3-form on the Calabi-Yau. But we know that this Ω is not unique: there is a whole moduli space of possible choices for Ω, with complex dimension $h^{2,1} + 1$. So (6.43) is not complete until we explain how to choose the appropriate Ω.

The crucial ingredient here is the "attractor mechanism" of $\mathcal{N} = 2$ supergravity [**60, 61**], which we now describe. Suppose we consider the supergravity theory obtained by compactifying Type IIB on X and look for classical solutions describing a spherically symmetric BPS black hole with charge C. The supergravity theory includes scalar fields corresponding to the moduli of X, and we can choose the expectation values of those scalar fields at infinity arbitrarily. Studying the evolution of the scalar fields as we move in from infinity toward the black hole horizon, one finds a remarkable phenomenon: the vector multiplet scalars and the graviphoton field strength approach fixed values at the horizon, independent of the boundary condition at infinity, depending only on the charge C of the black hole.[27]

Since we are in Type IIB, the vector multiplet scalars determine the holomorphic 3-form Ω on X up to an overall rescaling; this overall rescaling is determined by the graviphoton field strength. So the attractor mechanism can be viewed as the statement that the charge C determines Ω at the horizon. It is not easy to describe the map from C to Ω, but the inverse map is straightforward: choosing a basis of 3-cycles and corresponding electric-magnetic splitting $C = (P^I, Q_J)$, the relation is

$$P^I = \int_{A^I} \operatorname{Re}\Omega, \qquad Q_J = \int_{B_J} \operatorname{Re}\Omega, \tag{6.44}$$

[27]This statement needs to be slightly qualified: the moduli at the horizon are *locally* independent of the moduli at infinity, but there can be multiple basins of attraction [**62**].

or more invariantly, $\operatorname{Re}\Omega \in H^3(X,\mathbb{R})$ is the Poincarè dual of $C \in H_3(X,\mathbb{Z})$. Note that the counting of parameters works out correctly: the complex structure moduli, when augmented to include the choice of overall scaling of Ω, make up $2b_3(X)$ real parameters, and this is also the number of possible black hole charges.

So given this prescription for Ω, (6.43) is a sensible formula for the black hole entropy. Note that it has the expected scaling with the size of the black hole: namely, from (6.44) we see that a rescaling $C \mapsto \lambda C$ (which also rescales the size of the black hole by λ thanks to the BPS relation between mass and charge) rescales the attractor moduli by $\Omega \mapsto \lambda\Omega$, and hence $S \mapsto \lambda^2 S$. This is the expected behavior for the entropy of a black hole in four dimensions.

Now we want to highlight a connection between (6.43) and the topological string. To do so, we begin by noting that if we choose an electric-magnetic splitting, we can use the Riemann bilinear identity (4.33) to rewrite (6.43) as

$$S(P,Q) = \frac{\mathrm{i}\,\pi}{4}(X^I\overline{F}_I - \overline{X}^I F_I). \tag{6.45}$$

This expression is quadratic in the periods of Ω, which is reminiscent of the tree level B model free energy F_0. Indeed, it is very close to being the imaginary part of F_0,

$$\frac{\mathrm{i}\,\pi}{2}\operatorname{Im} F_0(X) = \frac{\mathrm{i}\,\pi}{4}(X^I F_I - \overline{X}^I\overline{F}_I). \tag{6.46}$$

Now (6.45) and (6.46) are not quite equal, but they are related, as explained in [**63**]: namely, beginning with (6.46), one can introduce the notation $\Phi^I = X^I - \overline{X}^I$, and then make a Legendre transform from Φ^I to a dual variable Q_I which we identify as the black hole electric charge. According to (6.44) this charge is given by $Q_I = F_I + \overline{F}_I$, and substituting this for Q_I in the Legendre transform of (6.46), one recovers (6.45)!

So the black hole entropy is the Legendre transform of $\operatorname{Im} F_0$, to leading order in the charge. There is a natural extension of this formula beyond the leading order, obtained by noting that the Legendre transform is the leading approximation to a Fourier transform:

$$\sum_Q \rho(P,Q)e^{-Q_I\Phi^I} = |Z_B(P + \mathrm{i}\,\Phi)|^2. \tag{6.47}$$

On the left side of (6.47), $\rho(P,Q)$ is the number of BPS black holes with electric and magnetic charges (P,Q), while on the right side $Z_B(P + \mathrm{i}\,\Phi)$ is the B model partition function, evaluated at the Ω determined by the A cycle periods $X^I = P^I + \mathrm{i}\,\Phi^I$. (Note that this formula for X^I determines even the overall scaling of Ω. This corresponds to fixing the coupling in the topological string; the expansion around large black hole charge corresponds to the genus expansion.) In other words, the partition function of the mixed ensemble of black holes where we fix the magnetic charges P and the electric potential Φ, then sum over all electric charges, is given by $|Z_B(P + \mathrm{i}\,\Phi)|^2$! This is a beautiful relation and it is very natural to conjecture that it indeed holds to all orders in the charge [**63**].

What is the evidence for this conjecture beyond leading order? The major source of evidence comes from a reconsideration of the corrections to $\mathcal{N} = 2$ supergravity computed by the topological string, which we wrote in (6.2):

$$\int \mathrm{d}^4 x \int \mathrm{d}^4\theta \, F_g(X^I)(\mathcal{W}^2)^g + \text{c.c.} \tag{6.48}$$

In the background of the charged black hole, the graviphoton field $\mathcal{W}$ and $\overline{\mathcal{W}}$ both have nonzero expectation values (and create a nontrivial gravitational backreaction.) The terms (6.48) therefore lead to a contribution to the free energy proportional to $F_g(X^I)$. These corrections were studied in [**64, 65, 66**], and found to give a correction to the black hole entropy which is consistent with the conjecture. There are also formulas for the one-loop correction to the black hole entropy [**67**] which agree with the conjecture. Finally, there is one example in which the conjecture can be checked exactly, studied in [**68**], which we now describe.

Example 6.7 (Black hole counting and 2-dimensional Yang-Mills)**.** For this example it is convenient to switch to the Type IIA language: namely, one considers the Type IIA string on the Calabi-Yau threefold geometry $X = \mathcal{L} \oplus \mathcal{L}^{-1} \to T^2$, where $\mathcal{L}$ is a particular complex line bundle over T^2.

First consider the counting of black hole BPS states in this geometry. The relevant black holes are obtained by wrapping D4-branes on $\mathcal{L}^{-1} \to T^2$ as well as wrapping D2-D0 bound states on T^2. One can then argue that the theory on the N D4-branes is a topological $U(N)$ gauge theory, and furthermore that it localizes to a bosonic $U(N)$ Yang-Mills theory on T^2. Bound states of D2 and D0 branes with the D4-branes can be realized as configurations of the gauge field on the D4-brane. The counting of BPS black hole states, summing over all D2 and D0 brane charges but fixing D4-brane charge N, is thus determined by the $U(N)$ Yang-Mills partition function on T^2. This theory was studied in detail in [**69**], where it was shown that the exact partition function can be obtained as a sum over representations of $U(N)$:

$$Z_{YM} = \sum_R e^{-\lambda C_2(R) + \mathrm{i}\,\theta |R|}, \tag{6.49}$$

where $|R|$ is the number of boxes in the Young diagram representing R.

Expanding around the large N limit, one finds that this Z_{YM} is the square of a holomorphic function to all orders in $1/N$, $Z_{YM} = |Z|^2$. (This splitting into "chiral" and "anti-chiral" parts is obtained by splitting up the Young diagrams R into short diagrams, with a finite number of boxes, and large diagrams, for which the size of each column differs from N by a finite number; from this description it is manifest that the splitting only makes sense in the large N limit.) So the partition function on the D4-branes is indeed of the form $|Z|^2$. Furthermore, X is a simple enough geometry that one can explicitly compute the A model partition function, and one finds that $Z = Z_A(X)$!

The example of Yang-Mills on T^2 thus provides a striking confirmation of the conjecture that $|Z_A|^2$ counts BPS black hole states in the mixed ensemble. It also gives us some insight into the difficulties one should expect to face in trying to define a nonperturbative topological string (i.e. to define Z_A as an honest function rather than as a formal power series in g_s.) Namely, as we noted above, the factorization of Z_{YM} into $|Z_A|^2$ is only valid to all orders in $1/N$, which in the topological string language means the expansion around $g_s = 0$. But whatever the nonperturbative topological string is, we want it to count BPS states and hence to agree with Z_{YM}. Therefore we might expect that Z_A itself probably only makes sense as a power series in g_s in general — the object that has a chance to have a nonperturbative completion is $|Z_A|^2$, but the nonperturbative completion probably is *not* generally factorized into chiral and anti-chiral parts.

7. Topological M-theory

In the last subsection, we described a conjectural relation between the square of the topological string partition function and the counting of black hole microstates. This relation, once fully understood, could be expected to lead to a proper nonperturbative understanding of the topological string. What kind of theory should we expect to find?

In the context of the physical string the answer to this question is rather surprising: it turns out that the proper nonperturbative description of the theory involves the dynamical emergence of an extra dimension, in other words, at strong coupling the theory is not 10-dimensional but 11-dimensional. While the fundamental degrees of freedom of this 11-dimensional "M-theory" are not known, we do know its low-energy description: it is 11-dimensional supergravity. It is natural to ask whether something similar might be true in the topological context; could the 6-dimensional theory be a shadow of some 7-dimensional lift? There are some tantalizing clues that this may be the case, which have been explored in [**70, 71**] (see also [**72**]); here we briefly summarize a few aspects of this story emphasized in [**70**].

The topological string theory is naturally related to backgrounds which preserve some supersymmetry when they appear in string compactifications — namely Calabi-Yau threefolds. From the target space point of view, these backgrounds should be understood as the solutions to a six-dimensional gravity theory, or more precisely two such gravity theories: the B model should be a theory whose classical solutions are complex threefolds equipped with a holomorphic 3-form, while the A model should have classical solutions which are symplectic manifolds. Giving both structures together is equivalent to giving the Ricci-flat metric on the Calabi-Yau threefold.[28] What about topological M-theory? There is a natural class of 7-dimensional Riemannian manifolds Y which preserve supersymmetry, namely manifolds of G_2 holonomy. Furthermore, given a Calabi-Yau manifold X, $Y = X \times S^1$ has holonomy contained in G_2. Hence it is natural to conjecture that the classical solutions of topological M-theory should be manifolds of G_2 holonomy.

With this guess for the classical solutions, one now has to ask: is there an action for which these are the extrema? Indeed there is a very natural candidate, recently discussed by Hitchin in [**73, 74**]. In Hitchin's theory the fundamental field is a locally defined 2-form β, which plays the role of an abelian gauge potential, from which one constructs the field strength Φ with some fixed flux (in other words, the field space consists of all closed 3-forms Φ in a fixed cohomology class.) If this Φ is suitably generic, then it defines a "G_2 structure" on Y, for which Φ is the "associative 3-form"; this just means that Φ picks out a privileged set of coordinate transformations, namely those which leave Φ invariant, and the group of such transformations at each point of Y is isomorphic to G_2. Indeed, these transformations also leave invariant a natural metric on Y, which we write as g_Φ; it can be given concretely in terms of Φ as

$$g_{jk} = B_{jk} \det(B)^{-1/9} \tag{7.1}$$

where

$$B_{jk} = -\frac{1}{144} \Phi_{ji_1i_2} \Phi_{ki_3i_4} \Phi_{i_5i_6i_7} \epsilon^{i_1 \dots i_7}. \tag{7.2}$$

Of course, for general Φ this metric need not have G_2 holonomy. To get this condition Hitchin now writes a rather remarkable action: it is simply the volume of Y in the metric g_Φ!

$$V_7(\Phi) = \int_Y \sqrt{g_\Phi}. \tag{7.3}$$

The extrema of this V_7, when Φ varies over a fixed cohomology class, turn out to give metrics of G_2 holonomy. In this sense V_7 is a candidate action for the proposed topological M-theory.

Some support for the conjecture that V_7 is related to topological strings comes from studying the theory on $Y = X \times S^1$. In this case it is natural to write Φ as

$$\Phi = \rho + k \wedge dt, \tag{7.4}$$

where t is a coordinate along S^1, and ρ and k are a 3-form and 2-form respectively on X. Then similarly expanding

$$*_\Phi \Phi = \sigma + \hat{\rho} dt, \tag{7.5}$$

[28]Strictly speaking, one has to require a compatibility condition: the symplectic form has to be of type $(1,1)$ in the complex structure so that it can be a Kähler form.

one finds[29]

$$V_7(\Phi) = V_H(\rho) + V_S(\sigma), \tag{7.6}$$

where V_H and V_S are two volume functionals in six dimensions analogous to V_7. Upon extremization these two functionals lead to complex and symplectic manifolds respectively; so they are candidate actions for the target space description of the B and A model topological string theories. To investigate this a little further one can try to compare the *partition function* of the theory with action V_H (call it Z_H) to the B model partition function Z_B. At the classical level (comparing the classical value of V_H to the genus zero part of Z_B) one finds that the proper conjecture is not $Z_H = Z_B$ but rather

$$Z_H = |Z_B|^2. \tag{7.7}$$

This is an encouraging result, since we have already seen in Section 6.4 some evidence that it is $|Z_B|^2$ rather than Z_B which is a natural candidate to be nonperturbatively completed; now we are finding that it is also the object which is naturally related to the 7-dimensional theory. One can also study Z_H at one loop; this was done in [**75**], which found that (7.7) is violated, but one can restore the agreement by replacing V_H with an action for which the extrema are *generalized* complex structures instead of ordinary ones, as given in [**76**]. In fact, the topological string makes sense on generalized Calabi-Yau manifolds, so it is very natural to use this modified V_H as its target space description; presumably this means that in 7 dimensions we should also replace V_7 by an action describing generalized G_2 manifolds as defined in [**77**].

So V_H passes a few basic checks as a candidate description of the B model. Similarly one can argue that the classical value of V_S agrees with the expectation from the A model; for this comparison one uses the "quantum foam" reformulation of that theory which we reviewed briefly in Section 5.6. These checks give some support to the conjecture that V_7 is indeed an appropriate description of a topological M-theory in 7 dimensions, since it is related to these reformulations of the topological string in 6 dimensions. Intriguingly, the reformulations of the topological string given by V_H and V_S seem to be naturally adapted to the problem of counting black hole microstates.

Even without understanding all of the details of the 7-dimensional theory, its existence can already shed some light on some properties of the 6-dimensional topological string. For example, it provides a natural interpretation of the fact that the topological string partition function behaves like a wavefunction, as we mentioned in Section 4.7; this is what one generally expects for the partition function of a 7-dimensional theory considered on a 7-manifold with a 6-dimensional boundary. In addition, the 7-dimensional description seems to unify the A and B model degrees of freedom in a natural

[29] Again here we are assuming some compatibility conditions between ρ and k.

way — the holomorphic and symplectic structures in 6 dimensions are naturally combined into the associative 3-form in 7 dimensions. Interestingly, looking at the canonical quantization on a 6-manifold one seems to find that the A model and B model degrees of freedom appear as canonical conjugates — this seems to suggest that in some sense these degrees of freedom cannot be specified simultaneously in the quantum theory, and also may be related to the conjectured S-duality between the two models.

Acknowledgements. We would like to thank the 2004 Simons Workshop on Mathematics and Physics, which was the motivation for this review. We are especially grateful to Martin Roček for his hospitality during the workshop and for preparing many of the figures in this paper, and to Luboš Motl, Boris Pioline and Marcel Vonk for helpful comments on earlier versions.

This work was supported in part by NSF grants PHY-0244821 and DMS-0244464.

References

[1] S.-T. Yau, *On the Ricci curvature of a compact Kaehler manifold and the complex Monge-Ampere equation*, I, Comm. Pure Appl. Math. **31** (1978), 339–411.

[2] P.A. Griffiths and J. Harris, *Principles of algebraic geometry*, John Wiley & Sons Inc., New York, 1994, reprint of the 1978 original.

[3] P. Candelas and X.C. de la Ossa, *Comments on conifolds*, Nucl. Phys. **B342** (1990), 246–268.

[4] K. Hori, S. Katz, A. Klemm, R. Pandharipande, R.P. Thomas, C. Vafa, R. Vakil, and E. Zaslow, *Mirror symmetry*, Clay Mathematics Monographs, **1**, American Mathematical Society, Providence, RI, 2003.

[5] E. Witten, *Phases of $\mathcal{N} = 2$ theories in two dimensions*, Nucl. Phys. **B403** (1993), 159–222, hep-th/9301042.

[6] E. Witten, *Topological sigma models*, Commun. Math. Phys. **118** (1988), 411.

[7] E. Witten, *Mirror manifolds and topological field theory*, in 'Mirror symmetry, I' (S.-T. Yau, ed.), 121–160.

[8] M. Bershadsky, S. Cecotti, H. Ooguri, and C. Vafa, *Holomorphic anomalies in topological field theories*, Nucl. Phys. **B405** (1993), 279–304, hep-th/9302103.

[9] M. Bershadsky, S. Cecotti, H. Ooguri, and C. Vafa, *Kodaira-Spencer theory of gravity and exact results for quantum string amplitudes*, Commun. Math. Phys. **165** (1994), 311–428, hep-th/9309140.

[10] D. Quillen, *Determinants of Cauchy-Riemann operators on Riemann surfaces*, Functional Anal. Appl. **19(1)** (1985), 31–34.

[11] E. Witten, *Quantum background independence in string theory*, hep-th/9306122.

[12] R. Dijkgraaf, E. Verlinde, and M. Vonk, *On the partition sum of the NS five-brane*, hep-th/0205281.

[13] A. Strominger, S.-T. Yau, and E. Zaslow, *Mirror symmetry is T-duality*, Nucl. Phys. **B479** (1996), 243–259, hep-th/9606040.

[14] K. Hori and C. Vafa, *Mirror symmetry*, hep-th/0002222.

[15] M. Aganagic and C. Vafa, *Mirror symmetry and supermanifolds*, hep-th/0403192.

[16] A. Schwarz, *Sigma models having supermanifolds as target spaces*, Lett. Math. Phys. **38** (1996), 91–96, hep-th/9506070.

[17] E. Witten, *Perturbative gauge theory as a string theory in twistor space*, hep-th/0312171.

[18] E. Witten, *An interpretation of classical Yang-Mills theory*, Phys. Lett. **B77** (1978), 394.

[19] A. Neitzke and C. Vafa, *$\mathcal{N} = 2$ strings and the twistorial Calabi-Yau*, hep-th/0402128.

[20] S. Katz, A. Klemm, and C. Vafa, *M-theory, topological strings and spinning black holes*, Adv. Theor. Math. Phys. **3** (1999), 1445–1537, hep-th/9910181.

[21] O. Aharony, S.S. Gubser, J.M. Maldacena, H. Ooguri, and Y. Oz, *Large N field theories, string theory and gravity*, Phys. Rept. **323** (2000), 183–386, hep-th/9905111.

[22] J.M. Maldacena, *The large N limit of superconformal field theories and supergravity*, Adv. Theor. Math. Phys. **2** (1998), 231–252, hep-th/9711200.

[23] R. Gopakumar and C. Vafa, *On the gauge theory/geometry correspondence*, Adv. Theor. Math. Phys. **3** (1999), 1415–1443, hep-th/9811131.

[24] H. Ooguri and C. Vafa, *Worldsheet derivation of a large N duality*, Nucl. Phys. **B641** (2002), 3–34, hep-th/0205297.

[25] E. Witten, *Chern-Simons gauge theory as a string theory*, Prog. Math. **133** (1995), 637–678, hep-th/9207094.

[26] E. Witten, *Noncommutative geometry and string field theory*, Nucl. Phys. **B268** (1986), 253.

[27] N.A. Nekrasov, H. Ooguri, and C. Vafa, *S-duality and topological strings*, hep-th/0403167.

[28] M. Aganagic, M. Marino, and C. Vafa, *All loop topological string amplitudes from Chern-Simons theory*, hep-th/0206164.

[29] H. Ooguri and C. Vafa, *Knot invariants and topological strings*, Nucl. Phys. **B577** (2000), 419–438, hep-th/9912123.

[30] E. Witten, *Quantum field theory and the Jones polynomial*, Commun. Math. Phys. **121** (1989), 351.

[31] M. Aganagic, A. Klemm, M. Marino, and C. Vafa, *The topological vertex*, hep-th/0305132.

[32] M. Aganagic, R. Dijkgraaf, A. Klemm, M. Marino, and C. Vafa, *Topological strings and integrable hierarchies*, hep-th/0312085.

[33] A. Okounkov, N. Reshetikhin, and C. Vafa, *Quantum Calabi-Yau and classical crystals*, hep-th/0309208.

[34] A. Iqbal, N.A. Nekrasov, A. Okounkov, and C. Vafa, *Quantum foam and topological strings*, hep-th/0312022.

[35] D. Maulik, N.A. Nekrasov, A. Okounkov, and R. Pandharipande, *Gromov-Witten theory and Donaldson-Thomas theory*, math.AG/0312059.

[36] M. Bershadsky and V. Sadov, *Theory of Kähler gravity*, Int. J. Mod. Phys. **A11** (1996), 4689–4730, hep-th/9410011.

[37] R. Hartshorne, *Algebraic geometry*, Graduate Texts in Mathematics, **52**, Springer-Verlag, New York, 1977.

[38] I. Antoniadis, E. Gava, K.S. Narain, and T.R. Taylor, *Topological amplitudes in string theory*, Nucl. Phys. **B413** (1994), 162–184, hep-th/9307158.

[39] S. Katz and C. Vafa, *Matter from geometry*, Nucl. Phys. **B497** (1997), 146–154, hep-th/9606086.

[40] N. Seiberg and E. Witten, *Electric-magnetic duality, monopole condensation, and confinement in $\mathcal{N} = 2$ supersymmetric Yang-Mills theory*, Nucl. Phys. **B426** (1994), 19–52, hep-th/9407087.

[41] S. Katz, A. Klemm, and C. Vafa, *Geometric engineering of quantum field theories*, Nucl. Phys. **B497** (1997) 173–195, hep-th/9609239.

[42] S. Kachru, A. Klemm, W. Lerche, P. Mayr, and C. Vafa, *Nonperturbative results on the point particle limit of $\mathcal{N} = 2$ heterotic string compactifications*, Nucl. Phys. **B459** (1996), 537–558, hep-th/9508155.

[43] S. Katz, P. Mayr, and C. Vafa, *Mirror symmetry and exact solution of 4d $\mathcal{N} = 2$ gauge theories*, I, Adv. Theor. Math. Phys. **1** (1998), 53–114, hep-th/9706110.

[44] D.R. Morrison and N. Seiberg, *Extremal transitions and five-dimensional supersymmetric field theories*, Nucl. Phys. **B483** (1997), 229–247, hep-th/9609070.

[45] A.D. Shapere and C. Vafa, *BPS structure of Argyres-Douglas superconformal theories*, hep-th/9910182.

[46] A.E. Lawrence and N.A. Nekrasov, *Instanton sums and five-dimensional gauge theories*, Nucl. Phys. **B513** (1998), 239–265, hep-th/9706025.

[47] C. Vafa, *Superstrings and topological strings at large N*, J. Math. Phys. **42** (2001), 2798–2817, hep-th/0008142.

[48] S. Gukov, C. Vafa, and E. Witten, *CFT's from Calabi-Yau four-folds*, Nucl. Phys. **B584** (2000), 69–108, hep-th/9906070.

[49] T.R. Taylor and C. Vafa, *RR flux on Calabi-Yau and partial supersymmetry breaking*, Phys. Lett. **B474** (2000), 130–137, hep-th/9912152.

[50] G. Veneziano and S. Yankielowicz, *An effective Lagrangian for the pure $\mathcal{N} = 1$ supersymmetric Yang-Mills theory*, Phys. Lett. **B113** (1982), 231.

[51] M. Aganagic and C. Vafa, *Mirror symmetry, D-branes and counting holomorphic discs*, hep-th/0012041.

[52] F.A. Cachazo, K.A. Intriligator, and C. Vafa, *A large N duality via a geometric transition*, Nucl. Phys. **B603** (2001), 3–41, hep-th/0103067.

[53] R. Dijkgraaf and C. Vafa, *Matrix models, topological strings, and supersymmetric gauge theories*, hep-th/0206255.

[54] R. Dijkgraaf and C. Vafa, *On geometry and matrix models*, hep-th/0207106.

[55] R. Dijkgraaf and C. Vafa, *A perturbative window into non-perturbative physics*, hep-th/0208048.

[56] E. Witten, *Constraints on supersymmetry breaking*, Nucl. Phys. **B202** (1982), 253.

[57] E. Witten, *On the Landau-Ginzburg description of $\mathcal{N} = 2$ minimal models*, Int. J. Mod. Phys. **A9** (1994), 4783–4800, hep-th/9304026.

[58] R. Gopakumar and C. Vafa, *M-theory and topological strings*, I, hep-th/9809187.

[59] R. Gopakumar and C. Vafa, *M-theory and topological strings*, II, hep-th/ 9812127.

[60] S. Ferrara, R. Kallosh, and A. Strominger, *$\mathcal{N} = 2$ extremal black holes*, Phys. Rev. **D52** (1995), 5412–5416, hep-th/9508072.

[61] A. Strominger, *Macroscopic Entropy of $\mathcal{N} = 2$ Extremal Black Holes*, Phys. Lett. **B383** (1996), 39–43, hep-th/9602111.

[62] G.W. Moore, *Arithmetic and attractors*, hep-th/9807087.

[63] H. Ooguri, A. Strominger, and C. Vafa, *Black hole attractors and the topological string*, hep-th/0405146.

[64] B. de Wit, G. Lopes Cardoso, and T. Mohaupt, *Corrections to macroscopic supersymmetric black-hole entropy*, Phys. Lett. **B451** (1999), 309–316, hep-th/9812082.

[65] B. de Wit, G. Lopes Cardoso, and T. Mohaupt, *Deviations from the area law for supersymmetric black holes*, Fortsch. Phys. **48** (2000) 49–64, hep-th/9904005.

[66] B. de Wit, G. Lopes Cardoso, and T. Mohaupt, *Macroscopic entropy formulae and non-holomorphic corrections for supersymmetric black holes*, Nucl. Phys. **B567** (2000), 87–110, hep-th/9906094.

[67] J.M. Maldacena, A. Strominger, and E. Witten, *Black hole entropy in m-theory*, JHEP **12** (1997), 002, hep-th/9711053.

[68] C. Vafa, *Two dimensional Yang-Mills, black holes and topological strings*, hep-th/ 0406058.

[69] D.J. Gross and W. Taylor, *Two-dimensional QCD is a string theory*, Nucl. Phys. **B400** (1993), 181–210, hep-th/9301068.

[70] R. Dijkgraaf, S. Gukov, A. Neitzke, and C. Vafa, *Topological M-theory as unification of form theories of gravity*, hep-th/0411073.
[71] N.A. Nekrasov, *Z theory*, hep-th/0412021.
[72] A.A. Gerasimov and S.L. Shatashvili, *Towards integrability of topological strings. I: Three- forms on Calabi-Yau manifolds*, hep-th/0409238.
[73] N.J. Hitchin, *The geometry of three-forms in six and seven dimensions*, math.dg/0010054.
[74] N.J. Hitchin, *Stable forms and special metrics*, math.dg/0107101.
[75] V. Pestun and E. Witten, *The Hitchin functionals and the topological B-model at one loop*, hep-th/0503083.
[76] N.J. Hitchin, *Generalized Calabi-Yau manifolds*, Quart. J. Math. Oxford Ser. **54** (2003), 281–308, math.dg/0209099.
[77] F. Witt, *Generalised G_2 manifolds*, math.dg/0411642.

Jefferson Physical Laboratory, Harvard University,
Cambridge, MA 02138
E-mail address: neitzke@post.harvard.edu
E-mail address: vafa@physics.harvard.edu

Notes on GIT and symplectic reduction for bundles and varieties

R.P. Thomas

1. Introduction

These notes give an introduction to Geometric Invariant Theory (GIT) and symplectic reduction, with lots of pictures and simple examples. We describe their applications to moduli of bundles and varieties, leaving the technical work on the analysis of the Hilbert-Mumford criterion in these situations to the final sections. We outline their infinite dimensional analogues (so called Hitchin-Kobayashi correspondences) in gauge theory and in the theory of constant scalar curvature Kähler (cscK) and Kähler-Einstein (KE) metrics on algebraic varieties. Donaldson's work on why these should be thought of as the classical limits of the original finite dimensional constructions – which are then their "quantisations" – is explained. The many analogies and connections between the bundle and variety cases are emphasised; in particular the GIT analysis of stability of bundles is shown to be a special case of the (harder) problem for varieties.

For GIT we work entirely over $\mathbb{C}$ and skip or only sketch many proofs. The interested reader should refer to the excellent [**Dl, GIT, Ne**] for more details. Throughout this survey we mention many names, but only include certain key papers in the references – apologies to those omitted but compiling a truly comprehensive bibliography would be fraught with danger.

Acknowledgements. I have learnt GIT from Simon Donaldson and Frances Kirwan over many years. Large parts of these notes are nothing but an account of Donaldson's point of view on GIT and symplectic reduction for moduli of varieties. The last sections describe joint work with Julius Ross, and I would like to thank him, Gábor Székelyhidi and Xiaowei Wang for useful conversations. Thanks also to Claudio Arezzo, Joel Fine, Daniel Huybrechts, Julien Keller, Dmitri Panov, Michael Singer and Burt Totaro for comments on the manuscript.

2. A brief review of affine and projective geometry

This section can be safely skipped by readers with any knowledge of algebraic geometry. We fix some notation and speedily review some standard theory of complex affine and projective varieties (and schemes). These are much simpler than arbitrary varieties in that they can be described by a single ring. Throughout "ring" means finitely generated commutative $\mathbb{C}$-algebra – i.e. a Noetherian ring with a scalar action of $\mathbb{C}$ making it into a $\mathbb{C}$-vector space and commuting with multiplication.

Affine varieties. *Affine varieties* X are just the irreducible components of the zero sets of finite collections of polynomials $p_1, \ldots, p_k$ in some affine space $\mathbb{C}^n$. They are in one-to-one correspondence with rings with no zero divisors (i.e., integral domains); in coordinates this is particularly simple:

$$\begin{array}{ccc} \mathbb{C}[x_1, \ldots, x_n] & \longleftrightarrow & \mathbb{C}^n, \\ \dfrac{\mathbb{C}[x_1, \ldots, x_n]}{(p_1, \ldots, p_k)} & \longleftrightarrow & (p_1 = 0 = \ldots = p_k) \subseteq \mathbb{C}^n, \\ \mathcal{O}_X & \longleftrightarrow & X. \end{array} \tag{2.1}$$

The arrow $\leftarrow$ replaces a variety by the ring of functions on it (i.e. the functions $\mathbb{C}[x_1, \ldots, x_n]$ on $\mathbb{C}^n$ divided out by the ideal of those that vanish on X). The arrow $\rightarrow$ recovers X from its ring of functions $\mathcal{O}_X$ by taking a finite number of generators $x_1, \ldots, x_n$ and a finite (by the Hilbert basis theorem) number of relations $p_1, \ldots, p_k$ (considered as polynomials in the generators) and setting X to be the affine variety in $\mathbb{C}^n$ cut out by the polynomials p_i.

The embedding is equivalent to the choice of generators: each is a map $X \to \mathbb{C}$ so n of them give the map $X \hookrightarrow \mathbb{C}^n$. (Invariantly, we embed X in the *dual* of the vector space on the generators,

$$X \ni x \mapsto ev_x \in (\mathbb{C}\langle x_1, \ldots, x_n\rangle)^*, \qquad ev_x(f) := f(x), \tag{2.2}$$

each point x of X mapping to the linear functional that evaluates functions at x.) The ideal of functions $(p_1, \ldots, p_k)$ vanishing on $X \subset \mathbb{C}^n$ is *prime* (if it contains fg then it contains one of f or g) reflecting the fact that the ring $\mathcal{O}_X$ has no zero divisors and that X is irreducible.

So really (2.1) is a correspondence between

(1) Affine varieties $X \subseteq \mathbb{C}^n$ with a fixed embedding into n-dimensional affine space;
(2) prime ideals $I \subseteq \mathbb{C}[x_1, \ldots, x_n]$, and
(3) rings without zero divisors plus a choice of n generators.

The coordinate-free approach (which also shows the above construction is independent of choices) is to note that the points x of X are in one-to-one correspondence with the maximal ideals $\mathscr{I}_x \subset \mathcal{O}_X$ of functions vanishing at x. So to any ring without zero divisors R we associate an affine variety

Spec R whose points are maximal ideals in R; the coordinate-independent version of (2.1):

$$\mathcal{O}_X = R \quad \longleftrightarrow \quad X = \operatorname{Spec} R.$$

This can be extended to a correspondence between arbitrary rings and affine *schemes*, which are allowed nilpotents in their ring of functions, corresponding to multiplicities or infinitesimal directions in the scheme.

Projective varieties. Now we just do everything $\mathbb{C}^*$-equivariantly. Recall that a $\mathbb{C}^*$-action on a vector space V is equivalent to a *grading*, i.e., a splitting into subspaces (the eigen- or weight spaces) V_k parameterised by the integers (the eigenvalues or weights). So we replace rings by graded rings, ideals by homogeneous ideals (those which are the sum of their graded pieces – i.e. which are $\mathbb{C}^*$-invariant), and get a correspondence between graded rings without zero divisors and projective varieties. This is easiest to express in coordinates when the ring is generated by its degree one piece:

$$\begin{aligned} \mathbb{C}[x_0, \ldots, x_n] \quad &\longleftrightarrow \quad \mathbb{P}^n, \\ \frac{\mathbb{C}[x_0, \ldots, x_n]}{(p_1, \ldots, p_k)} \quad &\longleftrightarrow \quad (p_1 = 0 = \ldots = p_k) \subseteq \mathbb{P}^n. \end{aligned} \tag{2.3}$$

Here $\mathbb{C}[x_0, \ldots, x_n]$ is given the standard grading it inherits from the scalar $\mathbb{C}^*$-action on $\mathbb{C}^{n+1}$ (i.e. the x_i have weight one), and the p_i are homogeneous polynomials (eigenvectors for the $\mathbb{C}^*$-action). They cut out an affine variety $\widetilde{X} = (p_1 = 0 = \ldots = p_k)$ in $\mathbb{C}^{n+1}$, which is $\mathbb{C}^*$-invariant and so a cone, determined by its set of lines through the origin $X \subseteq \mathbb{P}^n$. (X and $\mathbb{P}^n$ are of course the quotients of $\widetilde{X}\backslash\{0\}$ and $\mathbb{C}^{n+1}\backslash\{0\}$ by $\mathbb{C}^*$, but we have yet to develop a theory of quotients (!).)

This describes the arrow $\rightarrow$. For $\leftarrow$ we cannot simply take the ring of functions on X since this consists of just the constants; we have to take the ring of functions on the cone $\widetilde{X}$, which can be described on X in terms of a line bundle.

Since X is the space of lines in $\widetilde{X}$, it has a tautological line bundle $\mathcal{O}_X(-1) = \mathcal{O}_{\mathbb{P}^n}(-1)|_X$ over it whose fibre over a point in X is the corresponding line in $\widetilde{X} \subseteq \mathbb{C}^{n+1}$. The total space of $\mathcal{O}_X(-1)$ therefore has a tautological map to $\widetilde{X}$ which is an isomorphism away from the zero section $X \subset \mathcal{O}_X(-1)$, which is all contracted down to the origin in $\widetilde{X}$. In fact the total space of $\mathcal{O}_X(-1)$ is the *blow up* of $\widetilde{X}$ in the origin.

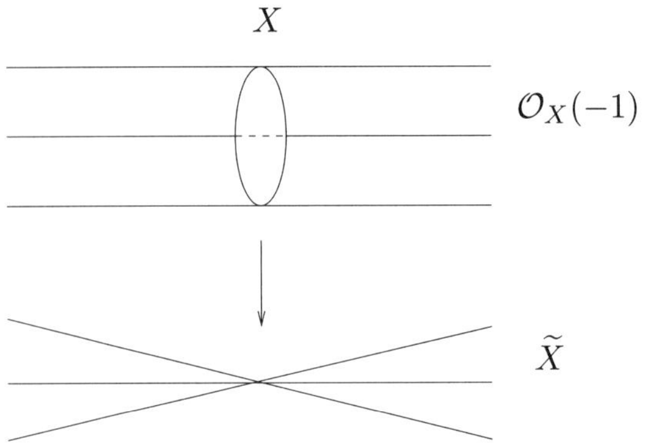

Linear functions on $\mathbb{C}^{n+1}$ like x_i, restricted to $\widetilde{X}$ and pulled back to the total space of $\mathcal{O}_X(-1)$, give functions which are linear on the fibres, so correspond to sections of the *dual* line bundle $\mathcal{O}_X(1)$. Similarly degree k homogeneous polynomials on $\widetilde{X}$ define functions on the total space of $\mathcal{O}_X(-1)$ which are of degree k on the fibres, and so give sections of the kth tensor power $\mathcal{O}_X(k)$ of the dual of the line bundle $\mathcal{O}_X(-1)$.

So the grading that splits the functions on $\widetilde{X}$ into homogeneous degree (or $\mathbb{C}^*$-weight spaces) corresponds to sections of different line bundles $\mathcal{O}_X(k)$ on X. So $\leftarrow$ takes the direct sum $\bigoplus_{k\geq 0} H^0(\mathcal{O}_X(k))$, considered a graded ring by tensoring sections $\mathcal{O}(k) \otimes \mathcal{O}(l) \xrightarrow{\simeq} \mathcal{O}(k+l)$. For the line bundle $\mathcal{O}_X(1)$ sufficiently positive, this ring will be generated in degree one. It is often called the (homogeneous) coordinate ring of the *polarised* (= endowed with an ample line bundle) variety $(X, \mathcal{O}_X(1))$.

The degree one restriction is for convenience and can be dropped (by working with varieties in weighted projective spaces), or bypassed by replacing $\mathcal{O}_X(1)$ by $\mathcal{O}_X(p)$, i.e. using the ring $R^{(p)} = \bigoplus_{k\geq 0} R_{kp}$; for $p \gg 0$ this will be generated by its degree one piece R_p.

The choice of generators of the ring is what gives the embedding in projective space. In fact the sections of any line bundle L over X define a (rational) map

$$(2.4) \qquad X \dashrightarrow \mathbb{P}(H^0(X,L)^*), \qquad x \mapsto ev_x, \qquad ev_x(s) := s(x),$$

(compare (2.2)) which in coordinates maps x to $(s_0(x) : \ldots : s_n(x)) \in \mathbb{P}^n$, where s_i form a basis for $H^0(L)$. This map is only defined for those x with $ev_x \neq 0$, i.e., for which $s(x)$ is not zero for every s.

It remains to describe $\rightarrow$ in a coordinate-free manner, by noting that the points of X, i.e. lines of $\widetilde{X}$, are $\mathbb{C}^*$-invariant subvarieties that are minimal among those which are not the origin in $\widetilde{X} \subseteq \mathbb{C}^{n+1}$; i.e. homogeneous ideals of the homogeneous coordinate ring that are maximal amongst all homogeneous ideals minus the one corresponding to the origin. So to any graded ring R we associate a projective variety $\operatorname{Proj} R$ whose points are the homogeneous ideals of R maximal amongst those except the *irrelevant* ideal $R^+ := \bigoplus_{k>0} R_k$. It comes equipped with an ample line bundle, the sections of whose kth power gives R_k for $k \gg 0$. This gives the coordinate-free (not quite one-to-one) correspondence (2.3):

$$\mathcal{O}_{\widetilde{X}} = R \quad \longleftrightarrow \quad (X, \mathcal{O}_X(1)) = \operatorname{Proj} R.$$

Replacing R by $R^{(p)}$ leaves the variety $\operatorname{Proj} R$ unaltered but changes the line bundle from $\mathcal{O}_X(1)$ to $\mathcal{O}_X(p)$.

Similarly arbitrary graded rings with zero divisors (finitely generated, as usual) correspond (not quite one-to-one) to polarised schemes.

Notation. Throughout G will be a connected **reductive complex linear algebraic group** with Lie algebra $\mathfrak{g}$. Reductive means that all (complex) representations split into sums of irreducibles, but equivalently it is the complexification of a compact real Lie group $K < G$ with Lie algebra $\mathfrak{k} < \mathfrak{g}$ such that $\mathfrak{g} = \mathfrak{k} + i\mathfrak{k}$. Therefore representations are also representations of the compact group, which preserve a hermitian inner product (by averaging any inner product on the representation using Haar measure on K) and so split into direct sums of irreducibles by taking orthogonal complements to invariant subspaces. These splittings are by complex subspaces, so are then also preserved by the complexification G.

For the purposes of these notes one can always assume that $K < G$ is $S^1 < \mathbb{C}^*$, $(S^1)^m < (\mathbb{C}^*)^m$ or $SU(m) < SL(m, \mathbb{C})$.

3. Geometric Invariant Theory

GIT is a way of taking quotients in algebraic geometry. This may sound like a dry and technical subject, but it is beautifully geometric (as we hope to show) and leads, through its link with symplectic reduction, to unexpected mathematics (some of which we describe later).

Suppose we are in the following situation, of G acting on a projective variety X through SL transformations of the projective space.

(3.1)
$$\begin{array}{ccc} G & \curvearrowright & X \\ \downarrow & & \cap \\ SL(n+1, \mathbb{C}) & \curvearrowright & \mathbb{P}^n. \end{array}$$

We would like to form a quotient X/G, ideally within the same category of projective varieties. There are a number of problems with this.

The topological quotient is not Hausdorff. Since X is compact but G is noncompact, a nontrivial G-action cannot be proper. There are nonclosed orbits (with lower dimensional orbits in their closures) so the topological quotient is not Hausdorff.

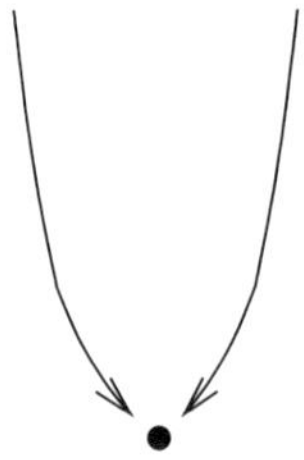

It is clear from the above simple illustration of 3 orbits, all of whose closures contain the smaller orbit, that we must remove some orbits to get a separated (Hausdorff) quotient. In the above case this would be the lower dimensional orbit; just as we would expect to remove the origin from $\mathbb{C}^{n+1}$ if we wanted to quotient by the scalar action of $\mathbb{C}^*$ to get $\mathbb{P}^n$ (an example to which we shall return).

Removing smaller orbits does not suffice. Another simple example shows that the quotient can still be nonseparated if we remove all lower dimensional orbits. Consider the action of $\mathbb{C}^*$ on $\mathbb{C}^2$ (or its projective completion $\mathbb{P}^2 \supset \mathbb{C}^2$) by matrices

$$\mathbb{C}^* \ni \lambda \mapsto \begin{pmatrix} \lambda & 0 \\ 0 & \lambda^{-1} \end{pmatrix} \in SL(2,\mathbb{C}) \tag{3.2}$$

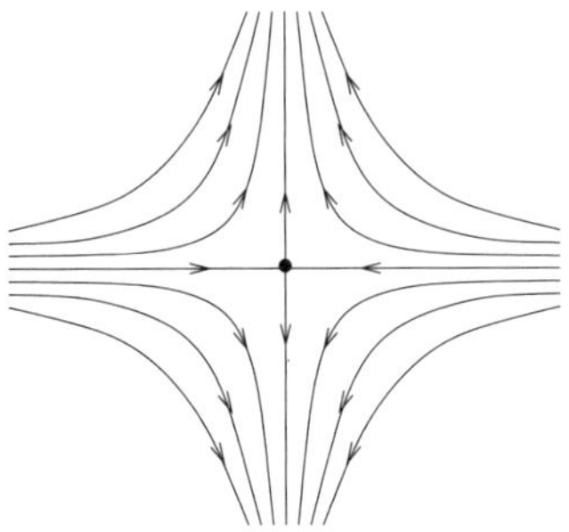

In this case removing the origin would make the topological quotient the affine line with nonseparated doubled point at the origin: it is clear that the punctured (origin removed) x- and y-axes are *both* orbits in the limit of the orbits $\{xy = \alpha\}$ as $\alpha \to 0$.

In this simple case it is clear that we would like the quotient to be $\mathbb{C}$, with $\alpha \in \mathbb{C}\backslash\{0\}$ representing the orbit $\{xy = \alpha\}$. But what then should the point $\alpha = 0$ of this "quotient" represent ? There are three standard solutions to the problem.

- Kapranov's **Chow quotient** (and the closely related **Hilbert quotient**, of which it is a contraction) parameterises the invariant conics in this example. For $\alpha \neq 0$ these are just the closed orbits of top dimension, while $\alpha = 0$ represents the invariant conic $\{xy = 0\}$, the *union* of all 3 bad orbits. We will not say more about the Chow quotient; instead see [**Hu**], for example (which, unsurprisingly, uses the same example for illustration).
- The **Geometric Invariant Theory quotient** gives the same for $\alpha \neq 0$, but $\alpha = 0$ represents any one of the 3 orbits: GIT identifies all 3 orbits with each other in an equivalence class.

- The **Symplectic reduction** throws away the two nonclosed orbits – the (punctured) x- and y-axes – and keeps the origin. So in this case the quotient represents the closed orbits, including the lower dimensional one.

In this simple case all three quotients give the same answer $\mathbb{C}$, but in general one expects the Chow quotient to dominate the GIT and symplectic quotients, which are isomorphic.

In general GIT and the symplectic quotient choose certain "unstable" orbits to remove to give a separated quotient. GIT also identifies some "semistable" orbits (those whose closures intersect each other) to compactify the quotient, resulting in a projective variety which we call X/G. The symplectic quotient compactifies by taking a distinguished representative of each semistable equivalence class above – the intersection of their closures, which turns out to be the unique *closed* semistable orbit in the equivalence class.

The construction of the GIT quotient. This is trivial, a formality. We consider X projective first, since although the affine case is often even easier, sometimes it is best embedded in the projective case, as we shall see below.

Since we have assumed that G acts through $SL(n+1,\mathbb{C})$ (rather than just its quotient $PSL(n+1,\mathbb{C})$), the action lifts from X to one covering it on $\mathcal{O}_X(-1)$. In other words we don't just act on the projective space (and X therein) but on the vector space overlying it (and the cone $\widetilde{X}$ on X therein). This is called a *linearisation* of the action. Thus G acts on each $H^0(\mathcal{O}_X(r))$.

Then, just as $(X,\mathcal{O}_X(1))$ is determined by its graded ring of sections of $\mathcal{O}(r)$ (i.e., the ring of functions on $\widetilde{X}$),

$$(X,\mathcal{O}(1)) \quad\longleftrightarrow\quad \bigoplus_r H^0(X,\mathcal{O}(r)),$$

we simply *construct* X/G (with a line bundle on it) from the ring of *invariant* sections:

$$X/G \quad\longleftrightarrow\quad \bigoplus_r H^0(X,\mathcal{O}(r))^G.$$

This is sensible, since if there is a good quotient then functions on it pullback to give G-invariant functions on X, i.e. functions constant on the orbits, the fibres of $X\to X/G$. For it to work we need

LEMMA 3.3. *$\bigoplus_r H^0(X,\mathcal{O}(r))^G$ is finitely generated.*

PROOF. Since $R:=\bigoplus_r H^0(X,\mathcal{O}(r))$ is Noetherian, Hilbert's basis theorem tells us that the ideal $R.\left(\bigoplus_{r>0} H^0(X,\mathcal{O}(r))^G\right)$ generated by $R_+^G := \bigoplus_{r>0} H^0(X,\mathcal{O}(r))^G$ is generated by a finite number of elements $s_0,\dots,s_k\in R_+^G$.

Thus any element $s \in H^0(X, \mathcal{O}(r))^G$, $r > 0$, may be written $s = \sum_{i=0}^k f_i s_i$, for some $f_i \in R$ of degree $< r$. To show that the s_i generate R_+^G as an algebra we must show that the f_i can be taken to lie in R^G.

We now use the fact that G is the complexification of the compact group K. Since K has an invariant metric, we can average over it and use the facts that s and s_i are invariant to give

$$s = \sum_{i=0}^{k} Av(f_i) s_i,$$

where $Av(f_i)$ is the (K-invariant) K-average of f_i. By complex linearity $Av(f_i)$ is also G-invariant (for instance, since G has a polar decomposition $G = K\exp(i\mathfrak{k})$). The $Av(f_i)$ are also of degree $< r$, and so we may assume, by an induction on r, that we have already shown that they are generated by the s_i in R_+^G. Thus s is also. $\square$

Thus we simply *define* X/G to be $\operatorname{Proj} \bigoplus_r H^0(X, \mathcal{O}(r))^G$. If X is a variety (rather than a scheme) then so is X/G, as its graded ring sits inside that of X and so has no zero divisors. Unfortunately this is not all there is to GIT, however. We have to work out what X/G is, which orbits points of X/G represent, and so on, which we tackle in the next section. Another important question, that we barely touch on, is how the quotient X/G changes with the linearisation. For some linearisations the quotient is empty, but if under a change of linearisation the moduli space remains the same dimension then it undergoes only a small birational transformation, a type of *flip* [**DH, Th**].

The affine case. The affine case is even easier; if G acts on $\operatorname{Spec} R$ we can form $\operatorname{Spec}(R^G)$ as a putative quotient. For instance in our example (3.2) the ring of invariants

$$R^G = \mathbb{C}[x,y]^{\mathbb{C}^*} = \mathbb{C}[xy]$$

is generated by xy, so that the quotient is $\operatorname{Spec} \mathbb{C}[xy] \cong \mathbb{C}$, as anticipated. The function xy does not distinguish between any of the "bad" orbits (the punctured x- and y- axes, and the origin), lumping them all in an equivalence class of orbits which get identified in the quotient.

In other cases this does not work so well; for instance under the scalar action of $\mathbb{C}^*$ on $\mathbb{C}^{n+1}$ the only invariant polynomials in $\mathbb{C}[x_0, \ldots, x_n]$ are the constants and this recipe for the quotient gives a single point. In the language of the next section, this is because there are no stable points in this example, and all semistable orbits' closures intersect (or equivalently, there is a unique polystable point, the origin). More generally in any affine case all points are always at least semistable (as the constants are always G-invariant functions) and so no orbits gets thrown away in making the quotient (though many may get identified with each other – those whose closures intersect which therefore cannot be separated by invariant functions). But for the scalar action of $\mathbb{C}^*$ on $\mathbb{C}^{n+1}$ we clearly need to remove at least the origin to get a sensible quotient.

So we should change the linearisation, from the trivial linearisation to a nontrivial one, to get a bigger quotient.

Example: $\mathbb{P}^n$ from GIT. That is, we consider the trivial line bundle on $\mathbb{C}^{n+1}$ but with a nontrivial linearisation, by composing the $\mathbb{C}^*$-action on $\mathbb{C}^{n+1}$ by a character $\lambda \mapsto \lambda^{-p}$ of $\mathbb{C}^*$ acting on the fibres of the trivial line bundle over $\mathbb{C}^{n+1}$. The invariant sections of this no longer form a ring; we have to take the direct sum of spaces of sections of *all powers* of this linearisation, just as in the projective case, and take Proj of the invariants of the resulting graded ring.

If $p < 0$ then there are no invariant sections and the quotient is empty. We have seen that for $p = 0$ the quotient is a single point. For $p > 0$ the invariant sections of the kth power of the linearisation are the homogeneous polynomials on $\mathbb{C}^n$ of degree kp. So for $p = 1$ we get the quotient

$$(3.4) \qquad \mathbb{C}^{n+1}/\mathbb{C}^* = \operatorname{Proj} \bigoplus_{k\geq 0} \big(\mathbb{C}[x_0, \dots, x_n]_k\big) = \operatorname{Proj} \mathbb{C}[x_0, \dots, x_n] = \mathbb{P}^n.$$

For $p \geq 1$ we get the same geometric quotient but with the line bundle $\mathcal{O}(p)$ on it instead of $\mathcal{O}(1)$.

Another way to derive this is to embed $\mathbb{C}^{n+1}$ in $\mathbb{P}^{n+1}$ as $x_{n+1} = 1$, act by $\mathbb{C}^*$ on the latter by $\operatorname{diag}\big(\lambda, \dots, \lambda, \lambda^{-(n+1)}\big) \in SL(n+2, \mathbb{C})$, and do projective GIT. This gives, on restriction to $\mathbb{C}^{n+1} \subset \mathbb{P}^{n+1}$, the $p = n+1$ linearisation above. The invariant sections of $\mathcal{O}((n+2)k)$ are of the form $x_{n+1}^k.f$, where f is a homogeneous polynomial of degree $(n+1)k$ in $x_1, \dots, x_n$. Therefore the quotient is

$$\operatorname{Proj} \bigoplus_{k\geq 0} \big(\mathbb{C}[x_1, \dots, x_n]_{(n+1)k}\big) = (\mathbb{P}^n, \mathcal{O}(n+1)).$$

In the language of the next section this is because the complement of $\mathbb{C}^{n+1} \subset \mathbb{P}^{n+1}$, and the origin $\{0\} \in \mathbb{C}^{n+1} \subset \mathbb{P}^{n+1}$, are unstable (either by noting that all of the nonconstant invariant polynomials above vanish on them, or by an easy exercise in using the Hilbert-Mumford criterion below – these loci are fixed points, but with a nontrivial action on the line above them). So these are removed and the projective quotient reduces to the affine case.

What are the points of a GIT quotient? By its very definition (and Lemma 3.3), for $r \gg 0$, X/G is just the image of X under the linear system $H^0(\mathcal{O}_X(r))^G$. That is, consider the Kodaira "embedding" of X/G (2.4),

$$(3.5) \qquad \begin{array}{ccc} X & \dashrightarrow & \mathbb{P}((H^0(X, \mathcal{O}(r))^G)^*), \\ x & \mapsto & ev_x \qquad\qquad (ev_x(s) := s(x)), \end{array}$$

that in coordinates takes x to $(s_0(x) : \dots : s_k(x)) \in \mathbb{P}^k$ (where the s_i form a basis for $H^0(X, \mathcal{O}(r))^G$).

This is only a rational map, since it is only defined on points for which $ev_x \not\equiv 0$ (equivalently the $s_i(x)$ are not all zero). That is, it is defined on the *semistable points* of X:

DEFINITION 3.6. $x \in X$ is **semistable** if and only if there exists $s \in H^0(X, \mathcal{O}(r))^G$ with $r > 0$ such that $s(x) \neq 0$.

Points which are not semistable we call (controversially) ***unstable***.

So semistable points are those that the G-invariant functions "see". The map (3.5) is well defined on the (Zariski open, though possibly empty) locus $X^{ss} \subseteq X$ of semistable points, and it is clearly constant on G-orbits, i.e. it factors through the set-theoretic quotient X^{ss}/G. But it may contract more than just G-orbits, so we need another definition.

DEFINITION 3.7. A semistable point x is **stable** if and only if $\bigoplus_r H^0(X, \mathcal{O}(r))^G$ separates orbits near x and the stabiliser of x is finite.

By "separates orbits near x" we mean the following. Since x is semistable there exists an $s \in H^0(\mathcal{O}_X(r))^G$ such that $s(x) \neq 0$. So now we work on the open set $U \subset X$ on which $s \neq 0$ and use s to trivialise $\mathcal{O}_U(r)$ (i.e., divide all sections of $\mathcal{O}_U(r)$ by s to consider them as functions). Then we ask that in U any orbit can be distinguished from $G.x$ by $H^0(X, \mathcal{O}(r))^G$. That is, there is an element of $H^0(X, \mathcal{O}(r))^G$ which takes different values on the two orbits, and this should also be true infinitesimally: given a vector $v \in T_xX \backslash T_x(G.x)$, there is an element of $H^0(X, \mathcal{O}(r))^G$ whose derivative down v is nonzero.

So we have a (surjective) map $X^{ss} \to X/G$ under which the line bundle on X/G (that arises from its Proj construction) pulls back to $\mathcal{O}_{X^{ss}}(1)$. This map has good geometric properties over the locus of stable points $X^s \subseteq X^{ss} \subseteq X$ (it only contracts single orbits, for instance; more properly it is a *geometric quotient* in the sense of [**GIT**, Definition 0.6]). The definitions of stable and semistable are the algebraic ones designed to make this true, but now we can relate them more to geometry.

Topological characterisation of (semi)stability. If we work upstairs in the vector space $\mathbb{C}^{n+1} \supset \widetilde{X}$ (or equivalently in the total space of $\mathcal{O}_X(-1)$) instead of in the projective space $\mathbb{P}^n \supset X$, we can get a beautiful topological characterisation of (semi)stability. Given our topological discussion about nonclosed orbits at the start of these notes, it is what one might guess, and the best one could possibly hope for. So for $x \in X$, pick $\tilde{x} \in \mathcal{O}_X(-1)$ covering it.

THEOREM 3.8.
x is semistable $\iff 0 \notin \overline{G.\tilde{x}}$.
x is stable $\iff$ *$G.\tilde{x}$ is closed in $\mathbb{C}^{n+1}$ and $\tilde{x}$ has finite stabiliser.*

When $G.\tilde{x}$ is closed, but not necessarily of full dimension, we call x **polystable**. (This is called "Kempf-stable" in [**Dl**], "weakly stable" in

[**Ti4**] and plain "stable" in [**Do5**]. Originally [**GIT**] the terms for stable and polystable were "properly stable" and "stable" respectively.)

In one direction the theorem is clear. G-invariant homogeneous functions of degree $r > 0$ on $\widetilde{X}$ are constant on orbits and so also their closures. So if the closure of the orbit of $\tilde{x}$ contains the origin then every such function is zero on $\widetilde{x}$ and x cannot be semistable. Similarly if the invariant functions separate orbits around the orbit of $\tilde{x}$ then it is the zero locus of a collection of invariant functions and so closed.

One can make the criterion (3.8) much simpler to calculate with, by considering one parameter subgroups instead of all of G.

The Hilbert-Mumford criterion. The key result is that x is (semi)-stable for G if and only if it is (semi)stable *for all one parameter subgroups* ($1 - PSs$) $\mathbb{C}^* < G$. We will outline a proof of this remarkable result once we have done some symplectic geometry.

So we may apply Theorem 3.8 to each of these 1-PS orbits, and determining the closedness of these one dimensional orbits is much easier by using their asymptotics. Setting $x_0 = \lim_{\lambda\to 0}\lambda.x$, this is a fixed point of the $\mathbb{C}^*$-action, so $\mathbb{C}^*$ acts on the line $\mathcal{O}_{x_0}(-1)$ in $\mathbb{C}^{n+1}$ that $x_0 \in \mathbb{P}^n$ represents. Letting $\rho(x) \in \mathbb{Z}$ denote the weight of this action (i.e. $\mathbb{C}^* \ni \lambda$ acts on $\mathcal{O}_{x_0}(-1)$ as $\lambda^{\rho(x)}$) we find the following, the Hilbert-Mumford criterion.

THEOREM 3.9.

- *If $\rho(x) < 0$ for all 1-PS then x is stable,*
- *If $\rho(x) \leq 0$ for all 1-PS then x is semistable,*
- *If $\rho(x) > 0$ for a 1-PS then x is unstable.*

The proof is the picture below; the 1-PS orbit is closed if and only if it is asymptotic to a negative weight $\mathbb{C}^*$-action on the limiting line at both $\lambda \to 0$ and $\lambda \to \infty$. But we can restrict to the former since the latter arises from the inverse 1-PS.

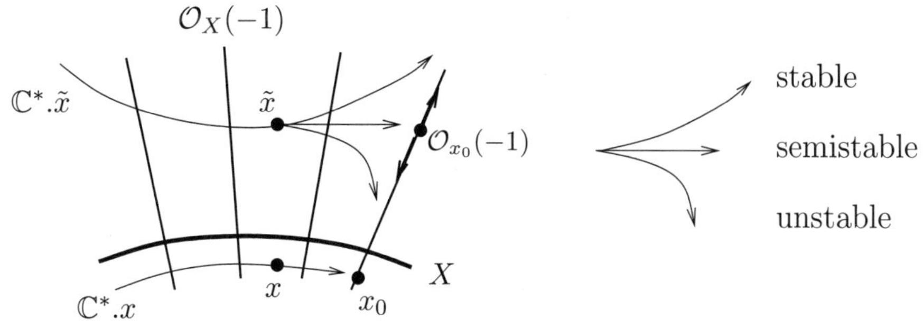

So we "just" have to compute the weight $\rho(x)$ for *all* $\mathbb{C}^* < SL(n+1, \mathbb{C})$; x is stable for G if and only if $\rho(x)$ is always < 0.

To sum up; the 1-PS orbits are not (in general) closed in the projective space, but they may be upstairs in the vector space. To decide if x is stable we first take the limit x_0 as we move through isomorphic objects $\lambda.x$; this limit is *not* (in general) isomorphic to x under the $\mathbb{C}^*$-action; it is only in the *closure* of the $\mathbb{C}^*$-orbit. This point x_0 represents a line $\mathcal{O}_{x_0}(-1)$ in the vector space, on which $\mathbb{C}^*$ acts. If this weight is negative then x (*not* x_0!) is stable for this 1-PS; if this is true for all 1-PSs then x is stable for G.

Fundamental example: points in $\mathbb{P}^1$. The standard example, from for instance [**GIT**], is to consider configurations of n (unordered) points in $\mathbb{P}^1$ up to the symmetries of $\mathbb{P}^1$. (This is of course a 0-dimensional algebraic variety, and so the easiest example of the stability of varieties that we shall study later.) In fact we allow multiplicities, i.e. we take length-n 0-dimensional subschemes of $\mathbb{P}^1$ – $S^n\mathbb{P}^1$ – modulo $SL(2,\mathbb{C})$.

To linearise the action we note that specifying any such n points is the same as specifying a degree n homogeneous polynomial on $\mathbb{P}^1$, unique up to scale, by taking the roots of the polynomial. That is, $S^n\mathbb{P}^1$ is the projectivisation of $H^0(\mathcal{O}_{\mathbb{P}^1}(n))$, so giving us a natural linearisation of the problem; we use the induced $SL(2,\mathbb{C})$-action on $H^0(\mathcal{O}_{\mathbb{P}^1}(n)) \cong S^n(\mathbb{C}^2)^*$.

We find that the configuration is stable unless it has a very singular point.

THEOREM 3.10. *A length-n subscheme of $\mathbb{P}^1$ is*

- *semistable if and only if each multiplicity $\leq n/2$,*
- *stable if and only if each multiplicity $< n/2$.*

PROOF. 1. Diagonalise a given $\mathbb{C}^* < SL(2,\mathbb{C})$:

$$\begin{pmatrix} \lambda^k & 0 \\ 0 & \lambda^{-k} \end{pmatrix} \text{ in } [x:y] \text{ coordinates on } \mathbb{P}^1. \ (k \geq 0.) \tag{3.11}$$

In these coordinates write our degree n homogeneous polynomial (whose roots give the n points) as $f = \sum_{i=0}^n a_i x^i y^{n-i}$. As $\lambda \to 0$ the first half (precisely the first $\lceil n/2 - 1 \rceil$) of these monomials tend to infinity (as there are more ys than xs in the monomial).

Thus $\lambda.f$ tends to ∞ and the orbit is closed about $\lambda \to 0$ unless $a_i = 0$ for $i \leq n/2$. That is, it is closed so long as f does not vanish to order $\geq n/2$ at $x = 0$.

Repeating over all 1-PS changes the coordinates $[x : y]$, so f is stable if and only if it does not vanish to order $\geq n/2$ at any point. □

Alternatively, we can use the Hilbert-Mumford criterion in terms of the weight on the limiting line.

PROOF. 2. Up to rescaling, under the action (3.11), $\lambda.f \to f_0 = a_j x^j y^{n-j}$, where j is smallest such that $a_j \neq 0$.

The weight of (3.11) on $\mathbb{C}.f_0$ is $k(j-(n-j)) = k(2j-n)$. So f is stable if and only if $k(2j-n) < 0 \iff j < n/2 \iff \mathrm{ord}_{x=0}(f) < n/2$ for all 1-PS (and so all points $x = 0$) as before. Semistability is similar. □

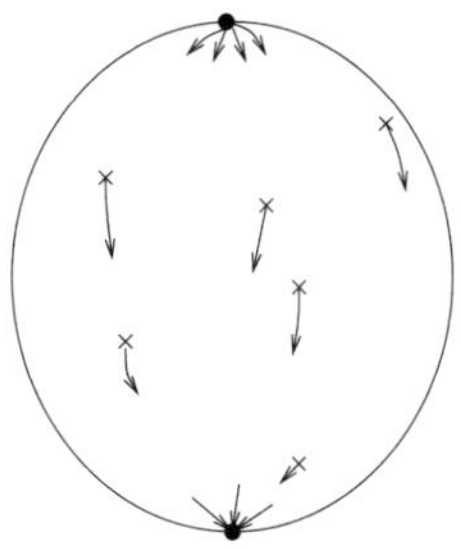

Geometrically what is happening is that the 1-PS moves almost all points to the "attractive" fixed point at $x = 0$ (weight $-k$), and this is the generic, stable, situation. Only those points stuck at "repulsive" fixed point $y = 0$ contribute positively ($+k$) to the weight. So the total weight is negative unless more than half of the points are at the repulsive fixed point.

More generally if we consider hypersurfaces $D \subset \mathbb{P}^n$ modulo the action of $SL(n+1, \mathbb{C})$, then the existence of the *discriminant* of the defining equation (which is automatically an invariant polynomial) means that

$$D \text{ smooth } \Rightarrow D \text{ semistable}, \tag{3.12}$$

since its discriminant does not vanish.

Another topological characterisation of stability. Instead of working in $\widetilde{X}$, we can give a topological characterisation of stability (not semi-stability) *downstairs* in X which, in the language of the first section, says that we have removed enough orbits to get a Hausdorff quotient.

Namely, polystability of x is equivalent to the orbit of x being *closed in the locus of semistable points.*

For stability, one direction is clear; by the original definition of stability of x, the *closed* locus where the invariant sections $\bigoplus_k H^0(L^k)^G$ take the value that they take on $G.x$ is precisely $G.x$.

For the converse we note that given any 1-PS orbit of a stable point x, its limit point x_0 is *unstable*, by the Hilbert-Mumford criterion applied to the inverse 1-PS (under which x_0 is fixed and $\mathcal{O}_{x_0}(-1)$ is acted on with positive weight). Thus once x_0 is removed, i.e. in the locus of semistable points, the 1-PS orbit of x is closed (at the $\lambda \to 0$ end; then one can consider the inverse 1-PS). Then one has to show that this enough to show that the whole of $G.x$ is closed in X^{ss}.

The polystable case follows by showing that for any semistable point x,

$$\bigcap \{\overline{G.y} \, : \, \overline{G.y} \cap \overline{G.x} \neq \emptyset\}$$

is a *single orbit*, which is therefore closed. This is the unique polystable representative of the semistable equivalence class of orbits which are identified together in the GIT quotient. This can be proved using the Kempf-Ness theorem in a later Section.

Another form of the Hilbert-Mumford criterion. The Hilbert-Mumford criterion can be recast in terms of $\tilde{x}$ only (with no mention of x_0) in a form that can be useful in calculations.

In our usual set-up of G acting on $X \subseteq \mathbb{P}^n$ linearised over $\mathcal{O}_X(-1)$, any 1-PS therefore gives a 1-PS of $SL(n+1,\mathbb{C})$ (in fact $GL(n+1,\mathbb{C})$, but it is an easy exercise to see that if it does not lie in $SL(n+1,\mathbb{C})$ then all points are unstable).

Diagonalising we can write the action as $\lambda \mapsto \operatorname{diag}(\lambda^{\rho_0},\dots,\lambda^{\rho_n})$, in which basis $\tilde{x} = (x_0,\dots,x_n)$. Then it is clear that the orbit $\{\lambda.\tilde{x}\}$ tends to ∞ as $\lambda \to 0$ if and only if there is an i such that

$$x_i \neq 0 \quad \text{and} \quad \rho_i < 0. \tag{3.13}$$

This is true for all 1-PSs if and only if $\tilde{x}$ is stable. The semi- and poly-stable cases are left as an exercise for the reader.

So the vector space $\mathbb{C}^{n+1} = W^+ \oplus W^0 \oplus W^-$ can be split into a sum of positive and negative weight spaces (with the sum of all weights, with multiplicities, zero) for the $\mathbb{C}^* < SL(n+1,\mathbb{C})$. The condition that $\tilde{x}$ be stable with respect to this 1-PS and its inverse is then that its components in both of W^+ and W^- be nonzero; i.e., that it have components of both positive and negative weight. Hence *generic* points are stable if the group acts effectively through $SL(n+1,\mathbb{C})$.

4. Symplectic reduction

In this section we use the compact subgroup $K < G$ to enlarge (3.1) to

$$\begin{array}{ccccc} K & < & G & \curvearrowright & X \\ \downarrow & & \downarrow & & \cap \\ SU(n+1) & < & SL(n+1,\mathbb{C}) & \curvearrowright & \mathbb{P}^n. \end{array} \tag{4.1}$$

So K acts on $\mathbb{P}^N$, preserving the complex structure J (as $SL(n+1,\mathbb{C})$ does) but also the Fubini-Study metric g. Thus K preserves the symplectic form $\omega = g(\,\cdot\,, J\,\cdot\,)$ and acts through *symplectomorphisms* in $\operatorname{Aut}(X,\omega)$.

Hamiltonian automorphisms. The Lie algebra of the symplectomorphism group consists of vector fields Y on X which preserve the symplectic form, i.e.

$$\mathcal{L}_Y\omega = d(Y \lrcorner\, \omega) = 0.$$

Since contraction with ω is an isomorphism $TX \xrightarrow{\simeq} T^*X$, the Lie algebra is isomorphic to the space of *closed 1-forms* $Z^1(X)$.

The subspace of *exact 1-forms* $dC^\infty(X,\mathbb{R})$ (which is all of them for simply connected spaces like $\mathbb{P}^N$) generate **hamiltonian** automorphisms – those which can be connected to the identity through a path of automorphisms whose flux homomorphism is zero (i.e., integration of ω over the cylinder traced out in this path by any loop in X is zero).

Since d has kernel the constants, the Lie algebra of the hamiltonian automorphisms is $C^\infty(X,\mathbb{R})/\mathbb{R}$; the function h generating the vector field

$$X_h \quad \text{such that} \quad X_h \lrcorner \omega = dh.$$

That is, using the metric, X_h is the symplectic gradient $J\nabla h$ of the *hamiltonian* h.

The Lie bracket of hamiltonian vector fields works out to be the **Poisson bracket** $\{f,g\}$ on functions, given by pairing df and dg using the (inverse of) the symplectic form and dividing out by constants. (Equivalently, this is the class of $X_f(g) = -X_g(f)$ in $C^\infty(X,\mathbb{R})/\mathbb{R}$.)

This bracket clearly lifts to $C^\infty(X,\mathbb{R})$, and can be checked to satisfy the Jacobi identity there too. The constants are central – they Poisson commute with all of $C^\infty(X,\mathbb{R})$ – so we get a *central extension* of the Lie algebra of hamiltonian automorphisms:

$$0 \to \mathbb{R} \to C^\infty(X,\mathbb{R}) \to C^\infty(X,\mathbb{R})/\mathbb{R} \to 0. \tag{4.2}$$

One might ask what $C^\infty(X,\mathbb{R})$ is the Lie algebra of, or what it acts on infinitesimally. Since $\mathbb{R}$ is the Lie algebra of isometries of the line $\mathbb{C}$, one might consider isometries of a line bundle $L \to X$ covering hamiltonian automorphisms on X .

This indeed can be made to work if ω is integral, i.e. its cohomology class lies in $H^2(X,\mathbb{Z})/\text{torsion} \le H^2(X,\mathbb{R})$ (as in our projective case, for instance). Then $2\pi i\omega$ is the curvature of a hermitian line bundle L with unitary connection, and we let $\widetilde{\text{Ham}}(X,\omega)$ be the isometries of L preserving its connection; these then cover hamiltonian automorphisms on X. Infinitesimally $h \in C^\infty(X,\mathbb{R})$ acts through vector fields on L given by

$$\tilde{X}_h + ih. \tag{4.3}$$

Here $\tilde{X}_h$ is the horizontal lift of the hamiltonian vector field X_h, and ih is the multiplication operator taking element of the line L to a perpendicular element in its tangent space (using the natural isomorphism between a line L and its tangent space). This action defines a homomorphism of Lie algebras (the Poisson bracket on $C^\infty(X,\mathbb{R})$ maps to the Lie bracket on vector fields on the total space of L), and the constants $\mathbb{R}$ act as the Lie algebra of global constant rotations $\{e^{i\theta}\} = U(1)$ of the fibres of L, yielding the exact sequence (4.2).

(This is often called *prequantisation*, giving a representation of the hamiltonian diffeomorphisms on the projectivisation of the Hilbert space of L^2-sections of L. This is considered too big a Hilbert space to be the set of quantum mechanical wave functions, and geometric quantisation attempts to replace it with holomorphic sections; about which more later.)

Moment maps and linearisations. Since K acts through symplectomorphisms of $\mathbb{P}^n$, which is simply connected (so that $Z^1(X) = dC^\infty(X)$), we get a Lie algebra homomorphism

$$\mathfrak{k} = \mathrm{Lie}(K) \to C^\infty(\mathbb{P}^n, \mathbb{R})/\mathbb{R} \to C^\infty(X, \mathbb{R})/\mathbb{R}, \qquad v \mapsto [m_v],$$

where any $v \in \mathfrak{k}$ generates a hamiltonian vector field X_v on X such that $X_v \lrcorner\, \omega = dm_v$, for some function m_v unique up to a constant.

We would like to choose these constants consistently, i.e., choose a lift

$$\begin{array}{ccc} & & C^\infty(X,\mathbb{R}) \\ & \nearrow & \downarrow \\ \mathfrak{k} & \longrightarrow & C^\infty(X,\mathbb{R})/\mathbb{R} \end{array} \qquad\qquad \begin{array}{ccc} & & m_v \\ & \nearrow & \downarrow \\ v & \longmapsto & [m_v] \cong X_v = dm_v \end{array} \tag{4.4}$$

which is a *homomorphism of Lie algebras*. One such always exists, since (4.2) is split by the Poisson subalgebra $C^\infty(X, \mathbb{R})_0 \cong C^\infty(X, \mathbb{R})/\mathbb{R}$ of functions of integral zero; i.e., we can choose each m_v to have integral zero. But we want to consider arbitrary lifts since from (4.3) we know that each is equivalent to a lift of $\mathfrak{k}$ to isometries of L preserving the connection and covering its hamiltonian action downstairs. This is the infinitesimal version of a linearisation, assigning to $v \in \mathfrak{k}$ the vector field

$$\tilde{X}_v + im_v \tag{4.5}$$

on the total space of L. It may or may not integrate up to an action of K on (X, L) covering that on X.

Said differently, we want to put together all of the hamiltonians m_v to give a **moment map**

$$m: X \to \mathfrak{k}^*, \tag{4.6}$$

such that $\langle m(x), v\rangle = m_v(x)$ for all $v \in \mathfrak{k}$. m is just a collection of $\dim K$ hamiltonians m_v, written invariantly. Then our lifting condition (4.4) becomes the condition that the undetermined constants in m_v be chosen such that (4.6) is K-*equivariant* (using the coadjoint action on the right hand side). Thus a moment map is unique up to the addition of a central element of $\mathfrak{k}^*$.

Yet another way of saying the same thing is that the derivative of the K-action maps $\mathfrak{k}$ to TX, so by contraction with the symplectic form $\lrcorner\,\omega\colon TX \cong T^*X$ is a section of $\mathfrak{k}^* \otimes T^*X$. It is closed and K-invariant, so we ask for it to be invariantly exact, i.e. d of a K-invariant section μ of $C^\infty(X, \mathfrak{k}^*)$.

(The name comes from the case of a cotangent bundle $X = T^*M$ with its canonical symplectic form and action induced from an action of K on M. Then the moment map really gives the momentum of the image $X_v \in TM$ of $v \in \mathfrak{k}$: $m_v(p, q) = \langle p, X_v\rangle$ at a point $q \in M$ and $p \in T_q^*M$. Hence for translations we get the usual linear momentum, and for rotations angular momentum.)

In the projective case that we have been considering, a natural m exists because we picked a linearisation. $SU(n+1) \curvearrowright (\mathbb{P}^n, \mathcal{O}(1))$ has a canonical moment map given by

$$\tilde{x} \mapsto \frac{i(\langle\,\cdot\,,\tilde{x}\rangle \otimes \tilde{x})_0}{||\tilde{x}||^2} \in \mathfrak{su}(n+1)^* \cong \mathfrak{su}(n+1), \tag{4.7}$$

where $(\ )_0$ denotes the trace-free part of an endomorphism. Restricting to X and projecting to $\mathfrak{k}^*$ by the adjoint of the map $\mathfrak{k} \to \mathfrak{su}(n+1)$ gives a moment map for the K-action on X.

The Kempf-Ness theorem. The key to the link between symplectic geometry and GIT is the following calculation. Suppose $(X, L = \mathcal{O}_X(1))$ is a polarised variety with a hermitian metric on L inducing a connection with curvature $2\pi i\omega$. Lift x to any $\tilde{x} \in \mathcal{O}_x(-1) = L_x^{-1}$ and consider the **norm functional** $||\tilde{x}||$. (If X is embedded in $\mathbb{P}(H^0(L)^*)$ then one way to get a metric on $\mathcal{O}(-1)$ is to induce it from one on $H^0(L)^*$ upstairs; then $||\tilde{x}||$ is just the usual norm in the vector space that $\widetilde{X}$ lives in.) As we move down a 1-PS orbit $\{\lambda.\tilde{x} \colon \lambda \in \mathbb{C}^*\}$ in the direction of $v \in \mathfrak{k}$ we see how $\log ||\tilde{x}||$ varies; for $\lambda \in U(1) < \mathbb{C}^*$ (which preserves the metric) not at all, but for λ in the complexified, radial direction $\lambda \in (0, \infty) < \mathbb{C}^*$ we get

$$m_v = \left.\frac{d}{d\lambda}\right|_{\lambda=1} \log ||\lambda\tilde{x}||_{\lambda\in(0,\infty)}. \tag{4.8}$$

That is, $X_v(\log ||\lambda\tilde{x}||) = 0$, but

$$(JX_v)(\log ||\lambda\tilde{x}||) = X_{iv}(\log ||\lambda\tilde{x}||) = m_v. \tag{4.9}$$

(This is just an unravelling of (4.5). For instance if x is a fixed point, then $\mathbb{C}^*$ acts on the line $\langle\tilde{x}\rangle$ with a weight ρ, and

$$m_v = \rho, \tag{4.10}$$

which is therefore an integer.)

Moreover, $\log ||\lambda\tilde{x}||$ is *convex* on $\mathbb{C}^*/U(1) \cong (0, \infty)$, as its second derivative is positive:

$$X_{iv}m_v = dm_v(JX_v) = \omega(X_v, JX_v) = ||X_v||^2.$$

It follows that the orbit tends to infinity at both ends, i.e., is closed, if and only if it contains a critical point (i.e. absolute minimum) of $\log ||\lambda\tilde{x}||$.

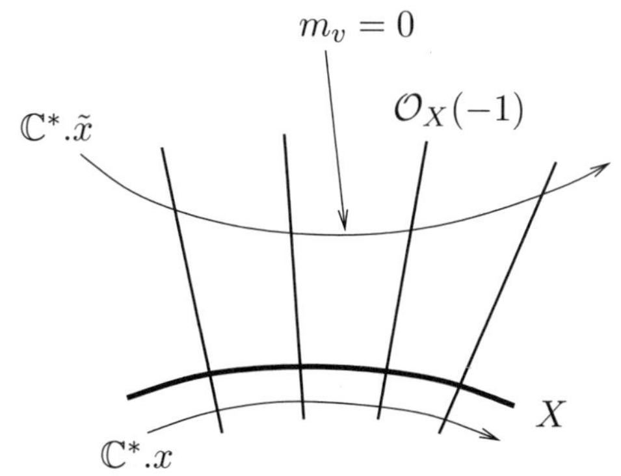

So a 1-PS orbit is polystable if and only if it contains a zero of the corresponding hamiltonian. That zero is then unique, up to the action of $U(1)$. This is the Kempf-Ness theorem for $\mathbb{C}^*$-actions.

Next we would like to consider a full G orbit, and find a zero of all the hamiltonians simultaneously, i.e. a zero of m. Pick v_i to form a basis for the Lie algebra of a maximal torus in K such that each generates a 1-PS. If an orbit is polystable then each 1-PS orbit is closed, so by the above there is a point with $m_{v_1} = 0$ in the first 1-PS. Now we move down the second 1-PS orbit of this point to a point with $m_{v_2} = 0$ *and* $m_{v_1} = 0$ since the two 1-PSs commute (i.e., $\{m_{v_1}, m_{v_2}\} = 0$). Inductively we find a point with $m_v = 0$ for all v in the Lie algebra of the torus, and so for all v conjugate to such (i.e. all v) by equivariance of the moment map. Thus the orbit contains a point with $m = 0$. Moreover, by the convexity of $\log ||\tilde{x}||$ on G/K, the zero is in fact unique up to the action of K.

(Alternatively, we could have proved this without using the Hilbert-Mumford criterion by noting that $\log ||g.\tilde{x}||$ is convex on the whole of G/K, instead of each $\mathbb{C}^*/U(1)$, so an orbit is closed if and only if this functional has a minimum, at which point $m = 0$ by (4.8).)

THEOREM 4.11. [Kempf-Ness] *A G-orbit contains a zero of the moment map if and only if it is polystable. It is unique up to the action of K.*

A G-orbit is semistable if and only if its closure contains a zero of the moment map; this zero is in the unique polystable orbit in the closure of the original orbit.

In particular, as sets,

$$\frac{X}{G} \cong \frac{m^{-1}(0)}{K} =: X/\!\!/K.$$

$X/\!\!/K := m^{-1}(0)/K$ is called the **symplectic reduction** of X, invented by Marsden-Weinstein and Meyer.

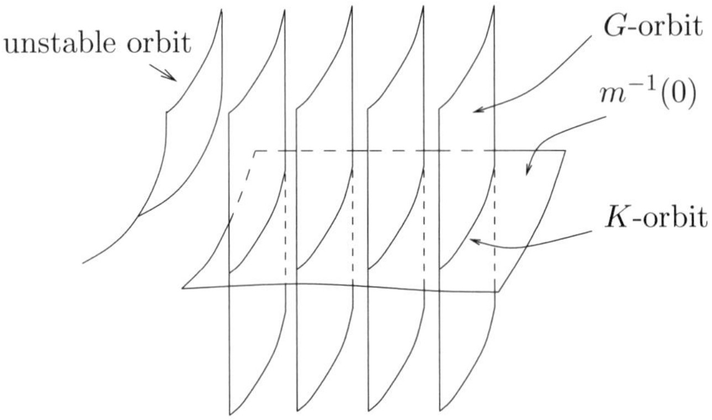

So on the locus of stable points $m^{-1}(0)$ provides a (K-equivariant) slice to the $i\mathfrak{k} < \mathfrak{g} = \mathfrak{k} + i\mathfrak{k}$ part of orbit; since this is topologically trivial (G retracts onto K) it makes topological sense that one could take a slice instead of a quotient. This leaves only the K-action to divide by to get the GIT quotient.

The Kempf-Ness theorem is a nonlinear generalisation of the isomorphism $V/W \cong W^{\perp}$ for vector spaces $W \leq V$. It works due to convexity, giving a unique distinguished K-orbit of points of least norm in each polystable G-orbit upstairs in $\widetilde{X}$.

When 0 is a regular value of m (which implies that $m^{-1}(0)$ is smooth and the $\mathfrak{k}$-action on it is injective, so the K-action has finite stabilisers and the quotient is a smooth orbifold at worst) then the restriction of ω to $m^{-1}(0)$ is degenerate precisely along the K-orbits, and so descends to a symplectic form on the quotient. This is in fact compatible with the complex (algebraic) structure on the GIT quotient, giving a Kähler form representing the first Chern class of the polarisation that X/G inherits from its Proj construction.

Example. $U(1) < \mathbb{C}^*$ acts on $\mathbb{C}^{n+1}$ with moment map $m = |\underline{z}|^2 - a$ for any constant $a \in \mathbb{R}$. For $a > 0$ this gives

$$\frac{\mathbb{C}^{n+1}\backslash\{0\}}{\mathbb{C}^*} \cong \frac{S^{2n+1} = \{\underline{z} : |\underline{z}|^2 = a\}}{U(1)} \cong \mathbb{P}^n.$$

$S^{2n+1} = m^{-1}(0)$ is a slice to the $(0,\infty)$-action, leaving the $U(1)$-action to divide by. The resulting Kähler form on $\mathbb{P}^n$ varies with the level a.

For $a = 0$ we get just a single point, while for $a < 0$ we get the empty set – as we showed already using GIT for different polarisations (3.4), where p played the role of a (but took integer values so that the lifted action of $\mathfrak{k}$ descended to an action of $K = U(1)$ on the trivial line bundle over $\mathbb{C}^{n+1}$).

Example: n points in $\mathbb{P}^1$ again. (Kirwan [**Ki**]) The moment map

$$SL(2,\mathbb{C}) \supset SU(2) \curvearrowright \mathbb{P}^1 \xrightarrow{m} \mathfrak{su}(2)^*$$

is just the inclusion of the unit sphere $S^2 \subset \mathbb{R}^3$.

Adding gives, for n points, the moment map $m = \sum_{i=1}^n m_i$:

$$S^n\mathbb{P}^1 \longrightarrow \mathbb{R}^3, \tag{4.12}$$

the sum of the n points in $\mathbb{R}^3$, i.e., (n times) their centre of mass.

So $m^{-1}(0)$ is the set of **balanced configurations** of points with centre of mass $0 \in \mathbb{R}^3$.

Since by Kempf-Ness polystability is equivalent to the existence of an $SL(2,\mathbb{C})$ transformation of $\mathbb{P}^1$ that balances the points, Theorem 3.10 yields

THEOREM 4.13. *A configuration of points with multiplicities in the unit sphere $S^2 \subset \mathbb{R}^3$ can be moved by an element of $SL(2,\mathbb{C})$ to have centre of mass the origin if and only if either each multiplicity is strictly less than half the total, or there are only 2 points and both have the same multiplicity.*

The first case is the stable case, the second the polystable case with a $\mathbb{C}^*$-stabiliser.

Example: Grassmannians from GIT and symplectic reduction. We have seen how to get $\mathbb{P}^n$ by GIT and symplectic reduction; we can do something similar for Grassmannians.

Consider $SL(r,\mathbb{C})$ acting on $\mathrm{Hom}(\mathbb{C}^r,\mathbb{C}^n)$, $r<n$, linearising the induced action on the projectivisation $\mathbb{P}$ of this vector space (we choose the left action of multiplying on the right by g^{-1}).

PROPOSITION 4.14. *$[A]\in\mathbb{P}$ is stable if $A\in\mathrm{Hom}(\mathbb{C}^r,\mathbb{C}^n)$ has full rank r, and unstable otherwise.*

PROOF. If $\mathrm{rank}(A)<r$ then we can pick a splitting $\mathbb{C}^r=\langle v\rangle\oplus W$ with $A(v)=0$. Then the 1-PS that acts as λ^{r-1} on v and λ^{-1} on W fixes $[A]\in\mathbb{P}$ and acts on the line $\mathbb{C}.A$ with weight $+1$. Therefore $[A]$ is unstable by the Hilbert-Mumford criterion.

Conversely, if A has full rank then, up to the action of $SL(r,\mathbb{C})$ some multiple of it is the inclusion of the first factor of some splitting $\mathbb{C}^n\cong\mathbb{C}^r\oplus\mathbb{C}^{n-r}$. Diagonalising a given 1-PS, we may assume further that in this basis we have the action

$$\mathrm{diag}(\lambda^{\rho_1},\dots,\lambda^{\rho_r}),\qquad \rho_1\geq\rho_2\geq\dots\geq\rho_r,\quad \sum_i\rho_i=0.$$

Ignoring the trivial 1-PS, there is some p such that $\rho_1=\rho_p>\rho_{p+1}$. Then the limit $[A_0]$ of $[A]$ under this 1-PS is the inclusion of $\mathbb{C}^p$ as the first p basis vectors of $\mathbb{C}^n$, with the 1-PS acting with weight $-\rho_1<0$ on $\mathbb{C}.A_0$. Therefore A is stable. □

So the points of the GIT quotient are the injections of $\mathbb{C}^r$ into $\mathbb{C}^n$ modulo the automorphisms of $\mathbb{C}^r$; i.e., they are the *images* of the injections – the Grassmannian $Gr(r,n)$ of r dimensional subspaces of $\mathbb{C}^n$.

For symplectic reduction, it is easier to consider the affine case of $U(r)<GL(r,\mathbb{C})$ acting on $\mathrm{Hom}(\mathbb{C}^r,\mathbb{C}^n)$, with all vector spaces endowed with their standard metrics. (Above, by working with $\mathbb{P}$, we had already divided out by the centre of $GL(r,\mathbb{C})$ but didn't describe it this way because, as we have seen, it is easier to deal with the linearisation issues in the symplectic picture, where it just amounts to changing the moment map by a central scalar.) The moment map is

$$A\mapsto i(A^*A-\mathrm{id}), \tag{4.15}$$

with zeros the orthogonal linear maps that embed $\mathbb{C}^r$ isometrically. Thus Kempf-Ness recovers the obvious fact that a linear map is congruent by $GL(r,\mathbb{C})$ to an isometric embedding if and only if it is injective. Dividing these isometric embeddings by $U(r)$ gives $Gr(r,n)$ again.

More affine examples. Our simple example (3.2) has moment map

$$(|x|^2-|y|^2)/2,$$

whose zero set intersects each good orbit $xy=\alpha\neq0$ in a unique $U(1)$ orbit $\sqrt{\alpha}(e^{i\theta},e^{-i\theta})$. It intersects the origin (another $U(1)$ orbit, corresponding to $\alpha = 0$) and misses the other two orbits (the punctured x- and y-axes). Therefore the symplectic quotient is a copy of $\mathbb{C}$ parameterised by α, representing the closed, *polystable* orbits, as anticipated.

If we chose the moment map $(|x|^2 - |y|^2 + a)/2$, $a > 0$, then we miss the x-axis and the origin, and gain a unique $U(1)$ orbit on the y-axis. So the symplectic quotient is isomorphic, but with a different interpretation. This corresponds in GIT to a different linearisation, in which the x-axis and the origin are unstable and the punctured y-axis is stable. (So this nonclosed orbit becomes closed upstairs in the new linearisation, and is closed in the locus of semistable points.)

Another standard example is to consider $n \times n$ complex matrices acted on by the adjoint action of $SL(n, \mathbb{C})$. The invariant polynomials are the symmetric functions in the eigenvalues of the matrix (by the denseness of the set of diagonalisable matrices) – i.e. functions in the coefficients of the characteristic polynomial. This reflects the fact that the matrices with non-diagonal Jordan canonical form have the corresponding diagonal matrices in the *closure* of their orbits – all matrices are semistable for this linearisation (the constant 1 does not vanish on any orbit!), with the diagonalisable matrices being polystable (their stabiliser is at least $(\mathbb{C}^*)^n$, after all).

The moment map (for the standard symplectic structure inherited from $\mathbb{C}^{n^2}$) for the induced action of $SU(n)$ is $A \mapsto \frac{1}{2}[A, A^*]$ with zeros the *normal* matrices. Since normal matrices are those that can be orthogonally diagonalised, the symplectic quotient $\{\text{normal matrices}\}/\ SU(n)$ is the set of diagonal matrices up to the action of the symmetric group, and so equal to the GIT quotient. (So in this case Kempf-Ness is the obvious fact that a matrix can be diagonalised if and only if it is similar to a matrix that can be orthogonally diagonalised.)

Back to the Hilbert-Mumford criterion. For simplicity of exposition we used the Hilbert-Mumford criterion to prove the Kempf-Ness theorem, to reduce everything to single hamiltonians. But as we noted there, we could have avoided this and proved it directly by noting that $\log ||g.\tilde{x}||$ is convex on the whole of G/K, so an orbit is closed if and only if this log-norm functional is proper, in which case it has a minimum, at which point $m = 0$ by (4.8).

We can then use this to go back and give a sketch proof (more of a discussion, really) of the Hilbert-Mumford criterion. That is we want to show that properness is equivalent to properness on 1-PSs. As usual one direction is trivial; for the other one can try to work on G/K as in, for instance, [**DK**]. The idea is that while 1-PSs cover very little of G, since K preserves the norm functional it descends to G/K, in which 1-PSs are dense (see the torus case below where the 1-PSs correspond to directions in $\mathfrak{g}/\mathfrak{k} \cong \mathfrak{k}$ of rational slope). Although it is not a priori clear that properness down each such rational direction is enough to give properness on all of G/K, it is clear by openness that if a G-orbit is strictly unstable then there will be a rational direction (1-PS) that detects it. So we see that (semi)stability of each 1-PS implies semistability for G.

So this leaves the hard part – that strict stability for each 1-PS implies strict stability for G. That is, we want to show that if a G-orbit is strictly semistable, then there is a 1-PS with zero weight; i.e. the non-properness is detected by a rational direction.

We first show this for G a torus $T^c = (\mathbb{C}^*)^r$. A T^c-action on a vector space splits it into a sum of weight spaces W_m, $m \in \mathfrak{t}^*$, on which $\exp(v) \in T^c$, $v \in \mathfrak{t}^c$, acts as the character $\exp(i\langle m, v\rangle)$.

Given any vector $\tilde{x}$, we let $\Delta_{\tilde{x}} \subset \mathfrak{t}^*$ denote the convex hull of only those weights m in whose weight spaces $\tilde{x}$ has nonzero components (i.e. the projection of $\tilde{x}$ to W_m is nonzero). Any 1-PS corresponds to an integral vector $v \in \mathfrak{t}$ and so a hyperplane $H_v \leq \mathfrak{t}^*$. The points of $\Delta_{\tilde{x}}$ on the negative side of this hyperplane correspond to negative weights in whose weight space $\tilde{x}$ has a nonzero component, so their existence implies that $\lambda.\tilde{x} \to \infty$ as $\lambda \to 0$ under this 1-PS, as in (3.13). Similarly the existence of points in $\Delta_{\tilde{x}}$ on the positive side of the hyperplane prove that $\lambda.\tilde{x} \to \infty$ as $\lambda \to \infty$.

Thus $\mathbb{C}^*.\tilde{x}$ is closed, and $\tilde{x}$ is stable for this 1-PS, if and only if its hyperplane $H_v \leq \mathfrak{t}^*$ cuts $\Delta_{\tilde{x}}$ through its interior. Applying this to all integral points $v \in \mathfrak{t}$ (including those whose hyperplanes H_v are parallel to the faces of $\Delta_{\tilde{x}}$) gives the first part of the following result, which was explained to me by Gábor Székelyhidi [**Sz2**].

THEOREM 4.16. *The point $\tilde{x}$ is stable for every 1-PS if and only if $0 \in \mathfrak{t}^*$ is in the interior of $\Delta_{\tilde{x}}$, if and only if $\tilde{x}$ is stable for T^c.*

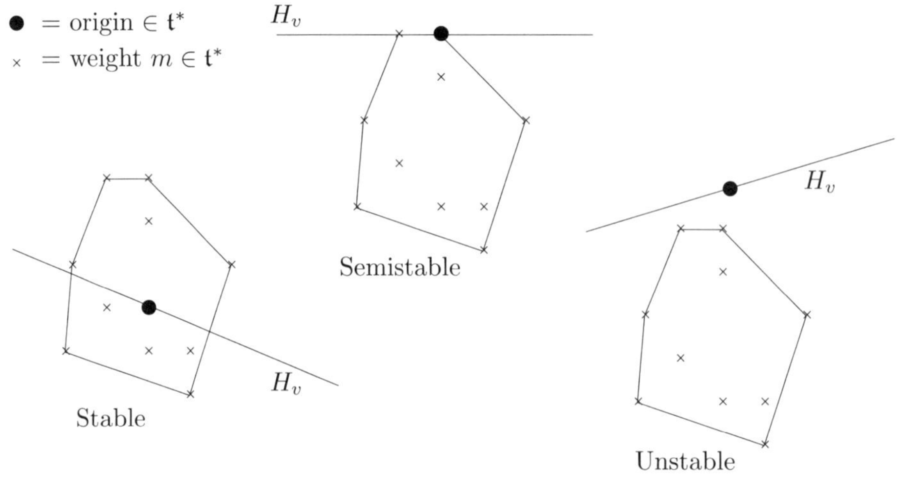

For the second result we cover the whole of T^c/T (where $T \leq T^c$ is the maximal compact subgroup $T \cong (S^1)^r$) by going in nonrational directions $v \in \mathfrak{t}$ too. And if the origin is in the interior, any such v has negative pairing with at least one of the weights, so the associated orbit (of an analytic $\mathbb{C}$-subgroup of T^c, if v is irrational) will go to infinity as we move along v. In fact the log-norm function will be proper with a growth that can be bounded below by the minimal $|\langle m, v\rangle|$. Thus the functional will be proper on all of T^c/T, and we can deduce the Hilbert-Mumford criterion.

Perhaps an easier proof, using the full Kempf-Ness theorem, comes from observing that the interior of $\Delta_{\tilde{x}}$ is the moment polytope of the orbit – the image of $T^c\tilde{x}$ under the moment map – so the moment map has a zero in this orbit if and only if the origin is in the interior.

But the first proof illustrates the key point, which is that the faces of $\Delta_{\tilde{x}}$ are *rational* – parallel to hyperplanes H_v. So if there is an irrational $v \in \mathfrak{t}$ that destabilises (has weight ≥ 0) then since it cannot be contained in a face there is in fact a rational v, giving rise to a 1-PS, with the same property. So we can avoid the situation of a sequence of stable 1-PSs of negative weight converging to an irrational "semistable" direction of weight zero lying in a face, making the T^c-orbit non-proper but without a 1-PS or rational direction to detect it.

For an arbitrary group G, we can try to reduce to the torus case by dividing G by K on both the left *and* right instead of considering just G/K. That is, by spectral theory we can write $G = KT^cK$ for any maximal torus $T^c \leq G$; then since the norm functional is invariant under the left hand action of K we are left with proving its properness on a *compact* family of T^c-actions – the conjugates of the original action by all $k \in K$. The result is then basically routine, the point being that in a compact family of polytopes each containing the origin in its interior, the distance of the origin to the boundary is bounded below by some $\epsilon > 0$.

As an application of Theorem 4.16, we can strengthen (3.12) to recover standard results [**GIT, Mu**] about which hypersurfaces of degree d in $\mathbb{P}^n$ are stable. Namely, forming the Newton polygon of degree d homogeneous polynomials in $(n+1)$ variables, a hypersurface $(f = 0)$ defines a subset of integral points of this polytope – those that appear in f with nonzero coefficient. Then $(f = 0)$ is semistable (or stable) if and only if these points do not lie to one side of (or strictly to one side of) any hyperplane through the centre of the Newton polytope.

5. Moduli of polarised algebraic varieties (X, L)

The GIT problem. This section is unnecessarily technical, and the squeamish reader can skip it once it is clear why forming moduli of algebraic varieties should be a GIT problem.

Suppose we want to form a moduli space of polarised algebraic varieties [**Mu**]. The polarisation allows us to embed X into a projective space

$$X \hookrightarrow \mathbb{P}(H^0(X, L^r)^*), \quad r \gg 0.$$

In fact for X smooth, a theorem of Matsusaka tells us that r can be chosen uniformly amongst all (X, L) with the same Hilbert polynomial $\mathcal{P}(r) = \chi(X, L^r)$. Moreover we can also assume that all higher cohomology groups $H^{\geq 1}(X, L^r)$ vanish so that $H^0(X, L^r)$ has dimension $\mathcal{P}(r)$, and that any two

(X_i, L_i) are isomorphic if and only if their embeddings $X_i \hookrightarrow \mathbb{P}^N$, $N+1 = \mathcal{P}(r)$, differ by a projective linear map.

Picking an isomorphism

$$H^0(X, L^r) \cong \mathbb{C}^{N+1}, \tag{5.1}$$

(X, L) defines a point in the *Hilbert scheme* of subvarieties (in fact subschemes) of $\mathbb{P}^N$. This moduli space is easy to construct; for instance as a subscheme of a Grassmannian of subspaces of $S^k(\mathbb{C}^{N+1})^*$; $X \subset \mathbb{P}^N$ corresponding to the subspace $H^0(\mathbb{P}^N, \mathscr{I}_X(k)) < H^0(\mathbb{P}^N, \mathcal{O}(k)) = S^k(\mathbb{C}^{N+1})^*$ of degree k polynomials vanishing on X. The natural Plücker line bundle then pulls back to give an anti-ample line bundle on Hilb whose fibre at a point (X, L) is

$$\Lambda^{\max} H^0(X, L^{rk})^* \otimes \Lambda^{\max} S^k H^0(X, L^r). \tag{5.2}$$

Then we must divide out the choice of isomorphism (5.1), i.e., take the GIT quotient of Hilb by $SL(N+1, \mathbb{C})$.

So by abstract GIT, any choice of $SL(N+1, \mathbb{C})$-equivariant (anti-)ample line bundle on Hilb gives rise to a notion of **stability for** (X, L). There are many such; we describe some of those whose associated weights can all be characterised in terms of weights on the line (5.2).

The Hilbert-Mumford criterion requires us to consider $\mathbb{C}^* < SL(N+1, \mathbb{C})$ orbits of $X \subset \mathbb{P}^N$. This gives rise to a $\mathbb{C}^*$-equivariant flat family, or **test configuration**, $(\mathscr{X}, \mathcal{L}) \to \mathbb{C}$.

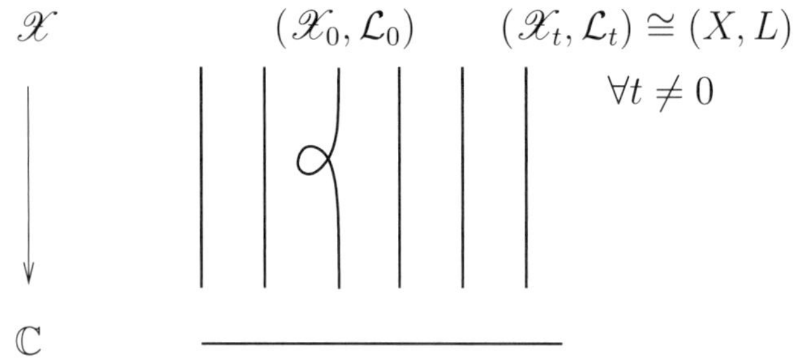

The weight $w_{r,k}$ of the $\mathbb{C}^*$-action on (5.2) is

$$w_{r,k} = a_{n+1}(r)k^{n+1} + a_n(r)k^n + \dots, \tag{5.3}$$

where

$$a_i(r) = a_{in}r^n + a_{i,n-1}r^{n-1} + \dots.$$

Then doing GIT on Hilb with the line (5.2), Mumford's Chow line, or Tian's CM line, gives rise to Hilbert-Mumford criteria that $\mathbb{C}^* < SL(N+1, \mathbb{C})$ destabilises (X, L) if $w_{r,k} \succ 0$ in the following senses:

- *HM(r)-unstable:* $w_{r,k} > 0$ *for all* $k \gg 0$.
- *Asymptotically HM-unstable: for all* $r \gg 0$, $w_{r,k} > 0$ *for all* $k \gg 0$.
- *Chow(r)-unstable: leading* k^{n+1}*-coefficient* $a_{n+1}(r) > 0$.
- *Asymptotically Chow unstable:* $a_{n+1}(r) > 0$ *for* $r \gg 0$.

- *K-unstable: leading coefficient* $a_{n+1,n} > 0$.

To make "if" into "iff" requires a few technicalities on the size of r; see [**RT1**]. In particular K-stability, which is Donaldson's refinement of Tian's original notion, requires one to pick a test configuration first, and then choose $r \gg 0$. The coefficient $a_{n+1,n}$ is Donaldson's version of the Futaki invariant of the $\mathbb{C}^*$-action on $(\mathscr{X}_0, L)$; see (5.20).

There are also notions semistability and polystability in all of these cases; both defined by nonstrict inequalities, the latter requiring also that whenever the inequality is an equality, the test configuration should arise from an automorphism of (X, L), i.e., it should be isomorphic as a scheme to the product $X \times \mathbb{C}$, but with a nontrivial $\mathbb{C}^*$-action.

In particular we have the following implications (see [**RT1**], where our a_i are denoted $-e_i$):

Asymptotically Chow stable $\Rightarrow$ Asymptotically Hilbert stable $\Rightarrow$ Asymptotically Hilbert semistable $\Rightarrow$ Asymptotically Chow semistable $\Rightarrow$ K-semistable.

The increasing number of test configurations that have to be tested as $r \to \infty$ currently prevents one from proving that K-stability implies asymptotic Chow stability.

The moment map problem. Fix a metric on $\mathbb{C}^{N+1}$ and so g_{FS} on $\mathbb{P}^N$ and an induced hermitian metric on $\mathcal{O}(-1)$. This induces the symplectic form ω_{FS} on a smooth $X \subset \mathbb{P}^N$. This induces a natural symplectic, in fact Kähler, structure on (any smooth subset of smooth points of) Hilb:

$$\Omega(v_1, v_2) := \int_X \omega_{FS}(v_1, v_2) \frac{\omega_{FS}^n}{n!}, \tag{5.4}$$

where the v_i are the normal components of holomorphic vector fields along $X \subset \mathbb{P}^N$. This is also (a multiple of) the first Chern class of a natural line bundle on Hilb coming from the "Deligne pairing" of $\mathcal{O}_X(1)$ with itself $(n+1)$-times [**Zh**].

Let $m\colon \mathbb{P}^N \hookrightarrow \mathfrak{su}(N+1)^*$ denote the usual moment map (4.7).

Then [**Do3**], just as for a finite number of points in $\mathbb{P}^1$ (4.12), the moment map for $SU(N+1) \curvearrowright (\text{Hilb}, \Omega)$ takes $X \subset \mathbb{P}^N$ to a multiple of its centre of mass in $\mathfrak{su}(N+1)^*$:

$$\mu(X) = \int_X m \frac{\omega_{FS}^n}{n!} \in \mathfrak{su}(N+1)^*. \tag{5.5}$$

So zeros of moment map correspond to **balanced** varieties $X \subset \mathbb{P}^N$.

The fact that Hilb is not smooth means there are complications in applying the Kempf-Ness theorem directly, but nonetheless the following is an essentially finite dimensional result. It was first proved by Zhang [**Zh**], and then rediscovered and reproved in different forms by Luo, Paul, Wang and Phong-Sturm.

THEOREM 5.6. *$X \subset \mathbb{P}^N$ can be balanced by an element of $SL(N+1,\mathbb{C})$ $\iff$ X is Chow polystable.*

The balanced condition can be re-expressed as follows. The metric on $\mathcal{O}_X(1) = L^r$ is the quotient metric induced from that on $H^0(\mathcal{O}_X(1)) = H^0(X, L^r)$ by the surjection of vector bundles

$$\underline{H^0(\mathcal{O}_X(1))} \to \mathcal{O}_X(1) \to 0$$

on X. So picking an orthonormal basis $\sigma_i \in H^0(X, L^r)$, we have the identity

$$\sum_i |\sigma_i(x)|^2 \equiv 1 \tag{5.7}$$

on X. (More generally, given an orthonormal basis e_i of an inner product space V and a surjection $V \xrightarrow{\pi} W$, we have the identity $\sum_i |\pi(e_i)|^2 = \dim W$ in the induced metric on W. Given any basis σ_i of $H^0(L^r)$ the above expression (5.7) is the pointwise ratio of the given metric on L^r and the Fubini-Study metric on L^r induced by embedding in $H^0(L^r)^*$ and pulling back the metric gotten by declaring the σ_i to be orthonormal. This is constant if and only if the metric on (X, L^r) really is such a Fubini-Study metric.)

But then in these coordinates, the moment map (5.5) constructed using (4.7), takes X to the matrix with (ij)th entry

$$i\left(\int_X \sigma_i(x)\sigma_j(x)^* \frac{\omega_{FS}^n}{n!} - \frac{1}{N+1}\delta_{ij}\right) \in \mathfrak{su}(N+1). \tag{5.8}$$

Thus the balanced condition is equivalent to the σ_i being orthonormal (up to a constant scale) in the induced L^2-metric on $H^0(\mathcal{O}_X(1))$. That is, up to scale, the original metric on $\mathbb{C}^{N+1} \cong H^0(\mathcal{O}_X(1)) = H^0(X, L^r)$ equals the L^2-metric given by integration against $g_{FS}|_X$. By (5.7) this is equivalent to

$$\sum_i |s_i(x)|^2 \equiv \text{const}, \tag{5.9}$$

where the s_i are now an orthonormal basis *with respect to the L^2-metric on* $H^0(X, L^r)$ (rather than the original metric). A final way of saying this is that starting with a metric on $\mathbb{C}^{N+1}$ we can induce another by first inducing the Fubini-Study metric on $X \subset \mathbb{P}^N$ and the hermitian metric on $\mathcal{O}_X(-1)$, and then using this to give, by integration, an L^2-metric on $\mathbb{C}^{N+1} = H^0(\mathcal{O}_X(1))$. Balanced metrics are then the fixed points of this operator.

Asymptotics as $r \to \infty$. Fix a metric on (X, L) (e.g., by picking a metric on $H^0(L)$ and then inducing the Fubini-Study metric on $X \subset \mathbb{P}(H^0(L^*))$ and $L = \mathcal{O}(1)|_X$). This then induces one on L^r for all r, and so L^2-metrics on $H^0(X, L^r)$ for all r.

Picking an L^2-orthonormal basis $s_i \in H^0(X, L^r)$, we can then define, for each r, the **Bergman kernel**

$$B_r(x_1, x_2) = \sum_i s_i(x_1) \otimes s_i(x_2)^* \tag{5.10}$$

on $X \times X$. This is the integral kernel for the L^2-orthogonal projection of C^∞ sections of L^r onto holomorphic sections. Restricting to the diagonal gives

$$B_r(x) := \sum_i |s_i(x)|^2, \tag{5.11}$$

so the balanced condition (5.9) is equivalent to B_r (5.11) being *constant on* X.

This expresses the finite dimensional balanced condition (a condition for a metric on $H^0(X, L^r)$) as a *pointwise* condition for a metric on (X, L) (a fact that will be explained later via Donaldson's double quotient construction) and we can look at the asymptotics of the "density of states" function $B_r(x)$ as $r \to \infty$ and expect it to only depend on local differential-geometric data. This is because, as is well known to quantum physicists and is made precise in [**Ti3**], one can form a basis of sections of $H^0(L^r)$ whose norms are approximately peaked Gaussians concentrated in balls of radius const$/\sqrt{r}$ and so volume const$/r^n$. (These are the *coherent states* of geometric quantisation; under the metric isomorphism $H^0(L^r) \cong H^0(L^r)^*$ they correspond to evaluating sections at points – the centres of the peaks.) The relationship between the volume of small balls about $x \in X$ and the scalar curvature $s(x)$ at x means that as $r \to \infty$ the number of peaked sections that can be packed into a fixed ball of volume ϵ about x is $\sim \epsilon(r^n + \frac{n}{2}s(x)r^{n-1} + \ldots)$. Globally this gives rise to

$$\mathrm{vol}(X)r^n + \frac{n}{2}\int_X s\frac{\omega^n}{n!}r^{n-1} + O(r^{n-2}) = \frac{L^n}{n!}r^n - \frac{K_X.L^{n-1}}{2(n-1)!}r^{n-1} + O(r^{n-2})$$

sections – approximating the Riemann-Roch formula.

In fact, as $r \to \infty$ ($\Rightarrow N \to \infty$) $B_r(x)$ has an asymptotic expansion (Tian, Zelditch, Catlin, W.-D. Ruan, Z. Lu)

$$B_r(x) \sim r^n + \frac{1}{2\pi}s(x)r^{n-1} + O(r^{n-2}). \tag{5.12}$$

More precisely, $||B_r(x) - r^n + \frac{1}{2\pi}s(x)r^{n-1}||_{C^\alpha} \leq Cr^{n-2}$ for $\alpha \geq 0$, where the constant C depends on both α *and* the metric – it can only be taken to be uniform for metrics in a *compact subset* of the space of metrics.

Roughly speaking then, balanced metrics should tend towards cscK metrics with $[\omega] = [c_1(L)]$. What we have seen so far should motivate the following results.

THEOREM 5.13. *If (X, L) admits a cscK metric in $[c_1(L)]$ and has finite automorphism group then (X, L^r) can be balanced in $\mathbb{P}(H^0(X, L^r)^*)$ for $r \gg 0$. Thus it is Chow stable, and so K-semistable.*

The metrics given by r^{-1} times by the pull backs of the balanced metrics converge to the cscK metric.

Conversely if $(X, L^r) \subset \mathbb{P}^{N(r)}$ is balanced for $r \gg 0$ and *the resulting $\omega_{FS,r}$ are convergent, then the limit metric is cscK.*

Finally, cscK metrics compatible with a fixed complex structure are unique up to holomorphic automorphisms of X.

This result is due to Donaldson [**Do3**]; we will discuss the proof of the balanced result in a later section. Tian had previously proved K-semistability for KE metrics [**Ti4**], and a related convergence result for sequences of Fubini-Study metrics [**Ti3**], following a suggestion of Yau [**Y2**]. Using [**Do3**] Mabuchi proved that cscK manifolds with automorphisms are Chow polystable if the automorphisms satisfy a certain stability condition [**Mb2**]. Donaldson [**Do6**] then showed that cscK $\Rightarrow$ K-semistable without any condition on automorphisms.

Uniqueness was originally proved by Bando-Mabuchi [**BM**] for KE metrics, by Chen [**Ch**] for cscK metrics when $c_1 \leq 0$, then by Donaldson in the general cscK case with finite automorphisms. Again the finite automorphisms condition was relaxed by Mabuchi, and, in the more general setting of extremal metrics and Kähler non-projective metrics, Chen-Tian [**CT**].

When our polarisation L is a power of the canonical bundle K_X, then cscK metrics are in fact KE: those with Ricci form (the induced curvature of K_X^{-1}) a constant multiple of the Kähler form. It is clear that KE metrics are cscK; the converse follows from the calculation

$$\Delta \operatorname{Ric} = -i\partial\bar{\partial}\Lambda \operatorname{Ric} = -i\partial\bar{\partial}s.$$

Here Λ is the adjoint of $\omega\wedge$, and we use the fact that Ric is a closed real $(1,1)$-form, so it is ∂- and $\bar{\partial}$-closed. $s = \Lambda \operatorname{Ric}$ is the scalar curvature. So for s a constant, Ric is harmonic, as is ω. But, after scaling, they represent the same cohomology class, and so are identically equal.

KE metrics were first proved to exist on compact Kähler manifolds with positive canonical bundle by Aubin [**Au**] and Yau [**Y1**], and with trivial canonical bundle by Yau [**Y1**]. It was Calabi [**Ca**] who initiated the study of cscK and *extremal* metrics: those which extremise the Calabi functional $\int_X s^2\omega^n$ over cohomologous Kähler forms; they are the metrics with ∇s a holomorphic vector field. Apart from Aubin and Yau's (nonconstructive) results, there are few compact examples of cscK or KE metrics. Siu [**S**], Tian and Nadel [**Na**] found examples with symmetry, Tian showed Fano surfaces with reductive automorphism groups admit KE metrics [**Ti2**], Burns-de Bartolomeis [**BdB**] and Hong [**Ho**] gave constructions of cscK metrics on certain projective bundles over cscK bases, and there are constructions for blow ups of these [**AP, LB, RS**] and smooth fibrations of cscK manifolds [**Fi**]. Bourguignon [**Bo**] and Biquard [**Bi**] have given excellent surveys of KE and cscK metrics respectively.

An example – blow ups of cscK manifolds. The results of [**AP**] give a beautiful illustration of the theory described here and the link between cscK and balanced metrics. Arezzo and Pacard consider a cscK manifold

(X, ω) and its blow up in some points p_i,

$$\begin{aligned} \pi \colon \widehat{X} &\longrightarrow X, \\ E_i &\rightarrow p_i. \end{aligned}$$

It is proved that there is a cscK metric in the class $\pi^*[\omega] - \epsilon \sum_i m_i [E_i]$ for $\epsilon, m_i > 0$ and ϵ sufficiently small, if the $m_i p_i$ satisfy two conditions with respect to $\mathrm{Aut}(X, \omega)$. Arezzo and Michael Singer observed that one of these conditions could be rewritten as a balanced condition. Namely there is a moment map $X \xrightarrow{\mu_a} \mathfrak{ham}(X, J, \omega)^*$ for the action of the hamiltonian isometry group of X, and the conditions are that

$$\sum_i m_i \mu_a(p_i) = 0, \text{ and the } \mu_a(p_i) \text{ generate } \mathfrak{ham}(X, J, \omega)^*. \tag{5.14}$$

We can interpret this in the projective case, where $\mathfrak{ham}(X, J, \omega)$ becomes $\mathfrak{aut}(X, L)$, as follows. Taking ϵ very small is equivalent to replacing the polarisation by a very large power $r \gg 0$, whereupon the cscK condition approximates the balanced condition (5.12) (for what follows we only need that the approximation is valid for the linearisation of the equations as $r \to \infty$). Then morally, in replacing (X, L^r) by $(\widehat{X}, \pi^* L^r(-\sum m_i E_i))$ we are perturbing a balanced $X \subset \mathbb{P}^N = \mathbb{P}(H^0(L^r)^*)$ only a little bit and so end up with a manifold that is nearly balanced. Slightly more precisely, set $I = H^0(L^r \otimes \mathscr{I}_{\cup_i m_i p_i})$ and split the exact sequence

$$0 \longrightarrow H^0(L^r \otimes \mathscr{I}_{\cup_i m_i p_i}) \longrightarrow H^0(L^r) \xrightarrow{\;\dashleftarrow\;} \bigoplus_i \mathbb{C}^{m_i}_{p_i} \longrightarrow 0$$

by picking peaked approximately Gaussian sections of L^r on X at the p_i, as in our discussion of (5.12). Away from the p_i, therefore, points in the image of $X \hookrightarrow \mathbb{P}(H^0(L^r)^*)$ almost annihilate this $\bigoplus_i \mathbb{C}^{m_i}_{p_i}$, i.e. they lie very close to $\mathbb{P}(I^*)$, as in the following diagram.

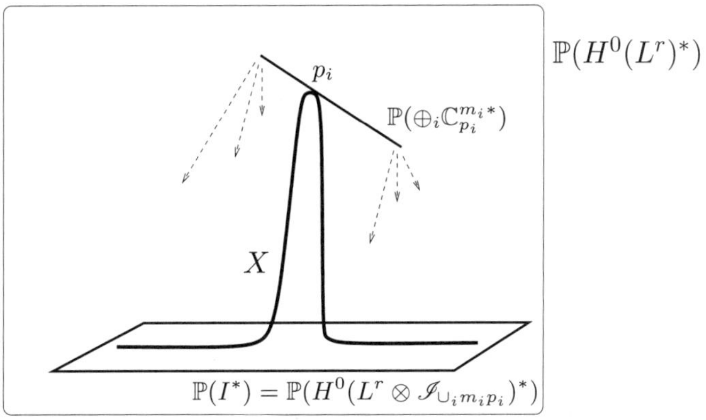

The dashed arrows denote the rational map $\mathbb{P}(H^0(L^r)^*) \dashrightarrow \mathbb{P}(I^*)$ blowing up $\mathbb{P}(\bigoplus_i \mathbb{C}^*_{p_i})$; on restriction to X this blows up the p_i and embeds the result in $\mathbb{P}(I^*)$.

So away from the p_i, the moment map $\mathbb{P}^N \to \mathfrak{su}(N+1)^*$ of (4.7), projected to $\mathfrak{su}(I)^*$, is very close (as $r \to \infty$) to the rational projection to $\mathbb{P}(I^*)$ followed by the moment map $\mathbb{P}(I^*) \to \mathfrak{su}(I)^*$.

Since the exceptional divisors are small, we can integrate over X (or its blow up in the p_i) to find that the centre of mass in $\mathfrak{su}(I)^*$ is close to the projection of that in $\mathfrak{su}(N+1)^*$. But this is zero, so $\widehat{X}$ is close to balanced, as claimed.

Now the exact sequence expressing the derivative D of the $SU(N+1)$ action on the Hilbert scheme of $\mathbb{P}(H^0(X,L^r)^*) \supset X$,

$$0 \to \mathfrak{aut}(X, L^r) \to \mathfrak{su}(N+1) \xrightarrow{D} T\,\mathrm{Hilb} \cong T^*\,\mathrm{Hilb}$$

(with the last isomorphism induced by the symplectic form), has dual

$$0 \leftarrow \mathfrak{aut}(X, L^r)^* \leftarrow \mathfrak{su}(N+1)^* \xleftarrow{d\mu} T\,\mathrm{Hilb}, \tag{5.15}$$

by the definition of the moment map $\mu = \int_X m\,\omega_{FS}/n!$ (5.5).

If the automorphism group of (X, L^r) is finite (so the condition (5.14) is vacuous) then D is injective and its adjoint $d\mu$ is *onto*. So we expect to be able to move a little in the orbit to move back to a balanced metric with $\mu = 0$ to correct the perturbation introduced by the p_i. This of course involves some estimates, which is what [**AP**] work out for the cscK problem, to show that for $\mathfrak{aut} = 0$ there is always a cscK metric on the blow up.

When the automorphism group is nontrivial this map $d\mu$ is *not* onto, so we must ensure that on perturbing as above we end up inside its image to apply the same argument. That is, by (5.15), the image of the moment map in $\mathfrak{aut}(X, L^r)^*$ should be zero. Since the moment map is the centre of mass, and since we have added masses m_i at the exceptional divisors E_i lying over p_i, we must ensure that, to first order, the $\cup_i m_i p_i$ should be balanced in $\mathfrak{aut}(X, L^r)^*$.

This recovers (5.14) as the necessary linearised condition. The second condition is a nondegeneracy condition that allows one to perturb the metric on and around the exceptional divisors to move the moment map enough to solve the equation to higher orders.

As pointed out by Donaldson, Hong's results [**Ho**] on when a cscK metric exists on the projectivisation of a HYM bundle over a cscK base involves a similar moment map condition for the action of the automorphism group of the base on the moduli of vector bundles.

These examples illustrate a general principle about moment map problems: that transverse (regular) points of $\mu^{-1}(0)$ have no automorphisms, whereas for nontransverse points x the cokernel of $d\mu$ is canonically $(\mathfrak{g}^x)^*$, the dual of the Lie algebra of the stabiliser subgroup of the point $x \in X$. Thus when one perturbs a solution x of $\mu = 0$ with stabiliser subgroup $G^x < G$, the obstruction to extending a first order deformation lies in $(\mathfrak{g}^x)^*$, and is nothing but the derivative of the moment map of the action of $G^x < G$.

This follows from the exact sequence

$$T_xX \xrightarrow{d\mu} \mathfrak{g}^* \to (\mathfrak{g}^x)^* \to 0,$$

the dual of

$$0 \to \mathfrak{g}^x \to \mathfrak{g} \to T_x^*X,$$

with the last map the composition of the $\mathfrak{g}$-action on T_xX and contraction with the symplectic form (cf. (5.15)).

The infinite dimensional setup. Instead of letting the dimension N of our quotient problem go to infinity, Donaldson [**Do1**] also gave a purely infinite dimensional formal symplectic quotient formulation.

The group of Hamiltonian diffeomorphisms acts on (X, ω) and so on the space of complex structures which make (X, ω) Kähler:

$$\mathrm{Ham}(X, \omega) \curvearrowright \mathcal{J} := \{\omega\text{-compatible complex structures on } X\}.$$

Acting by pullback, the infinitesimal action of a hamiltonian h, with hamiltonian vector field X_h, on a complex structure J is $\mathcal{L}_{X_h}J$. At the Lie algebra level this can be complexified so that ih acts as

$$J\mathcal{L}_{X_h}J = \mathcal{L}_{JX_h}J = \mathcal{L}_{X_{ih}}J,$$

by the integrability of J. Thus it acts through the vector field

$$X_{ih} := JX_h.$$

We note that the action of this vector field on ω is

$$\mathcal{L}_{JX_h}\omega = d(JX_h \lrcorner\, \omega) = d(Jdh) = d(-i\partial h + i\bar{\partial} h) = 2i\partial\bar{\partial} h,$$

changing ω within its cohomology class by the Kähler potential h to another form compatible with J.

We can contract these vector fields with ω to write them as one-forms. By Hodge theory,

$$\Omega^1(X) = dC^\infty(X) \;\oplus\; H^1(X, \mathbb{R}) \;\oplus\; d^*\Omega^2.$$

The first summand corresponds to the hamiltonian vector fields, the second to symplectomorphisms modulo those which are hamiltonian, and inside the third lies $d^*(C^\infty(X)\omega)$ as those which preserve the compatibility of ω with J (i.e. down which the Lie derivative of ω is of type (1,1)). These constitute the complexified hamiltonian action, by the Kähler identity

$$d^*(h\omega) = i(\bar{\partial} - \partial)h = Jdh,$$

whose contraction with (the inverse of) ω is $JX_h = X_{ih}$.

So, assuming $H^1(X, \mathbb{R}) = 0$ for simplicity, integrating up this complexified Lie algebra suggests defining the complexification of $\mathrm{Ham}(X, \omega)$ to be the set of diffeomorphisms of X such that the pullback of ω is compatible with J (i.e., of type (1,1)):

$$\big\{f\colon X \to X \;:\; \exists\, h \in C^\infty(X, \mathbb{R}) \text{ such that } f^*\omega = \omega + 2i\partial\bar{\partial} h\big\}. \tag{5.16}$$

While this description depends on J, it does formally complexify $\mathrm{Ham}(X,\omega)$: we have already seen that it has the right tangent space $C^\infty(X,\mathbb{R})_0\otimes\mathbb{C}$ at each point, and it is, crucially, *contractible onto* $\mathrm{Ham}\,(X,\ \omega)$ by Moser's theorem and the convexity of the space of Kähler forms.

The complexified orbits. Although (5.16) is not actually a group, its *orbits* on $\mathcal{J}$ (consisting of pullbacks of complex structures by the above diffeomorphisms) make perfect sense and complexify the $\mathrm{Ham}\,(X,\omega)$ orbits.

Since any two complex structures J, f^*J in such an orbit differ by a diffeomorphism, we consider them isomorphic. They are both, by construction, compatible with ω, but the Kähler structures (J,ω), (f^*J,ω) they define need not be isomorphic as the latter is only isomorphic to $(J,(f^{-1})^*\omega)$. Pulling back by the diffeomorphisms f in this way (i.e., fixing J and moving ω instead of the other way round) we get an exact sequence

$$(5.17)\quad \mathrm{Ham}(X,\omega)\to \mathrm{Ham}^c(X,\omega).J \twoheadrightarrow \big\{\text{compatible Kähler metrics on } (X,J) \text{ in the } H^2 \text{ class } [\omega]\big\}.$$

The last arrow is *onto* because any such ω' is of the form $\omega+2i\partial\bar\partial h$, and so diffeomorphic to ω (since by the convexity of the space of Kähler forms it is connected to ω through a family of Kähler forms $\omega+ti\partial\bar\partial h$ which are therefore all diffeomorphic by Moser's theorem). Thus the space of Kähler metrics on (X,J) is formally of the form G/K. This sequence should be compared to its (more familiar) bundle analogue in (6.1).

The set-theoretic "quotient" by the complexified group (i.e., the set of complexified orbits) is therefore the set of isomorphism classes of integrable complex structures on X (that are compatible with one of the symplectic forms $f^*\omega$).

Moment map = scalar curvature. The Kähler structure on X induces one on $\mathcal{J}$ by integration. This is preserved by $\mathrm{Ham}(X,\omega)$, and we can ask for a moment map. Considering $C^\infty(X,\omega)_0$ (the functions of integral zero) to lie in the dual of the Lie algebra $C^\infty(X,\mathbb{R})/\mathbb{R}$ by integration against ω^n, and setting s_0 to be the topological constant $\int_X c_1(X).\omega^{n-1}/\int_X\omega^n = \int s\omega^n/\int\omega^n$ (the average scalar curvature), Fujiki [**Fj**] and Donaldson [**Do1**] show that

$$(5.18)\qquad \text{Moment map } = s-s_0.$$

This should be no surprise, since we were looking for a function depending algebraically on the second derivatives of the metric, i.e., an invariant scalar derived from the curvature, which can only be a multiple of the scalar curvature. Thus zeros of the moment map correspond to cscK metrics.

Norm functional = Mabuchi's K-energy. The formula (4.9) for the change in the log-norm functional $M = \log ||\tilde{x}||$ along a complexified orbit, gives the following in this infinite dimensional set-up. Moving down the orbit of ih, $h \in C^\infty(X, \mathbb{R})$, i.e., in the family of Kähler forms $\omega_t = \omega + 2it\partial\bar{\partial}h$,

$$\frac{dM}{dt} = m_h, \tag{5.19}$$

where $m_h = \langle m, h\rangle$ is the hamiltonian function on $\mathcal{J}$ for the element of the Lie algebra $h \in C^\infty(X, \mathbb{R})$. Since the moment map $m = s - s_0$ (5.18), $m_h = \int_X (s - s_0) h\omega_t^n / n!$, and

$$M(\omega_s) = \int_0^s \int_X (s_t - s_0) h \frac{\omega_t^n}{n!}\, dt,$$

where s_t is the scalar curvature of ω_t. This is precisely the *Mabuchi functional* or *K-energy* [**Mb1**], defined up to a constant (equivalent to the ambiguity in the choice of a lift of a point to the line bundle above it). It can indeed be written as the log-norm functional for a Quillen metric on a line bundle over the space of Kähler metrics; see for example [**MW**]. Its critical points are cscK metrics, and one expects such a metric to exist on (X, J) if and only if M is proper on the space of Kähler metrics on (X, J) (which is the infinite dimensional analogue of G/K by (5.17)).

Weight = Futaki invariant. The formula (5.19) at a fixed point (e.g. the limit point of a 1-PS when this exists and is smooth), on the line over which $\mathbb{C}^*$ acts with weight ρ, is

$$\frac{dM}{dt} = \rho = m_h = \int_X (s - s_0) h \frac{\omega^n}{n!}\,. \tag{5.20}$$

Compare (4.5, 4.9, 4.10). This is the statement that "the derivative of the Mabuchi energy is the Futaki invariant" [**Mb1, DT**].

The right hand side is, up to a sign, the original definition of the Futaki invariant [**Fu**] for a smooth polarised manifold (X, L) with a $\mathbb{C}^*$-action. Noting as above that it is the weight of the induced action on a line led Donaldson to give the more general definition $a_{n+1,n}$ described earlier, for an arbitrary polarised scheme (X, L).

Approximation and quantisation. As Donaldson explains in [**Do4**], the finite dimensional problem of balanced metrics can be thought of as the quantisation of the infinite dimensional problem of cscK metrics, which emerges as the classical limit as $r, N \to \infty$.

As in quantum theory we think of the spaces of sections of the line bundle L^r as wave functions on X, with a basis of Gaussian sections, peaked around points on x. As $r \to \infty$ these peak more, largely supported in balls of radius const$/r$. Our $SL(N+1, \mathbb{C})$ group action moves these sections around the manifold, which may be thought of as moving quantised chunks of manifold of volume $\sim 1/r^n$ around X (thanks to Anton Gerasimov for this analogy). In the limit this is meant to approximate the classical limit of the diffeomorphisms (5.16) in the complexification of $\widetilde{\mathrm{Ham}}(X, \omega)$ moving points of the manifold around.

There is in fact a natural map $\mathfrak{su}(N+1) \to C^\infty(X, \mathbb{R})$, though it is only a homomorphism of Lie algebras to leading order in r [**CGR**]. A skew-adjoint endomorphism $iA \in \mathfrak{su}(N+1)$ gives an infinitesimal automorphism of $\mathbb{P}^N$ whose vector field v_A is hamiltonian with respect to the Fubini-Study symplectic form. Its hamiltonian is the function (Berezin symbol)

$$\mathbb{P}^N \ni x = [\tilde{x}] \mapsto \frac{\langle A\tilde{x}, \tilde{x}\rangle}{||\tilde{x}||^2} =: h_A. \tag{5.21}$$

On X, $h_A|_X$ induces a hamiltonian vector field which is the orthogonal projection of v_A from $T\mathbb{P}^N|_X$ to TX.

Using the fact that TX is invariant under the complex structure J, and working with complexified hamiltonian vector fields (of the form $X_h + JX_g =: X_{h+ig}$), the same working shows that the same formula defines a map from $\mathfrak{sl}(N+1, \mathbb{C})$ to the Lie algebra $C^\infty(X, \mathbb{C})$ of the complexification (5.16) of $\widetilde{\mathrm{Ham}}(X, \omega)$. Thus the change in metric on X induced by pulling back the metric along an $\mathfrak{sl}(N+1, \mathbb{C})$ vector field in $\mathbb{P}^N$ is the same as that induced by pulling back along its orthogonal projection *tangent to* X. (Thanks to Gábor Székelyhidi for this observation [**Sz2**].)

In this way algebraic 1-PS orbits, i.e. test configurations, give rise to curves in the complexification of the $\widetilde{\mathrm{Ham}}(X, \omega)$-action on $\mathcal{J}$ which approximate 1-PS orbits. Using the description of these orbits in terms of a fixed complex structure and varying Kähler form (5.17), this simply corresponds to restricting the Fubini-Study metric of $\mathbb{P}^N$ to the test configuration.

To get a map back we orthogonally project the prequantisation representation $\widetilde{\mathrm{Ham}}(X, \omega) \to \mathrm{Aut}(\Gamma(L^r))$ to $H^0(L^r) < \Gamma(L^r)$ using the Bergman kernel (5.10). That is $h \in C^\infty(X, \mathbb{R})$ maps to the infinitesimal automorphism $iA \in \mathfrak{su}(H^0(L^r))$ defined by

$$iA(s) = \sum_i \langle \nabla_{X_h} s + ihs, s_i \rangle_{L^2} s_i.$$

Again, this is not a homomorphism (except to leading order in r). The problem is that we had to use the Bergman kernel because quantisation is *not* a symplectic invariant (it cannot be done equivariantly with respect to symplectomorphisms or elements of $\widetilde{\mathrm{Ham}}(X, \omega)$). That is, it is not independent of choices of complex structure because the pullback of $s \in H^0(L^r)$ by $\widetilde{\mathrm{Ham}}(X, \omega)$ is not in general holomorphic.

Donaldson's double quotient construction. Because of this problem Donaldson [**Do3**] considers pairs of a complex structure $J \in \mathcal{J}$ and a section $s \in \Gamma(L^r)$ which is *holomorphic with respect to* J; these are clearly acted on by $\widetilde{\mathrm{Ham}}(X, \omega)$. In fact he considers $N + 1 = h^0(X, L^r)$-tuples of sections:

$$\mathcal{S} = \big\{(J, \{s_i\}) \in \mathcal{J} \times \Gamma(L^r)^{N+1} \, : \, \bar\partial_J s_i = 0, \ i = 1, \dots, N+1\big\}.$$

Here, as usual, L has a metric and hermitian connection, and $\bar\partial_J$ is the the $(0,1)$-part of the induced connection on L^r with respect to the complex structure J. Since the curvature $2\pi ri\omega$ is compatible with J (by the definition of $\mathcal{J}$), i.e. of type (1,1), $\bar\partial_J^2 = 0$ and $\bar\partial_J$ defines an integrable holomorphic structure on L^r by the Newlander-Nirenberg theorem.

We now have actions of $GL(N+1, \mathbb{C})$ *and* $\widetilde{\mathrm{Ham}}(X, \omega)^c$. These commute, and both have centre $\mathbb{C}^*$ acting by scalars on L, so we can quotient by $\widetilde{\mathrm{Ham}}(X, \omega)^c$ and then $SL(N+1, \mathbb{C})$, or by $GL(N+1, \mathbb{C})$ and then $\mathrm{Ham}(X, \omega)^c$. In this way we will see how an infinite dimensional moment map problem is equivalent to a finite dimensional one.

Dividing by $GL(N+1, \mathbb{C})$ leaves $\mathcal{J}$ (with a fibration over it by the Grassmannian of $(N+1)$-planes in $H^0(L^r, \bar\partial_J)$, by Proposition 4.14, but for L sufficiently ample $N+1 = h^0(L^r)$ and this is a single point). In turn the formal complex quotient of this by $\mathrm{Ham}(X, \omega)^c$, discussed above, is the space of complex structures on X (compatible with some symplectic structure in the diffeomorphism group orbit of ω). Taking symplectic reductions instead we end up with cscK metrics (together with orthonormal bases of $H^0(L^r)$ modulo the unitary group – i.e. just a point). So far then, we have just reproduced what we already knew.

However, we can put a different symplectic structure Ω_r on $\mathcal{J}$, and one that tends to Ω as $r \to \infty$. Namely, the fact that the s_i determine an embedding of X into $\mathbb{P}(H^0(L^r)^*)$ for $r \gg 0$ means that the natural projection

$$\mathcal{S} \to \Gamma(L)^{N+1}$$

is an embedding, and we can pullback the natural L^2-symplectic form from the latter to define Ω_r.

Now [**Do3, Do4**] the moment map for the $\mathrm{Ham}(X, \omega)$-action becomes

$$\Big(\frac{1}{2}\Delta + r\Big) \sum_i |s_i(x)|^2,$$

with zeros the solutions of $\sum_i |s_i(x)|^2 = \text{constant}$.

If we first take the symplectic reduction by $\mathrm{Ham}(X, \omega)$ then this involves solving $\sum_i |s_i(x)|^2 = \text{constant}$, which we have already observed in (5.7) says that the metric on X is the restriction of the Fubini-Study metric on $\mathbb{P}(H^0(L^r)^*) \supset X$ when we put the metric on $H^0(L^r)$ that makes the rs_i orthonormal (the scaling arising because we have ignored the central scalar action). But since this is a Kähler metric in the same class as ω, we have already observed (5.17) that we can solve this in a $\mathrm{Ham}(X, \omega)^c$ orbit, uniquely up to the action of $\mathrm{Ham}(X, \omega)$. Next we take the reduction by $SU(N+1, \mathbb{C})$, which by (5.5) means we try to balance $X \subset \mathbb{P}(H^0(L^r)^*)$ in the metric in which the rs_i are orthonormal. By Theorem 5.6 there is a solution to this problem in a $SL(N+1,\mathbb{C})$-orbit of X, unique up to $SU(N)$, if and only if $X \subset \mathbb{P}(H^0(L^r)^*)$ is Chow polystable.

So that gives us the finite dimensional problem of solving (5.8), (which, as observed there, is equivalent to the metric on $H^0(L^r)$ being the L^2-metric).

Taking the symplectic reduction in the opposite direction gives instead the pointwise description (5.9) of the balanced condition. Namely, first taking the reduction by $SU(N+1)$ gives us an *orthonormal* basis s_i (up to an overall scale which could be removed by putting back the central $\mathbb{C}^*$-action) in each $SL(N+1,\mathbb{C})$ orbit, unique up to $SU(N+1)$, if and only the original s_i were linearly independent (Proposition 4.14). Then taking the reduction by $\mathrm{Ham}(X,\omega)$ involves solving (5.9) $B_r(x) = \mathrm{const}$ for the metric. So we see how solving this infinite dimensional moment map problem has been reduced to the finite dimensional balanced moment map problem.

This latter equation has the advantage that it is asymptotically close to the cscK equation (5.12). If quantisation really "worked" it would be exactly the cscK equation, and proving Donaldson's result that cscK $\Rightarrow$ balanced would be trivial. Since it is only asymptotically close, Donaldson crucially uses the "failure" of quantisation to move from a cscK solution to a balanced solution, as we now describe.

CscK $\Rightarrow$ balanced. In [**Do3**], Donaldson proves a "quantitative" version of the Kempf-Ness theorem: if the moment map $m(x)$ at a point x is small, and the action of the Lie algebra at x is injective, with a sufficiently large lower bound on its smallest eigenvalue in a sufficiently large neighbourhood of x, then there exists a zero of m close to x in its complexified orbit. Flowing down the gradient of $-||m||^2$, i.e. down JX_{m^*} (where $m^* \in \mathfrak{k}$ is dual to $m \in \mathfrak{k}^*$ under the Killing form), the conditions ensure that X_{m^*} is sufficiently large and so $||m||^2$ decreases sufficiently fast for sufficiently long to converge to a zero of m.

He applies this to the $SU(N+1)$-action on the symplectic reduction of $\mathcal{S}$ by $\mathrm{Ham}(X,\omega)$. The cscK metric ensures that we are close to a balanced metric (zero of the moment map) as $r \to \infty$. Then to give a lower bound for the injectivity of the $\mathfrak{su}(N+1)$-action it is equivalent to give a bound for the orthogonal projection of its action perpendicular to the orbits of $\mathrm{Ham}(X,\omega)$ upstairs on $\mathcal{S}$.

Donaldson shows that the projection of the action of $iA \in \mathfrak{su}(N+1)$ onto the tangent to the $\mathrm{Ham}(X,\omega)$ orbits is just what one might expect from quantisation: it is the action of its Berezin symbol h_A (5.21). So the normal projection we require is given by the difference in the actions of iA and h_A on $\mathcal{S}$.

It is here is where the failure of quantisation to be equivariant with respect to $\mathrm{Ham}(X,\omega)$ is used – to show that this difference is sufficiently large in some sense. Of course quantisation is invariant with respect to *holomorphic* hamiltonian vector fields, i.e., those functions satisfying

$$\mathcal{D}h := \bar{\partial} X_h = \bar{\partial}(dh \lrcorner \omega^{-1}) = 0.$$

[**Do3**] assumes that $\mathrm{Aut}(X, J) = 0$, so that $\ker \mathcal{D}$ is just the constants. Then the (fixed) lowest eigenvalue of $\mathcal{D}$ gives the lower bound on the difference of the actions of iA and h_A. This gives the required estimates, as $r \to \infty$ and we get closer to a zero of the balanced moment map equation, to apply Donaldson's quantitative Kempf-Ness theorem.

So $SU(N+1)$ really "approximates" $\widetilde{\mathrm{Ham}}(X, \omega)$, in the sense that its finite dimensional moment map converges to the infinite dimensional one (5.12), the symplectic structures $\Omega_r \to \Omega$, and the natural norm functionals and weights tend to their infinite dimensional analogues (the Mabuchi functional and Futaki invariant) as $r \to \infty$; see [**Do4**]. Also the space of "algebraic metrics" (the restrictions of the Fubini-Study metrics $SL(N+1, \mathbb{C}).\omega_{FS}$ from $\mathbb{P}^N$) becomes dense in the space of all Kähler metrics as $r, N \to \infty$ [**Ti3**]. Thus the quantum picture tends to the classical one as $r \to \infty$.

The Yau-Tian-Donaldson conjecture. By analogy with the Kempf-Ness theorem in finite dimensions (and by taking the infinite limit of Theorem 5.6) it is natural to conjecture a *Hitchin-Kobayashi correspondence* (the name coming from the analogy with the bundle case in the next section). That is a variety should admit a cscK metric if and only if it is polystable in a certain sense.

In fact Yau [**Y3**] first suggested that there should be a relationship between stability and the existence of KE metrics. Tian [**Ti2**] proved this for surfaces, introduced his notion of K-stability, and, building on his work with Ding [**DT**], showed it was satisfied by Kähler-Einstein manifolds [**Ti4**]. The definition of K-stability was generalised to more singular test configurations by Donaldson [**Do5**] who also showed that cscK implies K-semistability [**Do3**]. So it was thought that **K-polystability**, as defined above, should be the right notion to be equivalent to cscK.

Recent explicit examples [**ACGT**] in the extremal metrics case (where there is a similar conjecture due to Székelyhidi [**Sz1**]) suggest that this should be strengthened to **analytic K-polystability**, allowing more general analytic (instead of just algebraic) test configurations. In particular one should allow the line bundle L over the test configuration to be an $\mathbb{R}$-line bundle: an $\mathbb{R}$-linear combination (by tensor product) of $\mathbb{C}^*$-linearised line bundles. So the most likely Yau-Tian-Donaldson conjecture as things stand at the end of 2005 is the following.

CONJECTURE 5.22. (X, L) is analytically K-polystable $\iff$ (X, L) admits a cscK metric. This is unique up to the holomorphic automorphisms of (X, L).

This would be the right higher dimensional generalisation of the uniformisation theorem for Riemann surfaces.

There is very little progress on this conjecture in the $\Rightarrow$ direction except for projective bundles [**BdB, Ho, RT2**] and Donaldson's deep work on toric surfaces [**Do5**]. In the KE case there are sufficient conditions for existence given by Tian's α-invariant [**Ti1**] and Nadel's multiplier ideal sheaf [**Na**], but no one has successfully related these to stability. Part of the problem, quite apart from the analytical difficulties, is that we do not have a good intrinsic understanding of stability for varieties – i.e., no one has successfully analysed the Hilbert-Mumford criterion for varieties.

Summarising the status of the whole theory for varieties, we have the infinite dimensional analogue of the balanced condition for points in $\mathbb{P}^1$ (i.e., cscK metrics) and part of the relationship to stability, but not the algebro-geometric description of stability. That is, the Hilbert-Mumford criterion, giving the analogue of the multiplicity $< n/2$ condition for points in $\mathbb{P}^1$, is missing.

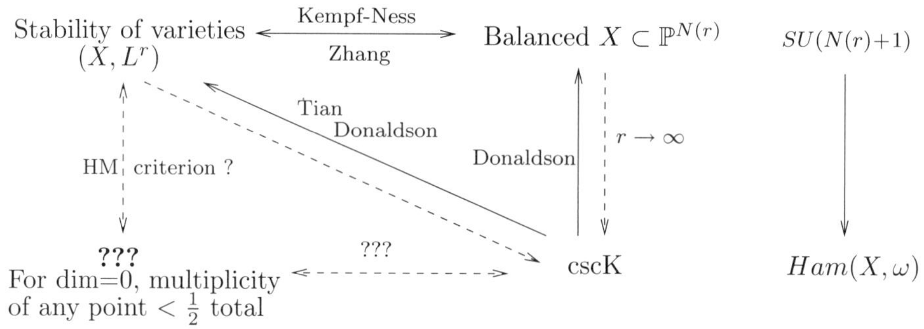

6. Moduli of bundles over (X, L)

For holomorphic bundles E over a polarised algebraic variety (X, L) there is a very similar story which is more-or-less completely worked out. Again there are subtleties due to different notions of stability, but for bundles for which Gieseker and slope stability coincide, for simplicity (such as those with coprime rank and degree, or bundles over curves), we have, for $r \gg 0$,

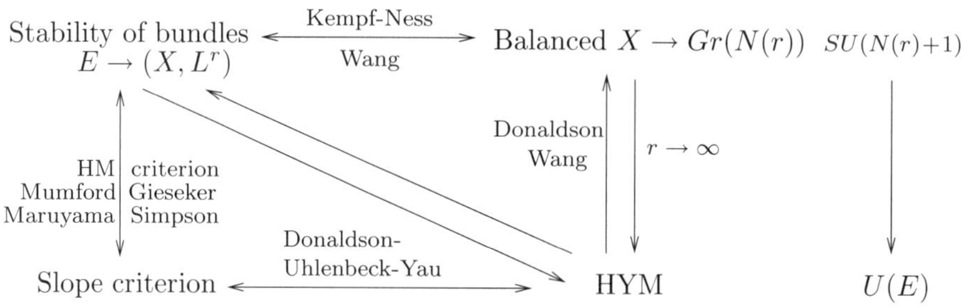

We now briefly explain this theory.

The gauge theory picture. The formal infinite dimensional picture was described by Atiyah-Bott [**AB**]. Fix a compatible hermitian metric on a C^∞-bundle E and consider the gauge group $U(E) = \{$unitary C^∞-maps

$E \to E\}$ and its (genuine) complexification $GL(E)$ of all C^∞ invertible bundle maps $E \to E$. These act on

$$\mathcal{A} = \{\text{unitary connections } A \text{ with } F_A^{0,2} = 0\}.$$

The $U(E)$-action is obvious; $GL(E)$ acts by pulling back the $(0,1)$-part $\bar\partial_A$ of the connection and then taking the unique Chern connection compatible with both this and the metric. The integrability condition $F_A^{0,2} = \bar\partial_A^2 = 0$ ensures that $\bar\partial_A$ defines a holomorphic structure on E. Thus any two $\bar\partial$-operators define isomorphic holomorphic structures on E if and only if they lie in the same $GL(E)$-orbit. So the formal complex quotient of $\mathcal{A}$ by $GL(E)$ is the moduli space of holomorphic vector bundles on X with topological type E. (Of course we expect to need a stability condition to form this quotient.)

Alternatively, fixing the $\bar\partial$-operator and pulling back the metric by $GL(E)$ gives the direct analogue of (5.17) for the complexified orbit of $\bar\partial_A$:

$$U(E) \to GL(E).\bar\partial_A \twoheadrightarrow \{\text{compatible metrics on } (E, \bar\partial_A)\}. \tag{6.1}$$

The last map is onto since $GL(E)$ acts transitively on the space of compatible hermitian metrics on E (the space of metrics being $GL(E)/U(E)$), so a complexified orbit can be thought of as giving all compatible metrics on a fixed holomorphic bundle $(E, \bar\partial_A)$, up to the action of $U(E)$.

Fix a compatible hermitian metric on L, inducing a Kähler form ω on X. Then $\mathcal{A}$ inherits a natural Kähler structure, with symplectic form given by $\Omega(a,b) = \int_X \mathrm{tr}(a \wedge b) \wedge \omega^{n-1}$ for $a, b \in \Omega^1(\operatorname{End} E)$ tangent vectors to $\mathcal{A}$. Atiyah-Bott show that $U(E) \curvearrowright \mathcal{A}$ has a moment map

$$A \mapsto F_A^{1,1} \wedge \omega^{n-1} - \lambda \, \mathrm{id} \, \omega^n \in \Omega^{2n}(\mathfrak{su}(E)),$$

thinking of the latter space as dual to $\Omega^0(\mathfrak{su}(E))$ by the trace pairing and integration. Here $\lambda = 2\pi i \mu(E) / \int_X \omega^n$ is a topological constant, where

$$\mu(E) = \frac{\int_X c_1(E).\omega^{n-1}}{\operatorname{rank} E} \tag{6.2}$$

is the *slope* of E.

Thus zeros of the moment map are **Hermitian-Yang-Mills** connections; solutions of $\Lambda F_A^{1,1} = \text{const.id}$. An infinite dimensional version of the Kempf-Ness theorem would be that in a polystable orbit of $GL(E)$ there should be a HYM connection (i.e. a metric whose associated Chern connection is HYM; we call this a HYM metric), unique up to the action of $U(E)$, as conjectured by Hitchin and Kobayashi.

THEOREM 6.3 (Donaldson-Uhlenbeck-Yau). *E slope polystable $\iff$ E admits a HYM metric. It is unique up to the automorphisms of E.*

The notion of stability that arises here (also called *Mumford stability*) comes from GIT.

The GIT picture. Suppose we wanted form an algebraic moduli space of bundles E over (X, L) of fixed topological type. (More generally, to get a compact moduli space, we have to consider coherent sheaves E of the same Hilbert polynomial $\chi(E(r))$.) We can twist $E(r) := E \otimes L^r$, $r \gg 0$ until (a bounded subset of) the Es have no higher cohomology and are generated by their holomorphic sections:

$$0 \to \ker \to \underline{H^0(E(r))} \to E(r) \to 0 \qquad \text{on } X. \tag{6.4}$$

Fixing an isomorphism $H^0(E(r)) \cong \mathbb{C}^N$, $N = \chi(E(r))$, we have expressed all such Es as quotients of $\mathcal{O}(-r)^{\oplus N}$. Such quotients are easily parameterised algebraically by a Quot scheme (for instance as a subset of the Grassmannian of subspaces $H^0(\ker(s)) \leq H^0(\mathcal{O}(s))^{\oplus N}$ given by taking sections of (6.4) tensored by L^s, $s \gg 0$.)

So we divide by choices of the isomorphism $H^0(E(r)) \cong \mathbb{C}^N$, i.e., by $SL(N, \mathbb{C})$, to get the moduli space of (semistable) sheaves.

In this case the Hilbert-Mumford criterion can be manipulated (Mumford, Takemoto, Gieseker, Maruyama, Simpson [**HL**]) to give an algebro-geometric understanding of stability. We describe this in a later section; the upshot is the following.

We write the Hilbert polynomial $\chi(E(r)) = a_0 r^n + a_1 r^{n-1} + \ldots$, where $a_0 = \operatorname{rank} E \int_X \omega^n/n!$, $a_1 = \int_X c_1(E).\omega^{n-1}/(n-1)! + \varepsilon(X)$, etc. are topological. We use its monic version, the reduced Hilbert polynomial

$$p_E(r) := \frac{\chi(E(r))}{a_0} = r^n + \frac{a_1}{a_0} r^{n-1} + \ldots. \tag{6.5}$$

Then E is stable if and only if *for all* coherent subsheaves $F \hookrightarrow E$, $p_F(r) \prec p_E(r)$ in the following sense (depending on the line bundle chosen on the Quot scheme):

- *Gieseker stable* if and only if $p_F(r) < p_E(r) \quad \forall r \gg 0$.
- *Slope stable* if and only if $\frac{a_1(F)}{a_0(F)} < \frac{a_1(E)}{a_0(E)}$ $\big(\iff \mu(F) < \mu(E)\big)$.

Here, as before, $\mu(E) = \int_X c_1(E).\omega^{n-1}/\operatorname{rank}(E)$ is the slope of E (6.2). Gieseker and slope stability coincide on curves X. Slope stability corresponds to taking a certain semi-ample line bundle on the Quot scheme (roughly speaking given by restricting sheaves to high degree complete intersection curves in X and using the usual line bundle for moduli of bundles on the curve). GIT needs amending for this situation; so far this has been carried out only for X a surface by Jun Li [**HL**].

Semistability is similar (replacing $<$ by $\leq$), while polystability is equivalent to semistability where the only destabilising subsheaves F are direct summands of E. Slope polystability then turns out to be the right stability notion for the infinite dimensional quotient and HYM of Theorem 6.3.

The symplectic reduction picture. For E a bundle, we can interpret the quotient $\underline{\mathbb{C}}^N \to E(r) \to 0$ (6.4) differently, via its classifying map $X \to Gr$ to the Grassmannian of quotients of $\mathbb{C}^N$.

Then, much as in the varieties case, we fix compatible hermitian metrics on L and E inducing a Kähler form on X and an L^2-metric on $\mathbb{C}^N \cong H^0(E(r))$. Then there are actions of $SU(N) < SL(N, \mathbb{C})$ on Gr, inducing a moment map $m\colon Gr \hookrightarrow \mathfrak{su}(N)^*$ and an action $SU(N) \curvearrowright \mathrm{Maps}(X, Gr)$. Its moment map is the integral of (the pullback of) m over X, so we can again talk about *balanced* $X \to Gr$ (those with centre of mass zero in $\mathfrak{su}(N)^*$) and asymptotics as $r, N \to \infty$.

Proving conjectures of Donaldson [**Do2**], Wang shows that the existence of a balanced map is equivalent to the Gieseker polystability of E [**Wa1**]. Slope stable bundles (which are therefore Gieseker stable) admit balanced maps $X \to Gr$ for $r \gg 0$ [**Wa2**], and pulling back the canonical quotient connection on Gr and taking $\lim_{r\to\infty}$ gives a conformally Hermitian-Yang-Mills connection on E (which is HYM after rescaling). (Unfortunately, this is not how the results are proved; Wang uses the Donaldson-Uhlenbeck-Yau theorem to give an a priori HYM connection which can be compared to the sequence of balanced metrics.)

7. Slope criteria for algebraic varieties

Slope for K-stability. So we have seen that the finite and infinite dimensional GIT and symplectic reduction pictures work for bundles, and tend to one another as $r \to \infty$. In particular the stability notion for bundles and sheaves involves only subsheaves $F < E$. So one might ask if the subvariety $\mathbb{P}(F) \subset \mathbb{P}(E)$ can destabilise the variety $\mathbb{P}(E)$. Or if, more generally, subschemes $Z \subset (X, L)$ can destabilise (X, L). (Cf. length $\geq n/2$ subschemes destabilising length-n schemes in $\mathbb{P}^1$ (3.10).)

We need some topological data for (X, L) analogous to that for bundles (6.5). Fixing $Z \subset (X, L)$, we have the Hilbert polynomial of $L = \mathcal{O}_X(1)$

$$h^0(\mathcal{O}_X(r)) = a_0 r^n + a_1 r^{n-1} + \dots ,$$

and the *Hilbert-Samuel* polynomial of $\mathscr{I}_Z$ (for $x \in \mathbb{Q}$ and $rx \in \mathbb{N}$):

$$h^0(\mathscr{I}_Z^{xr}(r)) = a_0(x) r^n + a_1(x) r^{n-1} + \dots .$$

Then by working on the blow up of X in Z one can see that the $a_i(x)$ are polynomials in $x \in \mathbb{Q} \cap [0, \epsilon(Z))$ for $r \gg 0$. (More precisely, there is a constant p and a polynomial which is equal to $a_i(x)$ for $xr > p$ or $x = 0$.) Here $\epsilon(Z)$ is the *Seshadri constant* of Z, the supremum of x such that $\mathscr{I}_Z^{xr}(r)$ is generated by global sections for $r \gg 0$.

For X normal, $a_0(0) = a_0$, and $a_1(0) = a_1$. All the $a_i(x)$ are given by topological formulae by Riemann-Roch, for instance

$$a_0 = \frac{\int_X \omega^n}{n!}, \qquad a_1 = \frac{\int_X c_1(X)\omega^{n-1}}{2(n-1)!}.$$

For any $c \leq \epsilon(Z)$, analogously to the definition of slope for bundles (6.2), we define the slope of Z to be

$$\mu_c(\mathscr{I}_Z) = \frac{\int_0^c a_1(x) + \frac{a_0'(x)}{2} dx}{\int_0^c a_0(x) dx}. \tag{7.1}$$

$Z = \emptyset$ gives

$$\mu(X) = \frac{a_1}{a_0}.$$

We then have the following [**RT2**].

THEOREM 7.2. *(X, L) K-semistable $\Rightarrow$ slope semistable: $\mu_c(\mathscr{I}_Z) \leq \mu(X)$ for all closed subschemes $Z \subset X$ and $c \leq \epsilon(Z)$.*

Removing the "semi" is a little more involved. If we use the algebraic K-stability of [**Do5, RT1**] then we define slope stability to mean

- $\mu_c(\mathscr{I}_Z) < \mu(X) \ \ \forall c \in (0, \epsilon(Z)) \cap \mathbb{Q}$, *and*
- $\mu_{\epsilon(Z)}(\mathscr{I}_Z) < \mu(X)$ if $\epsilon(Z) \in \mathbb{Q}$ and $\mathscr{I}_Z^{\epsilon(Z)r}(r)$ is saturated by global sections for $r \gg 0$.

(The quickest definition of saturated [**RT1**] is that on the blow up $\pi\colon \mathrm{Bl}_Z X \to X$ of X in Z with exceptional divisor E, $\pi^*\mathscr{I}_Z^{\epsilon(Z)r}(r) = \pi^* L^r(-\epsilon(Z)r)$ should be generated by global sections.) Similarly slope polystability is defined as slope stability except in the second part of the definition we allow $\mu_{\epsilon(Z)}(\mathscr{I}_Z)$ to equal $\mu(X)$ if on $\mathrm{Bl}_{Z\times\{0\}}(X \times \mathbb{C})$, $L(-\epsilon(Z)P)$ (where P is the exceptional divisor) is pulled back from a contraction $\mathrm{Bl}_{Z\times\{0\}}(X \times \mathbb{C}) \to X \times \mathbb{C}$ (which of course won't be the original blowup map).

If we use analytic K-stability [**RT2**], which is what should be relevant to the cscK problem (Conjecture 5.22), then the relevant definition of slope stability allows irrational c:

- $\mu_c(\mathscr{I}_Z) < \mu(X) \ \ \forall c \in (0, \epsilon(Z))$, *and*
- $\mu_{\epsilon(Z)}(\mathscr{I}_Z) < \mu(X)$ if $\epsilon(Z) \in \mathbb{Q}$ and $\mathscr{I}_Z^{\epsilon(Z)r}(r)$ is saturated by global sections for $r \gg 0$.

Again slope polystability is defined in the same way except for the second condition in which we allow $\mu_{\epsilon(Z)}(\mathscr{I}_Z) = \mu(X)$ if on $\mathrm{Bl}_{Z\times\{0\}}(X \times \mathbb{C})$, $L^r(-\epsilon(Z)rP)$ is pulled back from a contraction to $X \times \mathbb{C}$.

Then the analogue of Theorem 7.2, for the either notion of K-stability, is the following [**RT1**].

THEOREM 7.3. *(X, L) K-(poly)stable $\Rightarrow$ slope (poly)stable.*

As a corollary of Theorem 7.2 and the results of Donaldson and Chen-Tian mentioned in Theorem 5.13 we find the following.

COROLLARY 7.4. *If $\mu_c(\mathscr{I}_Z) > \mu(X)$ then X does not admit a cscK metric in the class of $c_1(L)$.*

We give some examples.

- If $F < E$ is a slope-destabilising subbundle of a vector bundle $E \to X$ the $\mathbb{P}(F) \subset \mathbb{P}(E)$ destabilises for suitable polarisations $\pi^* L^m \otimes \mathcal{O}_{\mathbb{P}(E)}(1)$, $m \gg 0$, on $\mathbb{P}(E)$. Conversely [**Ho**], if E is slope-stable and the base X is cscK with discrete automorphism group then $\mathbb{P}(E)$ is cscK for $m \gg 0$.
- When the base is a curve, we can do better [**RT2**]. In fact, for $\mathbb{P}(E)$ with discrete automorphism group and *any* polarisation,

$$\mathbb{P}(E) \text{ cscK} \iff E \text{ HYM} \iff E \text{ stable.}$$

  The converses follow from the Narasimhan-Seshadri theorem [**BdB**] that stable bundles admit projectively flat connections.
- -1-curves destabilise del Pezzo surfaces (X, L) for appropriate L. In particular one can find examples with reductive (even trivial) automorphism group, showing that the folklore conjecture Aut (X) reductive $\Rightarrow$ cscK does not hold for surfaces. (Tian showed that it does not hold for threefolds [**Ti4**], but that it does hold for anti-canonically polarised surfaces [**Ti2**].)
- For instance $\mathbb{P}^2$ blown up in one point cannot admit a cscK metric for any polarisation because Aut (X) is not reductive. From the above point of view this is because it is destabilised by the exceptional -1-curve for all polarisations.
- Generically stable varieties can specialise to unstable ones. For instance moving two -1-curves together on a stable del Pezzo gives a limiting -2-curve (if the -1-curves arise from blowing up distinct points, then blow up two "infinitely near" points – one point and then another on the exceptional divisor) which can destabilise for suitable L.
- Calabi-Yau manifolds, and varieties with canonical singularities and numerically trivial canonical bundle ($mK_X \sim 0$) are slope stable.
- Canonically polarised varieties with canonical singularities (i.e., the canonical models of Mori theory) are slope stable.
- Remarkably, the product of a (nongeneric) smooth curve with itself can admit polarisations which are slope unstable, giving surfaces of general type which are neither K- nor Chow stable, and which do not admit cscK metrics in that class [**Ro**].

We will sketch the algebro-geometric proof of the above results later. But differential-geometrically, what the proofs amount to is the following.

We know that the Mabuchi energy (or log-norm functional, in GIT language) is convex over the space of all Kähler metrics on our fixed complex manifold X. Intuitively, if it is proper in some sense (roughly, tends to $+\infty$ at infinity) then it should have a unique absolute minimum, which, modulo regularity issues, would be cscK. Conversely the manifold is strictly K-unstable, with no cscK metric, if the Mabuchi energy is unbounded below. If X is slope destabilised by some subscheme Z then there is a family of

Kähler metrics on X, given by "stretching the neck" around Z, along which the Mabuchi energy tends to $-\infty$, so X cannot be cscK.

Slope for Chow stability. Fix $Z \subset (X, \mathcal{O}_X(1)) \subset (\mathbb{P}^N, \mathcal{O}(1))$ embedded by sections of $\mathcal{O}_X(1)$, and, as before,

$$h^0(\mathcal{O}_X(r)) = a_0 r^n + a_1 r^{n-1} + \dots ,$$
$$h^0(\mathscr{I}_Z^{xr}(r)) = a_0(x) r^n + a_1(x) r^{n-1} + \dots .$$

Then, for all $c \le \epsilon(Z) \in \mathbb{N}$, define the *Chow slope* of Z to be

$$Ch_c(\mathscr{I}_Z) = \frac{\sum_{i=1}^{c} h^0(\mathscr{I}_Z^i(1))}{\int_0^c a_0(x)dx},$$

a discrete version of (7.1). $Z = \emptyset$ gives

$$Ch(X) = \frac{h^0(\mathcal{O}_X(1))}{a_0} = \frac{N+1}{a_0},$$

and we have the following [**RT1**].

THEOREM 7.5. *Chow (semi)stable $\Rightarrow$ slope (semi)stable:*

$$Ch_c(\mathscr{I}_Z) \underset{(\le)}{<} Ch(X) \ \ \forall Z \subset X.$$

To see where these results come from, and to explain how far one can get towards a converse, we need to analyse the Hilbert-Mumford criterion for varieties (X, L). We first warm up with a brief overview of the bundle case.

The Hilbert-Mumford criterion for vector bundles. Given a coherent sheaf E on X, recall (6.4) how an picking an identification $H^0(E(r)) \cong \mathbb{C}^N$ for $r \gg 0$, $N = \chi(E(r))$, realises $E(r)$ as a point of a Quot scheme of quotients

$$\mathcal{O}(-r)^{\oplus N} \to E \to 0. \tag{7.6}$$

Dividing out by the identification, i.e., quotienting the relevant subset of Quot by $SL(N, \mathbb{C})$, gives a moduli space of sheaves.

To apply the Hilbert-Mumford criterion, we consider the E-orbit of a 1-PS $\mathbb{C}^* < GL(N, \mathbb{C})$ on Quot [**HL**] (we shall normalise to $SL(N, \mathbb{C})$ later), whose eigenvalues we can assume are all positive, without loss of generality. The eigenspaces $V_\lambda < \mathbb{C}^N \cong H^0(E(r))$ give a weight filtration $V_{\le i} = \oplus_{\lambda \le i} V_\lambda$ of $\mathbb{C}^N$. Their images $F_i \le E$ under the map (7.6) give a filtration of E,

$$F_0 \le F_1 \le \dots \le F_p \le E,$$

and the orbit gives a sheaf over $X \times \mathbb{C}$ described in terms of the F_i as follows. Let $\mathbb{F}_i$, $\mathbb{E}$ denote the pullbacks of the sheaves F_i, E to $X \times \mathbb{C}$, so $\mathbb{F}_i = F_i \otimes \mathbb{C}[t]$ where t is the variable pulled back from $\mathbb{C}$. Then the orbit gives the following subsheaf of $\mathbb{E}$,

$$\mathbb{F}_0 + t.\mathbb{F}_1 + t^2.\mathbb{F}_2 + \dots + t^p.\mathbb{F}_p + t^{p+1}.\mathbb{E} < E \otimes \mathbb{C}[t]. \tag{7.7}$$

This is a degeneration of the general fibre E to

$$F_0 \oplus F_1/F_0 \oplus \ldots \oplus F_p/F_{p-1} \oplus E/F_p \tag{7.8}$$

over the central fibre $X \times \{0\}$. One can prove this inductively as follows.

Set $Q_i = E/F_i$, giving exact sequences $0 \to F_i/F_{i-1} \to Q_{i-1} \to Q_i \to 0$. In the $p = 0$ case, $\mathbb{E}_0 := \mathbb{F}_0 + t.\mathbb{E}$ is the kernel of the composition $\mathbb{E} \to E \to Q_0$, where the latter two sheaves are considered to be supported on the central fibre $X \times \{0\}$. (So $\mathbb{E}_0$ is the *elementary transform* of $\mathbb{E}$ in Q_0 on $X \times \{0\}$.) $\mathbb{E}_0$ has a map to F_0 induced by the diagram

$$\begin{array}{ccccccccc} 0 & \longrightarrow & \mathbb{E}_0 & \longrightarrow & \mathbb{E} & \longrightarrow & Q_0 & \longrightarrow & 0 \\ & & \downarrow & & \downarrow & & \| & & \\ 0 & \longrightarrow & F_0 & \longrightarrow & E & \longrightarrow & Q_0 & \longrightarrow & 0, \end{array} \tag{7.9}$$

and one to Q_0:

$$\begin{array}{ccccccccc} 0 & \longrightarrow & \mathbb{E}_0 & \longrightarrow & \mathbb{E} & \longrightarrow & Q_0 & \longrightarrow & 0 \\ & & \downarrow & & \downarrow & & \| & & \\ 0 & \longrightarrow & Q_0 & \xrightarrow{.t} & Q_0 \otimes \frac{\mathbb{C}[t]}{(t^2)} & \longrightarrow & Q_0 & \longrightarrow & 0. \end{array}$$

The pair make the central fibre of $\mathbb{E}_0$ isomorphic to $F_0 \oplus Q_0$, while the map $\mathbb{E}_0 \to Q_0$ can be composed with the surjection $Q_0 \to Q_1$ to continue the induction by defining $\mathbb{E}_1$ as its kernel. This is clearly just $\mathbb{F}_0 + t.\mathbb{F}_1 + t^2.\mathbb{E}$, and similar working shows it has central fibre $F_0 \oplus F_1/F_0 \oplus Q_1$, and so on.

Different 1-PSs can give the same filtration (if the ith piece of the weight filtration of $\mathbb{C}^N$ generates the same subsheaf $F_i \leq E$). But for every filtration F_i and sequence of weights there is a canonical *least stable* 1-PS, given by choosing the weight filtration to be $V_{\leq i} = H^0(F_i(r)) < H^0(E(r))$.

So we need only consider these canonical 1-PSs: they have the largest GIT weight in their class. Since (7.8) gives the weight space decomposition of the limiting sheaf over the central fibre, the weights of these 1-PSs $\sum_n nh^0\big((F_n/F_{n-1})(k)\big)$, $k \gg 0$, are positive linear combinations of weights of the canonical 1-PSs associated to the splittings

$$F_i \oplus E/F_i. \tag{7.10}$$

So in fact we need only control the weights of these simpler splittings. Calculating their weights (and then normalising into $SL(N, \mathbb{C})$ [**HL**]) gives the Gieseker stability condition for bundles described earlier, controlled by single subsheaves $F < E$ and their reduced Hilbert polynomials.

The Hilbert-Mumford criterion for varieties. We can now try to do the analogous thing for varieties, following [**Mu, RT1**].

Any test configuration $(\mathscr{X}, \mathcal{L})$ is $\mathbb{C}^*$-birational to $(X \times \mathbb{C}, L)$, so is (a contraction p of) the blow up of $X \times \mathbb{C}$ in a $\mathbb{C}^*$-invariant ideal I supported on the central fibre, with polarisation $p^*\mathcal{L} = (\pi^* L)(-cP) = \pi^*(L \otimes I^c)$, where P is the exceptional divisor.

(7.11)
$$\begin{array}{ccc} & \mathrm{Bl}_I(X \times \mathbb{C}) & \\ {}^{\pi}\swarrow & & \searrow^{p} \\ X \times \mathbb{C} & & \mathscr{X} \end{array}$$

Classifying such $\mathbb{C}^*$-invariant ideals, there exist subschemes

(7.12)
$$Z_{p-1} \subseteq \ldots \subseteq Z_1 \subseteq Z_0 \subseteq X$$

with ideal sheaves $\mathscr{I}_{p-1} \supseteq \ldots \supseteq \mathscr{I}_1 \supseteq \mathscr{I}_0$ such that [**Mu**]

(7.13)
$$I = \mathscr{I}_0 + t\mathscr{I}_1 + t^2\mathscr{I}_2 + \ldots + t^{p-1}\mathscr{I}_{p-1} + t^p.$$

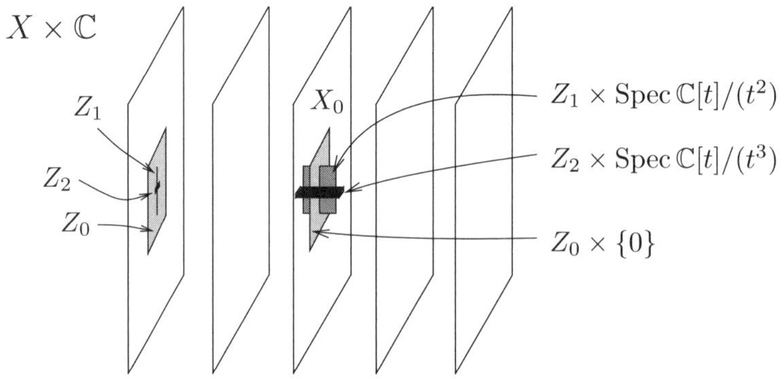

Firstly we have an analogue of the fact that, in the bundle case, one need only consider canonical 1-PSs. Namely, under any map of test configurations like p above, the weights are less stable (more positive) on the dominating test configuration. (Notice that the blow up map $\mathrm{Bl}_I(X \times \mathbb{C}) \to X \times \mathbb{C}$ above is not *such* a map of test configurations, since it does not preserve polarisations: the line bundle is not a pullback from downstairs.)

PROPOSITION 7.14. *Suppose X is normal. Given a test configuration $(\mathscr{X}, \mathcal{L})$ for (X, L) and another flat $\mathbb{C}^\times$-family $\mathscr{Y} \to \mathbb{C}$ with a birational $\mathbb{C}^*$-equivariant map $p\colon \mathscr{Y} \to \mathscr{X}$, there exists an $a \geq 0$ such that*

$$w(H^0_{\mathscr{X}_0}(\mathcal{L}^k)) = w\left(H^0_{\mathscr{Y}}(p^*\mathcal{L}^k)/tH^0_{\mathscr{Y}}(p^*\mathcal{L}^k)\right) - ak^n + O(k^{n-1}).$$

(Here w denotes the total weight of a $\mathbb{C}^*$-action – i.e. its weight on the determinant of the $\mathbb{C}^*$-module. The normality of X is required to equate $H^0(X, L^k)$ with sections of $p^*\mathcal{L}^k$ on a general fibre of $\mathscr{Y}$. The result is stated in rather more generality than we require here (in particular allowing $\mathscr{Y}$ to have general fibre some blow up of X, rather than just X) for future use. We are forced to use $H^0_{\mathscr{Y}}(p^*\mathcal{L}^k)/tH^0_{\mathscr{Y}}(p^*\mathcal{L}^k)$, rather than $H^0_{\mathscr{Y}_0}(p^*\mathcal{L}^k)$, because $p^*\mathcal{L}$ need not be ample on $\mathscr{Y}$.)

So we need only consider weights on the normalisation of the blow up of $X \times \mathbb{C}$ in I, as this is is itself a perfectly good test configuration. (In the Chow stability case this test configuration may not arise from a linear transformation of the given projective space – only from an embedding by a higher twist of L – but this does not concern us as Proposition 7.14 gives an equality in weights to $O(k^n)$, which is all that is required for Chow stability.)

Next we consider the easiest case of $p=1$, i.e., $I=\mathscr{I}_0+(t)$. So we blow up in $Z_0\times\{0\}\subset X\times\mathbb{C}$, giving the *deformation to normal cone of* Z_0. This is the analogue of the simplest degenerations of bundles and sheaves earlier; the canonical degenerations of a one-step filtration to a direct sum (7.10).

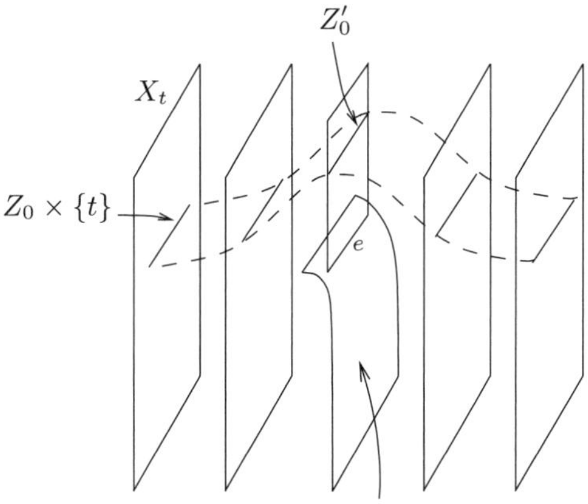

The exceptional divisor P is the *normal cone* of Z_0: if $Z_0\subset X$ is smooth then this is the projective completion $\mathbb{P}(\nu_{Z_0}\oplus\underline{\mathbb{C}})\to Z_0$ of the normal bundle $\nu_{Z_0}\to Z_0$.

$\mathbb{C}^*\ni\lambda$ acts on the blow up (as $[1:\lambda]=[\lambda^{-1}:1]$ on $\mathbb{P}(\nu_{Z_0}\oplus\underline{\mathbb{C}})$ in the smooth case) and on the line bundle $\pi^*L(-cP)$ over $\mathrm{Bl}_{Z_0\times\{0\}}(X\times\mathbb{C})$.

This deformation to the normal cone of Z replaces $H^0_X(L^r)$ (filtered by $H^0(L^r\otimes\mathscr{I}_Z^j)$) by the associated graded of the filtration on the central fibre:

$$\begin{aligned}H^0_X(\mathscr{I}_Z^{cr}(r))\oplus tH^0_X\left(\mathscr{I}_Z^{cr-1}(r)/\mathscr{I}_Z^{cr}(r)\right)\oplus\dots\\ \oplus\ t^{cr-1}H^0_X\left(\mathscr{I}_Z(r)/\mathscr{I}_Z^2(r)\right)\oplus t^{cr}H^0_X(\mathcal{O}_Z(r)).\end{aligned}$$

Here t is the coordinate on the $\mathbb{C}$-base. This is the splitting of sections of $L^r(-crP)$ on the central fibre into those on the proper transform of the original central fibre X (the first term) plus the polynomials on P. In turn the latter can be split into their $\mathbb{C}^*$-weight spaces as t^j times by the homogeneous polynomials on the normal bundle of $Z\subset X$ of degree $cr-j$.

So this is the weight space decomposition, with $\mathbb{C}^*$ acting on t^j with weight $-j$, and the weight on top exterior power is

$$\begin{aligned}w_r &= -\sum_{j=1}^{cr} jh^0_X\left(\mathscr{I}_Z^{cr-j}(r)/\mathscr{I}_Z^{cr-j+1}(r)\right)\\ &= -\sum_{j=1}^{cr} h^0(\mathscr{I}_Z^j(r))-crh^0(\mathcal{O}_X(r)).\end{aligned}\tag{7.15}$$

This looks like a discrete approximation (Riemann sum) for

$$\int_0^c h^0(\mathscr{I}_Z^{xr}(r))dx,$$

so we estimate it by the trapezium rule, giving

$$-\left(\int_0^c a_0(x)dx\right)r^{n+1} + \int_0^c \left(a_1(x) + \frac{a_0'(x)}{2}\right)dx\ r^n + O(r^{n-1}).$$

Normalising (to make the 1-PS lie in $SL(N,\mathbb{C})$ instead of $GL(N,\mathbb{C})$) we find the K-stability slope criterion of (7.1) and Theorems 7.2 and 7.3.

Alternatively, taking the leading order term of (7.15) and normalising gives the Chow slope criterion of Theorem 7.5.

General test configurations. So we have the analogue of the result in the bundle case that one need only consider canonical 1-PS orbits – i.e. we need only consider test configurations that are normalisations of the form (7.11). And we have the analogue of the simple degenerations (7.10), given by the deformation to the normal cone of a subscheme $Z \subset X$ yielding the right analogue of slope stability. For bundles any canonical 1-PS weight turned out to be a positive linear combination of these simple weights, so ideally one would like to write the weight of a test configuration (7.11) as a positive linear combination of weights (7.15) – if so one could conclude that stability of varieties was equivalent to slope stability.

So we need the analogue of the induction (7.9) that we did in the bundle case to handle (7.7). In fact it turns out we can mirror it almost completely; moreover the bundle induction is a special case of what follows below when we consider the sheaf $\mathbb{E}$ of (7.7) to be the sections of the polarisation $\mathcal{O}_{\mathbb{P}(E^*)}(1)$ on the variety $\mathbb{P}(E^*) \times \mathbb{C}$. The correspondence between (7.7) and (7.13) is clear, and the elementary transformations we did in the bundle case become, on projectivisation, the blowups below.

The proper transform $\overline{Z_0 \times \mathbb{C}}$ of $Z_0 \times \mathbb{C}$ is *flat* over the base $\mathbb{C}$. It defines a copy Z_0' of Z_0 in the central fibre of the deformation to the normal cone of Z_0:

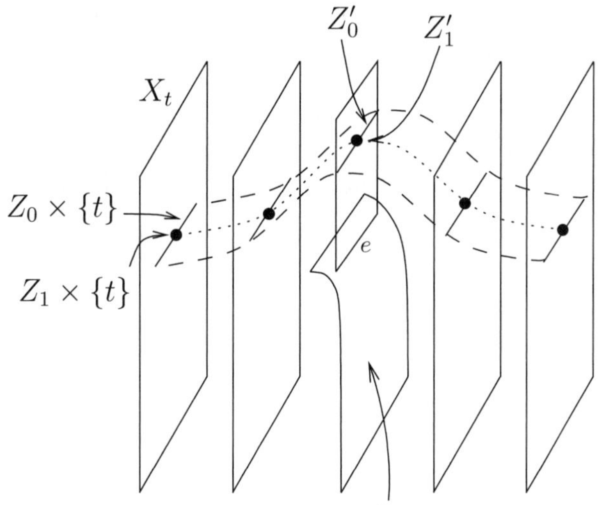

By (7.12) this defines subschemes $Z'_{p-1} \subseteq \ldots \subseteq Z'_1 \subset Z'_0$. So next we blow up in Z'_1, giving $Z''_{p-1} \subseteq \ldots \subseteq Z''_1$; next blow up Z''_2, and so on inductively up to $Z^{(p-1)}_{p-1}$.

(For $X = \mathbb{P}(E^*)$ and the degeneration (7.7), we are blowing up the subschemes $\mathbb{P}(Q_0^*)$ (in the central fibre), then the central fibre of the proper transform of $\mathbb{P}(Q_1^*)$, and so on; but this is just the projectivisation of the elementary transformations (7.9).)

THEOREM 7.16. *The blow up of $X \times \mathbb{C}$ in $I = \mathscr{I}_0 + t\mathscr{I}_1 + \ldots + t^{p-1}\mathscr{I}_{p-1} + t^p$ is a contraction of this iterated blow up.*

This is meant in the $\mathbb{C}^*$-equivariant polarised sense – the polarisation $L(-cP)$ on $\mathrm{Bl}_I(X \times \mathbb{C})$ is the pullback of the natural polarisation on the iterated blow up given by starting with L and, at each stage, pulling back and subtracting c times the exceptional divisor.

Thus, by our result (Proposition 7.14) that one need only calculate weights on dominating test configurations, we are left with calculating the amount that each blow up in $Z_i^{(i)}$ adds to the weight of the resulting $\mathbb{C}^*$-action on the determinant of the space of sections of the rth power of the polarisation on the central fibre.

THEOREM 7.17. *Consider the ith step, when we blow up $Z_i^{(i)}$. If all thickenings of $\overline{(Z_i \times \mathbb{C})}$ are flat over $\mathbb{C}$ then this adds $w(Z_i)$ to the weight, to $O(r^n)$.*

(Here $w(Z_i)$ is weight on deformation to normal cone of Z_i.)

In fact under certain conditions one can get the result to $O(r^{n-1})$ [**RT1**].

So if this flatness condition holds, the total (normalised) weight is $w(Z_0) + \cdots + w(Z_{p-1})$. So X is stable if and only if

$$w(Z_0) + \ldots + w(Z_{p-1}) \prec 0 \iff w(Z) \prec 0 \;\forall Z. \tag{7.18}$$

So if this flatness condition held in general, then stability and slope stability would be equivalent.

The flatness problem. In fact $\overline{(Z_i \times \mathbb{C})}$ *is* flat over $\mathbb{C}$, but since blow ups use all powers of an ideal, we require all of its scheme theoretic thickenings $k\overline{(Z_i \times \mathbb{C})}$ (defined by the ideal $\mathscr{I}^k_{\overline{Z_i \times \mathbb{C}}}$) to be flat too. The idea of the proof of Theorem 7.17 is then that a formal neighbourhood of $Z_i^{(i)}$ in the test configuration looks sufficiently like a formal neighbourhood of $Z_i \times \{0\} \subset X \times \mathbb{C}$ for the weights added by the two blow ups to be comparable to $O(r^n)$ – the corrections being due to estimates on the sizes of nonvanishing cohomology groups.

This flatness condition holds for $Z_i \subset X$ smooth, or reduced simple normal crossing (snc) divisors.

In general, one can use resolution of singularities

$$(X \supset Z_i) \xleftarrow{\ \pi\ } (\widehat{X} \supset m_i D_i), \quad D_i \text{ snc divisors,}$$

to replace (X, L) by $(\widehat{X}, \pi^* L)$. Test configurations for the latter dominate those of the former, so Proposition 7.14 allows us to obtain (7.18) for X *normal*, so long as $m_i = 1$ for all i.

So finally we need to be able to deal with the snc divisors D_i being possibly nonreduced. We can attempt to do this by basechange [**RT1**], which we illustrate with an example.

Consider the case where $Z_0 \subset X$ is a double point in a smooth curve. Locally then $\mathscr{I}_0 = (x^2)$, and the deformation to normal cone of Z_0 is the blow up of $X \times \mathbb{C}$ in (x^2, t).

Now consider squaring the $\mathbb{C}^*$-action. This is trivial from a GIT point of view (it just doubles the weight, which we can later halve). But it fundamentally alters our geometric description of the test configuration, blowing up $X \times \mathbb{C}$ in the ideal (x^2, t^2).

Taking the integral closure of this ideal corresponds to normalising the blow up, which we can deal with by Proposition 7.14. That is, we get a more unstable test configuration by blowing up in $(x^2, xt, t^2) = (x, t)^2$, and it suffices to control the weights of this test configuration.

But this is now a much nicer ideal, and corresponds to blowing up in (x, t) and using a different line bundle (squaring the exceptional divisor $P \mapsto 2P$ or $c \mapsto 2c$). So modulo doubling c and the weight, we have removed the multiplicity 2 of the double point Z_0.

In this way we can deal with D_i with multiplicities m_i when they *all have the same support.* This is enough to prove that (K- and Chow) stability coincides with (K- and Chow) slope stability for smooth curves, and indeed gives probably the "right" geometric proof, rather than the old combinatorial (for Chow stability) and analytical (for K-stability) proofs.

Of course, for higher dimensions, one would like to combine the two approaches to deal with snc divisors with intersecting components of different multiplicities, for instance, $D_0 = (x^2 y = 0)$, $D_1 = (x = 0)$. This is still work in progress, but its difficulty suggests that slope stability is not enough to describe stability (unlike for the more linear bundle case) except in one dimension or on projective bundles over stable bases.

References

[AB] M.F. Atiyah and R. Bott, *The Yang-Mills equations over Riemann surfaces*, Phil. Trans. R. Soc. Lond. Ser. A **308** (1982), 523–615.

[ACGT] V. Apostolov, D. Calderbank, P. Gauduchon, and W. Tønnesen-Friedman, *Hamiltonian 2-forms in Kähler geometry,* III *Extremal metrics and stability,* math.DG/0511118, 2005.

[AP] C. Arezzo and F. Pacard, *Blowing up Kähler manifolds with constant scalar curvature*, II, math.DG/0504115, 2004.

[Au] T. Aubin, *Équations du type Monge-Ampère sur les variétés kähleriennes compactes*, C.R. Acad. Sci. Paris Sér. A-B **283** (1976), A119–A121.

[BM] S. Bando and T. Mabuchi, *Uniqueness of Einstein-Kähler metrics modulo connected group actions*, in 'Algebraic Geometry', Advanced Studies in Pure Math. **10** (1987), 11–40.

[Bi] O. Biquard, *Métriques kählériennes à courbure scalaire constante*, Séminaire Bourbaki **938**, 2004.

[Bo] J.-P. Bourguignon, *Métriques d'Einstein-Kähler sur les variétés de Fano: obstructions et existence (d'après Y. Matsushima, A. Futaki, S.-T. Yau, A. Nadel et G. Tian)*, Séminaire Bourbaki, Astérisque **245** (1997), 277–305.

[BdB] D. Burns and P. De Bartolomeis, *Stability of vector bundles and extremal metrics*, Invent. Math. **92** (1988), 403–407.

[CGR] M. Cahen, S. Gutt, and J. Rawnsley, *Quantization of Kähler manifolds*, II, Trans. Amer. Math. Soc. **337** (1993), 73–98.

[Ca] E. Calabi, *Extremal Kähler metrics*, in 'Seminar on differential geometry' (S.-T. Yau, ed.), Princeton Univ. Press, New Jersey, 259–290, 1982.

[Ch] X. Chen, *The space of Kähler metrics*, Jour. Differential Geom. **56** (2000), 189–234.

[CT] X. Chen and G. Tian, *Geometry of Kähler metrics and holomorphic foliation by discs*, math.DG/0409433, 2004.

[DT] W. Ding and G. Tian, *Kähler-Einstein metrics and the generalized Futaki invariant*, Invent. Math. **110** (1992), 315–335.

[Dl] I. Dolgachev, *Lectures on invariant theory*, LMS Lecture Note Series, **296**, Cambridge University Press, 2003.

[DH] I. Dolgachev and Y. Hu, *Variation of Geometric Invariant Theory Quotients*, Publ. IHES **87** (1998), 5–56.

[Do1] S.K. Donaldson, *Remarks on gauge theory, complex geometry and 4-manifold topology*, in 'Fields Medallists Lectures', World Sci. Ser. 20th Century Math., **5**, World Sci. Publishing, River Edge, NJ, 384–403, 1997.

[Do2] S.K. Donaldson, *Geometry in Oxford c.* 1980–85, Sir Michael Atiyah: a great mathematician of the twentieth century, Asian J. Math. **3** (1999), xliii–xlvii.

[Do3] S.K. Donaldson, *Scalar curvature and projective embeddings*, I, Jour. Differential Geom. **59** (2001), 479–522.

[Do4] S.K. Donaldson, *Planck's constant in complex and almost-complex geometry*, XIIIth International Congress on Mathematical Physics (London, 2000), 63–72, Int. Press, Boston, 2001.

[Do5] S.K. Donaldson, *Scalar curvature and stability of toric varieties*, Jour. Differential Geom. **62** (2002), 289–349.

[Do6] S.K. Donaldson, *Lower bounds on the Calabi functional*, math.DG/0506501, 2005.

[DK] S.K. Donaldson and P. Kronheimer, *The geometry of four-manifolds*, Oxford Univ. Press, 1990.

[Fi] J. Fine, *Constant scalar curvature Kähler metrics on fibred complex surfaces*, Jour. Differential Geom. **68** (2004), 397–432.

[Fj] A. Fujiki, *Moduli space of polarized algebraic manifolds and Kähler metrics*, Sugaku **42** (1990), 231–243; English translation: Sugaku Expo. **5** (1992), 173–191.

[Fu] A Futaki, *An obstruction to the existence of Einstein-Kähler metrics*, Invent. Math. **73** (1983), 437–443.

[Ho] Y-J. Hong, *Gauge-fixing constant scalar curvature equations on ruled manifolds and the Futaki invariants*, Jour. Differential Geom. **60** (2002), 389–453.

[HL] D. Huybrechts and M. Lehn, *Geometry of moduli spaces of shaves*, Aspects in Mathematics, Vol. E31, Vieweg, 1997.

[Hu] Y. Hu, *Topological Aspects of Chow Quotients*, math.AG/0308027.

[Ki] F.C. Kirwan, *Cohomology of quotients in symplectic and algebraic geometry*, Mathematical Notes, **34**, Princeton University Press, Princeton, 1984.

[LB] C. LeBrun, *Anti-self-dual metrics and Kähler geometry*, Proceedings of the International Congress of Mathematicians (Zürich, 1994), 498–507, Birkhäuser, 1995.

[Mb1] T. Mabuchi, *K-energy maps integrating Futaki invariants*, Tohoku Math. Jour. **38** (1986), 245–257.

[Mb2] T. Mabuchi, *An energy-theoretic approach to the Hitchin-Kobayashi correspondence for manifolds*, I, Invent. Math. **159** (2004), 225–243.

[Mu] D. Mumford, *Stability of projective varieties*, Enseignement Math. (2) **23** (1977), 39–110.

[MW] W. Müller and K. Wendland, *Extremal Kähler metrics and Ray-Singer analytic torsion*, in 'Geometric aspects of partial differential equations' (Roskilde, 1998), 135–160, Contemp. Math., **242**, Amer. Math. Soc., Providence, RI, 1999.

[GIT] D. Mumford, J. Fogarty, and F. Kirwan, *Geometric Invariant Theory*, Third edition, Erg. Math., **34**, Springer-Verlag, Berlin, 1994.

[Na] A. Nadel, *Multiplier ideal sheaves and Kähler-Einstein metrics of positive scalar curvature*, Ann. of Math. **132** (1990), 549–596.

[Ne] P. Newstead, *Introduction to moduli problems and orbit spaces*, Tata Institute, Bombay. Springer-Verlag, 1978, out of print.

[RS] Y. Rollin and M. Singer, *Non-minimal scalar-flat Kähler surfaces and parabolic stability*, Invent. Math. **162** (2005), 235–270.

[Ro] J. Ross, *Unstable products of smooth curves*, to appear in Invent. Math., math.AG/0506447, 2005.

[RT1] J. Ross and R. Thomas, *A study of the Hilbert-Mumford criterion for the stability of projective varieties*, math.AG/0412519, 2004.

[RT2] J. Ross and R. Thomas, *An obstruction to the existence of constant scalar curvature Kähler metrics*, to appear in Jour. Differential Geom., math.DG/0412518, 2004.

[S] Y.T. Siu, *The existence of Kähler-Einstein metrics on manifolds with positive anticanonical line bundle and a suitable finite symmetry group*, Ann. of Math. **127** (1988), 585–627.

[Sz1] G. Székelyhidi, *Extremal metrics and K-stability*, math.AG/0410401, 2004.

[Sz2] G. Székelyhidi, Ph.D. thesis, Imperial College London, 2006, in preparation.

[Th] M. Thaddeus, *Geometric invariant theory and flips*, J. Amer. Math. Soc. **9** (1996), 691–723.

[Ti1] G. Tian, *On Kähler-Einstein metrics on certain Kähler manifolds with $C_1(M) > 0$*, Invent. Math. **89** (1987), 225–246.

[Ti2] G. Tian, *On Calabi's conjecture for complex surfaces with positive first Chern class*, Invent. Math. **101** (1990), 101–172.

[Ti3] G. Tian, *On a set of polarized Kähler metrics on algebraic manifolds*, Jour. Differential Geom. **32** (1990), 99–130.

[Ti4] G. Tian, *Kähler-Einstein metrics with positive scalar curvature*, Invent. Math. **130** (1997), 1–37.

[Wa1] X.-W. Wang, *Balance point and Stability of Vector Bundles Over a Projective Manifold*, Math. Res. Lett. **9** (2002), 393–411.

[Wa2] X.-W. Wang, *Canonical metrics on stable vector bundles*, Comm. Anal. Geom. **13** (2005), 253–285.

[Y1] S.-T. Yau, *On the Ricci curvature of a compact Kähler manifold and the complex Monge-Ampère equation*, I, Comm. Pure Appl. Math. **31** (1978), 339–411.

[Y2] S.-T. Yau, *Nonlinear analysis in geometry*, Enseign. Math. **33** (1987), 109–158.

[Y3] S.-T. Yau, *Open problems in geometry*, in 'Differential geometry: partial differential equations on manifolds' (Los Angeles, CA, 1990), 1–28, Proc. Sympos. Pure Math., **54**, AMS publications, 1993.

[Zh] S. Zhang, *Heights and reductions of semi-stable varieties*, Compositio Math. **104** (1996), 77–105.

Perspectives on geometric analysis

Shing-Tung Yau

This essay grew from a talk I gave on the occasion of the seventieth anniversary of the Chinese Mathematical Society. I dedicate the lecture to the memory of my teacher S.S. Chern who had passed away half a year before (December 2004).

During my graduate studies, I was rather free in picking research topics. I [**731**] worked on fundamental groups of manifolds with non-positive curvature. But in the second year of my studies, I started to look into differential equations on manifolds. However, at that time, Chern was very much interested in the work of Bott on holomorphic vector fields. Also he told me that I should work on Riemann hypothesis. (Weil had told him that it was time for the hypothesis to be settled.) While Chern did not express his opinions about my research on geometric analysis, he started to appreciate it a few years later. In fact, after Chern gave a course on Calabi's works on affine geometry in 1972 at Berkeley, S.Y. Cheng told me about these inspiring lectures. By 1973, Cheng and I started to work on some problems mentioned in Chern's lectures. We did not realize that the great geometers Pogorelov, Calabi and Nirenberg were also working on them. We were excited that we solved some of the conjectures of Calabi on improper affine spheres. But soon after we found out that Pogorelov [**563**] published his results right before us by different arguments. Nevertheless our ideas are useful in handling other problems in affine geometry, and my knowledge about Monge-Ampère equations started to broaden in these years. Chern was very pleased by my work, especially after I [**736**] solved the problem of Calabi on Kähler Einstein metric in 1976. I had been at Stanford, and Chern proposed to the Berkeley Math Department that they hire me. I visited Berkeley in 1977 for a year and gave a course on geometric analysis with emphasis on isometric embedding.

Chern nominated me to give a plenary talk at the International Congress in Helsinki. The talk [**737**] went well, but my decision not to stay at Berkeley

This research is supported by NSF grants DMS-0244464, DMS-0354737 and DMS-0306600.

did not quite please him. Nevertheless he recommended me for a position on the faculty at the Institute for Advanced Study. Before I accepted a faculty position at the Institute, I organized a special year on geometry in 1979 at the Institute at the invitation of Borel. That was an exciting year because most people in geometric analysis came.

In 1979, I visited China at the invitation of Professor L.K. Hua. I gave a series of talks on the bubbling process of Sacks-Uhlenbeck [**581**]. I suggested to the Chinese mathematicians that they apply similar arguments for a Jordan curve bounding two surfaces with the same constant mean curvature. I thought it would be a good exercise for getting into this exciting field of geometric analysis. The problem was indeed picked up by a group of students of Professor G.Y. Wang [**362**]. But unfortunately it also initiated some ugly fights during the meeting of the sixtieth anniversary of the Chinese Mathematical Society. Professor Wang was forced to resign, and this event hampered development of this beautiful subject in China in the past ten years.

In 1980, Chern decided to develop geometric analysis on a large scale. He initiated a series of international conferences on differential geometry and differential equations to be held each year in China. For the first year, a large group of the most distinguished mathematicians was gathered in Beijing to give lectures (see [**148**]). I lectured on open problems in geometry [**739**]. It took a much longer time than I expected for Chinese mathematicians to pick up some of these problems.

To his disappointment, Chern's enthusiasm about developing differential equations and differential geometry in China did not stimulate as much activity as he had hoped. Most Chinese mathematicians were trained in analysis but were rather weak in geometry. The goal of geometric analysis for understanding geometry was not appreciated. The major research center on differential geometry came from students of Chern, Hua and B.C. Su. The works of J.Q. Zhong (see, e.g., [**755, 527, 528**]) were remarkable. Unfortunately he died about twenty years ago. Q.K. Lu studied the Bergman metric extensively. C.H. Gu [**296**] studied gauge theory and considered harmonic map where the domain is $R^{1,1}$. J.X. Hong (see, e.g., [**345, 318**]) did some interesting work on isometric embedding of surfaces into $\mathbb{R}^3$. In the past five years, the research center at the Chinese University of Hong Kong, led by L.F. Tam and X.P. Zhu, has produced first class work related to Hamilton's Ricci flow (see, e.g., [**125, 126, 129, 130, 113**]).

In the hope that it will advance Chern's ambition to build up geometric analysis, I will explain my personal view to my Chinese colleagues. I will consider this article to be successful if it conveys to my readers the excitement of developments in differential geometry which have been taking place during the period when it has been my good fortune to contribute. I do not claim this article covers all aspects of the subject. In fact, I have given priority to those works closest to my personal experience, and, consequently, I have given insufficient space to aspects of differential geometry in which I have not participated. In spite of these shortcomings, I hope that my readers, particularly those too young to know the origins of geometric analysis, will be interested to learn how the field looks to someone who was there. I would like to thank comments given by R. Bryant, H.D. Cao, J. Jost, H. Lawson, N.C. Leung, T.J. Li, Peter Li, J. Li, K.F. Liu, D. Phong, D. Stroock, X.W. Wang, S. Scott, S. Wolpert, and S.W. Zhang. I am also grateful to J.X. Fu, especially for his help of tracking down references for the major part of this survey. When Fu went back to China, this task was taken up by P. Peng and X.F. Sun to whom I am grateful also.

In this whole survey, I follow the following:

Basic Philosophy:

Functions, tensors and subvarieties governed by natural differential equations provide deep insight into geometric structures. Information about these objects will give a way to construct a geometric structure. They also provide important information for physics, algebraic geometry and topology. Conversely it is vital to learn ideas from these fields.

Behind such basic philosophy, there are basic invariants to understand how space is twisted. This is provided by Chern classes [149], which appear in every branch of mathematics and theoretical physics. So far we barely understand the analytic meaning of the first Chern class. It will take much more time for geometers to understand the analytic meaning of the higher Chern forms. The analytic expression of Chern classes by differential forms has opened up a new horizon for global geometry. Professor Chern's influence on mathematics is forever.

An old Chinese poem says:

The reeds and rushes are abundant,
and the white dew has yet to dry.
The man whom I admire is on the bank of the river.
I go against the stream in quest of him,
But the way is difficult and turns to the right.
I go down the stream in quest of him,
and Lo! He is on the island in the midst of the water.

May the charm and beauty be always the guiding principle of geometry!

Contents

1. History and contributors of the subject

1.1. Founding fathers of the subject. Since the whole development of geometry depends heavily on the past, we start out with historical developments. The following are samples of work before 1970 which provided fruitful ideas and methods.

- **Fermat's principle of calculus of variation** (Shortest path in various media).
- **Calculus (Newton and Leibnitz)**: Path of bodies governed by law of nature.
- **Euler, Lagrange**: Foundation for the variational principle and the study of partial differential equations. Derivations of equations for fluids and for minimal surfaces.
- **Fourier, Hilbert**: Decomposition of functions into eigenfunctions, spectral analysis.
- **Gauss, Riemann**: Concept of intrinsic geometry.
- **Riemann, Dirichlet, Hilbert**: Solving Dirichlet boundary value problem for harmonic function using variational method.
- **Maxwell**: Electromagnetism, gauge fields, unification of forces.
- **Christoffel, Levi-Civita, Bianchi, Ricci**: Calculus on manifolds.
- **Riemann, Poincaré, Koebe, Teichmüller**: Riemann surface uniformization theory, conformal deformation.
- **Frobenius, Cartan, Poincaré**: Exterior differentiation and Poincaré lemma.
- **Cartan**: Exterior differential system, connections on fiber bundle.
- **Einstein, Hilbert**: Einstein equation and Hilbert action.
- **Dirac**: Spinors, Dirac equation, quantum field theory.
- **Riemann, Hilbert, Poincaré, Klein, Picard, Ahlfors, Beurling, Carlsson**: Application of complex analysis to geometry.
- **Kähler, Hodge**: Kähler metric and Hodge theory.
- **Hilbert, Cohn-Vossen, Lewy, Weyl, Hopf, Pogorelov, Efimov, Nirenberg**: Global surface theory in three space based on analysis.
- **Weierstrass, Riemann, Lebesgue, Courant, Douglas, Radó, Morrey**: Minimal surface theory.
- **Gauss, Green, Poincaré, Schauder, Morrey**: Potential theory, regularity theory for elliptic equations.
- **Weyl, Hodge, Kodaira, de Rham, Milgram-Rosenbloom, Atiyah-Singer**: de Rham-Hodge theory, integral operators, heat equation, spectral theory of elliptic self-adjoint operators.
- **Riemann, Roch, Hirzebruch, Atiyah-Singer**: Riemann-Roch formula and index theory.
- **Pontrjagin, Chern, Allendoerfer-Weil**: Global topological invariants defined by curvature forms.

- **Todd, Pontrjagin, Chern, Hirzebruch, Grothendieck, Atiyah**: Characteristic classes and K-theory in topology and algebraic geometry.
- **Leray, Serre**: Sheaf theory.
- **Bochner-Kodaira**: Vanishing of cohomology groups based on the curvature consideration.
- **Birkhoff, Morse, Bott, Smale**: Critical point theory, global topology, homotopy groups of classical groups.
- **De Giorgi-Nash-Moser**: Regularity theory for the higher dimensional elliptic equation and the parabolic equation of divergence type.
- **Kodaira, Morrey, Grauert, Hua, Hörmander, Bergman, Kohn, Andreotti-Vesentini**: Embedding of complex manifolds, $\bar{\partial}$-Neumann problem, L^2 method, kernel functions.
- **Kodaira-Spencer, Newlander-Nirenberg**: Deformation of geometric structures.
- **Federer-Fleming, Almgren, Allard, Bombieri, De Giorgi, Giusti**: Varifolds and minimal varieties in higher dimensions.
- **Eells-Sampson, Al'ber**: Existence of harmonic maps into manifolds with non-positive curvature.
- **Calabi**: Affine geometry and conjectures on Kähler Einstein metric.

1.2. Modern Contributors. The major contributors can be roughly mentioned in the following periods:

I. 1972 to 1982: M. Atiyah, R. Bott, I. Singer, E. Calabi, L. Nirenberg, A. Pogorelov, R. Schoen, L. Simon, K. Uhlenbeck, S. Donaldson, R. Hamilton, C. Taubes, W. Thurston, E. Stein, C. Fefferman, Y.T. Siu, L. Caffarelli, J. Kohn, S.Y. Cheng, M. Kuranishi, J. Cheeger, D. Gromoll, R. Harvey, H. Lawson, M. Gromov, T. Aubin, V. Patodi, N. Hitchin, V. Guillemin, R. Melrose, Colin de Verdière, M. Taylor, R. Bryant, H. Wu, R. Greene, Peter Li, D. Phong, S. Wolpert, J. Pitts, N. Trudinger, T. Hildebrandt, S. Kobayashi, R. Hardt, J. Spruck, C. Gerhardt, B. White, R. Gulliver, F. Warner, J. Kazdan.

Highlights of the works in this period include a deep understanding of the spectrum of elliptic operators, introduction of self-dual connections for four manifolds, introduction of a geometrization program for three manifolds, an understanding of minimal surface theory, Monge-Ampére equations and the application of the theory to algebraic geometry and general relativity.

II. 1983 to 1992: In 1983, Schoen and I started to give lectures on geometric analysis at the Institute for Advanced Study. J.Q. Zhong took notes on the majority of our lectures. The lectures were continued in 1985 in San Diego. During the period of 1985 and 1986, K.C. Chang and W.Y. Ding came to take notes of some part of our lectures. The book *Lectures on*

Differential Geometry was published in Chinese around 1989 [**606**]. It did have great influence for a generation of Chinese mathematicians to become interested in this subject. At the same time, a large group of my students made contributions to the subject. This includes A. Treibergs, T. Parker, R. Bartnik, S. Bando, L. Saper, M. Stern, H.D. Cao, B. Chow, W.X. Shi, F.Y. Zheng and G. Tian.

At the same time, D. Christodoulou, C.S. Lin, N. Mok, J.Q. Zhong, J. Jost, G. Huisken, D. Jerison, P. Sarnak, T. Ilmanen, C. Croke, D. Stroock, J. Bismut, Price, F. H. Lin, S. Zelditch, S. Klainerman, V. Moncrief, C.L. Terng, Michael Wolf, M. Anderson, C. LeBrun, M. Micallef, J. Moore, K. Fukaya, T. Mabuchi, John Lee, A. Chang, N. Korevaar were making contributions in various directions. One should also mention that in this period important work was done by the authors in the first group. For example, Donaldson, Taubes [**655**] and Uhlenbeck [**688, 689**] did spectacular work on Yang-Mills theory of general manifolds which led Donaldson [**195**] to solve the outstanding question on four manifold topology. Donaldson [**196**], Uhlenbeck-Yau [**691**] proved the existence of Hermitian Yang-Mills connection on stable bundles. Schoen [**590**] solved the Yamabe problem.

III. 1993 to now: Many mathematicians joined the subject. This includes P. Kronheimer, B. Mrowka, J. Demailly, T. Colding, W. Minicozzi, T. Tao, R. Thomas, Zworski, Y. Eliashberg, Toth, Andrews, L.F. Tam, N.C. Leung, Y.B. Ruan, W.D. Ruan, R. Wentworth, A. Grigor'yan, L. Saloff-Coste, J.X. Hong, X.P. Zhu, M. T. Wang, A.K. Liu, K.F. Liu, X.F. Sun, T.J. Li, X.J. Wang, J. Loftin, H. Bray, J.P. Wang, L. Ni, P.F. Guan, N. Kapouleas, P. Ozsváth, Z. Szabó and Y.I. Li. The most important event is of course the major breakthrough of Hamilton [**315**] in 1995 on the Ricci flow. I did propose to him in 1982 to use his flow to solve Thurston's conjecture. Only after this paper by Hamilton, it is finally realized that it is feasible to solve the full geometrization program by geometric analysis. (A key step was the estimates on parabolic equations initiated by Li-Yau [**445**] and accomplished by Hamilton for Ricci flow [**312, 313**].) In 2002, Perelman [**551, 552**] brought in fresh new ideas to solve important steps that remained in the program. Many contributors, including Colding-Minicozzi [**173**], Shioya-Yamaguchi [**616**] and Chen-Zhu [**129**], [**130**] have helped in filling gaps in the arguments of Hamilton-Perelman. Cao-Zhu has just finished a long manuscript which gives the first complete detailed account of the program. The paper appeared in Asian J. Math., **10**(**2**) (2006), 165–492 while the monograph will be published by International Press. In the other direction, we see the important development of Seiberg-Witten theory [**721**]. Taubes [**661, 662, 663, 664**] was able to prove the remarkable theorem for counting pseudo-holomorphic curves in terms of his invariants. Kronheimer-Mrowka [**402**] were able to solve the Thom conjecture that holomorphic curves provide the lowest genus surfaces in representing homology in algebraic surfaces. (Ozsváth-Szabó had a symplectic version [**548**].)

2. Construction of functions in geometry

The following is the basic principle [**737**]:

> **Linear or non-linear analysis is developed to understand the underlying geometric or combinatorial structure. In the process, geometry will provide deeper insight of analysis. An important guideline is that space of special functions defined by the structure of the space can be used to define the structure of this space itself.**

Algebraic geometers have defined the Zariski topology of an algebraic variety using ring of rational functions. In differential geometry, one should extract information about the metric and topology of the manifolds from functions defined over it. Naturally, these functions should be defined either by geometric construction or by differential equations given by the underlying structure of the geometry. (Integral equations have not been used extensively as the idea of linking local geometry to global geometry is more compatible with the ideas of differential equations.) A natural generalization of functions consists of the following: differential forms, spinors, and sections of vector bundles.

The dual concepts of differential forms or sections of vector bundles are submanifolds or foliations. From the differential equations that arise from the variational principle, we have minimal submanifolds or holomorphic cycles. Naturally the properties of such objects or the moduli space of such objects govern the geometry of the underlying manifold. A very good example is Morse theory on the space of loops on a manifold (see [**518**]).

I shall now discuss various methods for constructing functions or tensors of geometric interest.

2.1. Polynomials from ambient space.

If the manifold is isometrically embedded into Euclidean space, a natural class of functions are the restrictions of polynomials from Euclidean space. However, isometric embedding in general is not rigid, and so functions constructed in such a way are usually not too useful.

On the other hand, if a manifold is embedded into Euclidean space in a canonical manner and the geometry of this submanifold is defined by some group of linear transformations of the Euclidean space, the polynomials restricted to the submanifold do play important roles.

2.1.1. *Linear functions being the harmonic function or eigenfunction of the submanifold.* For minimal submanifolds in Euclidean space, the restrictions of linear functions are harmonic functions. Since the sum of the norm square of the gradient of the coordinate functions is equal to one, it is fruitful to construct classical potentials using coordinate functions. This principle was used by Cheng-Li-Yau [**140**] in 1982 to give a comparison theorem for the heat kernel of minimal submanifolds in Euclidean space, sphere and hyperbolic space. Li-Tian [**439**] also considered a similar estimate for complex submanifolds of $\mathbb{C}P^n$. But this follows from [**140**] as such submanifolds can be lifted to a minimal submanifold in S^{2n+1}.

Another very important property of a linear function is that when it is restricted to a minimal hypersurface in a sphere S^{n+1}, it is automatically an eigenfunction. When the hypersurface is embedded, I conjectured that the first eigenfunction is linear and the first eigenvalue of the hypersurface is equal to n (see [**739**]). While this conjecture is not completely solved, the work of Choi-Wang [**155**] gives strong support. They proved that the first eigenvalue has a lower bound depending only on n. Such a result was good enough for Choi-Schoen [**153**] to prove a compactness result for embedded minimal surfaces in S^3.

2.1.2. *Support functions.* An important class of functions that are constructed from the ambient space are the support functions of a hypersurface. These are functions defined on the sphere and are related to the Gauss map of the hypersurface. The famous Minkowski problem reduces to solving some Monge-Ampère equation for such support functions. This was done by Nirenberg [**540**], Pogorelov [**560**], Cheng-Yau [**144**]. The question of prescribed symmetric functions of principal curvatures has been studied by many people: Pogorelov [**564**], Caffarelli-Nirenberg-Spruck [**92**], P.F. Guan and his coauthors (see [**298, 297**]), Gerhardt [**249**], etc. It is not clear whether one can formulate a useful Minkowski problem for higher codimensional submanifolds.

The question of isometric embedding of surfaces into three space can also be written in terms of the Darboux equation for the support function. The major global result is the Weyl embedding theorem for convex surfaces, which was proved by Pogorelov [**561, 562**] and Nirenberg [**540**]. The rigidity part was due to Cohn-Vossen and an important estimate was due to Weyl himself. For local isometric embeddings, there is work by C.S. Lin [**455, 456**], which are followed by Han-Hong-Lin [**318**]. The global problem for surfaces with negative curvature was studied by Hong [**345**]. In all these problems, infinitesimal rigidity plays an important role. Unfortunately they are only well understood for a convex hypersurface. It is intuitively clear that generically, every closed surface is infinitesimally rigid. However, significant works only appeared for very special surfaces. Rado studied the set of surfaces that are obtained by rotating a curve around an axis. The surfaces constructed depend on the height of the curve. It turns out that such surfaces are infinitesimally rigid except on a set of heights which form part of a spectrum of some Sturm-Liouville operator.

2.1.3. *Gradient estimates of natural functions induced from ambient space.* A priori estimates are the basic tools for nonlinear analysis. In general the first step is to control the ellipticity of the problem. In the case of the Minkowski problem, we need to control the Hessian of the support function. For minimal submanifolds and other submanifold problem, we need gradient estimates which we shall discuss in Chapter 4. In 1974 and 1975, S.Y. Cheng and I [**143, 147**] developed several gradient estimates for linear or quadratic polynomials in order to control metrics of submanifolds in Minkowski spacetime or affine space. This kind of idea can be used to deal with many different metric problems in geometry.

The first theorem concerns a spacelike hypersurface M in the Minkowski space $\mathbb{R}^{n,1}$. The following important question arose: Since the metric on $\mathbb{R}^{n,1}$ is $\sum(dx^i)^2 - dt^2$, the restriction of this metric on M need not be complete even though it may be true for the induced Euclidean metric. In order to prove the equivalence of these two concepts for hypersurfaces whose mean curvatures are controlled, Cheng and I proved the gradient estimate of the function

$$\langle X, X\rangle = \sum_i (x^i)^2 - t^2$$

restricted on the hypersurface.

By choosing a coordinate system, the function $\langle X, X\rangle$ can be assumed to be positive and proper on M. For any positive proper function f defined on M, if we prove the following gradient estimate

$$\frac{|\nabla f|}{f} \leq C$$

where C is independent of f, then we can prove the induced metric on M is complete. This is obtained by integrating the inequality to get

$$|\log f(x) - \log f(y)| \leq Cd(x, y)$$

so that when $f(y) \to \infty$, $d(x, y) \to \infty$. Once we knew the metric was complete, we proved the Bernstein theorem which says that maximal spacelike hypersurface must be linear. Such work was then generalized by Treibergs [**685**], C. Gerhardt [**248**] and R. Bartnik [**40**] for hypersurfaces in more general spacetime. (It is still an important problem to understand the behavior of a maximal spacelike hypersurface foliation for general spacetime when we assume the spacetime is evolved by Einstein equation from a nonsingular data set.)

Another important example is the study of affine hypersurfaces M^n in an affine space A^{n+1}. These are the improper affine spheres

$$\det(u_{ij}) = 1$$

where u is a convex function or the hyperbolic affine spheres

$$\det(u_{ij}) = \left(-\frac{1}{u}\right)^{n+2}$$

where u is convex and zero on $\partial\Omega$ and Ω is a convex domain. Note that these equations describe hypersurfaces where the affine normals are either parallel or converge to a point.

For affine geometry, there is an affine invariant metric defined on M which is

$$(\det h_{ij})^{-\frac{1}{n+2}} \sum h_{ij} dx^i dx^j$$

where h_{ij} is the second fundamental form of M. A fundamental question is whether this metric is complete or not.

A coordinate system in A^{n+1} is chosen so that the height function is a proper positive function defined on M. The gradient estimate of the height function gives a way to prove completeness of the affine metric. Cheng and I [**147**] did find such an estimate which is similar to the one given above.

Once completeness of the affine metric is known, it is straight forward to obtain important properties of the affine spheres, some of which were conjectured by Calabi. For example we proved that an improper affine sphere is a paraboloid and that every proper convex cone admits a foliation of hyperbolic affine spheres. The statement about improper affine sphere was first proved by Jörgens [**364**], Calabi [**94**] and Pogorelov [**563**]. Conversely, we also proved that every hyperbolic affine sphere is asymptotic to a convex cone. (The estimate of Cheng-Yau was reproduced again by a Chinese mathematician who claimed to prove the result ten years afterwards.) Much more recently, Trudinger and X. J. Wang [**687**] solved the Bernstein problem for an affine minimal surface, thereby settling a conjecture by Chern. They found a counterexample for dim$\geq$ 10. These results are solid contributions to fourth order elliptic equations.

The argument of using gradient estimates for some naturally defined function was also used by me to prove that the Kähler Einstein metric constructed by Cheng and myself is complete for any bounded pseudo-convex domain [**145**]. (It appeared in my paper with Mok [**526**] who proved the converse statement which says that if the Kähler Einstein metric is complete, the domain is pseudo-convex.)

It should be noted that in most cases, gradient estimates amount to control of ellipticity of the nonlinear elliptic equation.

> **Comment**: To control a metric, find functions that are capable of describing the geometry and give gradient or higher order estimates for these functions.

2.2. Geometric construction of functions.

2.2.1. *Distance function and Busemann function.* When manifolds cannot be embedded into the linear space, we construct functions adapted to the metric structure. Obviously the distance function is the first major function to be constructed. A very important property of the distance function is that when the Ricci curvature of the manifold is greater than the Ricci curvature of a model manifold which is spherical symmetric at one point, the Laplacian of the distance function is not greater than the Laplacian of the distance function of the model manifold in the sense of distribution. This fact was used by Cheeger-Yau [**124**] to give a sharp lower estimate of the heat kernel of such manifolds. An argument of this type was also used by Perelman in his recent work.

Gromov [**284**] developed a remarkable Morse theory for the distance function (a preliminary version was developed by K. Grove and K. Shiohama [**295**]) to compare the topology of a geodesic ball to that of a large ball, thereby obtaining a bound on the Betti numbers of compact manifolds with nonnegative sectional curvature. (He can also allow the manifolds to have negative curvature. But in this case the diameter and the lower bound of the curvature will enter into the estimate.)

We can also take the distance function from a hypersurface and compute the Hessian of the distance function. In general, one can prove comparison theorems, and the principle curvatures of the hypersurface will come into the estimates. However, the upper bound of the Laplacian of the function depends only on the Ricci curvature of the ambient manifold and the mean curvature of the hypersurface. This kind of calculation was used in the sixties by Penrose and Hawking to study the focal locus of a closed surface under the assumption that the surface is "trapped," which means the mean curvatures are negative in both the ingoing and the outgoing null directions. This information allowed them to prove the first singularity theorem in general relativity (which demonstrates that the black hole singularity is stable under perturbation). The distance to hypersurfaces can be used as barrier functions to prove the existence of a minimal surface as was shown by Meeks-Yau [**507**], [**508**]. T. Frankel used the idea of minimizing the distance between two submanifolds to detect the topology of minimal surfaces. In particular, two maximal spacelike hypersurfaces in spacetime which satisfy the energy condition must be disjoint if they are parallel at infinity.

Out of the distance function, we can construct the Busemann function in the following way:

Given a geodesic ray $\gamma : [0,\infty) \to M$ *so that*

$$\textit{distance}(\gamma(t_1),\gamma(t_2)) = t_2 - t_1,$$

where $\| \frac{d\gamma}{dt} \| = 1$, *one defines*

$$B_\gamma(x) = \varliminf_{t\to\infty} (d(x,\gamma(t)) - t).$$

This function generalizes the notion of a linear function. For a hyperbolic space form, its level set defines horospheres. For manifolds with positive curvature, it is concave. Cohn-Vossen (for surface) and Gromoll-Meyer [**279**] used it to prove that a complete noncompact manifold with positive curvature is diffeomorphic to $\mathbb{R}^n$.

A very important property of the Busemann function is that it is superharmonic on complete manifolds with nonnegative Ricci curvature in the sense of distribution. This is the key to prove the splitting principle of Cheeger-Gromoll [**119**]. Various versions of this splitting principle have been important for applications to the structure of manifolds. When I [**736**] proved the Calabi conjecture, the splitting principle was used by me and others to prove the structure theorem for Kähler manifolds with a nonnegative first Chern Class. (The argument for the structure theorem is due to Calabi [**93**] who knew how to handle the first Betti number. Kobayashi [**387**] and Michelsohn [**516**] wrote up the formal argument and Beauville [**45**] had a survey article on this development.)

In 1974, I was able to use the Busemann function to estimate the volume of complete manifolds with nonnegative Ricci curvature [**734**]. After long discussions with me, Gromov [**285**] realized that my argument of Busemann function amounts to compare volumes of geodesic balls. The comparison theorem of Bishop-Gromov had been used extensively in metric geometry.

If we consider $\inf_\gamma B_\gamma$, where γ ranges from all geodesic rays from a point on the manifold, we may be able to obtain a proper exhaustion of the manifold. When M is a complete manifold with finite volume and its curvature is pinched by two negative constants, Siu and I [**634**] did prove that such a function gives a concave exhaustion of the manifold. If the manifold is also Kähler, we were able to prove that one can compactify the manifold by adding a point to each end to form a compact complex variety. In the other direction, Schoen-Yau [**603**] was able to use the Busemann function to construct a barrier for the existence of minimal surfaces to prove that any complete three dimensional manifold with positive Ricci curvature is diffeomorphic to Euclidean space.

The Busemann function also gives a way to detect the angular structure at infinity of the manifold. It can be used to construct the Poisson kernel of hyperbolic space form. For a simply connected complete manifold with bounded and strongly negative curvature, it is used as a barrier to solve the Dirichlet problem for bounded harmonic functions, after they are mollified at infinity. This was achieved by Sullivan [**647**] and Anderson [**8**]. Schoen and Anderson [**9**] obtained the Harnack inequality for a bounded harmonic function and identified the Martin boundary of such manifolds. W. Ballmann [**27**] then studied the Dirichlet problem for manifolds of non-positive curvature. Schoen and I [**606**] conjectured that nontrivial bounded harmonic function exists if the manifold has bounded geometry and a positive first eigenvalue. Many important cases were settled in [**606**]. Lyons-Sullivan [**487**] proved the existence of nontrivial bounded harmonic functions using the non-amenability of groups acting on the manifold.

The abundance of bounded harmonic functions on the universal cover of a compact manifold should mean that the manifold is "hyperbolic". Hence if the Dirichlet problem is solvable on the universal cover, one expects the Gromov volume of the manifold to be greater than zero.

The Martin boundary was studied by L. Ji and MacPherson (see [**303, 361**]) for the compactification of various symmetric spaces. For product of manifolds with negative curvature, it was determined by A. Freire [**232**]. For rank one complete manifolds with non-positive curvature, work has been done by Ballman-Ledrappier [**28**] and Cao-Fan-Ledrappier [**105**]. It should be nice to generalize the work of L.K. Hua on symmetric spaces with higher rank to general manifolds with non-positive curvature. Hua found that bounded harmonic functions satisfy extra equations (see [**348**]).

2.2.2. *Length function defined on loop space.* If we look at the space of loops in a manifold, we can take the length of each loop and thereby define a natural function on the space of loops. This is a function for which Morse theory found rich application. Bott [**71**] made use of it to prove his periodicity theorem. Bott [**68, 70**] and Morse also developed a formula for computing index of a geodesic. Bott showed that the index of a closed geodesic and its linearized Poincaré map determine the indexes of iterates of this geodesic. Starting from the famous works of Poincaré, Birkhoff, Morse and Ljusternik-Shnirel'man, there has been extensive work on proving the existence of a closed geodesic using Morse theory on the space of loops. Klingenberg and his students developed powerful tools (see [**386**]). Gromoll-Meyer [**280**] did important work in which they proved the existence of infinitely many closed geodesics assuming the Betti number of the free loop space of the manifold grows unboundedly. They used the results of Bott [**70**], Serre and some version of degenerate Morse theory. There was also later work by Ballmann, Ziller, G. Thorbergsson, Hingston and Kramer (see, e.g., [**30, 328, 401**]), who improved the Gromoll-Meyer theorem to give a low estimate of the growth of the number of geometrically distinct closed geodesics of length $\leq t$. In most cases, they grow at least as fast as the prime numbers. The classical important question that every metric on S^2 supports an infinite number of closed geodesics was also solved affirmatively by Franks [**228**], Bangert [**35**] and Hingston [**329**]. An important achievement was made by Vigué-Poirrier and D. Sullivan [**695**] who proved that the Gromoll-Meyer condition for the existence of infinite numbers of closed geodesics is satisfied if and only if the rational cohomology algebra of the manifold has at least two generators. They made use of Sullivan's theory of the rational homotopic type. When the metric is Finsler, the most recent work of Victor Bangert and Yiming Long [**36**] showed the existence of two closed geodesics on the two dimensional sphere. (Katok [**377**] produced an example which shows that two is optimal.) Length function is a natural concept in Finsler geometry. In the last fifty years, Finsler geometry has not been popular in western world. But under the leadership of Chern, David Bao, Z. Shen, X. H. Mo and M. Ji did develop Finsler geometry much further (see, e.g., [**37**]).

A special class of manifolds, all of whose geodesics are closed, has occupied quite a lot of interest of distinguished geometers. It started from the work of Zoll (1903) for surfaces where Guillemin did important contributions. Bott [**69**] has determined the cohomology ring of these manifolds. The well known Blaschke conjecture was proved by L. Green [**270**] for two dimension and by M. Berger and J. Kazdan (see [**51**]) for higher dimensional spheres. Weinstein [**713**] and C.T. Yang [**727, 728, 729**] made important contributions to the conjecture for other homotopic types.

2.2.3. *Displacement functions.* When the manifold has negative curvature, the length function of curves is related to the displacement function defined in the following way:

If γ is an element of the fundamental group acting on the universal cover of a complete manifold with non-positive curvature, we consider the function $d(x, \gamma(x))$: The study of such a function gives rise to properties of compact manifolds with non-positive curvature. For example, in my thesis, I generalized the Preissmann theorem to the effect that every solvable subgroup of the fundamental group must be a finite extension of an Abelian group which is the fundamental group of a totally geodesic flat sub-torus [**731**]. Gromoll-Wolf [**281**] and Lawson-Yau [**412**] also proved that if the fundamental group of such a manifold has no center and splits as a product, then the manifold splits as a metric product. Strong rigidity result for a discrete group acting on product of manifolds irreducibly was obtained by Jost-Yau [**372**] where they proved that these manifolds are homogeneous if the discrete group also appears as fundamental group of compact manifolds with nonpositive curvature.

When the manifold has bounded curvature, Margulis studied those points where $d(x, \gamma(x))$ is small and proved the famous Margulis lemma which was used extensively by Gromov [**282**] to study the structure of manifolds with non-positive curvature.

Comment: The lower bound of sectional curvature (or Ricci curvature) of a manifold gives upper estimate of the Hessian (or the Laplacian) of the distance functions. Since most functions constructed in geometry come from distance functions, we have partial control of the Hessian of these functions. The information provides us with basic tools to construct barrier functions for harmonic analysis or to produce convex functions. The Hessian of distance functions come from computations of second variation of geodesics. If we consider the second variation of closed geodesic loops, we get information about the Morse index of the loop, which enable us to link global topology to the existence of many closed geodesics or curvatures of the manifold.

We always look for canonical objects through geometric constructions and deform them to find their global properties.

2.3. Functions and tensors defined by linear differential equations. Direct construction of functions or tensors based on geometric intuitions alone is not rich enough to handle the very complicated geometric world. One should produce global geometric objects based on global differential equations. Often the construction depends on the maximal principle, integration by part, or the method of contradictions, and they are not necessarily geometric intuitive. On the other hand, the basic principle of global differential equations does fit well with modern geometry in relating local data to global behavior. In order for the theory to be effective, the global differential operator has to be constructed from a geometric structure naturally.

The key to understanding any self-adjoint linear elliptic differential operator is to understand its spectral resolution and the detail of the structure of objects in the process of the resolution: eigenvalues or eigenfunctions are particularly important for their relation to geometry. Low eigenvalues and low eigenfunctions give deep information about global geometry such as topology or isoperimetric inequalities. High eigenvalues and high eigenfunctions are related to local geometry such as curvature forms or characteristic forms. Semiclassical analysis in quantum physics give a way to relate these two ends. This results in using either the heat equation or the hyperbolic equation.

There are many important first order differential operators: d, δ, $\bar{\partial}$, Dirac operator. All these operators have contributed to a deeper understanding of geometry. They form systems of equations. Our understanding of them is not as deep as our understanding of the Laplacian acting on functions. The future of geometry will rest on an understanding of global systems of equations and their relation to global topology. The index theorem has given many important contributions as it provides significant information about the dimension of the kernel (or cokernel). However, a deeper understanding of the spectrum of these operators is still needed.

2.3.1. *Laplacian.*

(a). *Harmonic functions.* The spectral resolution of the Laplacian gives rise to eigenfunctions. Harmonic functions are therefore the simplest functions that play important roles in geometry.

If the manifold is compact, the maximum principle shows that harmonic functions are constant. However, when we try to understand the singularities of compact manifolds, we may create noncompact manifolds by scaling and blowing up processes, at which point harmonic functions can play an important role.

The first important question about harmonic functions on a complete manifold is the Liouville theorem. I started my research on analysis by understanding the right formulation of the Liouville theorem. In 1971, I thought that it is natural to prove that for complete manifolds with a nonnegative Ricci curvature, there is no nontrivial harmonic function [**732**]. I also thought that in the opposite case, when a complete manifold has strongly negative curvature and is simply connected, one should be able to solve Dirichlet problem for bounded harmonic functions.

The gradient estimates [**732**] that I derived for a positive harmonic function come from a suitable interpretation of the Schwarz lemma in complex analysis. In fact, I generalized the Ahlfors Schwarz lemma before I understood how to work out the gradient estimates for harmonic functions. The generalized Schwarz lemma [**738**] says that holomorphic maps, from a complete Kähler manifold with Ricci curvature bounded from below to a Hermitian manifold with holomorphic bisectional curvature bounded from above by a negative constant, are distance decreasing with constants depending only on the bound on the curvature. This generalization has since found many applications such as the study of the geometry of moduli spaces by Liu-Sun-Yau [**471, 472**]. They used it to prove the equivalence of the Bergman metric with the Kähler-Einstein metric on the moduli space. They also proved that these metrics are equivalent to the Teichmüller metric and the McMullen metric.

The classical Liouville theorem has a natural generalization: Polynomial growth harmonic functions are in fact polynomials. Motivated by this fact and several complex variables, I asked whether the space of polynomial growth harmonic functions with a fixed growth rate is finite dimension with the upper bound depending only on the growth rate [**741**]. This was proved by Colding-Minicozzi [**168**] and generalized by Peter Li [**437**]. (Functions can be replaced by sections of bundles). In a beautiful series of papers (see, e.g., [**440, 441**]), P. Li and J.P. Wang studied the space of harmonic functions in relation to the geometry of manifolds. In the case when harmonic functions are holomorphic, they form a ring. I am curious about the structure of this ring. In particular, is it finitely generated when the manifold is complete and has a nonnegative Ricci curvature? A natural generalization of such a question is to consider holomorphic sections of line bundles, especially powers of canonical line bundles. This is part of Mori's minimal model program.

(b). *Eigenvalues and eigenfunctions.* Eigenvalues reflect the geometry of manifolds very precisely. For domains, estimates of them date back to Lord Rayleigh. Hermann Weyl [**711**] solved a problem of Lorentz's on the asymptotic behavior of eigenvalues in relation to the volume of the domain and hence initiated a new subject of spectral geometry. Pólya-Szegö, Faber, Krahn and Levy gave estimates of eigenvalues of various geometric problems. On a general manifold, Cheeger [**114**] was the first person to relate a lower estimate of the first eigenvalue with the isoperimetric constant (now called the Cheeger constant). One may note that many questions on the eigenvalue for domains are still unsolved. The most noted one is the Pólya conjecture which gave a sharp lower estimate of the Dirichlet problem in terms of volume. Li-Yau [**444**] did settle the average version of the Pólya conjecture.

The gradient estimate that I found for harmonic functions can be generalized to cover eigenfunctions and Peter Li [**436**] was the first one to apply it to finding estimates for eigenvalues for manifolds with positive Ricci curvature. (If the Ricci curvature has a positive lower bound, this is due to Lichnerowicz.) Li-Yau [**442**] then solved the well-known problem of estimating eigenvalues of manifolds in terms of their diameter and the lower bound on their Ricci curvature. Li-Yau conjectured the sharp constant for their estimates, and Zhong-Yang [**755**] were able to prove this conjecture by sharpening Li and Yau's arguments. A probabilistic argument was later developed by Chen and Wang [**133**] to derive these inequalities. The precise upper bound for the eigenvalue was first obtained by S. Y. Cheng [**137**] also in terms of diameter and lower bound of the Ricci curvature. Cheng's theorem gives a very good demonstration of how the analysis of functions provides information about geometry. As a corollary of his theorem, he proved that if a compact manifold M^n has a Ricci curvature $\geq n-1$ and the diameter is equal to π, then the manifold is isometric to the sphere. He used a lower estimate for eigenvalues based on the work of Lichnerowicz and Obata. Colding [**167**] was able to use functions with properties close to those of the first eigenfunction to prove a pinching theorem which states that: When the Ricci curvature is bounded below by $n-1$ and the volume is close to that of the unit sphere, the manifold is diffeomorphic to the sphere. There is extensive work by Colding-Cheeger [**116, 117, 118**] and Perelman (see, e.g., [**88**]) devoted to the understanding of Gromov's theory of Hausdorff convergence for manifolds. The tools they used include the comparison theorem, the splitting theorem of Cheeger and Gromoll, and the ideas introduced earlier by Colding.

A very precise estimate of eigenvalues of the Laplacian has been important in many areas of mathematics. For example, the idea of Szegö [**651**]-Hersch [**327**] on the upper bound of the first eigenvalue in terms of the area alone was generalized by me to the higher genus in joint works with P. Yang [**730**] and P. Li [**443**]. For genus one, this was Berger's conjecture, as I was informed by Cheng. After Cheng showed me the paper of Hersch, I realized how to create trial functions by taking the branched conformal cover of S^2. While the constant in the paper of Yang-Yau [**730**] for torus is not the best possible, the recent work of Jakobson, Levitin, Nadirashvili, Nigam and Polterovich [**358**] demonstrated that the constant for a genus two surface may be the best possible and may be achieved by Bolza's surface.

Shortly afterwards, I applied the argument of [**730**] to prove that a Riemann surface defined by an arithmetic group must have a relative high degree when it is branched over the sphere. This observation of using Selberg's estimate coupled with Li-Yau [**443**] was made in 1985 when I was in San Diego, where I also used similar idea to estimate genus of mini-max surface in three dimensional manifolds and also to prove positivity of Hawking mass. After I arrived in Harvard, I discussed the idea with my colleague N. Elkies and B. Mazur. The paper was finally written up and published in 1995 [**745**]. In the meanwhile, ideas of using my work on eigenvalue coupled with Selberg's work to study congruence subgroup was generalized by D. Abramovich [**1**] (my idea was conveyed by Elkies to him) and by P. Zograf [**758**] to the case where the curve has cusps. Most recently Ian Agol [**2**] also used a similar idea to study arithmetic Kleinian reflection groups.

In a beautiful article, N. Korevaar [**397**] gave an upper bound, depending only on genus and n, for the n-th eigenvalue λ_n of a Riemann surface. His result answered a challenge of mine (see [**739**]) when I met him in Utah in 1989. Grigor'yan, Netrusov and I [**278**] were able to give a simplified proof and apply the estimate to bound the index of minimal surfaces. There are also works by P. Sarnak (see, e.g., [**586, 357**]) on understanding eigenfunctions for such Riemann surfaces. Iwaniec-Sarnack [**357**] showed that the estimate of the maximum norm of the n-th eigenfunction on an arithmetic surface has significant interest in number theory. Wolpert [**725**] analyzes perturbation stability of embedded eigenvalues and applies asymptotic perturbation theory and harmonic map theory to show that stability is equivalent to the non-vanishing of certain standard quantities in number theory. There was also the work of Schoen-Wolpert-Yau [**595**] on the behavior of eigenvalues $\lambda_1, \cdots, \lambda_{2g-3}$ for a compact Riemann surface of genus g. These are eigenvalues that may tend to zero for metrics with curvature -1. However, $\lambda_{2g-2}, \lambda_{2g-1}, \cdots, \lambda_{4g-1}$ always appear in $[c_g, \frac{1}{4}]$ where $c_g > 0$ depends only on g. It will be nice to find the optimal c_g.

In this regard, one may mention the very deep problem of Selberg on lower estimate of λ_1 for surfaces defined by an arithmetic group. Selberg proved that it is greater than $\frac{3}{16}$ and it was later improved by Luo-Rudnick-Sarnak [**482**]. For a higher dimensional locally symmetric space, there is a similar question of Selberg and results similar to Selberg's were found by J.S. Li [**425**] and Cogdell-Li-Piatetski-Shapiro-Sarnak [**166**]. Many researchers attempt to use Kazdhan's property T for discrete groups to study Selberg's problem.

There are many important properties of eigenfunctions that were studied in the seventies. For example, Cheng [**138**] found a beautiful estimate of multiplicities of eigenvalues of Riemann surfaces based only on genus. The idea was used by Colin de Verdière [**175**] to embedded graphes into $\mathbb{R}^3$ when they satisfy nice combinatorial properties. The connectivity and the topology of nodal domains are very interesting questions. Melas [**510**] did prove that for a convex planar domain, the nodal line of second eigenfunctions must intersect the boundary in exactly two points. Very little is known about the number of nodal domains except the famous theorem of Courant that the number of nodal domains of the m-th eigenfunction is less than m.

There are several important questions related to the size of nodal sets and the number of critical points of eigenfunctions. I made a conjecture (see [**739**]) about the area of nodal sets, and significant progress toward its resolution was made by Donnelly-Fefferman [**207**], Dong [**206**] and F.H. Lin [**458**]. The number of critical points of an eigenfunction is difficult to determine. I [**746**] managed to prove the existence of a critical point near the nodal set. Jakobson and Nadirashvili [**359**] gave a counterexample to my conjecture that the number of critical points of the n-th eigenfunction is unbounded when n tends to infinity. I believe the conjecture is true for generic metrics and deserves to be studied extensively. Nadirashvili and his coauthors [**344, 319**] were also the first to show that the critical sets of eigenfunctions in n-dimensional manifold have a finite H^{n-2}-Hausdorff measure. Afterwards, Han-Hardt-Lin [**317**] gave an explicit estimate.

When there is potential V, the eigenvalues of $-\triangle+V$ are also important. When V is convex, with Singer, Wong and Stephen Yau, I applied the argument that I had with Peter Li to estimate the gap $\lambda_2-\lambda_1$ [**624**]. When V is arbitrary, I [**747**] observed how this gap depends on the lower eigenvalue of the Hessian of $-\log\psi$, where ψ is the ground state. The method of continuity was used by me in 1980 to reprove the work of Brascamp-Lieb [**79**] on the convexity of $-\log\psi$ when V is convex. (This work appeared in the appendix of [**624**].) When V is the scalar curvature, this was studied by Schoen and myself extensively. In fact, in [**604**], we found an upper estimate of the first Dirichlet eigenvalue of the operator $-\triangle+\frac{1}{2}R$ in terms of $\frac{3\pi^2}{2r^2}$ where r is a certain concept of radius related to loops in a three dimensional manifold. (If we replace loops with higher dimensional spheres, one can define a similar concept of radius. It will be nice if such a concept can be related to eigenvalues of differential forms.) This operator is naturally related to conformal deformation, stability of minimal surfaces, etc. (The works of D. Fischer-Colbrie and Schoen [**223**], Micallef [**512**], Schoen-Yau [**597, 603**] on stable minimal surfaces all depend on an understanding of spectrum of this operator.) The parabolic version appears in the recent work of Perelman.

If there is a closed non-degenerate elliptic geodesic in the manifold, Babič [**25**], Guillemin and Weinstein [**302**] found a sequence of eigenvalues of the Laplacian which can be expressed in terms of the length, the rotation angles and the Morse index of the geodesic.

> **Comment**: It is important to understand how harmonic functions or eigenfunctions oscillate. Gradient estimate is a good tool to achieve this. Gradient estimate for the log of the eigenfunction can be used to prove the Louville theorem or give a good estimate of eigenvalues. For higher eigenfunctions, it is important to understand its zero set and its growth. By controlling this information, one can estimate the dimension of these functions. A good upper estimate for eigenvalues depends on geometric intuition which may lead to construction of trial functions that are more adaptive to geometry. It should be emphasized that a clean and sharp estimate for the linear operator is key to obtaining good estimates for the nonlinear operator.

(c). *Heat kernel.* Most of the work on the heat kernel over Euclidean space can be generalized to those manifolds where Sobolev and Poincaré inequalities hold. (It should be noted that Aubin [**22, 24**] and Talenti [**652**] did find the best constant for various Sobolev inequalities on Euclidean space.) These inequalities are all related to isoperimetric inequalities. C. Croke [**178**] was able to follow my work [**733**] on Poincaré inequalities to prove the Sobolev inequality depending only on volume, diameter and the lower bound of Ricci curvature. Arguments of John Nash were then used by Cheng-Li-Yau [**139**] to give estimates of the heat kernel and its higher derivatives. In this paper, an estimate of the injectivity radius was derived and this estimate turns out to play a role in Hamilton's theory of Ricci flow. A year later, Cheeger-Gromov-Taylor [**122**] made use of the wave kernel to reprove this estimate. In other direction, D. Stroock (see [**538**]) used Malliavin's calculus to give remarkable estimates for the heat kernel at a pair of points where one point is at the cut locus of another point.

The estimate of the heat kernel was later generalized by Davies [**185, 186**], Saloff-Coste [**582**] and Grigor'yan [**276, 277**] to complete manifolds with polynomial volume growth and volume doubling property. Since these are quasi-isometric invariants, their analysis can be applied to graphs or discrete groups. See Grigor'yan's survey [**277**] and Saloff-Coste's survey [**583**].

On the other hand, the original gradient estimate that I derived is a pointwise inequality that is much more adaptable to nonlinear theory. Peter Li and I [**445**] were able to find a parabolic version of it in 1984. We observed its significance for estimates on the heat kernel and its relation to the variational principle for paths in spacetime. Coupled with the work of Cheeger-Yau [**124**], this gives a very precise estimate of the heat kernel. Such ideas turn out to provide fundamental estimates which are crucial for the analysis of Hamilton's Ricci flow [**312, 313**].

Not much is known about the heat kernel on differential forms or differential forms with twisted coefficients. The fundamental idea of using the heat equation to prove the Hodge theory came from Milgram-Rosenbloom. The heat kernel for differential forms with twisted coefficients does play an important role in the analytic proof of the index theorem, as was demonstrated by Atiyah-Bott-Patodi [**13**]. It is the alternating sum that exhibits cancellations and gives rise to index of elliptic operators. When t is small, the alternating sum reduces to a calculation of curvature forms. When t is large, it gives global information on harmonic forms. Since the index of the operator is independent of t, we can relate the index to characteristic forms.

If a compact manifold is the quotient of a non-compact manifold by a discrete group and if the heat kernel of the non-compact manifold can be computed explicitly, it can be averaged to give the heat kernel of the quotient manifold. Since the integral of the later kernel on the diagonal can be computed by the spectrum to be $\sum e^{-t\lambda_i}$, one can relate the displacement function of the discrete group to the spectrum. This is the Selberg trace formula relating length of closed geodesics to the spectrum of the Laplacian.

Heat kernel converges to delta function when t approaches zero. This property was used by Kefeng Liu [**469, 470**] in an elegant way to obtain various localization formulas on the moduli space of bundles. Liu's idea was used later by Bismut to treat the formula of E. Verlinde [**54**].

> **Comment**: Understanding the heat kernel is almost the same as understanding the heat equation. However, heat kernel satisfies semi-group properties, which enables one to give a good estimate of the maximum norm or higher order derivative norms as long as the Sobolev inequality can be proved. It is useful to look at the heat equation in spacetime where the Li-Yau gradient estimate is naturally defined. The estimate provides special pathes in spacetime for the estimate of the kernel. However, the effects of closed geodesics have not been found in the heat equation approach. A sharp improvement of the Li-Yau estimate may lead to such information.

(d). *Isoperimetric inequalities.* Isoperimetric inequality is a beautiful subject. It has a long history. Besides its relation to eigenvalues, it reviews the deep structure of manifolds. The best constant for the inequality is important. Pólya-Szegö [**565**], G. Faber (1923), E. Krahn (1925) and P. Lévy (1951) made fundamental contributions. Gromov generalized the idea of Lévy to obtain a good estimate of Cheeger's constant (see [**289**]). C. Croke [**179**] and Cao-Escobar [**104**] have worked on domains in a simply connected manifold with non-positive curvature. It is assumed that the inequality holds for any minimal subvariety in Euclidean space. But it is not known to be true for the best constant. Li-Schoen-Yau [**438**] did prove it in the case of a minimal surface with a connected boundary, and E. Lutwak, Deane Yang and G.Y. Zhang did some beautiful work in the affine geometry case (see, e.g., [**484, 485**]). In Hamilton's proof of Ricci flow convergent to the round metric on S^2, he demonstrated that the isoperimetric constant of the metric is improving and geometry of the manifold is controlled.

> **Comment**: The variational principle has been the most important method in geometry since the Greek mathematicians. Fixing the area of the domain and minimizing the length of the boundary is the most classical form of isoperimetric inequality. This principle has been generalized to much more general situations in geometry and mathematical physics. In most cases, one tries to prove existence of the extremal object and establish isoperimetric inequalities by calculating corresponding quantities for the extremal object. There is also the idea of rearrangement or symmetrization to prove isoperimetric inequalities. In the other direction, there is the duality

principle in the calculus of variation: instead of minimizing the length of the boundary, one can fix it and maximize the area it encloses. The principle can be effective in complicated variational problems.

(e). *Harmonic analysis on discrete geometry.* There are many other ideas in geometric analysis that can be discretized and applied to graph theory. This is especially true for the theory of spectrum of graphs. Some of these were carried out by F.Chung, Grigor'yan and myself (see the reference of Chung's survey [**164**]). But the results in [**164**] are far away from being exhaustive. On the other hand, Margulis [**491**] and Lubotzky-Phillips-Sarnak [**480**] were able to make use of discrete group and number theory to construct expanding graphs. Methods to construct and classify these expanding graphs are important for application in computer science. It should be noted that Kazhdan's property (T) [**380**] did play an important role in such discussions. It is also important to see how to give a good decomposition of any graph using the spectral method.

The most important work for the geometry of a finitely presented group was done by Gromov [**284**]. He proved the fundamental structure theorem of groups where volume grows at most polynomially. These groups must be virtually nilpotent. Geometric ideas were developed by Varopoulos and his coauthors [**693, 38**] on the precise behaviors of the heat kernel in terms of volume growth. As an application of the theory of amenable groups, R. Brooks [**81**] was able to prove that if a manifold covers a compact set by a discrete group Γ, then it has positive eigenvalue if and only if Γ is non-amenable.

Gromov [**283**] also developed a rich theory of hyperbolic groups using concepts of isoperimetric type inequalities. It would be nice to characterize these groups that are fundamental groups of compact manifolds with non-positive curvature or locally symmetric spaces.

Comment: The geometry of a graph or complex can be used as a good testing ground for geometric ideas. They can be important in understanding smooth geometric structures. Many rough geometric concepts, such as isoperimetric inequalities, can be found on graphs, and in fact they play some roles in computer network theory. On the other hand, many natural geometric concepts should be generalized to graphs: for example, the concept of the fiber bundle, bundle theory over graphs and harmonic forms. It is likely that one needs to have a good way to define the concept of equivalence between such objects. When we approximate a smooth manifold by a graph or complex, we only care about the limiting object and therefore some equivalence relations should be allowed. In the case of Cayley graph of a finitely generated group, it depends on the choice of the generating set, and properties independent of this generating set are preferable if we are only interested in the group itself. In the other direction, computer networks and other practical subjects have independent interest in graph theory. A close collaboration between geometer and computer scientists would be fruitful.

(f). *Harmonic analysis via hyperbolic operators.* There are important works of Fefferman, Phong, Lieb, Duistermaat, Guillemin, Melrose, Colin de Verdier, Taylor, Toth, Zelditch and Sarnak on understanding the spectrum of the Laplacian from the point of view of semi-classical analysis (see, e.g. [**221, 210, 325, 587**]). Some of their ideas can be traced back to the geometric optics analysis of J. Keller. The fundamental work of Duistermaat and Hörmander [**209**] on propagation of singularities was also used extensively. There has been a lot of progress on the very difficult question of determining when one "Can hear the shape of a drum" by, among others, Melrose (see [**511**]), Guillemin [**299**] and Zelditch [**753**]. (Priori to this, Guillemin and Kazhdan [**300**] proved that no negatively curved closed surface can be isospectrally deformed.) The first counterexample for closed manifolds was given by J. Milnor [**519**] on a 16 dimensional torus. The idea was generalized by Sunada [**649**], Gordon-Wilson [**263**]. For domains in Euclidean spaces, there were examples by Urakawa in three dimensions. Two dimensional counterexamples were given by Gordon-Webb-Wolpert [**262**], Wilson and Szabó [**650**]. Most of the ideas for counterexamples are related to the Selberg trace formula discussed in the section of heat kernel. The semi-classical analysis based on the hyperbolic operator also gives a very precise estimate or relation between the geodesic and the spectrum. The support of the singularities of the trace of the wave kernel $\sum e^{\sqrt{-1}t\sqrt{\lambda_i}}$ is a subset of the set of the lengths of closed geodesics. It is difficult to achieve such results by elliptic theory. However, most results are asymptotic in nature. It will be remarkable if both methods can be combined.

Comment: Fourier expansion has been a very powerful tool in analysis and geometry. Practically any general theorem in classical Fourier analysis should have a counterpart in analysis of the spectrum of the Laplacian. The theory of geometric optics and the propagation of a singularity gives deep understanding of the singularity of a wave kernel. Geodesic and closed geodesic becomes an important means to understand eigenvalues. However, the theory has not been fruitful for the Laplacian acting on differential forms. Should areas of minimal submanifolds play a role? In the case of Kähler manifolds, holomorphic cycles or the volume of special Lagrangian cycles should be important, as the length of close geodesics appear in the exponential decay term of the heat kernel. It would be useful to sharpen the heat equation method to capture this lower order information.

(g). *Harmonic forms.* Natural generalizations of harmonic or holomorphic functions are harmonic or holomorphic sections of bundles with connections. The most important bundles are the exterior power of cotangent bundles. Using the Levi-Civita connection, harmonic sections are harmonic forms which, by the theory of de Rham and Hodge, give canonical representation of cohomology classes. The major research on harmonic forms comes from Bochner's vanishing theorem [**59**]. But our understanding is still poor except for 1-forms or when the curvature operator is positive, in which case the Bochner argument proved the manifold to be a homology sphere. If there is any nontrivial operator which commutes with the Laplacian, the eigenforms split accordingly. Making use of special structures of such splitting, the Bochner method can be more effective. For example, when the manifold is Kähler, differential forms can be decomposed further to (p, q)-forms and the Kodaira vanishing theorem [**390**] yields much more powerful information, when the (p, q) forms are twisted with a line bundle or vector bundles. Similar arguments can be applied to manifolds with a special holonomy group depending on the representation theory of the holonomy group. When the complex structure moves holomorphically, the subbundles of (p, q) forms in the bundle of $(p + q)$ forms do not necessarily deform holomorphically. The concept of Hodge filtration is therefore introduced. When we deform the complex structure around a point where the complex structure degenerates, there is a monodromy group acting on the Hodge filtration. The works of Griffiths-Schmid [**275**] and Schmid's $SL_2(\mathbb{R})$ theorem [**588**] give powerful control on the degeneration of the Hodge structure. Deligne's theory of mixed Hodge structure [**187**] plays a fundamental role for studying singular algebraic varieties. The theory of variation of Hodge structures is closely related to the study of period of the differential forms. This theory also appears in the subject of mirror symmetry. It is desirable to give a precise generalization of these works to higher dimensional moduli spaces where Kaplan-Cattani-Schmid made important contributions.

Harmonic forms give canonical representation to de Rham cohomology. However, the wedge product of harmonic forms need not be harmonic. The obstruction comes from secondary cohomology cooperation. K.T. Chen [**132**] studied the case of 1-forms and Sullivan [**646**] studied the general case and gave a minimal model theory for a rational homotopic type of a manifold. Using $\partial\bar{\partial}$-lemma of Kähler manifolds, Deligne-Griffiths-Morgan-Sullivan [**188**] showed that the rational homotopic type is formal for Kähler manifolds.

The importance of harmonic forms is that they give canonical representation to the de Rham cohomology which is isomorphic to singular cohomology over real numbers. It gives a powerful tool to relate local geometry to global topology. In fact the vanishing theorem of Bochner-Kodaira-Lichnerowicz allows one to deduce from sign of curvature to vanishing of cohomology.

This has been one of the most powerful tools in geometry in the past fifty years.

The idea of harmonic forms came from fluid dynamics and Maxwell equations. The non-Abelian version is the Yang-Mills theory. Most of the works on Yang-Mills theory have been focused on these gauge fields where the absolute minimum is achieved by some (topological) characteristic number. (These are called BPS state in physics literature.) When the dimension of the manifold is four, the star operator maps two form to two form and it makes sense to require the curvature form to be self-dual or anti-self-dual depending on whether the curvature form is invariant or anti-invariant under the star operator. These curvature forms can be interpreted as non-Abelian harmonic forms. The remarkable fact is that when the metric is Kähler, the anti-self-dual connections give rise to holomorphic bundles. The moduli space of such bundles can often be computed using tools from algebraic geometry.

On the product space $M \times \mathcal{M}$ where M is the four dimensional manifold and $\mathcal{M}$ is the moduli space of anti-self-dual connections, there is a universal bundle V over $M \times \mathcal{M}$. By studying the slant product and the Chern classes of V, we can construct polynomials on the cohomology of M that are invariants of the differentiable structure of M. These are Donaldson polynomials (see [**204**]). In general $\mathcal{M}$ is not compact and Donaldson has to construct cycles in $\mathcal{M}$ for such operations. Donaldson invariants are believed to be equivalent to Seiberg-Witten invariants, where the vanishing theorem can apply and powerful geometric consequences can be found. Kronheimer and Mrowka [**402**] built an important concept of simple type for Donaldson invariants. It is believed that Donaldson invariants of algebraic surfaces of general type are of simple type.

If the manifold is symplectic, we can look at the moduli space of pseudo-holomorphic curves. (These are J-invariant maps from Riemann surfaces to the manifold. J is an almost complex structure that is tame to the symplectic form.) Symplectic invariants can be created and they are called Gromov-Witten invariants. Y. Ruan [**579**] has observed that they need not be diffeomorphic invariants. It may still be interesting to know whether Gromov-Witten invariants are invariants of differentiable structures for Calabi-Yau manifolds.

De Rham cohomology can only capture the non-torsion part of the singular cohomology. Weil [**710**] and Allendoerfer-Eells [**5**] attempted to use differential forms with poles to compute cohomology with integer coefficients. Perhaps one should study Chern forms of a complex bundle with a connection that satisfies the Yang-Mills equation and whose curvature is square integrable. The singular set of the connection may be allowed to be minimal submanifolds. The moduli space of such objects may give information about integral cohomology. It should be noted that Cheeger-Simons [**123**] did develop a rich theory of differential character with values in $\mathbb{R}/\mathbb{Z}$.

It depends on the connections of the bundle. Witten managed to integrate the Chern-Simons forms [**152**] on the space of connections to obtain the knot invariants of Jones [**363**].

When we look for different operators acting on different forms, we may have to look into different kinds of harmonic forms. For example, if we are interesting in $\partial\bar{\partial}$ cohomology, we may look for the operator $\left(\partial\bar{\partial}\right)^* \partial\bar{\partial} + \partial\partial^* + \bar{\partial}\bar{\partial}^*$. It would be interesting to see how super-symmetry may be used to generalize the concept of harmonic forms.

> **Comment**: The theory of harmonic form is tremendously powerful because it provides a natural link between global topology, analysis, geometry, algebraic geometry and arithmetic geometry. However, our analytic understanding of high degree forms is poor. For one forms, we can integrate along paths. For two forms, we can take an interior product with a vector field to create a moment map. For closed $(1,1)$-forms in a Kähler manifold, we can express them locally as $\partial\bar{\partial} f$. However, we do not have good ways to reduce a high degree form to functions which are easier to understand. Good estimates of higher degree forms will be very important.

2.3.2. *$\bar{\partial}$-operator.* Construction of holomorphic functions or holomorphic sections of vector bundles and holomorphic curves are keys to understanding complex manifolds.

In order to demonstrate the idea behind the philosophy of determining the structure of manifolds by function theory, I was motivated to generalize the uniformization theory of a Riemann surface to higher dimensions when I was a graduate student. During this period, I was influenced by the works of Greene-Wu [**273**] in formulating these conjectures. Greene and Wu were interested in knowing whether the manifolds are Stein or not.

When the complete Kähler manifold is compact with positive bisectional curvature, this is the Frankel conjecture, as was proved independently by Mori [**530**] and Siu-Yau [**633**]. Both arguments depend on the construction of rational curves of low degree. Mori's argument is stronger, and it will be good to capture his result by the analytic method. When the manifold has nonnegative bisectional curvature and positive Ricci curvature, Mok-Zhong [**527**] and Mok [**523**], using ideas of Bando [**31**] in his thesis on Hamilton's Ricci flow, proved that the manifold is Hermitian symmetric unless it is biholomorphic to projective space.

When the complete Kähler manifold is noncompact with positive bisectional curvature, I conjectured that it must be biholomorphic to $\mathbb{C}^n$ (see [**739**]). Siu-Yau [**632**] made the first attempt to prove such a conjecture by using the L^2-method of Hörmander [**347**] to construct holomorphic functions with slow growth. (Note that Hörmander's method goes back to Kodaira, which was also generalized by Calabi-Vesentini [**95**].) Singular weight functions were used in this paper and later much more refined arguments were developed by Nadel [**535**] and Siu [**630**] using what is called the multiplier ideal sheaf method. Siu found important applications of this method in algebraic geometry and also related the idea to the powerful work of J. Kohn on weakly pseudo-convex domain.

This work of Siu-Yau was followed by Siu-Mok-Yau [**524**] and Mok [**521, 522**] under assumptions about the decay of curvature and volume growth. Shi [**611, 612, 613**] introduced Hamilton's Ricci flow to study my conjecture, and his work is fundamental. This was followed by beautiful works of Cao [**101, 102**], Chen-Zhu [**127, 128**] and Chen-Tang-Zhu [**125**]. Assuming the manifold has maximal Euclidean volume growth and bounded curvature, Chen-Tang-Zhu [**125**] (for complex dimension two) and then Ni [**539**] (for all higher dimension) were able to prove the manifold can be compactified as a complex variety. Last year, Albert Chau and Tam [**113**] were finally able to settle the conjecture assuming maximal Euclidean volume growth and bounded curvature. An important lemma of L. Ni [**539**] was used, where a conjecture of mine (see [**742**] or the introduction of [**539**]) was proved. The conjecture says that maximal volume growth implies scalar curvature decays quadratically in the average sense.

While we see great accomplishments for Kähler manifolds with positive curvature, very little is known for Kähler manifolds, which are complete simply connected with strongly negative curvature. It is conjectured to be a bounded domain in $\mathbb{C}^n$. (Some people told me that Kodaira considered a similar problem. But I cannot find the appropriate reference.) The major problem is to construct bounded holomorphic functions.

The difficulty of construction of bounded holomorphic functions is that the basic principle of the L^2-method of Hörmander comes from Kodaira's vanishing theorem. It is difficult to obtain elegant results by going from weighted L^2 space to bounded functions. In this connection, I was able to show that non-trivial bounded holomorphic functions do not exist on a complete manifold with non-negative Ricci curvature [**738**].

If the manifold is the universal cover of a compact Kähler manifold M which has a homotopically nontrivial map to a compact Riemann surface with genus > 1, then one can construct a bounded holomorphic function, using arguments of Jost-Yau [**370**]. In particular, if M has a map to a product of Riemann surfaces with genus > 1 with nontrivial topological degree, the universal cover should have a good chance to be a bounded domain.

Of course, this kind of construction is based on the fact that holomorphic functions are harmonic. Certain rigidity based on curvature forced the converse to be true. For functions, the target space has no topology and rigidity is not expected. Bounded holomorphic functions can not be constructed by solving the Dirichlet problem unless some boundary condition is assumed. This would make good sense if the boundary has a nice CR structure. Indeed, for odd dimensional real submanifold in $\mathbb{C}^n$ which has maximal complex linear subspace on each tangent plane, Harvey-Lawson [**321, 322**] proved the remarkable theorem that they bound complex submanifolds. Unfortunately the boundary of a complete simply connected manifold with bounded negative curvature does not have a smooth boundary. It will be nice to define a CR structure on such a singular boundary. One may mention the remarkable work of Kuranishi [**405, 406, 407**] on embedding of an abstract CR structure.

Historically a motivation for the development of the $\bar{\partial}$ operator came from the Levi problem, which was solved by Morrey, Grauert and greatly improved by Kohn and Hörmander. Their methods are powerful in studying pseudoconvex manifolds.

In this regard, one may mention the conjecture of Shafarevich that the universal cover of an algebraic manifold is pseudoconvex. Many years ago, I conjectured that if the second homotopy group of the manifold is trivial, its universal cover can be embedded into a domain of some algebraic manifold where the covering transformations act on the domain by birational transformations. One may also mention the work of S. Frankel [**227**] on proving that an algebraic manifold is Hermitian symmetric if the universal cover is a convex domain in complex Euclidean space.

> **Comment**: The $\bar{\partial}$ operator is the fundamental operator in complex geometry. Classically it was used to solve the uniformization theorem, the Levi problem and the Corona problems. We have seen much progress on the higher dimensional generalizations of the first two problems. However, due to poor understanding of the construction of bounded holomorphic functions, we are far away from understanding the Corona problem in higher dimensional manifolds and many related geometric questions.

2.3.3. *Dirac operator.* A very important bundle is the bundle of spinors. The Dirac operator acting on spinors is the most mysterious but major geometric operator. Atiyah-Singer were the first mathematicians to study it in detail in geometry and by thoroughly understanding the Dirac operator, they were able to prove their celebrated index theorem [**20**]. On a Kähler manifold, the Dirac operator can be considered as a $\bar{\partial} + \bar{\partial}^*$ operator acting on differential forms with coefficients on the square root of the canonical line bundle. Atiyah-Singer's original proof can be traced back to the celebrated Riemann-Roch-Hirzebruch formula and the Hirzebruch index formula. The formulas of Gauss-Bonnet-Chern and Atiyah-Singer-Hirzebruch should certainly be considered as the most fundamental identities in geometry. The vanishing theorem of Lichnerowicz [**453**] on harmonic spinors over spin manifolds with positive scalar curvature gives strong information. Through the Atiyah-Singer index theorem, it gives the vanishing theorem for the $\hat{A}$-genus and the α invariants for spin manifolds with positive scalar curvature. The method was later sharpened by Hitchin [**331**] to prove that every Einstein metric over $K3-$surfaces must be Kähler and Ricci flat. An effective use of Lichnerowicz formula for a spin$_{\mathbb{C}}$ structure for a four dimensional manifold is important for Seiberg-Witten theory, which couples the Dirac operator with a complex line bundle. Lawson-Yau [**413**] were able to use Lichnerowicz's work coupled with Hitchin's work to prove a large class of smooth manifolds have no smooth non-Abelian group action and, by using modular forms, K.F. Liu proved a loop space analogue of the Lawson-Yau's theorem for the vanishing of the Witten genus in [**467**].

On the basis of the surgery result of Schoen-Yau [**597, 600**] and Gromov-Lawson [**290, 291**], one expects that a suitable converse to Lichnerowicz's theorem exists. The chief result is that surgery on spheres with codimension ≥ 3 preserves a class of metrics with positive scalar curvature. Once geometric surgery is proved, standard works on cobordism theory allow one to deduce existence results for simply connected manifolds with positive scalar curvature. The best work in this direction is due to Stolz [**641**] who gave a complete answer in the case of simply connected manifolds with dimension greater then 4. I also suggested the possibility of performing surgery on an asymptotic hyperbolic manifold with conformal boundary whose scalar curvature is positive. This is related to the recent work of Witten-Yau [**722**] on the connectedness of the conformal boundary.

The study of metrics with positive scalar curvature is the first important step in understanding the positive mass conjecture in general relativity. Schoen-Yau [**598, 602**] gave the first proof using ideas of minimal surfaces. Three years later, Witten [**716**] gave a proof using harmonic spinors. Both approaches have been fundamental to questions related to mass and other conserved quantities in general relativity. In the other direction, Schoen-Yau [**600**] generalized their argument in 1979 to find topological obstructions for higher dimensional manifolds with positive scalar curvature. Subsequently Gromov-Lawson [**290, 291**] observed that the Lichnerowicz theorem can be coupled with a fundamental group and give topological obstructions for a metric with positive scalar curvature. This work was related to the Novikov conjecture where many authors, including Lusztig [**483**], Rosenberg [**571**], Weinberger [**712**] and G.L. Yu [**751**] made contributions.

Besides its importance in demonstrating the stability of Minkowski spacetime, the positive mass conjecture was used by Schoen [**590**] in a remarkable manner to finish the proof of the Yamabe problem where Trudinger [**686**] and Aubin [**21**] made substantial contributions.

> **Comment**: The Dirac operator is perhaps one of the most mysterious operators in geometry. When it is twisted with other bundles, it gives the symbol of all first order elliptic operators. When it couples with a complex line bundle it gives the Seiberg-Witten theory which provides powerful information for four manifolds. On the other hand, there were two different methods to study metrics with positive scalar

curvature. It should be fruitful to combine both methods: the method of Dirac operator and the method of minimal submanifolds.

2.3.4. *First order operator twisted by vector fields or endomorphisms of bundles.* Given a vector field X on a manifold, we can consider the complex of differential forms ω so that $L_X \omega = 0$. On such a complex, $d + \iota_X$ defines a differential and the resulting cohomology is called equivariant cohomology.

During the seventies, Bott [**72**] and Atiyah-Bott [**12**] developed the localization formula for equivariant cohomology. Both the concepts of a moment map and equivariant cohomology have become very important tools for computations of various geometric quantities, especially Chern numbers of natural bundles. The famous work of Atiyah, Guillemin-Sternberg on the convexity of the image of the moment map gives a strong application of equivariant cohomology to toric geometry. The formula of Duistermaat-Heckman [**208**] played an important role in motivation for evaluation of path integrals. These works have been used by Jeffrey and Kirwan [**360**] and by K.F. Liu and his coauthors on several topics: the mirror principle (Lian-Liu-Yau [**449, 450, 451, 452**]), topological vertex (Li-Liu-Liu-Zhou [**430**]), etc. The idea of applying localization to enumerative geometry was initiated by Kontsevich [**393**] and later by Givental [**257**] and Lian-Liu-Yau [**449**] independently. (Lian-Liu-Yau [**449**] formulated a functorial localization formula which has been fundamental for various calculations in mirror geometry.) These works solve the identities conjectured by Candelas et al [**99**] based on mirror symmetry, and provide good examples of the ways in which conformal field theory can be a source of inspiration when looking at classical problems in mathematics.

If we twist the $\bar\partial$ operator with an endomorphism valued holomorphic one form s so that $s \circ s = 0$, it gives rise to a complex $\bar\partial + s$. This was the Higgs theory initiated by Hitchin [**332**] and studied extensively by Simpson [**621**]. There is extensive work of Zuo Kang and Jost-Zuo (see [**759**]) on Higgs theory and representation of fundamental groups of algebraic manifold.

In string theory, there is a three form H and the cohomology of $d^c + H$ has not been well understood. It would be interesting to develop a deeper understanding of such twisted cohomology and its localization.

Comment: The idea of deforming a de Rham operator by twisting with some other zero order operators has given powerful information to geometry. Witten's idea of the analytic proof of Morse theory is an example. Equivariant cohomology is another example. We expect to see more works in such directions.

2.3.5. *Spectrum and global geometry.* Weyl made a famous address in the early fifties. The title of his talk was *The Eigenvalue Problem Old and New.* He was excited by the work of Minakshisundaram and Pleijel which asserts that the zeta function $\zeta(s) = \sum_\lambda \lambda^{-s}$, where λ are eigenvalues of the

Laplacian, not only makes sense for Re's large, but also has meromorphic extension to the whole complex s-plane, the position of whose poles could be described explicitly. In particular, it is analytic near $s = 0$. Formally $\frac{d\zeta(s)}{ds}|_{s=0}$ can be viewed as $-\log\det(\Delta)$. This gives a definition of determinant of Laplacian which entered into the fundamental work of Ray-Singer relating Reidemeister's combinational invariant of a manifold with analytic torsion defined by determinants of the Laplacians acting on differential forms of various degrees. Other application of zeta function expressed in terms of kernel is the calculation of the asymptotic growth of eigenvalues in terms of volume of the manifold. Tauberian type theorem is needed.

This initiated the subject of finding a formula to relate spectrum of manifolds with their global geometry. Atiyah and Singer [**20**] were the most important contributors to this beautiful subject. Atiyah-Bott-Patodi [**13**] applied the heat kernel expansion to a proof of the local index theorem where Gilkey [**256**] also made an important contributions. Atiyah-Patodi-Singer [**17, 18, 19**] initiated the study of spectrum flow and gave important global spectral invariants on odd dimensional manifolds. These global invariants become boundary terms for the L^2-index theorem developed by Atiyah-Donnelly-Singer [**14**] and Mark Stern [**640**]. (A method of Callias [**96**] has been used for such calculations.) Witten [**717, 718**] has introduced supersymmetry and analytic deformation of the de Rham complex to Morse theory, and thereby revealed a new aspect of the connection between global geometry and theoretical physics. Witten's work has been generalized by Demailly [**190**] and Bismut-Zhang [**56, 57**] to study the holomorphic Morse inequality and analytic torsion. Novikov [**541**] also studied Morse theory for one forms. Witten's work on Morse theory inspired the work of Floer (see, e.g., [**224, 225, 226**]) who used his ideas in Floer cohomology to prove Arnold's conjecture in the case where the manifold has vanishing higher homotopic group. Floer's theory is related to knot theory (through Chern-Simon's theory [**152**]) on three manifolds. Atiyah, Donaldson, Taubes, Dan Freed, P. Braam, and others (see, e.g., [**10, 658, 77, 229**]) all contributed to this subject. Fukaya-Ono [**241**], Oh [**544**], Kontsevich [**394**], Hofer-Wysocki-Zehnder [**341**], G. Liu-Tian [**466**], all studied such a theory in symplectic geometry. Some part of Arnold's conjecture on fixed points of groups acting on symplectic manifolds was claimed to be proven. But a satisfactory proof has not been forthcoming.

One should also mention here the very important work of Cheeger [**115**] and Müller [**533**] in which they verify the conjecture of Ray-Singer equating analytic torsion with the combinational torsion of the manifold. The fundamental idea of Ray-Singer [**567**] on holomorphic torsion is still being vigorously developed. It appeared in the beautiful work of Vafa et al [**50**]. Many more works on analytic torsion were advanced by Quillen, Todorov, Kontsevich, Borcherds, Bismut, Lott, Zhang, and Z.Q. Lu (see [**55**] and its reference, [**365**], [**64, 65**]). The local version of the index theorem by

Atiyah-Bott-Patodi [**13**] was later extended by Bismut [**53**] to an index theorem for a family of elliptic operators. However, the pushed forward Chern forms have not been calculated and the formula has not been used effectively. (The local index argument dates back to the foundational work of McKean-Singer [**499**] where methods were developed to calculate coefficients of heat kernel expansion.) The study of elliptic genus by Witten [**719**], Bott-Taubes [**73**], Taubes [**657**], K.F. Liu [**468**] and M. Hopkins [**346**] has built a bridge between topology and modular form.

> **Comment**: The subject of relating the spectrum to global topology is extremely rich. It is likely that we have only touched part of this rich subject. The deformation of spectrum associated with the deformation of geometric structure is always a fascinating subject. Global invariants are created by spectral flows. Determinants of elliptic operators are introduced to understand measures of infinite dimensional space. Geometric invariants that are created by asymptotic expansion of heat or wave kernels are in general not well understood. It will be a long time before we have a much better understanding of the global behavior of spectrum.

3. Mappings between manifolds and rigidity of geometric structures

There is a need to exhibit a geometric structure in a simpler space: hence we embed algebraic manifolds into complex projective space, we isometrically embed a Riemannian manifold into Euclidean space and we classify structures such as bundles by studying maps into Grassmannian.

We are also interested in probing the structure of a manifold by mapping Riemann surfaces inside the manifold, an important example being holomorphic curves in algebraic manifolds. Of course, we are also interested in maps that can be used to compare the geometric structures of different manifolds.

3.1. Embedding.

3.1.1. *Embedding theorems.* Holomorphic sections of holomorphic line bundles have always been important in algebraic geometry. The Riemann-Roch formula coupled with vanishing theorems gave very powerful existence results for sections of line bundles. The Kodaira embedding theorem [**391**] which said that every Hodge manifold is projective has initiated the theory of holomorphic embedding of Kähler manifolds. For example, Hirzebruch-Kodaira [**330**] proved that every odd (complex) dimensional Kähler manifold diffeomorphic to projective space is biholomorphic to projective space. (I proved the same statement for even dimensional Kähler manifolds based on Kähler Einstein metric.)

Given an orthonormal basis of holomorphic sections of a very ample line bundle, we can embed the manifold into projective space. The induced metric is the Bergman metric associated with the line bundle. Note that the original definition of the Bergman metric used the canonical line bundle and L^2-holomorphic sections.

In the process of understanding the relation between stability of a manifold and the existence of the Kähler Einstein metric, I [**741**] proposed that every Hodge metric can be approximated by the Bergman metric as long as we allow the power of the line bundle to be large. Following the ideas of the paper of Siu-Yau [**632**], Tian [**676**] proved the C^2 convergence in his thesis under my guidance. My other student W. D. Ruan [**575**] then proved C^∞ convergence in his thesis. This work was followed by Lu [**479**], Zelditch [**752**] and Catlin [**110**] who observed that the asymptotic expansion of the kernel function follows from some rather standard expressions of the Szegö kernel, going back to Fefferman [**220**] and Boutet de Monvel-Sjöstrand [**75**] on the circle bundle associated with the holomorphic line bundle over the Kähler manifold. Recently, Dai, Liu, Ma and Marinescu [**182**] [**488**] obtained the asymptotic expansion of the kernel function by using the heat kernel method, and gave a general way to compute the coefficients, thus also extended it to symplectic and orbifold cases.

Kodaira's proof of embedding Hodge manifolds by the sufficiently high power of a positive line bundle is not effective. Matsusaka [**493**] and later Kollár [**392**], Siu [**628**] were able to provide effective estimate of the power. Demailly [**192, 193**] and Siu [**628, 630**] made a significant contribution toward the solution of the famous Fujita conjecture [**237**] (see also Ein and Lazarsfeld [**213**]). Siu's powerful method also leads to a proof of the deformation invariance of plurigenera of algebraic manifolds [**629**]. It should be noticed that the extension theorem of Ohsawa-Takegoshi played an important role in this last work of Siu.

> **Comment**: The idea of embedding a geometric structure is clearly important because once they are put in the same space, we can compare them and study the moduli space of the geometric structure. For example, one can define Chow coordinate of a projective manifold and we can study various concepts of geometric stability of these structures. However, there is no natural universal space of Kähler manifolds or complex manifolds as we may not have a positive holomorphic line bundle over such manifolds to embed into complex projective space. In a similar vein, it will be nice to find a universal space for symplectic manifolds.

3.1.2. *Compactification.* The problem of compactification of the manifold dates back to Siegel, Satake, Baily-Borel [**26**] and Borel-Serre [**67**]. They are important for representation theory, for algebraic geometry and for number theory.

For geometry of non-compact manifolds, we like to control behavior of differentiable forms at infinity. A good exhaustion function is needed.

Construction of a proper exhaustion function with a bounded Hessian on a complete manifold with bounded curvature was achieved by Schoen-Yau [**606**] in 1983 in our lectures in Princeton. Based on this exhaustion function, M. Dafermos [**180**] was able to give a transparent proof of the theorem of Cheeger-Gromov [**121**] that such manifolds admit an exhaustion by compact hypersurfaces with bounded second fundamental form. Such exhaustions are useful to understand characteristic forms on noncompact manifolds as the boundary term can be controlled by the second fundamental form of the hypersurfaces.

After my work with Siu [**634**] on compactification of a strongly, negatively curved Kähler manifold with finite volume, I proposed that every complete Kähler manifold with bounded curvature, finite volume and finite topology should be compactifiable to be a compact complex variety. I suggested this problem to Mok and Zhong in 1982 who did significant work [**528**] in this direction. (The compactification by Mok-Zhong is not canonical and it is desirable to find an algebraic geometric analogue of Borel-Baily compactification [**26**] so that we can study the L^2-cohomology in terms of the intersection cohomology of the compactification.) Recall that the important conjecture of Zucker on identifying L^2-cohomology with the intersection cohomology of the Borel-Baily compactification was settled by Saper-Stern [**585**] and Looijenga [**476**]. (Intersection cohomology was introduced by Goresky-MacPherson [**265, 266**]. It is a topological concept and hence the Zucker conjecture gives a topological meaning of the L^2-cohomology.) It would be nice to find compactification for algebraic varieties so that a suitable form of intersection cohomology can be used to understand L^2 cohomology. Goresky-Harder-MacPherson [**264**] and Saper [**584**] have contributed a lot toward this kind of question. For moduli space of bundles, or polarized projective structures, compactification means studying of degeneration of these structures in a suitable canonical manner. For algebraic curves, there is Deligne-Mumford compactification [**189**] which has played a fundamental role in understanding algebraic curves. Geometric invariant theory (see [**534**]) gives a powerful method to introduce the concept of stable structures. Semi-stable structures can give points at infinity. The compactification based on the geometric invariant theory for moduli space of surfaces of the general type was done by Gieseker [**254**]. For a higher dimension, this was done by Viehweg [**694**]. Detailed analysis of the divisors at infinity is still missing.

> **Comment**: Compactification of a manifold is very much related to the embedding problem. One needs to construct functions or sections of bundles near infinity. For the moduli space of geometric structures, it amounts to study of degeneration of the structures canonically, e.g., the degeneration of Hermitian Yang-Mills connections and Kähler Einstein metrics.

3.1.3. *Isometric embedding.* Given a metric tensor on a manifold, the problem of isometric embedding is equivalent to find enough functions $f_1, \ldots,$ f_N so that the metric can be written as $\sum(df_i)^2$. Much work was accomplished for two dimensional surfaces as was mentioned in section 2.1.2. Isometric embedding for the general dimension was solved by the famous work of J. Nash [**536, 537**]. Nash used his famous implicit function theorem which depends on various smoothing operators to gain derivatives. In a remarkable work, Günther [**307**] was able to avoid the Nash procedure. He used only the standard Hölder regularity estimate for the Laplacian to reproduce the Nash isometric embedding with the same regularity result. In his book [**287**], Gromov was able to lower the codimension of the work of Nash. He called his method the h-principle.

When the dimension of the manifold is n, the expected dimension of the Euclidean space for the manifold to be isometrically embedded is $\frac{n(n+1)}{2}$. It is important to understand manifolds isometrically embedded into Euclidean space with this optimal dimension. Only in such a dimension does it make sense to talk about rigidity questions. It remains a major open problem whether one can find a nontrivial smooth family of isometric embeddings of a closed manifold into Euclidean space with an optimal dimension. Such a nontrivial family was found for a polyhedron in Euclidean three space by Connelly [**176**]. For a general manifold, it is desirable to find a canonical isometric embedding into a given Euclidean space by minimizing the L^2 norm of its mean curvature within the space of isometric embeddings.

Chern told me that he and H. Lewy studied local isometric embedding of a three manifold into six dimensional Euclidean space. But they didn't have any publication on it. The major work was done by E. Berger, Bryant, Griffiths and Yang [**85**], [**47**]. They showed that a generic three dimensional embedding system is strictly hyperbolic, and the generic four dimensional system is a real principal type. Local existence is true for a generic metric using a hyperbolic operator and the Nash-Moser implicit function theorem.

If the target space of isometric embedding is a linear space with indefinite metric, it is possible that the problem is easier. For example, by a theorem of Pogorelov [**561, 562**], any metric on the two dimensional sphere can be isometrically embedded into a three dimensional hyperbolic space-form (where the sectional curvature may be a large negative constant). Hence it can always be embedded into the hyperboloid of the Minkowski spacetime. This statement may also be true for surfaces with higher genus. The fundamental group may cause obstruction, hence the first step should be an attempt to canonically embed any complete metric (with bounded curvature) on a simply connected surface into a three dimensional hyperbolic space form. It should be also very interesting to study the rigidity problem of a space-like surface in Minkowski spacetime. Besides requesting the metric to be the induced metric, we shall need one more equation. Such an equation should be related to the second fundamental form. A candidate appeared in the work of M. Liu-Yau [**464, 465**] on the quasi-local mass in general relativity.

In the other direction, Calabi found the condition for a Kähler metric to be isometrically and holomorphically embedded into Hilbert space with an indefinite signature. In the course of his investigation, he introduced some kind of distance function that can be defined by the Kähler potential and enjoys many interesting properties. Calabi's work in this direction which should be relevant to the flat coordinate appeared in the recent works of Vafa et al [**50**].

> **Comment**: The theory of isometric embedding is a classical subject. But our knowledge is still rather limited, especially in dimension greater than three. Many difficult problems are related to nonlinear mixed type equation or hyperbolic differential systems over a closed manifold.

3.2. Rigidity of harmonic maps with negative curvature. One can define the energy of maps between manifolds and the critical maps are called harmonic maps. In 1964, Eells-Sampson [**212**] and Al'ber [**3**] independently proved the existence of such maps in their homotopy class if the image manifold has a non-positive curvature.

When I was working on manifolds with non-positive curvature, I realized that it is possible to use harmonic map to reprove some of the theorems in my thesis. I was convinced that it is possible to use harmonic maps to study rigidity questions in geometry such as Mostow's theorem [**531**]. In 1976, I proved the Calabi conjecture and applied the newly proved existence of the Kähler Einstein metric and the Mostow rigidity theorem to prove uniqueness of a complex structure on the quotient of the ball [**735**]. Motivated by this theorem, I proposed to use the harmonic map to prove the rigidity of a complex structure for Kähler manifolds with strongly negative sectional curvature. I proposed this to Siu who carried out the idea when the image manifold satisfies a stronger negative curvature condition [**625**]. Jost-Yau [**368**] proved that for harmonic maps into manifolds with non-positive curvature, the fibers give rise to holomorphic foliations even when the map is not holomorphic. Such a work was found to be useful in the work of Corlette, Simpson et al.

A further result was obtained by Jost-Yau [**371**] and Mok-Siu-Yeung [**525**] on the proof of the superrigidity theorem of Margulis [**490**], improving an earlier result of Corlette [**177**] who proved superrigidity for a certain rank one locally symmetric space. Complete understanding of superrigidity for the quotient of a complex ball is not yet available. One needs to find more structures for harmonic maps which reflect the underlying structure of the manifold. The analytic proof of super-rigidity was based on an argument of Matsushima [**495**] as was suggested by Calabi. (This was a topic discussed by Calabi in the special year on geometry in the Institute for Advanced Study.)

The discrete analogue of harmonic maps is also important. When the image manifold is a metric space, there are works by Gromov-Schoen [**292**], Korevaar-Schoen [**399**] and Jost [**367**]. Margulis knew that the super-rigidity for both the continuous and the discrete case is enough to prove Selberg's conjecture for the arithmeticity of lattices in groups with rank ≥ 2. Unfortunately, the analytic argument mentioned above only works if the lattices are cocompact as it is difficult to find a degree one smooth map with finite energy for non-cocompact lattices. Harmonic maps into a tree have given interesting applications to group theory. When the domain manifold is a simplicial complex, there are articles by Ballmann-Śwwiątkowski [**29**] and M.T. Wang [**703, 704**], where they introduce maps from complices which are generalizations of buildings. They also generalized the work of H. Garland [**245**] on the vanishing of the cohomology group for p-adic buildings.

Using the concept of the center of gravity, Besson-Courtois-Gallot [**52**] give a metric rigidity theorem for rank one locally symmetric space. They also proved a rigidity theorem for manifolds with negative curvature: if the fundamental group can be split as a nontrivial free product over some other group C, the manifold can be split along a totally geodesic submanifold with the fundamental group C.

> **Comment**: The harmonic map gives the first step in matching geometric structures of different manifolds. Eells-Sampson derived it from the variational principle. One can also use different elliptic operators to define maps which satisfy elliptic equations. Higher dimensional applications are mostly based on the assumption that the image manifold has a metric with non-positive curvature. In such a case, existence is easier and uniqueness (as shown by Hartman) is also true. Up to now, significant results on higher dimensional harmonic maps are based on such assumptions. Generalization to Kähler manifold should be reasonable. The second homotopic group should play a role as one may look at it as a generalization of the work of Sacks-Uhlenbeck. It may be possible to use harmonic maps to study the moduli of geometric structure on a fixed manifold as was done by Michael Wolf for Riemann surfaces. It will also be nice to see how a harmonic map can be used to compare graphs.

3.3. Holomorphic maps. The works of Liouville, Picard, Schwarz-Pick and Ahlfors show the importance of hyperbolic complex analysis. Grauèrt-Reckziegel [**268**] generalized this kind of analysis to higher dimensional complex manifolds. Kobayashi [**388**] and H. Wu [**726**] put this theory in an elegant setting. Kobayashi introduced the concept of hyperbolic complex manifolds. Its elegant formulation has been influential. An important application of the negative curvature metric is the extension theorem for holomorphic maps, as was achieved by the work of Griffiths-Schmid [**275**] on maps to a period domain and by the extension theorem of Borel [**66**] on compactification of Hermitian symmetric space. A major question was Lang's conjecture: on an algebraic manifold of a general type, there exists a proper subvariety such that the image of any holomorphic map from $\mathbb{C}$ must be a subset of this subvariety. It has deep arithmetic geometric meaning. In terms of the Kobayashi metric, it says that the Kobayashi metric is nonzero on a Zariski open set. Many works were done towards subvarieties of Abelian variety by Bloch, Green-Griffiths, Kobayashi-Ochiai, Voitag and Faltings. For generic hypersurfaces in $\mathbb{C}P^n$, there is work by Siu [**631**]. They developed the tool of jet differentials and meromorphic connections. For algebraic surfaces with $C_1^2 > 2C_2$, Lu-Yau [**477**] proved Lang's conjecture, based on the ideas of Bogomolov.

> **Comment**: Holomorphic maps have been studied for a long time. There is no general method to construct such maps based on the knowledge of topology alone, except the harmonic map approach proposed by me and carried out by Siu, Jost-Yau and others. But the approach is effective only for manifolds with negative curvature. For rigidity questions, the most interesting manifolds are Kähler manifolds with non-positive Ricci curvature, which give the major chunk of algebraic manifolds of a general type. The Kähler-Einstein metric should provide tools to study such problems. Is there any intrinsic way, based on the metric, to find the largest subvariety where the image of all holomorphic maps from the complex line lie? Deformation theory of such a subvariety should be interesting. There is also the question of when the holomorphic image of the complex line will intersect a divisor. Cheng and I did find good conditions for the complement of a divisor to admit the complete Kähler-Einstein metric. For such a geometry, the holomorphic line should either intersect the divisor or a subset of some subvarieties. These kinds of questions are very much related to arithmetic questions if the manifolds are defined over number fields.

3.4. Harmonic maps from two dimensional surfaces and pseudo-holomorphic curves. Harmonic maps behave especially well for Riemann surface. Morrey was the first one who solved the Dirichlet problem for energy minimizing harmonic map into any Riemannian manifold.

Another major breakthrough was made by Sacks-Uhlenbeck [**581**] in 1978 where they constructed minimal spheres in Riemannian manifolds representing elements in the second homotopy group using a beautiful extension theorem of a harmonic map at an isolated point. By pushing their method further, Siu-Yau [**633**] studied the bubbling process for the harmonic map and made use of it to prove a stable harmonic map must be holomorphic under curvature assumptions. As a consequence, they proved the famous conjecture of Frankel that a Kähler manifold with positive bisectional curvature is $\mathbb{C}P^n$, as was discussed in Section [2.3.2].

Gromov [**286**] then realized that a pseudoholomorphic curve for an almost complex structure can be used in a similar way to prove rigidity of a symplectic structure on $\mathbb{C}P^n$. The bubbling process mentioned above was sharpened further to give compactification of the moduli space of pseudoholomorphic maps by Ye [**749**] and Parker-Wolfson [**549**]. Based on these ideas, Kontsevich [**393**] introduced the concept of stable maps and the compactification of their moduli spaces.

The formal definitions of Gromov-Witten invariants and quantum cohomology were based on these developments and the ideas of physicists. For example, quantum cohomology was initiated by Vafa (see, e.g., [**692**]) and his coauthors (the name was suggested by Greene and me). Associativity in quantum cohomology was due to four physicists WDVV [**720, 194**]. The mathematical treatment (done by Ruan [**579**] and subsequently by Ruan-Tian [**580**]) followed the gluing ideas of the physicists. Ruan-Tian made use of the ideas of Taubes [**656**]. But important points were overlooked. A. Zinger [**756, 757**] has recently completed these arguments.

In close analogy with Donaldson's theory, one needs to introduce the concept of virtual cycle in the moduli space of stable maps. The algebraic setting of such a concept is deeper than the symplectic case and is more relevant to the development for algebraic geometry. The major idea was conceived by Jun Li who also did the algebraic geometric counterpart of Donaldson's theory (see [**426, 431**]). (The same comment applies to the concept of the relative Gromov-Witten invariant, where Jun Li made the vital contribution in the algebraic setting [**428, 429**].) The symplectic version of Li-Tian [**432**] ignores difficulties, many of which were completed recently by A. Zinger [**756, 757**].

Sacks-Uhlenbeck studied harmonic maps from higher genus Riemann surfaces. Independently, Schoen-Yau [**601**] studied the concept of the action of an L_1^2 map on the fundamental group of a manifold. It was used to prove the existence of a harmonic map with prescribed action on the fundamental group. Jost-Yau [**369**] generalized such action on fundamental group to a more general setting which allows the domain manifold to be higher dimensional. Recently F. H. Lin developed this idea further [**460**]. He studied extensively geometric measure theory on the space of maps (see, e.g., [**457, 459**]). The action on the second homotopy group is much more difficult to understand. I think there should exist a harmonic map with nontrivial action on the second homotopic group if such a continuous map exists. Such an existence theorem will give interesting applications to Kähler geometry.

There is a supersymmetric version of harmonic maps studied by string theorists. This is obtained by coupling the map with Dirac spinors in different ways (which corresponds to different string theories). While this kind of world sheet theory is fundamental for the development of string theory, geometers have not paid much attention to the supersymmetric harmonic map. Interesting applications may be found. The most recent paper of Chen, Jost, Li and Wang [**134**] does address to a related problem where they studied the regularity and energy identities for Dirac-Harmonic maps.

> **Comment**: Maps from circle or Riemann surfaces into a Riemannian manifold give a good deal of information about the manifold. The capability to construct holomorphic or pseudo-holomorphic maps from spheres with low degree was the major reason that Mori, Siu-Yau and Taubes were able to prove the rigidity of algebraic or symplectic structures on the complex projective space. It will be desirable to find more ways to construct such maps from low genus curves to manifolds that are not of a rational type. Their moduli space can be used to produce various invariants. An outstanding problem is to understand the invariants on counting curves of a higher genus which appeared in the fundamental paper of Vafa et al [**50**].

3.5. Morse theory for maps and topological applications. The energy functional for maps from S^2 into a manifold does not quite give rise to Morse theory. But the perturbation method of Sacks-Uhlenbeck did provide enough information for Micallef-Moore [**513**] to prove some structure theorem for manifolds with positive isotropic curvature. (Micallef and Wang [**514**] then proved the vanishing of second Betti number in the even dimensional case. If the manifold is irreducible, has non-negative isotropic curvature and non-vanishing second Betti number, then they proved that its second Betti number equals to one and it is Kähler with positive first Chern class.)

If the image manifold has negative curvature, the theorem of Eells-Sampson [**212**] says that any map can be canonically deformed by the heat flow to a unique harmonic map. Hence the topology of the space of maps is given by the space of homomorphism between the fundamental groups of the manifolds. This gives some information of the topology of manifolds with negative curvature. Farrell and Jones [**218**] have done much deeper analysis on the differentiable structure of manifolds with negative curvature.

Schoen-Yau [**601**] exploited the uniqueness theorem for harmonic maps to demonstrate that only finite groups can act smoothly on a manifold which admits a non-zero degree map onto a compact manifold with negative curvature. The size of the finite group can also be controlled. If the image manifold has non-positive curvature, then the only compact continuous group actions are given by the torus.

The topology of the space of maps into Calabi-Yau manifolds should be very interesting for string theory. Sullivan [**648**] has developed an equivariant homology theory for loop space. It will be interesting to link such a theory with quantum cohomology when the manifold has a symplectic structure.

> **Comment**: Morse theory has been one of the most powerful tools in geometry and topology as it connects local to global information. One does not expect full Morse theory for harmonic maps as we have difficulty even proving their existence. However, if their existence can be proven, the perturbation technique may be used and powerful conclusions may be drawn.

3.6. Wave maps. In the early eighties, C.H. Gu [**296**] studied harmonic maps when the domain manifold is the two dimensional Minkowski spacetime. They are called wave maps. Unfortunately, good global theory took much longer to develop as there were not many good a priori estimates. This subject was studied extensively by Christodoulou, Klainerman, Tao, Tataru and M. Struwe (see, e.g., [**161, 385, 653, 654, 610**]). It is hoped that such theory may shed some light on Einstein equations.

> **Comment**: The geometric or physical meaning of wave maps should be studied. The problem of vibrating membrane gives a good motivation to study time-like minimal hypersurface in a Minkowski spacetime. One can study the vibration of a submanifold by looking into the minimal time-like hypersurface with the boundary given by the submanifold. It is a mystery how such vibrations can be related to the eigenvalues of the submanifold.

3.7. Integrable system. Classically, Bäcklund (1875) was able to find a nonlinear transformation to create a surface with constant curvature in $\mathbb{R}^3$ from another one. The nonlinear equation behind it is the Sine-Gordon equation. Then in 1965, Kruskal and Zabusky (see [**403**]) discovered solitons and subsequently in 1967, Gardner, Greene, Kruskal and Miura [**244**] discovered the inverse scatting method to solve the KdV equations. The subject of a completely integrable system became popular.

Uhlenbeck [**690**] used techniques from integrable systems to construct harmonic maps from S^2 to $U(n)$, Bryant [**83**] and Hitchin [**334**] also contributed to related constructions using twistor theory and spectral curves. These inspired Burstall, Ferus, Pedit and Pinkall [**89**] to construct harmonic maps from a torus to any compact symmetric space. In a series of papers, Terng and Uhlenbeck [**668, 669**] used loop group factorizations to solve the inverse scattering problem and to construct Bäcklund transformations for soliton equations, including Schrödinger maps from $\mathbb{R}^{1,1}$ to a Hermitian symmetric space. There have been recent attempts by Martin Schmidt [**589**] to use an integrable system to study the Willmore surface.

The integrable system also appeared naturally in several geometric questions such as the Schottky problem (see Mulase [**532**]) and the Witten conjecture on Chern numbers of bundles over moduli space of curves.

Geroch found the Backlund transformation for axially symmetric stationary solutions of Einstein equations. It will be nice to find such nonlinear transformations for more general geometric structures.

> **Comment**: It is always important to find an explicit solution to a nonlinear problem. Hopefully an integrable system can help us to understand general structures of geometry.

3.8. Regularity theory. The major work on regularity theory of harmonic maps in higher dimensions was done by Schoen-Uhlenbeck [**592, 593**]. (There is a weaker version due to Giaquinta-Giusti [**252**] and also the earlier work of Ladyzhenskaya-Ural'ceva and Hildebrandt-Kaul-Widman where the image manifolds for the maps are more restrictive.) Leon Simon (see [**620**]) made a deep contribution to the structure of harmonic maps or minimal subvarieties near their singularity. This was followed by F.H. Lin [**459**]. The following is still a fundamental problem: Are singularities of harmonic maps or minimal submanifolds stable when we perturb the metric of the manifolds? Presumably some of them are. Can we characterize them? How big is the codimension of generic singularities?

In the other direction Schoen-Yau [**596**] also proved that degree one harmonic maps are one to one if the image surface has a non-positive curvature. Results of this type work only for two dimensional surfaces. It will be nice to study the set where the Jacobian vanishes.

> **Comment**: There is a very rich theory of stable singularity for smooth maps. However, in most problems, we can only afford to deform certain background geometric structures, while the extremal objects are still constrained by the elliptic variational problem. Understanding this kind of stable singularity should play fundamental roles in geometry.

4. Submanifolds defined by variational principles

4.1. Teichmüller space. The totality of the pair of polarized Kähler manifolds with a homotopic equivalence to a fixed manifold gives rise to the Teichmüller space. For an algebraic curve, this is the classical Teichmüller space. This space is important for the construction of the mapping problem for minimal surfaces of a higher genus.

In fact, given a conformal structure on a Riemann surface Σ, a harmonic map from Σ to a fixed Riemannian manifold may minimize energy within a certain homotopy class. However, it may not be conformal and may not be a minimal surface. In order to obtain a minimal surface, we need to vary the conformal structure on Σ also. Since the space of conformal structures on a surface is not compact, one needs to make sure the minimum can be achieved.

If the map f induces an injection on the fundamental group of the domain surface, Schoen-Yau [**597**] proved the energy of the harmonic map is proper on the moduli space of conformal structure on this surface by making use of a theorem of Linda Keen [**381**]. Based on a theory of topology of the L_1^2 map, they proved the existence of incompressible minimal surfaces. As a product of this argument, it is possible to find a nice exhaustion function for the Teichmüller space. Michael Wolf [**723**] was able to use harmonic maps to give a compactification of Teichmüller space which he proved to be equivalent to the Thurston compactification. S. Wolpert studied extensively the behavior of the Weil-Petersson metric (see Wolpert's survey [**724**]). A remarkable theorem of Royden [**574**] says that the Teichmüller metric is the same as the Kobayashi metric. C. McMullen [**502**] introduced a new Kähler metric on the moduli space which can be used to demonstrate that the moduli space is hyperbolic in the sense of Gromov [**288**]. The great detail of comparison of various intrinsic metrics on the Teichmüller space had been a major problem [**741**]. It was accomplished recently in the works of Liu-Sun-Yau [**471, 472**]. Actually Liu-Sun-Yau introduced new metrics with bounded negative curvature and geometry and found the stability of the logarithmic cotangent bundle of the moduli spaces. Recently L. Habermann and J. Jost [**308, 309**] also studied the geometry of the Weil-Petersson metric associated to the Bergmann metric on the Riemann surface instead of the Poincaré metric.

> **Comment**: For a conformally invariant variational problem, Teichmüller space plays a fundamental role. It covers the moduli space of curves and in many ways behaves like a Hermitian symmetric space of noncompact type. Unfortunately, there is no good canonical realization of it as a pseudo-convex domain in Euclidean space. For example, we do not know whether it can be realized as a smooth domain or not.
>
> There is also Teichmüller space for other algebraic manifolds, such as Calabi-Yau manifolds. It is an important question in understanding their global behavior.

4.2. Classical minimal surfaces in Euclidean space. There is a long and rich history of minimal surfaces in Euclidean space. Recent contributions include works by Meeks, Osserman, Lawson, Gulliver, White, Hildebrandt, Rosenberg, Collin, Hoffman, Karcher, Ros, Colding, Minicozzi, Rodríguez, Nadirashvili and others (see the reference in Colding and Minicozzi's survey [**174**]) on embedded minimal surfaces in Euclidean space. They come close to classifying complete embedded minimal surfaces and give a good understanding of complete minimal surface in a bounded domain. For example, Meeks-Rosenberg [**503**] proved that the plane and helicoid are the only properly embedded simply connected minimal surfaces in $\mathbb{R}^3$.

Calabi also initiated the study of isometric embedding of Riemann surfaces into S^N as minimal surfaces. The geometry of minimal spheres and minimal torus was then pursued by many geometers [**107**], [**150**], [**83**], [**334**], [**410**].

> **Comment**: This is one of the most beautiful subjects in geometry where Riemann made important contributions. Classification of complete minimal surface is nearly accomplished. However, a similar problem for compact minimal surfaces in S^3 is far from being solved. It is also difficult to detect which set of disjoint Jordan curves can bound a connected minimal surface. The classification of moduli space of complete minimal surfaces with finite total curvature should be studied in detail.

4.3. Douglas-Morrey solution, embeddedness and application to topology of three manifolds. In a series of papers started in 1978, Meeks-Yau [**505, 506, 507, 508**] settled a classical conjecture that the Douglas solution for the Plateau problem is embedded if the boundary curve is a subset of a mean convex boundary. (One should note that Osserman [**546**] had already settled the old problem of non-existence of branched points for the Douglas solution while Gulliver [**306**] proved non-existence of false branched points.) We made use of the area minimizing property of minimal surfaces to prove these surfaces are equivariant with respect to the group action. Embedded surfaces which are equivariant play important roles for finite group actions on manifolds. Coupling with a theorem of Thurston, we can then prove the Smith conjecture [**748**] for cyclic groups acting on the spheres: that the set of fixed points is not a knotted curve.

The Douglas-Morrey solution of the Plateau problem is obtained by fixing the genus of the surfaces. However, it is difficult to minimize the area when the genus is allowed to be arbitrary large. This was settled by Hardt-Simon [**320**] by proving the boundary regularity of the varifold solution of the Plateau problem. In the other direction, Almgren-Simon [**7**] succeeded in minimizing the area among embedded disks with a given boundary in Euclidean space. The technique was used by Meeks-Simon-Yau [**504**] to prove the existence of embedded minimal spheres enclosing a fake ball. This theorem has been important to prove that the universal covering of an irreducible three manifold is irreducible. They also gave conditions for the existence of embedding minimal surfaces of a higher genus. This work was followed by topologists Freedman-Hass-Scott [**231**]. Pitts [**557**] used the mini-max argument for varifolds to prove the existence of an embedded minimal surfaces. Simon-Smith (unpublished) managed to prove the existence of an embedded minimax sphere for any metric on the three sphere. J. Jost [**366**] then extended it to find four mini-max spheres. Pitts-Rubinstein (see, e.g., [**558**]) continued to study such mini-max surfaces. Since such mini-max surfaces have Morse index one, I was interested in representing such a minimal surface as a Heegard splitting of the three manifolds. I estimated its genus based on the fact that the second eigenvalue of the stability operator is nonnegative. This argument (dates back to Szego-Hersch) is to map the surface conformally to S^2. Hence we can use three coordinate functions, orthogonal to the first eigenfunction, to be trial functions. The estimate gave an upper bound of the genus for mini-max surfaces in compact manifolds with positive scalar curvature. About twenty years ago, I was hoping to use such an estimate to control a Heegard genus as a way to prove Poincaré conjecture. While the program has not materialized, three manifold topologists did adapt the ideas of Meeks-Yau to handle combinational type minimal surfaces and gave applications in three manifold topology.

The most recent works of Colding and Minicozzi [**169, 170, 171, 172**] on lamination by minimal surfaces and estimates of minimal surfaces without the area bound are quite remarkable. They [**173**] made contributions to Hamilton's Ricci flow by bounding the total time for evolution. Part of the idea came from the above mentioned inequality.

> **Comment**: The application of minimal surface theory to three manifold topology is a very rich subject. However, one needs to have a deep understanding of the construction of minimal surfaces. For example, if minimal surfaces are constructed by the method of mini-max, one needs to know the relation of their Morse index to the dimension of the family of surfaces that we use to perform the procedure of mini-max. A detailed understanding may lead to a new proof of the Smale conjecture, as we may construct a minimal surface by a homotopic group of embeddings of surfaces. Conversely, topological methods should help us to classify closed minimal surfaces.

4.4. Surfaces related to classical relativity. Besides minimal surfaces, another important class of surfaces are surfaces with constant mean curvature and also surfaces that minimize the L^2-norm of the mean curvature. It is important to know the existence of such surfaces in a three dimensional manifold with nonnegative scalar curvature, as they are relevant to the questions in general relativity.

The existence of minimal spheres is related to the existence of black holes. The most effective method was developed by Schoen-Yau [**604**] where they [**599**] proved the existence theorem for the equation of Jang. It should be nice to find new methods to prove existence of stable minimal spheres. The extremum of the Hawking mass is related to minimization of the L^2 norm of mean curvature. Their existence and behavior have not been understood.

For surfaces with constant mean curvature, we have the concept of stability. (Fixing the volume it encloses, the second variation of area is nonnegative.) Making use of my work on eigenvalues with Peter Li, I proved with Christodoulou [**162**] that the Hawking mass of such a surface is positive. (This was part of my contribution to the proposed joint project with

Christodoulou-Klainerman which did not materialize.) This fact was used by Huisken and me [**354**] to prove uniqueness and the existence of foliation by constant mean curvature spheres for a three dimensional asymptotically flat manifold with positive mass. (We initiated this research in 1986. Ye studied our work and proved existence of similar foliations under various conditions, see [**750**].)

This foliation was used by Huisken and Yau [**354**] to give a canonical coordinate system at infinity. It defines the concept of center of gravity where important properties for general relativity are found. The most notable is that total linear momentum is equal to the total mass multiple with the velocity of the center of the mass. One expects to find good asymptotic properties of the tensors in general relativity along these canonical surfaces. We hope to find a good definition of angular momentum based on this concept of center of gravity so that global inequality like total mass can dominate the square norm of angular momentum.

The idea of using the foliation of surfaces satisfying various properties (constant Gauss curvature, for example) to study three manifolds in general relativity was first developed by R. Bartnik [**41**]. His idea of quasi-spherical foliation gives a good parametrization of a large class of metrics with positive scalar curvature.

Some of these ideas were used by Shi-Tam [**614**] to study quantities associated to spheres which bound three manifolds with positive scalar curvature. Such a quantity is realized to be the quasi-local mass of Brown-York [**82**]. At the same time, Melissa Liu and Yau [**464, 465**] were able to define a new quasi-local mass for general spacetimes in general relativity, where some of the ideas of Shi-Tam were used. Further works by M. T. Wang and myself generalized Liu-Yau's work by studying surfaces in hyperbolic space-form.

My interest in quasi-local mass dates back to the paper that I wrote with Schoen [**604**] on the existence of a black hole due to the condensation of matter. It is desirable to find a quasi-local mass which includes the effect of matter and the nonlinear effect of gravity. Hopefully one can prove that when such a mass is larger than a constant multiple of the square root of the area, a black hole forms. This has not been achieved.

> **Comment**: When surfaces theory appears in general relativity, we gain intuitions from both geometry and physics together. This is a fascinating subject.

4.5. Higher dimensional minimal subvarieties. Higher dimensional minimal subvarieties are very important for geometry. There are works by Federer-Fleming [**219**], Almgren [**6**] and Allard [**4**]. The attempt to prove the Bernstein conjecture, that minimal graphs are linear, was a strong drive for its development. Bombieri, De Giorgi and Giusti [**63**] found the famous counterexample to the Bernstein problem. It initiated a great deal of interest in the area minimizing cone (as a graph must be area minimizing).

Schoen-Simon-Yau [**591**] found a completely different approach to the proof of Bernstein problem in low dimensions. This paper on stable minimal hypersurfaces initiated many developments on curvature estimates for the codimension one stable hypersurfaces in higher dimension. There are also works by L. Simon with Caffarelli and Hardt [**91**] on constructing minimal hypersurfaces by deforming stable minimal cones. Recently N. Wickramasekera [**714, 715**] did some deep work on stable minimal (branched) hypersurfaces which generalizes Schoen-Simon-Yau.

Michael-Simon [**515**] proved the Sobolev inequality and mean valued inequalities for such manifolds. This enables one to apply the classical argument of harmonic analysis to minimal submanifolds. For a minimal graph, Bombieri-Giusti [**62**] used ideas of De Giorgi-Nash to prove gradient estimates of the graph. N. Korevaar [**396**] was able to reprove this gradient estimate based on the maximal principle.

The best regularity result for higher codimension was done by F. Almgren [**6**] when he proved that for any area minimizing variety, the singular set has the codimension of at least two. How such a result can be used for geometry remains to be seen.

It was observed by Schoen-Yau [**597**] that for a closed stable minimal hypersurface in a manifold with positive scalar curvature, the first eigenfunction of the second variational operator can be used to conformally deform the metric so that the scalar curvature is positive. This provides an induction process to study manifolds with a positive scalar curvature. For example, if the manifold admits a nonzero degree map to the torus, one can then construct stable minimal hypersurfaces inductively until we find a two dimensional surface with higher genus which cannot support a metric with positive scalar curvature. At this moment, the argument encounters difficulty for dimensions greater than seven as we may have problems of singularity. In any case, we did apply the argument to prove the positive action conjecture in general relativity. The question of which type of singularities for minimal subvariety are generic under metric perturbation remains a major one for the theory of minimal submanifolds.

Perhaps the most important possible application of the theory of minimal submanifolds is the Hodge conjecture: whether a multiple of a (p,p) type integral cohomology class in a projective manifold can be represented by an algebraic cycle. Lawson made an attempt by combining a result of Lawson-Simons [**411**] and work of J. King [**383**] and Harvey-Shiffman [**324**]. (Lawson-Simons proved that currents in $\mathbb{C}P^n$ which are minimizing with respect to the projective group action are complex subvarieties.) The problem of how to use the hypothesis of (p,p) type has been difficult. In general, the algebraic cycles are not effective. This creates difficulties for analytic methods. The work of King [**383**] and Shiffman [**615**] on complex currents may be relevant.

Perhaps one should generalize the Hodge conjecture to include general (p,q) classes, as it is possible that every integral cycle in $\bigoplus_{i=-k}^{k} H^{p-i,p+i}$ is rationally homologous to an algebraic sum of minimal varieties such that there is a $p-k$ dimensional complex space in the tangent space for almost every point of the variety: it may be important to assume the metric to be canonical, e.g. the Kähler Einstein metric.

A dual question is how to represent a homology class by Lagrangian cycles which are minimal submanifolds also. When the manifold is Calabi-Yau, these are special Lagrangian cycles. Since they are supposed to be dual to holomorphic cycles, there should be an analogue of the Hodge conjecture. For example, if $\dim_{\mathbb{C}} M = n$ is odd, any integral element in $\bigoplus_{i+j=n} H^{i,j}$ should be representable by special Lagrangian cycles up to a rational multiple provided the cup product of it with the Kähler class is zero.

A very much related question is: if the Chern classes of a complex vector bundle are of (p,p) type, does the vector bundle, after adding a holomorphic vector bundle, admit a holomorphic structure? If the above generalization of the Hodge conjecture holds, there should be a similar generalization for the vector bundle. It should also be noted that Voisin [**696**] observed that Chern classes of all holomorphic bundles do not necessarily generate all rational (p,p) classes. On the other hand, the Kähler manifold that she constructed is not projective.

These questions had a lot more success for four dimensional symplectic manifolds by the work of Taubes both on the existence of pseudoholomorphic curves [**665**] and on the existence of anti-self-dual connections [**655, 656**]. On a Kähler surface, anti-self-dual connections are Hermitian connections for a holomorphic vector bundle. In particular, Taubes gave a method to construct holomorphic vector bundles over Kähler surfaces. Unfortunately this theorem does not provide much information on the Hodge conjecture as it follows from Lefschetz theorem in this dimension.

Another important class of minimal varieties is the class of special Lagrangian cycles in Calabi-Yau manifolds. Such cycles were first developed by Harvey-Lawson [**323**] in connection to calibrated geometry. Major works were done by Schoen-Wolfson [**594**], Yng-Ing Lee [**417**] and Butscher [**90**]. One expects Lagrangian cycles to be mirror to holomorphic bundles and special Lagrangian cycles to be mirror to Hermitian-Yang-Mills connections. Hence by the Donaldson-Uhlenbeck-Yau theorem, it is related to stability. The concept of stability for Lagrangian cycles was discussed by Joyce and Thomas. Since the Yang-Mills flow for Hermitian connection exists for all time, Thomas-Yau [**671**] suggested an analogy with the mean curvature flow for Lagrangian cycles. For stable Lagrangian cycles, mean curvature flow should converge to special Lagrangian cycles. See M.T. Wang [**705, 706**], Smoczyk [**636**] and Smoczyk-Wang [**637**]. The geometry of mirror symmetry was explained by Strominger-Yau-Zaslow in [**643**] using a family of special Lagrangian tori. There are other manifolds with special holonomy group. They have similar calibrated submanifolds. Conan Leung has contributed to studies of such manifolds and their mirrors (see, e.g., [**420, 421**]).

Submanifolds of space forms are called isoparametric if the normal bundle is flat and the principal curvatures are constants along parallel normal fields. These were studied by E. Cartan [**108**]. Minimal submanifolds with constant scalar curvature are believed to be isoparametric surfaces. There is work done by Lawson [**409**], Chern-de Carmo-Kobayashi [**151**] and Peng-Terng [**550**]. Recently there have been works by Terng and Thorbergsson (see Terng's survey [**667**] and Thorbergsson [**672**]). Terng [**666**] related isometric embedded hyperbolic spaces in Euclidean space to soliton theory. A theory of Lax pair and loop groups related to geometry has been developed.

> **Comment**: The theory of higher dimensional minimal submanifolds is one of the deepest subjects in geometry. Unfortunately our knowledge of the subject is not mature enough to give applications to solve outstanding problems in geometry, such as the Hodge conjecture. But the future is bright.

4.6. Geometric flows. The major geometric flows are flows of submanifold driven by mean curvature, gauss curvature, inverse mean curvature. Flows that change geometric structures are Ricci flows and Einstein flow.

Mean curvature flow for varifolds was initiated by Brakke [**78**]. The level set approach was studied by many people: S. Osher, L. Evans, Giga, etc (see [**545, 217, 136**]). Huisken [**349, 350**] did the first important work when the initial surface is convex. His recent work with Sinestrari [**352, 353**] on mean convex surfaces is remarkable and gives a good understanding of the structure of singularities of mean curvature flow. Mean curvature flow has many geometric applications. For example, the work of Huisken-Yau mentioned in 4.4 was achieved by mean curvature flow. Mean curvature flow for spacelike hypersurfaces in Lorentzian manifolds should be very interesting. Ecker [**211**] did interesting work in this direction. It will be nice to find the Li-Yau type estimate for such flows.

The inverse mean curvature was proposed by Geroch [**250**] to understand the Penrose conjecture relating the mass with the area of the black hole. Such a procedure was finally carried out by Huisken-Ilmanen [**351**] when the scalar curvature is non-negative. There was a different proof by H. Bray [**80**] subsequently.

Ricci flow has had spectacular successes in recent years. However, not much progress has been made on the Calabi flow (see Chang's survey [**111**]) for Kähler metrics. They are higher order problems where the maximal principle has not been effective. An important contribution was made by Chrúsciel [**163**] for Riemann surface. Inspired by the concept of the Bondi mass in general relativity, Chruściel was able to give a new estimate for the

Calabi flow. Unfortunately, a higher dimensional analogue had not been found.

Natural higher order elliptic problems are difficult to handle. Affine minimal surfaces and Willmore surfaces are such examples. L. Simon [**619**] made an important contribution to the regularity of the Willmore surfaces. The corresponding flow problem should be interesting.

The dynamics of Einstein equations for general relativity is a very difficult subject. The Cauchy problem was considered by many people: A. Lichnerowicz, Y. Choquet-Bruhat, J. York, V. Moncrief, H. Friedrich, D. Christodoulou, S. Klainerman, H. Lindblad, M. Dafermos (see, e.g., [**454**], [**158**], [**157**], [**234**], [**159**], [**384**], [**461**], [**181**]). But the global behavior is still far from being understood. The major unsolved problem is to formulate and prove the fundamental question of Penrose on Cosmic censorship. I suggested to Klainerman and Christodoulou to consider small initial data for the Einstein system. The treatment of stability of Minkowski spacetime was accomplished by Christodoulou-Klainerman [**160**] under small perturbation of flat spacetime and fast fall off conditions. Recently Lindblad and Rodnianski [**461**] gave a simpler proof. A few years ago, N. Zipser (Harvard thesis) added Maxwell equation to gravity and still proved stability of Minkowski spacetime. There is remarkable progress on the problem of Cosmic censorship by M. Dafermos [**181**]. He made an important contribution for the spherical case. Stability for Schwarzschild or Kerr solutions is far from being known. Finster-Kamran-Smoller-Yau [**222**] had studied decay properties of Dirac particles with such background. The work does indicate the stability of these classical spacetimes.

The no hair theorem for stationary black holes is a major theorem in general relativity. It was proved by W. Israël [**355**], B. Carter [**109**], D. Robinson [**570**] and S. Hawking [**326**]. But the proof is not completely rigorous for the Kerr metric. In any case, the existing uniqueness theorem does assume regularity of the horizon of the black hole. It is not clear to me whether a nontrivial asymptotically flat solitary solution of a vacuum Einstein equation has to be the Schwarzschild solution. There is a possibility that the Killing field is spacelike. In that case, there may be a new interesting vacuum solution.

There is extensive literature on spacelike hypersurfaces with constant mean curvature. The foliation defined by them gives interesting dynamics of Einstein equation. These surfaces are interesting even for $\mathbb{R}^{n,1}$. A. Treiberges studied it extensively [**685**]. Li, Choi-Treibergs [**154**] and T. Wan [**699**] observed that the Gauss maps of such surfaces give very nice examples of harmonic maps mapping into the disk. Recently Fisher and Moncrief used them to study the evolution equation of Einstein in $2+1$ dimension.

> **Comment**: The dynamics of submanifolds and geometric structures reveal the true nature of these geometric objects deeply. In the process of arriving at a stationary object or a

solitary solution, it encounters singularities. Understanding the structures of such singularity will solve many outstanding conjectures in topology such as Shoenflies conjecture.

5. Construction of geometric structures on bundles and manifolds

A fundamental question is how to build geometric structures over a given manifold. In general, the group of topological equivalences that leaves this geometric structure invariant should be a special group. With the exception of symplectic structures, these groups are usually finite dimensional. When the geometric structure is unique (up to equivalence), it can be used to produce key information about the topological structure.

The study of special geometric structures dates back to Sophus Lie, Klein and Cartan. In most cases, we like to be able to parallel transport vectors along paths so that we can define the concept of holonomy group.

5.1. Geometric structures with noncompact holonomy group. When the holonomy group is not compact, there are examples of projective flat structure, affine flat structure and conformally flat structure. It is not a trivial matter to determine which topological manifolds admit such structures. Since the structure is flat, there is a unique continuation property and hence one can construct a developing map from a suitable cover of the manifold to the real projective space, the affine space and the sphere respectively. The map gives rise to a representation of the fundamental group of the manifold to the real projective group, the special linear group and the Möbius group respectively. This holonomy representation gives a great deal of information for the geometric structure. Unfortunately, the map is not injective in general. In the case where it is injective, the manifold can be obtained as a quotient of a domain by a discrete subgroup of the corresponding Lie group. In this case, a lot more can be said about the manifold as the theories of partial differential equations and discrete groups can play important roles.

5.1.1. *Projective flat structure.* If a projective flat manifold can be projectively embedded as a bounded domain, Cheng-Yau [**145**] were able to construct a canonical metric from the real Monge-Ampère equation which generalizes the Hilbert metric. When the manifold is two dimensional, there are works of C.P. Wang [**700**] and J. Loftin [**473**] on how to associate such metrics to a conformal structure and a holomorphic section of the cubic power of a canonical bundle. This is a beautiful theory related to the hyperbolic affine sphere mentioned in chapter one.

There are fundamental works by S.Y. Choi, W. Goldman (see the reference of Choi-Goldman [**156**]), N. Hitchin [**335**] and others on the geometric decomposition and the moduli of flat projective structures on Riemann surfaces. It should be interesting to extend them to three or four dimensional manifolds.

5.1.2. *Affine flat structure.* It is a difficult question to determine which manifolds admit flat affine structures. For example, it is still open whether the Euler number of such spaces is zero, although great progress was made by D. Sullivan [**644**]. W. Goldman [**259**] has also found topological constraints on three manifolds in terms of fundamental groups. The difficulty arises as there is no useful metric that is compatible with the underlining affine structure. This motivated Cheng-Yau [**146**] to define the concept of affine Kähler metric.

When Cheng and I considered the concept of affine Kähler metric, we thought that it was a natural analogue of Kähler metrics. However, compact nonsingular examples are not bountiful. Strominger-Yau-Zaslow [**643**] proposed the construction of mirror manifolds by constructing the quotient space of a Calabi-Yau manifold by a special Lagrangian torus. At the limit of the large Kähler class, it was pointed out by Hitchin [**336**] that the quotient space admits a natural affine structure with a compatible affine Kähler structure. But in general, we do expect singularities of such a structure. It now becomes a deep question to understand what kind of singularity is allowed and how we build the Calabi-Yau manifold from such structures. Loftin-Yau-Zaslow [**474**] have initiated the study of the structure of a "Y" type singularity. Hopefully one can find an existence theorem for affine structures over compact manifolds with prescribed singularities along codimension two stratified submanifolds.

5.1.3. *Conformally flat structure.* Construction of conformally flat manifolds is also a very interesting topic. Similar to projective flat or affine flat manifolds, there are simple constraints from curvature representation for the Pontrjagin classes. The deeper problem is to understand the fundamental group and the developing map. When the structure admits a conformal metric with positive scalar curvature, Schoen-Yau [**605**] proved the rather remarkable theorem that the developing map is injective. Hence such a manifold must be the quotient of a domain in S^n by a discrete subgroup of Möbius transformations. It would be interesting to classify such manifolds. In this regard, the Yamabe problem as was solved by Schoen [**590**] did provide a conformal metric with constant scalar curvature. One hopes to be able to use such metrics to control the conformal structure. Unfortunately the metric is not unique and a deep understanding of the moduli space of conformal metrics with constant scalar curvature is needed.

Kazdan-Warner [**379**] and Korevaar-Mazzeo-Pacard-Schoen [**398**] developed a conformal method to understand Nirenberg's problem on prescribed scalar curvature. It was followed by Chen-Lin [**131**], Chang-Gursky-Yang [**112**]. Chen-Lin have related this problem to mean field theory. Their computations in the relevant degree theory involve deep analysis. One should be able to generalize their works to functions which are sections of a flat line bundle because it is related to the previously mentioned work of Loftin-Yau-Zaslow [**474**]. In any case, the integrability condition of Kazdan and Warner is still not fully understood.

It is curious that while bundle theory was used extensively in Riemannian geometry, it has not been used in the study of these geometries. One can construct real projective space bundles, affine bundles or sphere bundles by mapping coordinate charts projectively, affinely or conformally to the corresponding model spaces (possible with dimensions different from the original manifold) and gluing the target model spaces together to form natural bundles. Perhaps one may study their associated Chern-Simons forms [**152**].

Many years ago, H.C. Wang [**702**] proved the theorem that if a compact complex manifold has trivial holomorphic tangent bundle, it is covered by a complex Lie group. It will be nice to generalize and interpret such a theorem in terms of Hermitian connections on the manifold with a special holonomy group and torsion.

This program was discussed in my paper [**744**] on algebraic characterization of locally Hermitian symmetric spaces. For a holomorphic stable vector bundle V, we can form a stable vector bundles from V by taking irreducible representation of $GL(n, \mathbb{C})$ from decomposition of the tensor product representation $\bigotimes_p V \bigotimes_q V^*$. By twisting with powers of canonical line bundle, we can form irreducible stable bundles with trivial determinant line bundle. In general, such bundles may not have holomorphic sections. If they do, the section must be parallel with respect to the Hermitian-Yang-Mills connection on the bundle, and the structure group of V can be reduced to a smaller group. Hermitian-Yang-Mills connections with reduced holonomy group have good geometric properties. We may formulate a principle: For stable holomorphic bundles, existence of nontrivial holomorphic invariants implies the existence of parallel tensors and therefore the reduction of structure group. If the holonomy group is reduced to a discrete group, the bundle will provide representations of the fundamental group into unitary group. This should compare with Wang's theorem when the bundle is the tangent bundle.

> **Comment**: Geometric structures with a noncompact holonomy group are less intuitive than Riemannian geometry. Perhaps we need to deepen our intuitions by relating them to other geometric structures, especially those structures that may carry physical meaning.

5.2. Uniformization for three manifolds. An important goal of geometry is to build a canonical metric associated to a given topology. Besides the uniformization theorem in two dimensions, the only (spectacular) work in higher dimensions is the geometrization program of Thurston (see [**674**]).

W. Thurston made use of ideas from Riemann surface theory, W. Haken's work [**310**] on three-manifolds, G. Mostow's rigidity [**531**] to build his manifolds. Many mathematicians have contributed to the understanding of this program of Thurston's (e.g., J. Morgan [**529**], C. McMullen [**500, 501**], J. Otal [**547**], J. Porti [**566**]). Thurston's orbifold program was finally settled by M. Boileau, B. Leeb and J. Porti [**61**]. However, one needs to assume that an incompressible surface (or corresponding surface in case of orbifold) exists. When R. Hamilton [**311**] had his initial success on his Ricci flow, I suggested (around 1981) to him to use his flow to break up the manifold and prove Thurston's conjecture. His generalization of the theory of Li-Yau [**445**] to Ricci flow [**312, 313**] and his seminal paper in 1996 [**315**] on breaking up the manifold mark a cornerstone of the remarkable program. Perelman's recent idea [**551, 552**] built on these two works and has gone deeply into the problem. Detailed discussions have been pursued by Hamilton, Zhu, Cao, Colding-Minicozzi, Shioya-Yamaguchi, and Huisken in the past two years. Hopefully it may lead to the final settlement of the geometrization program. This theory of Hamilton and Perelman should be considered as a crowning achievement of geometric analysis in the past thirty years. Most ideas developed in this period by geometric analysts are used.

Let me now explain briefly the work of Hamilton and Perelman.

In the early 90's, Hamilton [**313, 314, 315**] developed methods and theorems to understand the structure of singularities of the Ricci flow. Taking up my suggestions, he proved a fundamental Li-Yau type differential inequality (now called the Li-Yau-Hamilton estimate) for the Ricci flow with non-negative curvature in all dimensions. He gave a beautiful interpretation of the work of Li-Yau and observed the associated inequalities should be equalities for solitary solutions. He then established a compactness theorem for (smooth) solutions to the Ricci flow, and observed (also independently by T. Ivey[**356**]) a pinching estimate for the curvature for three-manifolds. By imposing an injectivity radius condition, he rescaled the metric to show that each singularity is asymptotic to one of the three singularity models. For type I singularities in dimension three, Hamilton established an isoperimetric ratio estimate to verify the injectivity radius condition and obtained spherical or neck-like structures. Based on the Li-Yau-Hamilton estimate, Hamilton showed that any type II model is either a Ricci soliton with a neck-like structure or the product of the cigar soliton with the real line. Similar characterization for type III model was obtained by Chen-Zhu [**126**]. Hence Hamilton had already obtained the canonical neighborhood structures (consisting of spherical, neck-like and cap-like regions) for the singularities of three-dimensional Ricci flow.

But two obstacles remained: one is the injectivity radius condition and the other is the possibility of forming a singularity modelled on the product of the cigar soliton with a real line which could not be removed by surgery.

Recently, Perelman [**551**] removed these two stumbling blocks in Hamilton's program by establishing a local injectivity radius estimate (also called "Little Loop Lemma" by Hamilton in [**314**]). Perelman proved the Little Loop Lemma in two ways, one with an entropy functional he introduced in [**551**], the other with a reduced distance function based on the same idea as Li-Yau's path integral in obtaining their inequality [**551**]. This reduced distance question gives rise to a Gaussian type integral which he called reduced volume. The reduced volume satisfies monotonicity property. Furthermore, Perelman [**552**] developed a refined rescaling argument (by considering local limits and weak limits in Alexandrov spaces) for singularities of the Ricci flow on three-manifolds to obtain a uniform and global version of the canonical neighborhood structure theorem.

After obtaining the canonical neighborhoods for the singularities, one performs geometric surgery by cutting off the singularities and continuing the Ricci flow. In [**315**], Hamilton initiated such a surgery procedure for four-manifolds with a positive isotropic curvature. Perelman [**552**] adapted Hamilton's geometric surgery procedure to three-manifolds. The most important question is how to prevent the surgery times from accumulations and make sure there are only a finite number of surgeries on each finite time interval. When one performs the surgeries with a given accuracy at each surgery time, it is possible that the errors may add up, which causes the surgery times to accumulate. Hence at each step of surgery one is required to perform the surgery more accurately than the former one. In [**553**], Perelman presented a clever idea on how to find "fine" necks, how to glue "fine" caps and how to use rescaling arguments to justify the discreteness of the surgery times. In the process of rescaling for surgically modified solutions, one encounters the difficulty of how to use Hamilton's compactness theorem, which works only for smooth solutions. The idea to overcome such difficulty consists of two parts. The first part, due to Perelman [**552**], is to choose cutoff radius (in neck-like regions) small enough to push the surgical regions far away in space. The second part, due to Chen-Zhu [**130**] and Cao-Zhu [**103**], is to show that the solutions are smooth on some uniform small time intervals (on compact subsets) so that Hamilton's compactness theorem can be used.

Once surgeries are known to be discrete in time, one can complete Schoen-Yau's classification [**603**] for three-manifolds with positive scalar curvature. For simply connected three manifolds, if one can show solution to the Ricci flow with surgery extincts in finite time, Poincaré conjecture will be proved. Recently, such a finite extinction time result was proposed by Perelman [**553**] and a proof appeared in Colding-Minicozzi [**173**].

For the full geometrization program, one still needs to find the long time behavior of surgically modified solutions. In [**316**], Hamilton studied the long time behavior of the Ricci flow on a compact three-manifold for a special class of (smooth) solutions called "nonsingular solutions". Hamilton proved that any (three-dimensional) nonsingular solution either collapses or subsequently converges to a metric with constant curvature on the compact manifold, or at large time it admits a thick-thin decomposition where the thick part consists of a finite number of hyperbolic pieces and the thin part collapses. Moreover, by adapting Schoen-Yau's minimal surface arguments to a parabolic version, Hamilton showed that the boundary of hyperbolic pieces are incompressible tori. Then by combining with the collapsing results of Cheeger-Gromov [**120**], any nonsingular solution to the Ricci flow is geometrizable.

In [**551, 552**], Perelman modified Hamilton's arguments to analyze the long-time behavior of arbitrary solutions to the Ricci flow and solutions with surgery in dimension three. Perelman also argued by showing a thick-thin decomposition, except that he can only show the thin part has (local) lower bound on sectional curvature. For the thick part, based on Li-Yau-Hamilton estimate, Perelman established a crucial elliptic type Harnack estimate to conclude that the thick part consists of hyperbolic pieces. For the thin part, he announced a new collapsing result which states that if a three-manifold collapses with a (local) lower bound on the sectional curvature, then it is a graph manifold. However, the proof of the new collapsing result has not been published. Shioya and Yamaguchi [**616, 617**] offered a proof for compact manifolds. Very recently, Cao-Zhu claimed to have a complete proof for compact manifolds based only on the Shioya-Yamaguchi's collapsing result.

Hopefully all these arguments can be checked thoroughly in the near future. It should also be interesting to see whether other famous problems in three manifold can be settled by analysis: Does every three dimensional hyperbolic manifold admit a finite cover with nontrivial first Betti number?

Hyperbolic metrics have been used by topologists to give invariants for three dimensional manifolds. Thurston [**673**] observed that the volume of a hyperbolic metric is an important topological invariant. The associated Chern-Simons [**152**] invariant, which is defined by mod integers, can be looked upon as a phase for such manifolds. These invariants appeared later in Witten's theory of $2+1$ dimensional gravity [**719**] and S. Gukov [**304**] was able to relate them to fundamental questions in knot theory.

> **Comment**: This is the most spectacular development in the last thirty years. Once the three manifold is hyperbolic, Ricci flow does not give much more information. Perhaps, one may obtain further information by performing reduction from four dimension Ricci flow to three dimension by circle action. Is there any effective way to understand the totality of all hyperbolic manifolds with finite volume by constructing flows that may break up topology?

5.3. Four manifolds. The major accomplishment of Thurston, Hamilton, Perelman et al is the ability to create a canonical structure on three manifolds. Such a structure has not even been conjectured for four manifolds despite the great success of Donaldson invariants and Seiberg-Witten invariants. Taubes [**659**] did prove a remarkable existence theorem for self-dual metrics on a rather general class of four dimensional manifolds. Unfortunately their moduli space is not understood and their topological implication is not clear at this moment. Since the twistor space of Taubes metric admits integrable complex structure, ideas from complex geometry may be helpful. Prior to the construction of Taubes, Donaldson-Friedman [**205**] and LeBrun [**414**] have used ideas from twistor theory to construct self-dual metrics on the connected sum of $\mathbb{C}P^2$.

The problem of the four manifold is the lack of good diffeomorphic invariants. Donaldson or Seiberg-Witten provide such invariants. But they are not powerful enough to control the full structure of the manifold. A true understanding of four manifolds probably should come from understanding the question of existence of the integrable complex structures. The Riemann-Roch-Hirzebruch formula has been the basic tool to find the integrability condition. In the last twenty-five years, there are nonlinear methods from Käher-Einstein metrics, harmonic maps, anti-self-dual connections and Seiberg-Witten invariants. However, one needs an existence theorem to find a canonical way to deform an almost complex structure to an integrable complex structure. What kind of obstructions do we expect? The work of Donaldson [**199, 201**] and Gompf [**261**] gave a good characterization of symplectic manifolds in terms of Lefschetz fibration. It may be useful to know under which condition such fibration will give rise to complex structures. I did ask several of my students to work on it. But no definite answer is known. J Jost and I [**370**] studied the rigidity part: if a Kähler surface has a topological map to a Riemann surface with higher genus, it can be deformed to be a holomorphic map by changing the complex structure of the Riemann surface. One can derive from the work of Griffiths [**274**] that every algebraic surface has a Zariski open set which admits a complete Kähler-Einstein metric with finite volume and is covered by a contractible pseudo-convex domain. Perhaps one can classify these manifolds by topological means.

While the Donaldson invariant gave the first counterexample to the h-cobordism theorem and irreducibility (nontrivial connected sum with manifolds not homotopic to $\mathbb{C}P^2$) of four manifolds, the Seiberg-Witten invariant gave the remarkable result that an algebraic surface of general type can not be diffeomorphic to rational or elliptic surfaces. It also solves the famous Thom conjecture that holomorphic curves realize the lowest genus for embedded surface in a Kähler surface (Kronheimer-Mrowka [**402**] and Ozsváth-Szabó [**548**]). One wonders whether one can construct a diffeomorphic invariant based on metrics which are a generalization of Kähler-Einstein metrics.

Comment: A good conjectural statement needs to be made on the topology of four manifolds that may admit an integrable complex structure. Pseudo-holomorphic curve and fibration by Riemann surfaces should provide important information. Geometric flows may still be the major tool.

5.4. Special connections on bundles. In the seventies, theoretic physicists were very much interested in the theory of instantons: self-dual connections on four manifolds. Singer was able to communicate the flavor of this excitement to the mathematical community which soon led to his paper with Atiyah and Hitchin [**16**] and also the complete solution of the problem over the four sphere by Atiyah-Hitchin-Drinfel'd-Manin [**15**] using twistor technique of Penrose.

While the paper of Atiyah-Hitchin-Singer [**16**] laid the algebraic and geometric foundation for self-dual connections, the analytic foundation was laid by Uhlenbeck [**688, 689**] where she established the removable singularity theorem and compactness theorem for Yang-Mills connections. This eventually led to the fundamental works of Taubes [**656**] and Donaldson [**195**] which revolutionized four manifold topology.

In the other direction, Atiyah-Bott [**11**] applied Morse theory to the space of connections over Riemann surface. They solved important questions on the moduli space of holomorphic bundles which was studied by Narasimhan, Seshadri, Ramanathan, Newstead and Harder. In the paper of Atiyah-Bott, Morse theory, moment map and localization of equivariant cohomology were introduced on the subject of vector bundle. It laid the foundation of works in the last twenty years.

The analogue of anti-self dual connections over Kähler manifolds are Hermitian Yang-Mills connections, which was shown by Donaldson [**196**] for Kähler surfaces and Uhlenbeck-Yau [**691**] for general Kähler manifolds to be equivalent to the polystability of bundles. (That polystability of bundle is a consequence of the existence of Hermitian Yang-Mills connection was first observed by Lübke [**486**]. Donaldson [**197**] was able to make use of the theorem of Mehta-Ramanathan [**509**] and ideas of the above two papers to prove the theorem for projective manifold). It was generalized by C. Simpson [**621**], using ideas of Hitchin [**333**], to bundles with Higgs fields. It has important applications to the theory of variation of a Hodge structure [**622, 623**]. G. Daskalopoulos and R. Wentworth [**184**] studied such a theory for moduli space of vector bundles over curves. Li-Yau [**433**] generalized the existence of Hermitian Yang-Mills connections to non-Kähler manifolds. (Buchdahl [**87**] subsequently did the same for complex surfaces.) Li-Yau-Zheng [**435**] then used the result to give a complete proof of Bogomolov's theorem for class VII_0 surfaces. The only missing parts of the classification of non-Kähler surfaces are those complex surfaces with a finite number of holomorphic curves. It is possible that the argument of Li-Yau-Zheng can be used. One may want to use Hermitian Yang-Mills connections with poles along such curves. I expect more applications of Donaldson-Uhlenbeck-Yau theory to algebraic geometry.

It should be noted that the construction of Taubes [**659**] on the anti-self-dual connection is achieved by singular perturbation after gluing instantons from S^4. The method is rather different from Donaldson-Uhlenbeck-Yau. While it applies to arbitrary four manifolds, it does require some careful choice of Chern classes for the bundle. It will be nice to find a concept of stability for a general complex bundle so that a similar procedure of Donaldson-Uhlenbeck-Yau can be applied. The method of singular perturbation has an algebraic geometric counterpart as was found by Gieseker-Li [**255**] and O'Grady [**542**], who proved that moduli spaces of algebraic bundles with fixed Chern classes over algebraic surfaces are irreducible. Li [**427**] also obtained information about Betti number of such moduli space. Not many general theorems are known for bundles over algebraic manifolds of a higher dimension. It will be especially useful for bundles over Calabi-Yau manifolds.

D. Gieseker [**253**] developed the geometric invariant theory for the moduli space of bundles and introduced the Gieseker stability of bundles. Conan Leung [**419**] introduced the analytic counterpart of such bundles in his thesis under my guidance. While it is a natural concept, there is still an analytic problem to be resolved. (He assumed the curvature of the bundles to be uniformly bounded.)

There were attempts by de Bartolomeis-Tian [**43**] to generalize Yang-Mills theory to symplectic manifolds and also by Tian [**679**] to manifolds with a special holonomy group, as was initiated by the work of Donaldson and Thomas. However, the arguments for both papers are not complete and still need to be finished.

For a given natural structure on a manifold, we can often fix a structure and linearize the equation to obtain a natural connection on the tangent bundle. Usually we obtain Yang-Mills connections with the extra structure given by the holonomy group of the original structure. It is interesting to speculate whether an iterated procedure can be constructed to find an interesting metric or not. In any case, we can draw analogous properties between bundle theory and metric theory. The concept of stability for bundles is reasonably well understood for the holomorphic category. I believed that for each natural geometric structure, there should be a concept of stability. Donaldson [**197**] was able to explain stability in terms of moment map, generalizing the work of Atiyah-Bott [**11**] for bundles over Riemann surfaces. It will be nice to find moment maps for other geometric structures.

> **Comment**: Bundles with anti-self-dual connections or Hermitian Yang-Mills connections have been important for geometry. However, we do not have good estimates of the curvature of such connections. Such an estimate would be useful

> to handle important problems such as the Hartshorne question (see, e.g., [**39**]) on the splitting of rank two bundle over high dimension complex projective space.

5.5. Symplectic structures. Symplectic geometry had many important breakthroughs in the past twenty years. A moment map was developed by Atiyah-Bott [**12**], Guillemin-Sternberg [**301**] who proved the image of the map is a convex polytope. Kirwan and Donaldson had developed such a theory to be a powerful tool. The Marsden-Weinstein [**492**] reduction has become a useful method in many branches of geometry. At around the same time, other parts of symplectic topology were developed by Donaldson [**200**], Taubes [**660**], Gompf [**260**], Kronheimer-Mrowka [**402**] and others.

The phenomenon of symplectic rigidity is manifested by the existence of symplectic invariants measuring the 2–dimensional size of a symplectic manifold. The first such invariant was discovered by Gromov [**286**] via pseudo-holomorphic curves. Hofer [**340**] then developed several symplectic invariants based on variational methods and successfully applied them to Weinstein conjectures. Ekeland-Hofer [**214**] introduced a concept of symplectic capacity and used it to provide a characterization of a symplectomorphism not involving any derivatives. The C^0-closed property of the symplectomorphism group as a subgroup of the diffeomorphism group then follows, which was independently established by Y. Eliashberg [**215**] via wave front methods. Hofer-Zehnder [**343**] introduced another capacity and discovered the displacement-energy on $\mathbb{R}^{2n}$. By relating the two invariants with the energy-capacity inequality, Hofer [**339**] found a bi-invariant norm on the infinite dimensional group of Hamiltonian symplectomorphisms of $\mathbb{R}^{2n}$. The existence of such a norm has now been established for general symplectic manifolds by Lalonde-McDuff [**408**] via pseudo-holomorphic curves and symplectic embedding techniques. The generalized Weinstein conjecture on the existence of periodic orbits of Reeb flows for many 3-manifolds including the 3-sphere was also established in Hofer [**340**] by studying the finite energy pseudo-holomorphic plane in the symplectization of contact 3-manifolds.

Eliashberg-Givental-Hofer [**216**] recently introduced the concept of symplectic field theory, which is about invariants of punctured pseudo-holomorphic curves in a symplectic manifold with cylindrical ends. Though it has not been rigorously established, some applications in contact and symplectic topology have been found.

By analyzing the singularities of pseudo-holomorphic curves in a symplectic 4–manifold, D. McDuff [**498**] established rigorously the positivity of intersections of two distinct curves and the adjunction formula of an irreducible curve. Applying these basic properties to symplectic 4-manifolds containing embedded pseudo-holomorphic spheres with self-intersections at least -1, she was able to construct minimal models of general symplectic 4–manifolds, and classify those containing embedded symplectic spheres with non-negative self-intersections.

A fundamental question in symplectic geometry is to decide which topological manifold admits a symplectic structure and how, as was pointed out by Smith-Thomas-Yau [**635**], mirrors of certain non-Kähler complex manifolds should be symplectic manifolds. Based on this point of view, they construct a large class of symplectic manifolds with trivial first Chern class by reversing the procedure of Clemens-Friedman on non-Kähler Calabi-Yau manifolds [**165, 233**]. In dimension four, the Betti numbers of such manifolds are determined by T. J. Li [**446**]. In the last ten years, there has been extensive work on symplectic manifolds, initiated by Gromov [**286**], Taubes [**661, 662, 663, 664**], Donaldson [**198, 199, 201**] and Gompf [**261**]. These works are based on the understanding of pseudo-holomorphic curves and Lefschetz fibrations. They are most successful for four dimensional manifolds. The major tools are Seiberg-Witten theory [**607, 608, 721**] and analysis. The work of Taubes on the existence of pseudo-holomorphic curves and the topological meaning of its counting is one of the deepest works in geometry. Based on this work, Taubes [**661**] was able to prove the old conjecture that there is only one symplectic structure on the standard $\mathbb{C}P^2$. However, the following question of mine is still unanswered: If M is a symplectic 4-manifold homotopic to $\mathbb{C}P^2$, is M symplectomorphic to the standard $\mathbb{C}P^2$? (The corresponding question for complex geometry was solved by me in [**735**].) On the other hand, based on the work of Taubes [**660**], T.J. Li and A.K. Liu [**447**] did find a wall crossing formula for four dimensional manifolds that admit metrics with a positive scalar curvature. Subsequently A. Liu [**462**] gave the classification of such manifolds. (The surgery result by Stolz [**641**] based on Schoen-Yau-Gromov-Lawson for manifolds with positive scalar curvature is not effective for four dimensional manifolds.) As another application of the general wall crossing formula in [**447**], it was proved by T.J. Li and A. Liu in [**448**] that there is a unique symplectic structure on S^2-bundles over any Riemann surface. A main result of D. McDuff in [**497**] is used here.

McDuff [**496**] also used a refined bordism type Gromov-Witten invariant to distinguish two cohomologous and deformation equivalent symplectic forms on $S^2 \times S^2 \times T^2$, showing that they are not isotopic. Notice that there are also cohomologous but non-deformation equivalent symplectic forms on $K3 \times S^2$ as shown by Y. Ruan [**578**]. In contrast, it is not known whether examples of this kind exist in dimension 4 or not. This phenomenon might be related to the special features of pseudo-holomorphic curves in a 4-manifold.

Fukaya and Oh [**239**] have developed an elaborate theory for symplectic manifolds with Lagrangian cycles. Pseudo-holomorphic disks appeared as trace of motions of curves according to Floer theory. Due to boundary bubbles, the Lagrangian Floer homology is not always defined. Oh [**543**] developed some works on pseudo-holomorphic curves with Lagrangian boundary conditions and extended the Lagrangian Floer homology to all monotone symplectic manifolds. In order to understand open string theory,

Katz-Liu [**378**] and Melissa Liu [**463**] developed the theory in analogue of the Gromov-Witten invariant for a holomorphic curve with boundaries on a given Lagrangian submanifold. Fukaya [**238**] discovered the underlying A^∞ structure of the Lagrangian Floer homology on the chain level, leading to the Fukaya category. By carefully analyzing this A^∞ structure, Fukaya, Oh, Ohta and Ono in [**240**] have constructed a sequence of obstruction classes which elucidate the rather difficult Lagrangian Floer homology theory to a great extent. Seidel-Thomas [**609**] and W.D. Ruan [**577**] discussed Fukaya's category in relation to Kontsevich's homological mirror conjecture [**395**]. One wonders whether Fukaya's theory can help to construct canonical metrics for symplectic structures. For symplectic manifolds that admit an almost complex structure with zero first Chern class, it would be nice to construct Hermitian metrics with torsion that admit parallel spinor. Such structures may be considered as a mirror to the system constructed by Strominger on non-Kähler complex manifolds. Perhaps one can also gain some knowledge by reduction of G_2 or Spin(7) structures to six dimensions.

> **Comment**: Geometry from the symplectic point of view has seen powerful development in the past twenty years. Its relation to Seiberg-Witten theory and mirror geometry is fruitful. More interesting development is expected.

5.6. Kähler structure. The most interesting geometric structure is the Kähler structure. There are two interesting pre-Kähler structures. One is the complex structure and the other is the symplectic structure. The complex structure is rather rigid for complex two dimensional manifolds. However, it is much more flexible in dimension greater than two. For example, the twistor space of anti-self-dual four manifolds admits complex structures. Taubes [**659**] constructed a large class of such manifolds and hence a large class of complex three manifolds. There is also the construction of Clemens-Friedman for non-Kähler Calabi-Yau manifolds which will be explained later.

For quite a long time, it was believed that every compact Kähler manifold can be deformed to a projective manifold until C. Voisin [**697, 698**] found many counterexamples. We still need to digest the distinction between these two categories.

Besides some obvious topological obstruction from Hodge theory and the rational homotopic type theory of Deligne-Griffiths-Morgan-Sullivan [**188**], it has been difficult to decide which complex manifolds admit Kähler structure. The harmonic map argument does give some information. But it requires the fundamental group to be large.

Many years ago, Sullivan [**645**] proposed to use the Hahn-Banach theorem to construct Kähler metrics. This involves the concept of duality and hence closed currents. P. Gauduchon [**246**] has proposed those Hermitian metrics ω which is $\partial\bar{\partial}\omega^{n-1} = 0$. Siu [**627**] was able to use these ideas to prove that every $K3$ surface is Kähler. Demailly [**191**] did some remarkable work on regularization of closed positive currents. Singular Kähler metrics have been studied and used by many researchers. In fact, in my paper on proving the Calabi conjecture, I proved the existence of the Kähler metrics singular along subvarieties with control on volume element. They can be used to handle problems in algebraic geometry, including Chern number inequalities, and possibly problems arising in the minimal model program.

> **Comment**: The Kähler structure is one of the richest structures in geometry. Deeper understanding may require some more generalized structure such as a singular Kähler metric or balanced metrics.

5.6.1. *Calabi-Yau manifolds.* The construction of Calabi-Yau manifolds was based on the existence of a complex structure which can support a Kähler structure and a pluriharmonic volume form.

A fundamental question is whether an almost complex manifold admits an integrable complex structure when complex dimension is greater than two. The condition that the first Chern class is zero is equivalent to the existence of pluriharmonic volume for Kähler manifolds. Such a condition is no more true for non-Kähler manifolds. It would be nice to find a topological method to construct an integrable complex structure with pluriharmonic volume form.

Once we have an integrable complex structure, we can start to search for Hermitian metrics with special properties. As was mentioned earlier, if we would like the geometry to have supersymmetries, then a Kähler metric is the only choice if the connection is torsion free. Further supersymmetry would then imply the manifold to be Calabi-Yau. However, if we do not require the connection to be torsion free, Strominger [**642**] did derive a set of equations that exhibit supersymmetries without requiring the manifold to be Kähler. It is a coupled system of Hermitian Yang-Mills connections with Hermitian metrics. Twenty years ago, I tried to develop such a coupled system. The attempt was unsuccessful as I restricted myself to Kähler geometry. My student Bartnik with Mckinnon [**42**] did succeed in doing so in the Lorentzian case. They found non-singular solutions for such a coupled system. (The mathematically rigorous proof was provided by Smoller-Wasserman-Yau-Mcleod [**639**] and [**638**]).

The Strominger's system was shown to be solvable in a neighborhood of a Calabi-Yau structure by Jun Li and myself [**434**]. Fu and I [**235**] were also able to solve it for many complex manifolds which admit no Kähler structure. These manifolds are balanced manifolds and were studied by M. Michelsohn [**517**]. These manifolds can be used to explain some questions of flux in string theory (see, e.g., [**46, 106**]). Since Strominger has shown such manifolds admits parallel spinors, I have directed my student C.C. Wu to decompose cohomology group of such manifolds correspondingly. It is expected that many theorems in Kähler geometry may have counterparts in such geometry.

Such a structure may help to understand a proposal of Reid [**568**] in connecting Calabi-Yau manifolds with a different topology. This was initiated by a construction of Clemens [**165**] who proposed to perform complex surgery by blowing down rational curves with negative normal bundles in a Calabi-Yau manifold to rational double points. Friedman [**233**] found the condition to smooth out such singularities. Based on this Clemens-Friedman procedure, one can construct a complex structure on connected sums of $S^3 \times S^3$. It would be nice to construct Strominger's system on these manifolds.

The Calabi-Yau structure was used by me and others to solve important problems in algebraic geometry before it appeared in string theory. For example, the proof of the Torelli theorem (by Piatetskii-Shapiro and Shafarevich [**559**]) for a $K3$ surface by Todorov [**683**]-Siu [**627**] and the surjectivity of the period map of a $K3$ surface (by Kulikov [**404**]) by Siu [**626**]-Todorov [**683**] are important works for algebraic surfaces. The proof of the Bogomolov [**60**]-Tian [**675**]-Todorov [**684**] theorem also requires the metric. (One needs to use the statement that the holomorphic n-form is parallel. This was overlooked in [**675**].) The last theorem helps us to understand the moduli space of Calabi-Yau manifolds. It is important to understand the global behavior of the Weil-Petersson geometry for Calabi-Yau manifolds. C.L. Wang [**701**] was able to characterize these points which have finite distance in terms of the degeneration of the Hodge structure.

In my talk [**737**] in the Congress in 1978, I outlined the program and the results of classifying noncompact Calabi-Yau manifolds. Some of this work was written up in Tian-Yau [**681, 682**] and Bando-Kobayashi [**32, 33**]. During the period of 1984, there was an urgent request by string theorists to construct Calabi-Yau threefolds with a Euler number equal to ± 6. During the Argonne Lab conference, I [**740**] constructed such a manifold with a $\mathbb{Z}_3$ fundamental group by taking the quotient of a bi-degree $(1, 1)$ hypersurface in the product of two cubics. Soon afterwards, more examples were constructed by Tian and myself [**680**]. However, it was pointed out by Brian Greene that all the manifolds constructed by Tian-Yau can be deformed to my original manifold. The idea of producing Calabi-Yau manifolds by the complete intersection of hypersurfaces in products of weighted projective space was soon picked up by Candelas et al [**97**]. By now, on the order of ten thousand examples of different homotopic types had been constructed. The idea of using toric geometry for construction was first performed by S. Roan and myself [**569**]. A few years later, the systematic study by Batyrev [**44**] on toric geometry allowed one to construct mirror pairs for a large class of Calabi-Yau manifolds, generalizing the construction of Greene-Plesser [**271**] and Candelas et al [**97**]. Tian and I [**680**] were also the first one to apply flop construction to change topology of Calabi-Yau manifolds. Greene-Morrison-Plesser [**272**] then made the remarkable discovery of isomorphic quantum field theory on two topological distinct Calabi-Yau manifolds. Most Calabi-Yau threefolds are a complete intersection of some toric varieties and they admit a large set of rational curves. It will be important to understand the reason behind it. Up to now all the Calabi-Yau manifolds that have a Euler number ± 6 and a nontrivial fundamental group can be deformed from the birational model of the manifold (or their mirrors) that I constructed. It would be important if one could give a proof of this statement.

The most spectacular advancement on Calabi-Yau manifolds come from the work of Greene-Plesser, Candelas et al on construction of pairs of mirror manifolds with isomorphic conformal field theories attached to them. It allows one to calculate Gromov-Witten invariants. Existence of such mirror pairs was conjectured by Lerche-Vafa-Warner [**418**] and rigorous proof of mirror conjecture was due to Givental [**258**] and Lian-Liu-Yau [**449**] independently. The deep meaning of the symmetry is still being pursued.

In [**643**], Strominger, Yau and Zaslow proposed a mathematical explanation for the mirror symmetry conjecture for Calabi-Yau manifolds. Roughly speaking, mirror Calabi-Yau manifolds should admit special Lagrangian tori fibrations and the mirror transformation is a nonlinear analog of the Fourier transformation along these tori.

This proposal has opened up several new directions in geometric analysis. The first direction is the geometry of special Lagrangian submanifolds in Calabi-Yau manifolds. This includes constructions of special Lagrangian submanifolds ([**417**] and others) and (special) Lagrangian fibrations by Gross [**293, 294**] and W.D. Ruan [**576**], mean curvature of Lagrangian submanifolds in Calabi–Yau manifolds by Thomas and Yau [**670**] [**671**], structures of singularities on such submanifolds by Joyce [**376**], and Fourier transformations along special Lagrangian fibration by Leung-Yau-Zaslow [**424**] and Leung [**422**].

The second direction is affine geometry with singularities. As explained in [**643**], the mirror transformation at the large structure limit corresponds to a Legendre transformation of the base of the special lagrangian fibration which carries a natural special affine structure with singularities. Solving these affine problems is not trivial in geometric analysis [**473**] [**474**] and much work is still needed to be done here.

The third direction is the geometry of special holonomy and duality and triality transformation in M-theory. In [**305**], Gukov, Yau and Zaslow proposed a similar picture to explain the duality in M-theory. The corresponding differential geometric structures are fibrations on G_2 manifolds by coassociative submanifolds. These structures are studied by Kovalev [**400**], Leung and others [**416**] [**423**].

> **Comment**: Although the first demonstration of the existence of Kähler Ricci flat metric was shown by me in 1976, it was not until the first revolution of string theory in 1984

> that a large group of researchers did extensive calculations and changed the face of the subject. It is a subject that provides a good testing ground for analysis, geometry, physics, algebraic geometry, automorphic forms and number theory.

5.6.2. *Kähler metric with harmonic Ricci form and stability.* The existence of a Kähler Einstein metric with negative scalar curvature was proved by Aubin [**23**] and me [**736**] independently. I [**735**] found many important applications for solving classical problems in algebraic geometry, e.g., the uniqueness of complex structure over $\mathbb{C}P^2$ [**735**], the Chern number inequality of Miyaoka [**520**]-Yau [**735**] and the rigidity of algebraic manifolds biholomorphic to Shimura varieties. The problem of existence of Kähler Einstein metrics with positive scalar curvature in the general case is not solved. However, my proof of the Calabi conjecture already provided all the necessary estimates except some integral estimate on the unknown. This of course can be turned into hypothesis. I conjectured that an integral estimate of this sort is related to the stability of manifolds. Tian [**678**] called it K-stability. Mabuchi's functional [**489**] made the integral estimate more intrinsic and it gave rise to a natural variational formulation of the problem. Siu has pointed out that the work of Tian [**677**] on two dimensional surfaces is not complete. The work of Nadel [**535**] on the multiplier ideal sheaf did give useful methods for the subject of the Kähler-Einstein metric.

For Kähler Einstein manifolds with positive scalar curvature, it is possible that they admit a continuous group of automorphisms. Matsushima [**494**] was the first one to observe that such a group must be reductive. Futaki [**242**] introduced a remarkable invariant for general Kähler manifolds and proved that it must vanish for such manifolds. In my seminars in the eighties, I proposed that Futaki's theorem should be generalized to understand the projective group acting on the embedding of the manifold by a high power of anti-canonical embedding and that Futaki's invariant should be relevant to my conjecture [**743**] relating the Kähler Einstein manifold to stability. Tian asked what happens when manifolds have no group actions. I explained that the shadow of the group action is there once it is inside the projective space and one should deform the manifold to a possibly singular variety to obtain more information. The connection of Futaki invariant to stability of manifolds has finally appeared in the recent work of Donaldson [**202, 203**]. Donaldson introduced a remarkable concept of stability based purely on concept of algebraic geometry. It is not clear that Donaldson's algebraic definition has anything to do with Tian's analytic definition of stability. Donaldson proved that the existence of Kähler-Einstein did imply his K-stability which in turn implies Hilbert stability and asymptotic Chow stability of the manifold. This theorem of Donaldson already gives nontrivial information for manifolds with negative first Chern class and Calabi-Yau manifolds, where existence of Kähler-Einstein metrics was known. Some part of the deep work of Gieseker [**253**] and Viehweg [**694**] can be recovered by these theorems. One should also mention the recent interesting work of Ross-Thomas [**572, 573**] on the stability of manifolds. Phong-Strum [**555**] also studied solutions of certain degenerate Monge-Ampére equations and [**556**] the convergence of the Kähler-Ricci flow.

A Kähler metric with constant scalar curvature is equivalent to the harmonicity of the first Chern form. The uniqueness theorem for harmonic Kähler metric was due to X. Chen [**135**], Donaldson [**202**] and Mabuchi for various cases. (Note that the most important case of the uniqueness of the Kähler Einstein metric with positive scalar curvature was due to the remarkable argument of Bando-Mabuchi [**34**].) My general conjecture for existence of harmonic Kähler manifolds based on stability of such manifolds is still largely unknown. In my seminar in the mid-eighties, this problem was discussed extensively. Several students of mine, including Tian [**676**], Luo [**481**] and Wang [**709**] had written a thesis related to this topic. Prior to them, my former students Bando [**31**] and Cao [**100**] had made attempts to study the problem of constructing Kähler-Einstein metrics by Ricci flow. The fundamental curvature estimate was due to Cao [**101**]. The Kähler Ricci flow may either converge to Kähler Einstein metric or Kähler solitons. Hence in order for the approach, based on Ricci flow, to be successful, stability of the projective manifold should be related to such Kähler solitons. The study of harmonic Kähler metrics with constant scalar curvature on toric variety was initiated by S. Donaldson [**203**], who proposed to study the existence problem via the real Monge-Ampère equation. This problem of Donaldson in the Kähler-Einstein case was solved by Wang-Zhu [**708**]. LeBrun and his coauthors [**382**] also have found special constructions, based on twistor theory, for harmonic Kähler surfaces. Bando was also interested in Kähler manifolds with harmonic i-th Chern form. (There should be an analogue of stability of algebraic manifolds associated to manifolds with harmonic i-th Chern form.)

In the early 90's, S.W. Zhang [**754**] studied heights of manifolds. By comparing metrics on Deligne pairings, he found that a projective variety is Chow semistable if and only if it can be mapped by an element of a special linear group to a balanced subvariety. (Note that a subvariety in $\mathbb{C}P^N$ is called balanced if the integral of the moment map with respect to $SU(N+1)$ is zero, where the measure for the integral is induced from the Fubini-Study metric.) Zhang communicated his results to me. It is clearly related to Kähler-Einstein metric and I urged my students, including Tian, to study this connection.

Zhang's work has a nontrivial consequence on the previous mentioned development of Donaldson [**202, 203**]. Assume the projective manifold is embedded by an ample line bundle L into projective space. If the manifold has a finite automorphism group and admits a harmonic Kähler metric in $c_1(L)$, then Donaldson showed that for k large, L^k gives rise to an embedding which is balanced. Furthermore, the induced Fubini-Study form divided by

k will converge to the harmonic Kähler form. Combining the work of Zhang and Luo, he then proved that the manifold given by the embedding of L^k is stable in the sense of geometric invariant theory. Recently, Mabuchi generalized Donaldson's theorem to certain case which allow nontrivial projective automorphism.

Donaldson considered the problem from the point of view of symplectic geometry (Kähler form is a natural symplectic form). The Hamiltonian group then acts on the Hilbert space H of square integrable sections of the line bundle L where the first Chern class is the Kähler form. For each integrable complex structure on the manifold compatible with the symplectic form, the finite dimensional space of holomorphic sections gives a subspace of H. The Hamiltonian group acts on the Grassmannian of such subspaces. The moment map can be computed to be related to the Bergman kernel $\sum_\alpha s_\alpha(x) \otimes s^*_\alpha(y)$ where s_α form an orthonormal basis of the holomorphic sections. On the other hand, Fujiki [**236**] and Donaldson [**200**] computed the moment map for the Hamiltonian group action on the space of integrable complex structure, which turns out to be the scalar curvature of the Kähler metric. These two moment maps may not match, but for the line bundle L^k with large k, one can show that they converge to each other after normalization. Lu [**479**] has shown the first term of the expansion (in terms of $1/k$) of the Bergman kernel gives rise to scalar curvature. Hence we see the relevance of constant scalar curvature for a Kähler metric to these with a constant Bergman kernel function. S.W. Zhang's result says that the manifold is Chow semistable if and only if it is balanced. The balanced condition implies that there is a Kähler metric where the Bergman kernel is constant. With the work of Zhang and Donaldson, what remains to settle my conjecture is the convergence of the balanced metric when k is large. In general, we should not expect this to be true. However, for toric manifolds, this might be the case.

It may be noted that in my paper with Bourguignon and P. Li [**74**] on giving an upper estimate of the first eigenvalue of an algebraic manifold, this balanced condition also appeared. Perhaps the first eigenfunction may play a role for questions of stability.

> **Comment**: Kähler metrics with constant scalar curvature is a beautiful subject as it is related to structure questions of algebraic varieties including the concept of stability of manifolds. The most effective application of such metrics to algebraic geometry are still restricted to the Kähler-Einstein metric. The singular Kähler-Einstein metric as was initiated by my paper on Calabi conjecture should be studied further in application to algebraic geometry.

5.7. Manifolds with special holonomy group. Besides Kähler manifolds, there are manifolds with special holonomy groups. Holonomy groups of Riemannian manifolds were classified by Berger [**48**]. The most important ones are $O(n)$, $U(n)$, $SU(n)$, G_2 and $Spin(7)$. The first two groups correspond to Riemannian and Kähler geometry respectively. $SU(n)$ corresponds to Calabi-Yau manifolds. A G_2 manifold is seven dimensional and a $Spin(7)$ is eight dimensional (assuming they are irreducible manifolds). These last three classes of manifolds have zero Ricci curvature. It may be noted that before I [**736**] proved the Calabi conjecture in 1976, there was no known nontrivial compact Ricci flat manifold. Manifolds with a special holonomy group admit nontrivial parallel spinors and they correspond to supersymmetries in the language of physics. The input of ideas from string theory did give a lot of help to understand these manifolds. However, the very basic question of constructing these structures on a given topological space is still not well understood. In the case of G_2 and $Spin(7)$, it was initiated by Bryant (see [**84, 86**]). The first set of compact examples was given by Joyce [**373, 374, 375**]. Recently Dai-Wang-Wei [**183**] proved the stability of manifolds with parallel spinors.

The nice construction of Joyce was based on a singular perturbation which is similar to the construction of Taubes [**655**] on anti-self-dual connections. However, it is not global enough to give a good parametrization of G_2 or $Spin(7)$ structures. A great deal more work is needed. The beautiful theory of Hitchin [**337, 338**] on three forms and four forms may lead to a resolution of these important problems.

> **Comment**: Recent interest in M-theory has stimulated a lot of activities on manifolds with special holonomy group. We hope a complete structure theorem for such manifolds can be found.

5.8. Geometric structures by reduction. One can also obtain new geometric structures by imposing some singular structures on a manifold with a special holonomy group. For example, if we require a metric cone to admit a G_2, $Spin(7)$ or Calabi-Yau structure, the link of the cone will be a compact manifold with special structures. They give interesting Einstein metrics. When the cone is Calabi-Yau, the structure on the odd dimensional manifold is called Sasakian Einstein metric.

There is a natural Killing field called the Reeb vector field defined on a Sasakian Einstein manifold. If it generates a circle action, the orbit space gives rise to a Kähler Einstein manifold with positive scalar curvature. However, it need not generate a circle action and J. Sparks, Gauntlelt, Martelli and Waldram [**247**] gave many interesting explicit examples of non-regular Sasakian Einstein structures. They have interesting properties related to conformal field theory. For quasi-regular examples, there was work by Boyer, Galicki and Kollár [**76**]. The procedure gave many interesting examples of Einstein metrics on odd dimensional manifolds.

Sparks, Matelli and I have been pursuing general theory of Sasakian Einstein manifolds. I would like to consider them as a natural generalization of Kähler manifolds.

Comment: The recent development of the Sasakian Einstein metric shows that it gives a natural generalization of the Kähler-Einstein metric. Its relation with the recent activities on ADS/CFT theory is exciting.

5.9. Obstruction for existence of Einstein metrics on general manifolds. The existence of Einstein metrics on a fixed topological manifold is clearly one of the most important questions in geometry. Any metrics with a compact special holonomy group are Einstein. Besides Kähler geometry, we do not know much of their moduli space. For an Einstein metric with no special structures, we know only some topological constraints on four dimensional manifolds. There is work by Berger [**49**], Gray [**269**] and Hitchin [**332**] in terms of inequalities linking a Euler number and the signature of the manifold. (This is of course based on Chern's work [**149**] on the representation of characteristic classes by curvature forms.) Gromov [**285**] made use of his concept of Gromov volume to give further constraint. LeBrun [**415**] then introduced the ideas from Seiberg-Witten invariants to enlarge such classes and gave beautiful rigidity theorems on Einstein four manifolds. Unfortunately it is very difficult to understand moduli space of Einstein metrics when they admit no special structures. For example, it is still an open question of whether there is only one Einstein metric on the four dimensional sphere. M. Wang and Ziller [**707**] and C. Boehm [**58**] did use symmetric reductions to give many examples of Einstein metrics for higher dimensional manifolds. There may be much more examples of Einstein manifolds with negative Ricci curvature than we expected. This is certainly true for compact manifolds, with negative Ricci curvature. Gao-Yau [**243**] was the first one to demonstrate that such a metric exists on the three sphere. A few years later, Lokhamp [**475**] used the h-principle of Gromov to prove such a metric exists on any manifold with a dimension greater than three. It would be nice to prove that every manifold with a dimension greater than 4 admits an Einstein metric with negative Ricci curvature.

Comment: The Einstein manifold without extra special structures is a difficult subject. Do we expect a general classification for such an important geometric structure?

5.10. Metric Cobordism. In the last five years, a great deal of attention was addressed by physicists on the holographic principle: boundary geometry should determine the geometry in the interior. The ADS/CFT correspondence studies the conformal boundary of the Einstein manifold which is asymptotically hyperbolic. Gauge theory on the boundary is supposed to be dual to the theory of gravity in the bulk. Much fascinating work was done in this direction. Manifolds with positive scalar curvature appeared as conformal boundary are important for physics. Graham-Lee [**267**] have studied a perturbation problem near the standard sphere which bounds the hyperbolic manifold. Witten-Yau [**722**] proved that for a manifold with positive scalar curvature to be a conformal boundary, it must be connected. It is not known whether there are further obstructions.

Cobordism theory had been a powerful tool in the classification of the topology of manifolds. The first fundamental work was done by Thom who determined the cobordism group. Characteristic numbers play important roles. When two manifolds are cobordant to each other, the theory of surgery helps us to deform one manifold to another. It is clear that any construction of surgery that may preserve geometric structures would play a fundamental role in the future of geometry.

There are many geometric structures that are preserved under a connected sum construction. This includes the category of conformally flat structures, metrics with positive isotropic curvature and metrics with positive scalar curvature. For the last category, there was work by Schoen-Yau-Gromov-Lawson where they perform surgery on spheres with a codimension greater than or equal to 3. A key part of the work of Hamilton-Perelman is to find a canonical neighborhood to perform surgery. If we can deform the spheres in the above SYGL construction to a more canonically defined position, one may be able able to create an extra geometric structure for the result of SYGL. In fact, the construction of Schoen-Yau did provide some information about the conformal structures of the manifold. In complex geometry, there are two important canonical neighborhoods given by the log transform of Kodaira and the operation of flop. There should be similar constructions for other geometric structures.

The theory of quasi-local mass mentioned in Section 4.4 is another example of how boundary geometry can be controlled by the geometry in the bulk. The work of Choi-Wang [**155**] on the first eigenvalue is also based on the manifold that it bounds. There can be interesting theory of metric cobordism.

In the other direction, there are also beautiful rigidity of inverse problems for metric geometry by Gerver-Nadirashvili [**251**] and Pestov-Uhlmann [**554**] on recovering a Riemannian metric when one knows the distance functions between a pair of points on the boundary, if the Riemannian manifold is reasonably convex.

> **Comment**: There should be a mathematical foundation of the holographic principle of physicists. Good understanding of metric cobordism may be useful.

References

[1] D. Abramovich, *A linear lower bound on the gonality of modular curves*, Internat. Math. Res. Notices **20** (1996), 1005–1011.

[2] I. Agol, *Finiteness of arithmetic Kleinian reflection groups*, math.GT/0512560.

[3] S. I. Al'ber, *Higher-dimensional problems in global variational calculus*, Dokl. Akad. Nauk SSSR **156** (1964), 727–730.

[4] W.K. Allard, *On the first variation of a varifold: boundary behavior*, Ann. of Math. (2) **101** (1975), 418–446.
[5] C. Allendoerfer and J. Eells, *On the cohomology of smooth manifolds*, Comment. Math. Helv. **32** (1958), 165–179.
[6] F.J. Almgren Jr., *Almgren's big regularity paper. Q-valued functions minimizing Dirichlet's integral and the regularity of area-minimizing rectifiable currents up to codimension* 2, With a preface by Jean E. Taylor and Vladimir Scheffer, World Scientific Monograph Series in Mathematics, **1**, World Scientific Publishing Co., Inc., River Edge, NJ, 2000.
[7] F. Almgren and L. Simon, *Existence of embedded solutions of Plateau's problem*, Ann. Scuola Norm. Sup. Pisa Cl. Sci. (4) **6(3)** (1979), 447–495.
[8] M.T. Anderson, *The Dirichlet problem at infinity for manifolds of negative curvature*, J. Differential Geom. **18(4)** (1983), 701–721.
[9] M.T. Anderson and R. Schoen, *Positive harmonic functions on complete manifolds of negative curvature*, Ann. of Math. (2) **121(3)** (1985), 429–461.
[10] M. Atiyah, *New invariants of* 3- *and* 4-*dimensional manifolds*, The mathematical heritage of Hermann Weyl, 285–299, Proc. Sympos. Pure Math., **48**, Amer. Math. Soc., Providence, RI, 1988.
[11] M.F. Atiyah and R. Bott, *The Yang-Mills equations over Riemann surfaces*, Philos. Trans. Roy. Soc. London Ser. A **308(1505)** (1983), 523–615.
[12] M.F. Atiyah and R. Bott, *The moment map and equivariant cohomology*, Topology **23(1)** (1984), 1–28.
[13] M. Atiyah, R. Bott, and V.K. Patodi, *On the heat equation and the index theorem*, Invent. Math. **19** (1973), 279–330.
[14] M.F. Atiyah, H. Donnelly, and I.M. Singer, *Eta invariants, signature defects of cusps, and values of L-functions*, Ann. of Math. (2) **118(1)** (1983), 131–177.
[15] M.F. Atiyah, N.J. Hitchin, V.G. Drinfel'd, and Y.I. Manin, *Construction of instantons*, Phys. Lett. A **65(3)** (1978), 185–187.
[16] M.F. Atiyah, N.J. Hitchin, and I.M. Singer, *Self-duality in four-dimensional Riemannian geometry*, Proc. Roy. Soc. London Ser. A **362(1711)** (1978), 425–461.
[17] M.F. Atiyah, V.K. Patodi, and I.M. Singer, *Spectral asymmetry and Riemannian geometry*, I. Math. Proc. Cambridge Philos. Soc. **77** (1975), 43–69.
[18] M.F. Atiyah, V.K. Patodi, and I.M. Singer, *Spectral asymmetry and Riemannian geometry*, II, Math. Proc. Cambridge Philos. Soc. **78(3)** (1975), 405–432.
[19] M.F. Atiyah, V.K. Patodi, and I.M. Singer, *Spectral asymmetry and Riemannian geometry*, III, Math. Proc. Cambridge Philos. Soc. **79(1)** (1976), 71–99.
[20] M.F. Atiyah and I.M. Singer, *The index of elliptic operators on compact manifolds*, Bull. Amer. Math. Soc. **69** (1963), 422–433.
[21] T. Aubin, *Équations différentielles non linéaires et problème de Yamabe concernant la courbure scalaire*, J. Math. Pures Appl. (9) **55(3)** (1976), 269–296.
[22] T. Aubin, *Problèmes isopérimétriques et espaces de Sobolev*, J. Differential Geometry **11(4)** (1976), 573–598.
[23] T. Aubin, *Équations du type Monge-Ampère sur les variétés kähleriennes compactes*, C.R. Acad. Sci. Paris Sér. A-B **283(3)** (1976), Aiii, A119–A121.
[24] T. Aubin, *Espaces de Sobolev sur les variétés riemanniennes*, Bull. Sci. Math. (2) **100(2)** (1976), 149–173.
[25] V. M. Babič, *Eigenfunctions which are concentrated in the neighborhood of a closed geodesic*, Zap. Naučn. Sem. Leningrad. Otdel. Mat. Inst. Steklov (LOMI) **9** (1968), 15–63.
[26] W.L. Baily Jr. and A. Borel, *Compactification of arithmetic quotients of bounded symmetric domains*, Ann. of Math. (2) **84** (1966), 442–528.
[27] W. Ballmann, *On the Dirichlet problem at infinity for manifolds of nonpositive curvature*, Forum Math. **1(2)** (1989), 201–213.

[28] W. Ballmann and F. Ledrappier, *The Poisson boundary for rank one manifolds and their cocompact lattices*, Forum Math. **6(3)** (1994), 301–313.
[29] W. Ballmann and J. Świątkowski, *On L^2-cohomology and property (T) for automorphism groups of polyhedral cell complexes*, Geom. Funct. Anal. **7(4)** (1997), 615–645.
[30] W. Ballmann, G. Thorbergsson, and W. Ziller, *Closed geodesics on positively curved manifolds*, Ann. of Math. (2) **116(2)** (1982), 213–247.
[31] S. Bando, *On the classification of three-dimensional compact Kaehler manifolds of nonnegative bisectional curvature*, J. Differential Geom. **19(2)** (1984), 283–297.
[32] S. Bando and R. Kobayashi, *Ricci-flat Kähler metrics on affine algebraic manifolds*, Geometry and analysis on manifolds (Katata/Kyoto, 1987), 20–31, Lecture Notes in Math., **1339**, Springer, Berlin, 1988.
[33] S. Bando and R. Kobayashi, *Ricci-flat Kähler metrics on affine algebraic manifolds*, II, Math. Ann. **287(1)** (1990), 175–180.
[34] S. Bando and T. Mabuchi, *Uniqueness of Einstein Kähler metrics modulo connected group actions*, Algebraic geometry, Sendai, 1985, 11–40, Adv. Stud. Pure Math., **10**, North-Holland, Amsterdam, 1987.
[35] V. Bangert, *On the existence of closed geodesics on two-spheres*, Internat. J. Math. **4(1)** (1993), 1–10.
[36] V. Bangert and Y.M. Long, *The existense of two closed geodesics on every Finsler 2-sphere*, preprint, 2005, submitted.
[37] D. Bao, S.-S. Chern and Z. Shen, *An introduction to Riemann-Finsler geometry*, Graduate Texts in Mathematics, **200**, Springer-Verlag, New York, 2000.
[38] M. Barlow, T. Coulhon and A. Grigor'yan, *Alexander Manifolds and graphs with slow heat kernel decay*, Invent. Math. **144(3)** (2001), 609–649.
[39] W. Barth and A. Van de Ven, *A decomposability criterion for algebraic 2-bundles on projective spaces*, Invent. Math. **25** (1974), 91–106.
[40] R. Bartnik, *Existence of maximal surfaces in asymptotically flat spacetimes*, Comm. Math. Phys. **94(2)** (1984), 155–175.
[41] R. Bartnik, *Quasi-spherical metrics and prescribed scalar curvature*, J. Differential Geom. **37(1)** (1993), 31–71.
[42] R. Bartnik and J. McKinnon, *Particlelike solutions of the Einstein-Yang-Mills equations*, Phys. Rev. Lett. **61(2)** (1988), 141–144.
[43] P. de Bartolomeis and G. Tian, *Stability of complex vector bundles*, J. Differential Geom. **43(2)** (1996), 231–275.
[44] V. V. Batyrev, *Dual polyhedra and mirror symmetry for Calabi-Yau hypersurfaces in toric varieties*, J. Algebraic Geom. **3(3)** (1994), 493–535.
[45] A. Beauville, *Variétés Kähleriennes dont la première classe de Chern est nulle*, J. Differential Geom. **18(4)** (1983), 755–782.
[46] K. Becker and K. Dasgupta, *Heterotic strings with torsion*, J. High Energy Phys. **11** (2002), 006, 30 pp.
[47] E. Berger, R. Bryant, and P. Griffiths, *The Gauss equations and rigidity of isometric embeddings*, Duke Math. J. **50(3)** (1983), 803–892.
[48] M. Berger, *Sur les groupes d'holonomie homogène des variétés à connexion affine et des variétés riemanniennes*, Bull. Soc. Math. France **83** (1955), 279–330.
[49] M. Berger, *Sur quelques variétés d'Einstein compactes*, Ann. Mat. Pura Appl. (4) **53** (1961), 89–95.
[50] M. Bershadsky, S. Cecotti, H. Ooguri, and C. Vafa, *Kodaira-Spencer theory of gravity and exact results for quantum string amplitudes*, Comm. Math. Phys. **165(2)** (1994), 311–427.
[51] A.L. Besse, *Manifolds all of whose geodesics are closed*, With appendices by D.B.A. Epstein, J.-P. Bourguignon, L. Bérard-Bergery, M. Berger and J.L. Kazdan, Ergebnisse der Mathematik und ihrer Grenzgebiete, **93**, Springer-Verlag, Berlin-New York, 1978.

[52] G. Besson, G. Courtois, and S. Gallot, *Entropies et rigidités des espaces localement symétriques de courbure strictement négative*, Geom. Funct. Anal. **5**(**5**) (1995), 731–799.

[53] J.M. Bismut, *The Atiyah-Singer index theorem for families of Dirac operators: two heat equation proofs*, Invent. Math. **83**(**1**) (1986), 91–151.

[54] J.M. Bismut and F. Labourie, *Symplectic geometry and the Verlinde formulas*, Surveys in differential geometry: differential geometry inspired by string theory, 97–311, Surv. Differ. Geom., **5**, Int. Press, Boston, MA, 1999.

[55] J.M. Bismut and S. Goette, *Equivariant de Rham torsions*, Ann. of Math. (2) **159**(**1**) (2004), 53–216.

[56] J.M. Bismut and W.P. Zhang, *An extension of a theorem by Cheeger and Müller*, With an appendix by François Laudenbach, Astérisque **205** (1992).

[57] J.M. Bismut and W.P. Zhang, *Milnor and Ray-Singer metrics on the equivariant determinant of a flat vector bundle*, Geom. Funct. Anal. **4**(**2**) (1994), 136–212.

[58] C. Boehm, *Inhomogeneous Einstein metrics on low-dimensional spheres and other low-dimensional spaces*, Inv. Math. **134** (1998), 145-176.

[59] S. Bochner, *Curvature and Betti numbers*, Ann. of Math. (2) **49** (1948), 379–390.

[60] F. A. Bogomolov, *Hamiltonian Kälerian manifolds*, Dokl. Akad. Nauk SSSR **243**(**5**) (1978), 1101–1104.

[61] M. Boileau, B. Leeb, and J. Porti, *Uniformization of small* 3*-orbifolds*, C.R. Acad. Sci. Paris Ser. I Math. **332**(**1**) (2001), 57–62.

[62] E. Bombieri and E. Giusti, *Harnack's inequality for elliptic differential equations on minimal surfaces*, Invent. Math. **15** (1972), 24–46.

[63] E. Bombieri, E. De Giorgi, and E. Giusti, *Minimal cones and the Bernstein problem*, Invent. Math. **7** (1969), 243–268.

[64] R.E. Borcherds, *Automorphic forms on* $O_{s+2,2}(R)$ *and infinite products*, Invent. Math. **120**(**1**) (1995), 161–213.

[65] R.E. Borcherds, *The moduli space of Enriques surfaces and the fake Monster Lie superalgebra*, Topology **35**(**3**) (1996), 699–710.

[66] A. Borel, *Some metric properties of arithmetic quotients of symmetric spaces and an extension theorem*, Collection of articles dedicated to S.S. Chern and D.C. Spencer on their sixtieth birthdays, J. Differential Geometry **6** (1972), 543–560.

[67] A. Borel and J.-P. Serre, *Corners and arithmetic groups. Avec un appendice: Arrondissement des variétés à coins, par A. Douady et L. Hérault*, Comment. Math. Helv. **48** (1973), 436–491.

[68] R. Bott, *Nondegenerate critical manifolds*, Ann. of Math. (2) **60** (1954), 248-261.

[69] R. Bott, *On manifolds all of whose geodesics are closed*, Ann. of Math. (2) **60** (1954), 375–382.

[70] R. Bott, *On the iteration of closed geodesics and the Sturm intersection theory*, Comm. Pure Appl. Math. **9** (1956), 171–206.

[71] R. Bott, *The stable homotopy of the classical groups*, Ann. of Math. (2) **70** (1959), 313–337.

[72] R. Bott, *A residue formula for holomorphic vector-fields*, J. Differential Geometry **1** (1967), 311–330.

[73] R. Bott and C.H. Taubes, *Clifford On the rigidity theorems of Witten*, J. Amer. Math. Soc. **2**(**1**) (1989), 137–186.

[74] J.-P. Bourguignon, P. Li, and S.-T. Yau, *Upper bound for the first eigenvalue of algebraic submanifolds*, Comment. Math. Helv. **69**(**2**) (1994), 199–207.

[75] L. Boutet de Monvel and J. Sjöstrand, *Sur la singularité des noyaux de Bergman et de Szegö*, Journées: Équations aux Dérivées Partielles de Rennes (1975), Asterisque, **34-35**, Soc. Math. France, Paris, 1976, 123–164.

[76] C.P. Boyer, K. Galicki, and J. Kollár, *Einstein Metrics on Spheresmath*, arXiv: math.DG/0309408.

[77] P.J. Braam and S.K. Donaldson, *Floer's work on instanton homology, knots and surgery*, The Floer memorial volume, 195–256, Progr. Math., **133**, Birkhäuser, Basel, 1995.
[78] K.A. Brakke, *The motion of a surface by its mean curvatur*, Mathematical Notes, **20**, Princeton University Press, Princeton, NJ, 1978.
[79] H. Brascamp and E. Lieb, *On extensions of the Brunn-Minkowski and Prékopa-Leindler theorems, including inequalities for log concave functions, and with an application to the diffusion equation*, J. Functional Analysis **22(4)** (1976), 366–389
[80] H.L. Bray, *Proof of the Riemannian Penrose inequality using the positive mass theorem*, J. Differential Geom. **59(2)** (2001), 177–267.
[81] R. Brooks, *The fundamental group and the spectrum of the Laplacian*, Comment. Math. Helv. **56(4)** (1981), 581C598.
[82] J.D. Brown and J.W. York, *Quasilocal energy and conserved charges derived from the gravitational action*, Phys. Rev. D (3) **47(4)** (1993), 1407–1419.
[83] R. Bryant, *Conformal and minimal immersions of compact surfaces into the 4-sphere*, J. Differential Geom. **17(3)** (1982), 455–473.
[84] R. Bryant, *Metrics with exceptional holonomy*, Ann. of Math. (2) **126(3)** (1987), 525–576.
[85] R. Bryant, P. Griffiths, and D. Yang, *Characteristics and existence of isometric embeddings*, Duke Math. J. **50(4)** (1983), 893–994.
[86] R. Bryant and S. Salamon, *On the construction of some complete metrics with exceptional holonomy*, Duke Math. J. **58** (1989), 829–850.
[87] N.P. Buchdahl, *Hermitian-Einstein connections and stable vector bundles over compact complex surfaces*, Math. Ann. **280(4)** (1988), 625–648.
[88] Y. Burago, M. Gromov, and G.A.D. Perelman, *Aleksandrov spaces with curvatures bounded below*, Uspekhi Mat. Nauk **47(2)(284)** (1992), 3–51, 222; translation in Russian Math. Surveys **47(2)** (1992), 1–58.
[89] F.E. Burstall, D. Ferus, F. Pedit, and U. Pinkall, *Harmonic tori in symmetric spaces and commuting Hamiltonian systems on loop algebras*, Ann. of Math. (2) **138(1)** (1993), 173–212.
[90] A. Butscher, *Regularizing a singular special Lagrangian variety*, Comm. Anal. Geom. **12(4)** (2004), 733–791.
[91] L. Caffarelli, R. Hardt, and L. Simon, *Minimal surfaces with isolated singularities*, Manuscripta Math. **48(1-3)** (1984), 1–18.
[92] L. Caffarelli, L. Nirenberg, and J. Spruck, *The Dirichlet problem for nonlinear second order elliptic equations,* III. *Functions of the eigenvalues of the Hessian*, Acta Math. **155** (1985), 261–301.
[93] E. Calabi, *On Kähler manifolds with vanishing canonical class*, Algebraic geometry and topology. A symposium in honor of S. Lefschetz, 78–89, Princeton University Press, Princeton, NJ, 1957.
[94] E. Calabi, *Improper affine hyperspheres of convex type and a generalization of a theorem by K. Jörgens*, Michigan Math. J. **5** (1958) 105–126.
[95] E. Calabi and E. Vesentini, *Sur les variétés complexes compactes localement symétriques*, Bull. Soc. Math. France **87** (1959), 311–317.
[96] C. Callias, *Axial anomalies and index theorems on open spaces*, Comm. Math. Phys. **62(3)** (1978), 213–234.
[97] P. Candelas, A.M. Dale, C.A. Lütken, and R.C. Schimmrigk, *Complete intersection Calabi-Yau manifolds*, Frontiers of high energy physics (London, 1986), 88–134, Hilger, Bristol, 1987.
[98] P. Candelas, G.T. Horowitz, A. Strominger, and E. Witten, *Vacuum configurations for superstrings*, Nuclear Phys. B **258(1)** (1985), 46–74.

[99] P. Candelas, X.C. de la Ossa, P. S. Green, and L. Parkes, *A pair of Calabi-Yau manifolds as an exactly soluble superconformal theory*, Nuclear Phys. B **359(1)** (1991), 21–74.

[100] H.D. Cao, *Deformation of Kähler metrics to Kähler-Einstein metrics on compact Kähler manifolds*, Invent. Math. **81(2)** (1985), 359–372.

[101] H.D. Cao, *On Harnack inequalities for the Kähler-Ricci flow*, Invent. Math. **109** (1992), 247–263.

[102] H.D. Cao, *Limits of solutions to the Kähler-Ricci flow*, J. Differential Geom. **45(2)** (1997), 257–272.

[103] H.D. Cao and X.P. Zhu, *Ricci flow and its application.*

[104] J.G. Cao and J.F. Escobar, *A new 3-dimensional curvature integral formula for PL-manifolds of non-positive curvature*, Comm. Anal. Geom. **11(3)** (2003), no. 3, 489–551.

[105] J.G. Cao, H.J. Fan, and F. Ledrappier, *Martin points on open manifolds of non-positive curvature*, to appear in Trans. AMS.

[106] G.L. Cardoso, G. Curio, G. Dall'Agata, and D. Lüst, *Heterotic string theory on non-Kähler manifolds with H-flux and gaugino condensate*, Proceedings of the 36th International Symposium Ahrenshoop on the Theory of Elementary Particles. Fortschr. Phys. **52(6-7)** (2004), 483–488.

[107] M. do Carmo and N. Wallach, *Minimal immersions of spheres into spheres*, Proc. Nat. Acad. Sci. U.S.A. **63** (1969), 640–642.

[108] E. Cartan, *Sur des familles remarquables d'hypersurfaces isoparamétriques dans les espaces sphériques*, Math. Z. **45** (1939), 335–367.

[109] B. Carter, *Global Structure of the Kerr Family of Gravitational Fields*, Phys. Rev. **174** (1968), 1559–1571.

[110] D. Catlin, *The Bergman kernel and a theorem of Tian*, Analysis and geometry in several complex variables (Katata, 1997), 1–23, Trends Math., Birkhäuser Boston, Boston, MA, 1999.

[111] S.C. Chang, *Recent developments on the Calabi flow*, Geometric evolution equations, 17–42, Contemp. Math., **367**, Amer. Math. Soc., Providence, RI, 2005.

[112] S.Y.A. Chang, M.J. Gursky, and P. Yang, *An a priori estimate for a fully nonlinear equation on four-manifolds*, Dedicated to the memory of Thomas H. Wolff, J. Anal. Math. **87** (2002), 151–186.

[113] A. Chau and L.-F. Tam, *On the complex structure of Kähler manifolds with non-negative curvature*, arXiv:math.DG/0504422.

[114] J. Cheeger, *A lower bound for the smallest eigenvalue of the Laplacian*, Problems in analysis (Papers dedicated to Salomon Bochner, 1969), 195–199, Princeton Univ. Press, Princeton, NJ, 1970.

[115] J. Cheeger, *Analytic torsion and Reidemeister torsion*, Proc. Nat. Acad. Sci. USA **74(7)** (1977), 2651–2654.

[116] J. Cheeger and T.H. Colding, *On the structure of spaces with Ricci curvature bounded below*, I, J. Differential Geom. **46(3)** (1997), 406–480.

[117] J. Cheeger and T.H. Colding, *On the structure of spaces with Ricci curvature bounded below*, II, J. Differential Geom. **54(1)** (2000), 13–35.

[118] J. Cheeger and T.H. Colding, *On the structure of spaces with Ricci curvature bounded below*, III, J. Differential Geom. **54(1)** (2000), 37–74.

[119] J. Cheeger and D. Gromoll, *The splitting theorem for manifolds of nonnegative Ricci curvature*, J. Differential Geometry **6** (1971/72), 119–128.

[120] J. Cheeger and M. Gromov, *Collapsing Riemannian manifolds while keeping their curvature bounded*, I, J. Differential Geom. **23(3)** (1986), 309–346.

[121] J. Cheeger and M. Gromov, *Chopping Riemannian manifolds*, Differential geometry, 85–94, Pitman Monogr. Surveys Pure Appl. Math., **52**, Longman Sci. Tech., Harlow, 1991.

[122] J. Cheeger, M. Gromov, and M. Taylor, *Finite propagation speed, kernel estimates for functions of the Laplace operator, and the geometry of complete Riemannian manifolds*, J. Differential Geom. **17(1)** (1982), 15–53.

[123] J. Cheeger and J. Simons, *Differential characters and geometric invariants*, Geometry and topology (College Park, Md., 1983/84), 50–80, Lecture Notes in Math., **1167**, Springer, Berlin, 1985.

[124] J. Cheeger and S.-T. Yau, *A lower bound for the heat kernel*, Comm. Pure Appl. Math. **34(4)** (1981), 465–480.

[125] B.L. Chen, S.H. Tang, and X.P. Zhu, *A uniformization theorem for complete non-compact Kähler surfaces with positive bisectional curvature*, J. Differential Geom. **67(3)** (2004), 519–570.

[126] B.L. Chen and X.P. Zhu, *Complete Riemannian manifolds with pointwise pinched curvature*, Invent. Math. **140(2)** (2000), 423–452

[127] B.L. Chen and Z.X. Ping, *On complete noncompact Kähler manifolds with positive bisectional curvature*, Math. Ann. **327(1)** (2003), 1–23.

[128] B. L. Chen and X.P. Zhu, *Volume growth and curvature decay of positive curved Kähler manifold*, arXiv:math.DG/0211374.

[129] B.L. Chen and X.P. Zhu, *Flow with Surgery on Four-manifolds with Positive Isotropic Curvature*, arXiv:math.DG/0504478.

[130] B.L. Chen and X.P. Zhu, *Uniqueness of the Ricci Flow on Complete Noncompact Manifolds*, arXiv:math.DG/0505447.

[131] C.C. Chen and C.S. Lin, *A spherical Harnack inequality for singular solutions of nonlinear elliptic equations*, Ann. Scuola Norm. Sup. Pisa Cl. Sci. (4) **30(3-4)** (2001), 713–738.

[132] K.-T. Chen, *Differential forms and homotopy groups*, J. Differential Geometry **6** (1971/72), 231–246.

[133] M.F. Chen and F.Y. Wang, *Application of coupling method to the first eigenvalue on manifold*, Progr. Natur. Sci. (English Ed.) **5(2)** (1995), 227–229.

[134] Q. Chen, J. Jost, G.F. Wang, and J.Y. Li, *Regularity theorems and energy identities for Dirac-harmonic maps*, Math.DG/0411327.

[135] X.X. Chen, *The space of Kähler metrics*, J. Differential Geom. **56(2)** (2000), 189–234.

[136] Y.G. Chen, Y. Giga, and S. Goto, *Uniqueness and existence of viscosity solutions of generalized mean curvature flow equations*, J. Differential Geom. **33(3)** (1991), 749–786.

[137] S.Y. Cheng, *Eigenvalue comparison theorems and its geometric applications* Math. Z. **143(3)** (1975), 289–297.

[138] S.Y. Cheng, *Eigenfunctions and nodal sets*, Comment. Math. Helv. **51(1)** (1976), 43–55.

[139] S.Y. Cheng, P. Li, and S.-T. Yau, *On the upper estimate of the heat kernel of a complete Riemannian manifold*, Amer. J. Math. **103(5)** (1981), 1021–1063.

[140] S.Y. Cheng, P. Li, and S.-T. Yau, *Heat equations on minimal submanifolds and their applications*, Amer. J. Math. **106(5)** (1984), 1033–1065.

[141] S.Y. Cheng and S.-T. Yau, *Differential equations on Riemannian manifolds and their geometric applications*, Comm. Pure Appl. Math. **28(3)** (1975), 333–354.

[142] S.Y. Cheng and S.-T. Yau, *On the regularity of the Monge-Ampère equation* $\det(\partial^2 u/\partial x_i \partial s x_j) = F(x,u)$, Comm. Pure Appl. Math. **30(1)** (1977), 41–68.

[143] S.Y. Cheng and S.-T. Yau, *Maximal space-like hypersurfaces in the Lorentz-Minkowski spaces*, Ann. of Math. (2) **104(3)** (1976), 407–419.

[144] S. Y. Cheng and S.-T. Yau, *On the regularity of the solution of the* n*-dimensional Minkowski problem*, Commun. Pure Appl. Math. **29** (1976), 495–516.

[145] S.Y. Cheng and S.-T. Yau, *On the existence of a complete Kähler metric on non-compact complex manifolds and the regularity of Fefferman's equation*, Comm. Pure Appl. Math. **33(4)** (1980), 507–544.
[146] S.Y. Cheng and S.-Y. Yau, *The real Monge-Ampère equation and affine flat structures*, in Proceedings of the 1980 Beijing Symposium on Differential Geometry and Differential Equations, Science Press, Beijing, China (1982), Gordon and Breach Science Publishing Company, New York, 339–370.
[147] S.Y. Cheng and S.-T. Yau, *Complete affine hypersurfaces.* I. *The completeness of affine metrics*, Comm. Pure Appl. Math. **39(6)** (1986), 839–866.
[148] *Proceedings of the* 1980 *Beijing Symposium on Differential Geometry and Differential Equations*, Vol. **1**, **2**, **3**, Held in Beijing, August 18–September 21, 1980, Edited by S.-S. Chern and Wen Tsün Wu [Wen Jun Wu], Science Press, Beijing; Gordon and Breach Science Publishers, New York, 1982.
[149] S.-S. Chern, *Characteristic classes of Hermitian manifolds*, Ann. of Math. (2) **47** (1946), 85–121.
[150] S.-S. Chern, *On minimal spheres in the four-sphere*, 1970 Studies and Essays (Presented to Yu-why chen on his 60th Birthday, April 1, 1970), 137–150, Math. Res. Center, Nat. Taiwan Univ., Taipei.
[151] S.-S. Chern, M. de Carmo, and S. Kobayashi, *Minimal submanifolds of a sphere with second fundamental form of constant length*, Functional Analysis and Related Fields (Proc. Conf. for M. Stone, Univ. Chicago, Chicago, Ill., 1968), Springer, New York, 1970, pp. 59C75.
[152] S.-S. Chern and J. Simons, *Characteristic forms and geometric invariants*, Ann. of Math. (2) **99** (1974), 48–69.
[153] H.I. Choi and R. Schoen, *The space of minimal embeddings of a surface into a three-dimensional manifold of positive Ricci curvature*, Invent. Math. **81(3)** (1985), 387–394.
[154] H.I. Choi and A. Treibergs, *Gauss maps of spacelike constant mean curvature hypersurfaces of Minkowski space*, J. Differential Geom. **32(3)** (1990), 775–817.
[155] H.I. Choi and A.N. Wang, *A first eigenvalue estimate for minimal hypersurfaces*, J. Differential Geom. **18(3)** (1983), 559–562.
[156] S.Y. Choi and W.M. Goldman, *The classification of real projective structures on compact surfaces*, Bull. Amer. Math. Soc. (N.S.) **34(2)** (1997), 161–171.
[157] Y. Choquet-Bruhat and V. Moncrief, *Future global in time Einsteinian spacetimes with* U(1) *isometry group*, Ann. Henri Poincare **2(6)** (2001), 1007–1064.
[158] Y. Choquet-Bruhat and J.W. York, *On H. Friedrich's formulation of the Einstein equations with fluid sources*, Topol. Methods Nonlinear Anal. **18(2)** (2001), 321–335.
[159] D. Christodoulou, *The instability of naked singularities in the gravitational collapse of a scalar field*, Ann. of Math. (2) **149(1)** (1999), 183–217.
[160] D. Christodoulou and S. Klainerman, *The global nonlinear stability of the Minkowski space*, Princeton Mathematical Series, **41**, Princeton University Press, Princeton, NJ, 1993.
[161] D. Christodoulou and A. Tahvildar-Zadeh, *On the regularity of spherically symmetric wave maps*, Comm. Pure Appl. Math. **46(7)** (1993), 1041–1091.
[162] D. Christodoulou and S.-T. Yau, *Some remarks on the quasi-local mass*, Mathematics and general relativity, 9–14, Contemp. Math., **71**, Amer. Math. Soc., Providence, RI, 1988.
[163] P.T. Chrúsciel, *Semi-global existence and convergence of solutions of the Robinson-Trautman* (2-*dimensional Calabi*) *equation*, Comm. Math. Phys. **137(2)** (1991), 289–313.
[164] F. Chung, *Discrete Isoperimetric Inequalities*, Surveys in Differential Geometry, IX, Inrernational Press, 53–82.
[165] C.H. Clemens, *Double solids*, Adv. in Math. **47(2)** (1983), 107–230.

[166] J. Cogdell, J.S. Li, I. Piatetski-Shapiro, and P. Sarnak, Poincaré series for SO$(n, 1)$, Acta Math. **167(3-4)** (1991), 229–285.
[167] T.H. Colding, *Shape of manifolds with positive Ricci curvature*, Invent. Math. **124(1-3)** (1996), 175–191.
[168] T.H. Colding and W.P. Minicozzi II, *Harmonic functions on manifolds*, Ann. of Math. (2) **146(3)** (1997), 725–747.
[169] T.H. Colding and W.P. Minicozzi II, *The space of embedded minimal surfaces of fixed genus in a 3-manifold.* I. *Estimates off the axis for disks*, Ann. of Math. (2) **160(1)** (2004), 27–68.
[170] T.H. Colding and W.P. Minicozzi II, *The space of embedded minimal surfaces of fixed genus in a 3-manifold.* II. *Multi-valued graphs in disks*, Ann. of Math. (2) **160(1)** (2004), 69–92.
[171] T.H. Colding and W.P. Minicozzi II, *The space of embedded minimal surfaces of fixed genus in a 3-manifold.* III. *Planar domains*, Ann. of Math. (2) **160(2)** (2004), 523–572.
[172] T.H. Colding and W.P. Minicozzi II, *The space of embedded minimal surfaces of fixed genus in a 3-manifold.* IV. *Locally simply connected*, Ann. of Math. (2) **160(2)** (2004), 573–615.
[173] T.H. Colding and W.P. Minicozzi II, *Estimates for the extinction time for the Ricci flow on certain 3-manifolds and a question of Perelman*, J. Amer. Math. Soc. **18(3)** (2005), 561–569.
[174] T.H. Colding and W.P. Minicozzi II, *An excursion into geomtric analysis*, Surveys in Differential Geomtry, **IX**, International Press, 83–146.
[175] Y. Colin de Verdière, *Sur la multiplicitë de la première valeur propre non nulle du laplacien*, Comment. Math. Helv. **61(2)** (1986), 254–270.
[176] R. Connelly, *The rigidity of suspensions*, J. Differential Geom. **13(3)** (1978), 399–408.
[177] K. Corlette, *Archimedean superrigidity and hyperbolic geometry*, Ann. of Math. (2) **135(1)** (1992), 165–182.
[178] C.B. Croke, *Some isoperimetric inequalities and eigenvalue estimates*, Ann. Sci. ecole Norm. Sup. (4) **13(4)** (1980), 419–435.
[179] C.B. Croke, *A sharp four-dimensional isoperimetric inequality*, Comment. Math. Helv. **59(2)** (1984), 187–192.
[180] M. Dafermos, *Exhaustions of complete manifolds of bounded curvature*, Tsing Hua lectures n geometry and analysis, 319–322, Internat. Press, Cambridge, MA, 1997.
[181] M. Dafermos, *Stability and instability of the Cauchy horizon for the spherically symmetric Einstein-Maxwell-scalar field equations*, Ann. of Math. (2) **158(3)** (2003), 875–928.
[182] X.Z. Dai, K.F. Liu, and X.N. Ma, *On the asymptotic expansion of Bergman kernel*, C.R. Math. Acad. Sci. Paris **339(3)** (2004), 193-198, The full version: J. Differential Geom., to appear, math.DG/0404494.
[183] X.Z. Dai, X.D. Wang, and G.F. Wei, *On the stability of Riemannian manifold with parallel spinors*, Invent. Math. **161** (2005), 151 - 176.
[184] G.D. Daskalopoulos and R.A. Wentworth, *Geometric quantization for the moduli space of vector bundles with parabolic structure*, Geometry, topology and physics, 119–155, de Gruyter, Berlin, 1997.
[185] E.B. Davies, *Gaussian upper bounds for the heat kernels of some second-order operators on Riemannian manifolds*, J. Funct. Anal. **80(1)** (1988), 16–32.
[186] E.B. Davies, *An eigenvalue bound for compact manifolds with varying curvature*, Ann. Global Anal. Geom. **7(2)** (1989), 107–114.
[187] P. Deligne, *Théorie de Hodge*, II, Inst. Hautes études Sci. Publ. Math. No. **40** (1971), 5–57.

[188] P. Deligne, P. Griffiths, J. Morgan, and D. Sullivan, *Real homotopy theory of Kähler manifolds*, Invent. Math. **29(3)** (1975), 245–274.
[189] P. Deligne and D. Mumford, *The irreducibility of the space of curves of given genus*, Inst. Hautes études Sci. Publ. Math. No. **36** (1969), 75–109.
[190] J. Demailly, *Holomorphic Morse inequalities. Several complex variables and complex geometry*, 93–114, Proc. Sympos. Pure Math., **52**, Part 2, Amer. Math. Soc., Providence, RI, 1991.
[191] J. Demailly, *Regularization of closed positive currents and intersection theory*, J. Algebraic Geom. **1(3)** (1992), 361–409.
[192] J. Demailly, *A numerical criterion for very ample line bundles*, J. Differential Geom. **37(2)** (1993), 323–374.
[193] J. Demailly, *Effective bounds for very ample line bundles*, Invent. Math. **124(1-3)** (1996), 243–261.
[194] R. Dijkgraaf, H. Verlinde, and E. Verlinde, *Topological strings in $d < 1$*, Nuclear Phys. B **352(1)** (1991), 59–8.
[195] S.K. Donaldson, *An application of gauge theory to four-dimensional topology*, J. Differential Geom. **18(2)** (1983), 279–315.
[196] S.K. Donaldson, *Anti self-dual Yang-Mills connections over complex algebraic surfaces and stable vector bundles*, Proc. London Math. Soc. (3) **50(1)** (1985), 1–26.
[197] S.K. Donaldson, *Infinite determinants, stable bundles and curvature*, Duke Math. J. **54(1)** (1987), 231–247.
[198] S.K. Donaldson, *The Seiberg-Witten equations and 4-manifold topology*, Bull. Amer. Math. Soc. (N.S.) **33(1)** (1996), 45–70.
[199] S.K. Donaldson, *Symplectic submanifolds and almost-complex geometry*, J. Differential Geom. **44(4)** (1996), 666–705.
[200] S.K. Donaldson, *Remarks on gauge theory, complex geometry and 4-manifold topology*, Fields Medallists' lectures, 384–403, World Sci. Ser. 20th century Math., **5**, World Sci. Publishing, River Edge, NJ, 1997.
[201] S.K. Donaldson, *Lefschetz pencils on symplectic manifolds*, J. Differential Geom. **53(2)** (1999), 205–236.
[202] S.K. Donaldson, *Scalar curvature and projective embeddings*, I, J. Differential Geom. **59(3)** (2001), 479–522.
[203] S.K. Donaldson, *Scalar curvature and stability of toric varieties*, J. Differential Geom. **62(2)** (2002), 289–349.
[204] S.K. Donaldson and P.B. Kronheimer, *The geometry of four-manifolds*, Oxford Mathematical Monographs, Oxford Science Publications, The Clarendon Press, Oxford University Press, New York, 1990.
[205] S.K. Donaldson and R.D. Friedman, *Connected Sums of Self-Dual Manifolds and Deformations of Singular Spaces*, Nonlinearity **2** (1989), 197–239.
[206] R.T. Dong, *Nodal sets of eigenfunctions on Riemann surfaces*, J. Differential Geom. **36(2)** (1992), 493–506.
[207] H. Donnelly and C. Fefferman, *Nodal sets of eigenfunctions on Riemannian manifolds*, Invent. Math. **93(1)** (1988), 161–183.
[208] J.J. Duistermaat and G.J. Heckman, *On the variation in the cohomology of the symplectic form of the reduced phase space*, Invent. Math. **69(2)** (1982), 259–268.
[209] J.J. Duistermaat and L. Hörmander, *Fourier integral operators*, II, Acta Math. **128(3-4)** (1972), 183–269.
[210] J.J. Duistermaat and V.W. Guillemin, *The spectrum of positive elliptic operators and periodic bicharacteristics*, Invent. Math. **29(1)** (1975), 39–79.
[211] K. Ecker, *Interior estimates and longtime solutions for mean curvature flow of noncompact spacelike hypersurfaces in Minkowski space*, J. Differential Geom. **46(3)** (1997), 481–498.

[212] J. Eells, Jr. and J.H. Sampson, *Harmonic mappings of Riemannian manifolds*, Amer. J. Math. **86** (1964), 109–160.

[213] L. Ein and R. Lazarsfeld, *Global generation of pluricanonical and adjoint linear series on smooth projective threefolds*, J. Amer. Math. Soc. **6**(**4**) (1993), 875–903.

[214] I. Ekeland and H. Hofer, *Symplectic topology and Hamiltonian dynamics*, Math. Z. **200** (1989), 355–378.

[215] Y. Eliashberg, *The theorem on structure of wave fronts and its appilcations in Symplectic geometry*, Func. Ana. and its App. **21** (1987), 65-72.

[216] Y. Eliashberg, A. Giventhal, and H. Hofer, *Introduction to symplectic field theory*, GAFA (2000), 560–673.

[217] L.C. Evans and J. Spruck, *Motion of level sets by mean curvature*, I, J. Differential Geom. **33**(**3**) (1991), 635–681.

[218] F. Farrell and L. Jones, *Negatively curved manifolds with exotic smooth structures*, J. Amer. Math. Soc. **2**(**4**) (1989), 899–908.

[219] H. Federer and W.H. Fleming, *Normal and integral currents*, Ann. of Math. (2) **72** (1960), 458–520.

[220] C. Fefferman, *The Bergman kernel and biholomorphic mappings of pseudoconvex domains*, Invent. Math. **26** (1974), 1–65.

[221] C. Fefferman and D.H. Phong, *The uncertainty principle and sharp Gårding inequalities*, Comm. Pure Appl. Math. **34**(**3**) (1981), 285–331.

[222] F. Finster, N. Kamran, J. Smoller, and S.-T. Yau, *Decay rates and probability estimates for massive Dirac particles in the Kerr-Newman black hole geometry*, Comm. Math. Phys. **230**(**2**) (2002), 201–244.

[223] D. Fischer-Colbrie and R. Schoen, *The structure of complete stable minimal surfaces in 3-manifolds of nonnegative scalar curvature*, Comm. Pure Appl. Math. **33**(**2**) (1980), 199–211.

[224] A. Floer, *An instanton-invariant for 3-manifolds*, Comm. Math. Phys. **118**(**2**) (1988), 215–240.

[225] A. Floer, *Morse theory for Lagrangian intersections*, J. Differential Geom. **28**(**3**) (1988), 513–547.

[226] A. Floer, *Witten's complex and infinite-dimensional Morse theory*, J. Differential Geom. **30**(1989), no. 1, 207–221.

[227] S. Frankel, *Complex geometry of convex domains that cover varieties*, Acta Math. **163**(**1-2**) (1989), 109–149.

[228] J. Franks, *Geodesics on S^2 and periodic points of annulus homeomorphisms*, Invent. Math. **108**(**2**) (1992), 403–418.

[229] D.S. Freed, *Characteristic numbers and generalized path integrals*, Geometry, topology and physics, 126–138, Conf. Proc. Lecture Notes Geom. Topology, **IV**, Internat. Press, Cambridge, MA, 1995.

[230] M. Freedman, *The topology of four-dimensional manifolds*, J. Differential Geom. **17**(**3**) (1982), 357–453.

[231] M. Freedman, J. Hass, and P. Scott, *Least area incompressible surfaces in 3-manifolds*, Invent. Math. **71**(**3**) (1983), 609–642.

[232] A. Freire, *On the Martin boundary of Riemannian products*, J. Differential Geom. **33**(**1**) (1991), 215–232.

[233] R. Friedman, *Simultaneous resolution of threefold double points*, Math. Ann. **274**(**4**) (1986), 671–689.

[234] H. Friedrich, *Hyperbolic reductions for Einstein's equations*, Classical Quantum Gravity **13**(**6**) (1996), 1451–1469.

[235] J.X. Fu and S.-T. Yau, *Existence of Supersymmetric Hermitian Metrics with Torsion on Non-Kaehler Manifolds*, arXiv:hep-th/0509028.

[236] A. Fujiki, *The moduli spaces and Kähler metrics of polarized algebraic varieties*, Sugaku **42**(**3**) (1990), 231–243.

[237] T. Fujita, *On polarized manifolds whose adjoint bundles are not semipositive*, Algebraic geometry, Sendai, 1985, 167–178, Adv. Stud. Pure Math., **10**, North-Holland, Amsterdam, 1987.
[238] K. Fukaya, *Morse homotopy, A^∞-category, and Floer homologies*, Proceedings of GARC Workshop on Geometry and Topology '93 (Seoul, 1993), 1–102, Lecture Notes Ser., **18**, Seoul Nat. Univ., Seoul, 1993.
[239] K. Fukaya and Y.G. Oh, *Zero-loop open strings in the cotangent bundle and Morse homotopy*, Asian J. Math. **1(1)** (1997), 96–180.
[240] K. Fukaya, Y.-G Oh, H. Ohta, and K. Ono, *Lagrangian Intersection Theory–Anomaly and Obstruction*, book in preparation.
[241] K. Fukaya and K. Ono, *Arnold conjecture and Gromov-Witten invariant*, Topology **38(5)** (1999), 933–1048.
[242] A. Futaki, *An obstruction to the existence of Einstein-Kähler metrics*, Inv. Math., **73** (1983), 437–443.
[243] Z.Y. Gao and S.-T. Yau, *The existence of negatively Ricci curved metrics on three-manifolds*, Invent. Math. **85(3)** (1986), 637–652.
[244] C.S. Gardner, J.M. Greene, M.D. Kruskal, and R.M. Miura, *Korteweg-deVries equation and generalization.* VI. *Methods for exact solution*, Comm. Pure Appl. Math. **27** (1974), 97–133.
[245] H. Garland, *p-adic curvature and the ohomology of discrete subgroups of p-adic groups*, Ann. of Math. (2) **97** (1973), 375–423.
[246] P. Gauduchon, 1-*forme de torsion d'une variété hermitienne compacte*, Math. Ann. **267(4)** (1984), 495–518.
[247] J.P. Gauntlett, D. Martelli, J. Sparks, and D. Waldram, *Sasaki-Einstein metrics on* $S^2 \times S^3$, Adv. Theor. Math. Phys. **8(4)** (2004), 711–734.
[248] C. Gerhardt, *H-surfaces in Lorentzian manifolds*, Comm. Math. Phys. **89(4)** (1983), 523–553.
[249] C. Gerhardt, *Hypersurfaces of prescribed scalar curvature in Lorentzian manifolds*, J. Reine Angew. Math. **554** (2003), 157–199.
[250] R. Geroch, *Energy extraction*, Ann. New York Acad. Sci., **224**, 108–117.
[251] M.L. Gerver and N.S. Nadirashvili, *A condition for isometry of Riemannian metrics in the disc*, Dokl. Akad. Nauk SSSR **275(2)** (1984), 289–293.
[252] M. Giaquinta and E. Giusti, *The singular set of the minima of certain quadratic functionals*, Ann. Scuola Norm. Sup. Pisa Cl. Sci. (4) **11(1)** (1984), 45–55.
[253] D. Gieseker, *On moduli of vector bundles on an algebraic surface*, Ann. of Math. **106** (1977) 45–60.
[254] D. Gieseker, *Global moduli for surfaces of general type*, Invent. Math. **43(3)** (1977), 233–282.
[255] D. Gieseker and J. Li, *Irreducibility of moduli of rank-2 vector bundles on algebraic surfaces*, J. Differential Geom. **40(1)** (1994), 23–104.
[256] P.B. Gilkey, *Curvature and the eigenvalues of the Laplacian for elliptic complexes*, Advances in Math. **10** (1973), 344–382.
[257] A.B. Givental, *Equivariant Gromov-Witten invariants*, Internat. Math. Res. Notices **13** (1996), 613–663.
[258] A. Givental, *A mirror theorem for toric complete intersections*, Topological field theory, primitive forms and related topics (Kyoto, 1996), 141–175, Progr. Math., **160**, Birkhüser Boston, Boston, MA, 1998.
[259] W.M. Goldman, *On the polynomial cohomology of affine manifolds*, Invent. Math. **65(3)** (1981/82), 453–457.
[260] R. Gompf, *A new construction of symplectic manifolds*, Ann. of Math. (2) **142(3)** (1995), 527–595.
[261] R. E. Gompf, *Toward a topological characterization of symplectic manifolds*, J. Symplectic Geom. **2(2)** (2004), 177–206.

[262] C. Gordon, D. Webb, and S. Wolpert, *Isospectral plane domains and surfaces via Riemannian orbifolds*, Invent. Math. **110(1)** (1992), 1–22.

[263] C. Gordon and E. Wilson, *Continuous families of isospectral Riemannian metrics which are not locally isometric*, J. Differential Geom. **47(3)** (1997), 504–529.

[264] M. Goresky, G. Harder, and R. MacPherson, *Weighted cohomology*, Invent. Math. **116(1-3)** (1994), 139–213.

[265] M. Goresky and R. MacPherson, *Intersection homology theory*, Topology **19(2)** (1980), 135–162.

[266] M. Goresky and R. MacPherson, *Intersection homology*, II, Invent. Math. **72(1)** (1983), 77–129.

[267] C. Graham and J. Lee, *Einstein metrics with prescribed conformal infinity on the ball*, Adv. Math. **87(2)** (1991), 186–225.

[268] H. Grauert and H. Reckziegel, *Hermitesche Metriken und normale Familien holomorpher Abbildungen* (German), Math. Z. **89** (1965), 108–125.

[269] A. Gray, *Invariants of curvature operators of four-dimensional Riemannian manifolds*, Proceedings of the Thirteenth Biennial Seminar of the Canadian Mathematical Congress, **2**, 42–65, Canad. Math. Congr., Montreal, Que., 1972.

[270] L.W. Green, *Proof of Blaschke's sphere conjecture*, Bull. Amer. Math. Soc. **67** (1961), 156–158.

[271] B. Greene and M. Plesser, *Duality in Calabi-Yau moduli space*, Nuclear Phys. B **338(1)** (1990), 15–37.

[272] B. Greene, D. Morrison, and M. Plesser, *Mirror manifolds in higher dimension*, Comm. Math. Phys. **173(3)** (1995), 559–597.

[273] R.E. Greene and H. Wu, *Analysis on noncompact Kähler manifolds*, Proc. Sympos. Pure Math., Vol. **XXX**, Part 2, 69–100, Amer. Math. Soc., Providence, RI, 1977.

[274] P.A. Griffiths, *Complex-analytic properties of certain Zariski open sets on algebraic varieties*, Ann. of Math. (2) **94** (1971), 21–51.

[275] P.A. Griffiths and W. Schmid, *Locally homogeneous complex manifolds*, Acta Math. **123** (1969) 253–302.

[276] A. Grigor'yan, *The heat equation on noncompact Riemannian manifolds*, Mat. Sb. **182(1)** (1991), 55–87; translation in Math. USSR-Sb. **72(1)** (1992), 47–77

[277] A. Grigor'yan, *Heat kernel of a noncompact Riemannian manifold*, Proc. Sympos. Pure Math., **57**, 239–263, Amer. Math. Soc., Providence, RI, 1995.

[278] A. Grigor'yan, Y. Netrusov, and S.-T. Yau, *Eigenvalues of elliptic operators and geometric applications*, in 'Servey in Differential Geometry', **IX**, 147-217, A. Grigor'yan and S.-T. Yau, eds., International Press.

[279] D. Gromoll and W. Meyer, *On complete open manifolds of positive curvature*, Ann. of Math. (2) **90** (1969) 75–90.

[280] D. Gromoll and W. Meyer, *Periodic geodesics on compact riemannian manifolds*, J. Differential Geometry **3** (1969), 493–510.

[281] D. Gromoll and J.A. Wolf, *Some relations between the metric structure and the algebraic structure of the fundamental group in manifolds of nonpositive curvature*, Bull. Amer. Math. Soc. **77** (1971), 545–552.

[282] M. Gromov, *Manifolds of negative curvature*, J. Differential Geom. **13(2)** (1978), 223–230.

[283] M. Gromov, *Hyperbolic groups*, Essays in group theory, 75–263, Math. Sci. Res. Inst. Publ., **8**, Springer, New York, 1987.

[284] M. Gromov, *Curvature, diameter and Betti numbers*, Comment. Math. Helv. **56(2)** (1981), 179–195.

[285] M. Gromov, *Volume and bounded cohomology*, Inst. Hautes Études Sci. Publ. Math. **56** (1982), 5–99.

[286] M. Gromov, *Pseudoholomorphic curves in symplectic manifolds*, Invent. Math. **82(2)** (1985), 307–347.

[287] M. Gromov, *Partial differential relations*, Ergebnisse der Mathematik und ihrer Grenzgebiete (3), **9**, Springer-Verlag, Berlin, 1986.

[288] M. Gromov, *Kähler hyperbolicity and L_2-Hodge theory*, J. Differential Geom. **33(1)** (1991), 263–292.

[289] M. Gromov, *Metric structures for Riemannian and non-Riemannian spaces*, Based on the 1981 French original, With appendices by M. Katz, P. Pansu and S. Semmes; Translated from the French by Sean Michael Bates, Progress in Mathematics, **152**, Birkhäuser Boston, Inc., Boston, MA, 1999.

[290] M. Gromov and H.B. Lawson Jr., *Spin and scalar curvature in the presence of a fundamental group*, I, Ann. of Math. (2) **111(2)** (1980), 209–230.

[291] M. Gromov and H.B. Lawson Jr., *The classification of simply connected manifolds of positive scalar curvature*, Ann. of Math. (2) **111(3)** (1980), 423–434.

[292] M. Gromov and R. Schoen, *Harmonic maps into singular spaces and p-adic superrigidity for lattices in groups of rank one*, Inst. Hautes Études Sci. Publ. Math. **76** (1992), 165–246.

[293] M. Gross, *Special Lagrangian fibrations. I. Topology*, Integrable systems and algebraic geometry (Kobe/Kyoto, 1997), 156–193, World Sci. Publishing, River Edge, NJ, 1998.

[294] M. Gross, *Special Lagrangian fibrations. II. Geometry*, A survey of techniques in the study of special Lagrangian fibrations, Surveys in differential geometry: differential geometry inspired by string theory, 341–403, Surv. Differ. Geom., **5**, Int. Press, Boston, MA, 1999.

[295] K. Grove and K. Shiohama, *A generalized sphere theorem*, Ann. Math. (2) **106(2)** (1977), 201–211.

[296] C.H. Gu, *On the initial-boundary value problem for harmonic maps from the 2-dimensional Minkowski space*, Manuscripta Math. **33(1)** (1980/81), 51–58.

[297] B. Guan and P. Guan, *Hypersurfaces with prescribed curvatures*, Ann. Math. **156** (2002), 655–674.

[298] P.F. Guan and X.N. Ma, *The Christoffel-Minkowski problem. I. Convexity of solutions of a Hessian equation*, Invent. Math. **151(3)** (2003), 553–577.

[299] V. Guillemin, *Wave-trace invariants*, Duke Math. J. **83(2)** (1996), 287–352.

[300] V. Guillemin and D. Kazhdan, *Some inverse spectral results for negatively curved 2-manifolds*, Topology **19(3)** (1980), 301–312.

[301] V. Guillemin and S. Sternberg, *Convexity properties of the moment mapping*, Invent. Math. **67(3)** (1982), 491–513.

[302] V. Guillemin and A. Weinstein, *Eigenvalues associated with a closed geodesic*, Bull. Amer. Math. Soc. **82(1)** (1976), 92–94 and Bull. Amer. Math. Soc. **82(6)** (1976), 966.

[303] Y. Guivarc'h, L.Z. Ji, and J.C. Taylor, *Compactifications of symmetric spaces*, Progress in Mathematics, **156**, Birkhäuser Boston, Inc., Boston, MA, 1998.

[304] S. Gukov, *Three-dimensional quantum gravity, Chern-Simons theory, and the A-polynomial*, Comm. Math. Phys. **255(3)** (2005), 577–627.

[305] X. Gukov, S.-T. Yau, and E. Zaslow, *Duality and fibrations on G_2 manifolds*, Turkish J. Math. **27(1)** (2003), 61–97.

[306] R. Gulliver, *Regularity of minimizing surfaces of prescribed mean curvature*, Ann. of Math. (2) **97** (1973), 275–305.

[307] M. Günther, *On the perturbation problem associated to isometric embeddings of Riemannian manifolds*, Ann. Global Anal. Geom. **7(1)** (1989), 69–77.

[308] L. Habermann and J. Jost, *Riemannian metrics on Teichmüller space*, preprint.

[309] L. Habermann and J. Jost, *Metrics on Riemann surfaces and the geometry of moduli space*, preprint.

[310] W. Haken, *Various aspects of the three-dimensional Poincarè problem*, 1970 Topology of Manifolds (Proc. Inst., Univ. of Georgia, Athens, Ga., 1969), 140–152, Markham, Chicago, **Ill** .

[311] R.S. Hamilton, *Three-manifolds with positive Ricci curvature*, J. Differential Geom. **17**(**2**) (1982), 255–306.

[312] R.S. Hamilton, *The Ricci flow on surfaces*, Mathematics and general relativity, 237–262, Contemp. Math., **71**, Amer. Math. Soc., Providence, RI, 1988.

[313] R.S. Hamilton, *The Harnack estimate for the Ricci flow*, J. Differential Geom. **37**(**1**) (1993), 225–243.

[314] R.S. Hamilton, *The formation of singularities in the Ricci flow*, Surveys in differential geometry, Vol. II (Cambridge, MA, 1993), 7–136, Internat. Press, Cambridge, MA, 1995.

[315] R.S. Hamilton, *Four-manifolds with positive isotropic curvature*, Comm. Anal. Geom. **5**(**1**) (1997), 1–92.

[316] R.S. Hamilton, *Non-singular solutions of the Ricci flow on three-manifolds*, Comm. Anal. Geom. **7**(**4**) (1999), 695–729.

[317] Q. Han, R. Hardt, and F.H. Lin, *Geometric measure of singular sets of elliptic equations*, Comm. Pure Appl. Math. **51**(**11-12**) (1998), 1425–1443.

[318] Q. Han, J.X. Hong, and C.S. Lin, *Local isometric embedding of surfaces with non-positive Gaussian curvature*, J. Differential Geom. **63**(**3**) (2003), 475–520.

[319] R. Hardt, M. Hoffmann-Ostenhof, T. Hoffmann-Ostenhof, and N. Nadirashvili, *Critical sets of solutions to elliptic equations*, J. Differential Geom. **51**(**2**) (1999), 359–373.

[320] R. Hardt and L. Simon, *Boundary regularity and embedded solutions for the oriented Plateau problem*, Ann. of Math. (2) **110**(**3**) (1979), 439–486.

[321] F.R. Harvey and H.B. Lawson Jr., *On boundaries of complex analytic varieties*, I, Ann. of Math. (2) **102**(**2**) (1975), 223–290.

[322] F. R. Harvey and H.B. Lawson Jr., *On boundaries of complex analytic varieties*, II, Ann. Math. (2) **106**(**2**) (1977), 213–238.

[323] R. Harvey and H.B. Lawson Jr., *Calibrated geometries*, Acta Math. **148** (1982), 47–157.

[324] R. Harvey and B. Shiffman, *A characterization of holomorphic chains*, Ann. of Math. (2) **99** (1974), 553–587.

[325] A. Hassell, R. Melrose, and A. Vasy, *Spectral and scattering theory for symbolic potentials of order zero*, Adv. Math. **181**(**1**) (2004), 1–87.

[326] S. W. Hawking, *Black hole in general relativity*, Commun. Math. Phys. **25** (1972), 152–166.

[327] J. Hersch, *Quatre propriétés isopérimétriques de membranes sphériques homogènes*, C.R. Acad. Sci. Paris Sér. A-B **270** (1970), A1645–A1648.

[328] N. Hingston, *Equivariant Morse theory and closed geodesics*, J. Differential Geom. **19**(**1**) (1984), 85–116.

[329] N. Hingston, *On the growth of the number of closed geodesics on the two-sphere*, Internat. Math. Res. Notices **9** (1993), 253–262.

[330] F. Hirzebruch and K. Kodaira, *On the complex projective spaces*, J. Math. Pures Appl. (9) **36** (1957), 201–216.

[331] N. Hitchin, *Harmonic spinors*, Advances in Math. **14** (1974), 1–55.

[332] N. Hitchin, *Compact four-dimensional Einstein manifolds*, J. Differential Geometry **9** (1974), 435–441.

[333] N.J. Hitchin, *The self-duality equations on a Riemann surface*, Proc. London Math. Soc. (3) **5**(**1**) (1987), 59–126.

[334] N.J. Hitchin, *Harmonic maps from a 2-torus to the 3-sphere*, J. Differential Geom. **31**(**3**) (1990), 627–710.

[335] N.J. Hitchin, *Lie groups and Teichmüller space*, Topology **31**(**3**) (1992), 449–473.

[336] N.J. Hitchin, *The moduli space of special Lagrangian submanifolds*, Ann. Scuola Norm. Sup. Pisa Cl. Sci. (4) **25(3-4)** (1997), 503–515.
[337] N. Hitchin, *The geometry of three-forms in six dimensions*, J. Differential Geom. **55(3)** (2000), 547–576.
[338] N. Hitchin, *Stable forms and special metrics*, Global differential geometry: the mathematical legacy of Alfred Gray (Bilbao, 2000), 70–89, Contemp. Math., **288**, Amer. Math. Soc., Providence, RI, 2001.
[339] H. Hofer, *On the topological properties of symplectic maps*, Proc. Royal Soc. Edinburgh, Sect. A **115** (1990), 25–38.
[340] H. Hofer, *Pseudo-holomorphic curves and Weinstein conjecture in dimension three*, Invent. math. **114(3)** (1993), 515–563.
[341] H. Hofer, K. Wysocki, and E. Zehnder, *The dynamics on three-dimensional strictly convex energy surfaces*, Ann. of Math. (2) **148(1)** (1998), 197–289.
[342] H. Hofer, K. Wysocki, and E. Zehnder, *Finite energy foliations of tight three-spheres and Hamiltonian dynamics*, Ann. of Math. (2) **157** (2003), 125–255.
[343] H. Hofer and E. Zehnder, *Symplectic capacities and Hamiltonian dynamics*, Birkhauser, Basel, 1994.
[344] M. Hoffmann-Ostenhof, T. Hoffmann-Ostenhof, and N. Nadirashvili, *Critical sets of smooth solutions to elliptic equations in dimension* 3, Indiana Univ. Math. J. **45** (1996), 15–37.
[345] J.X. Hong, *Realization in R^3 of complete Riemannian manifolds with negative curvature*, Comm. Anal. Geom. **1(3-4)** (1993), 487–514.
[346] M.J. Hopkins, *Topological modular forms, the Witten genus, and the theorem of the cube*, Proceedings of the International Congress of Mathematicians, Vol. **1**, **2** (Zürich, 1994), 554–565, Birkhäuser, Basel, 1995.
[347] L. Hörmander, *An introduction to complex analysis in several variables*, Third edition, North-Holland Mathematical Library, **7**, North-Holland Publishing Co., Amsterdam, 1990.
[348] L.K. Hua, *Harmonic analysis of functions of several variables in the classical domains*, American Mathematical Society, Providence, RI, 1963.
[349] G. Huisken, *Flow by mean curvature of convex surfaces into spheres*, J. Differential Geom. **20(1)** (1984), 237–266.
[350] G. Huisken, *Contracting convex hypersurfaces in Riemannian manifolds by their mean curvature*, Invent. Math. **84(3)** (1986), 463–480.
[351] G. Huisken and T. Ilmanen, *The inverse mean curvature flow and the Riemannian Penrose inequality*, J. Differential Geom. **59(3)** (2001), 353–437.
[352] G. Huisken and C. Sinestrari, *Mean curvature flow singularities for mean convex surfaces*, Calc. Var. Partial Differential Equations **8(1)** (1999), 1–14.
[353] G. Huisken and C. Sinestrari, *Convexity estimates for mean curvature flow and singularities of mean convex surfaces*, Acta Math. **183(1)** (1999), 45–70.
[354] G. Huisken and S.-T. Yau, *Definition of center of mass for isolated physical systems and unique foliations by stable spheres with constant mean curvature*, Invent. Math. **124(1-3)** (1996), 281–311.
[355] W. Israël, *Event Horizons in Static Vacuum Space-Times*, Phys. Rev. **164** (1967), 1776–1779.
[356] T. Ivey, *Ricci solitons on compact* 3*-manifolds*, Differential Geom. Appl. **3** (1993) 301–307.
[357] H. Iwaniec and P. Sarnak, *L^∞ norms of eigenfunctions of arithmetic surfaces*, Ann. of Math. (2) **141(2)** (1995), 301–320.
[358] D. Jakobson, M. Levitin, N. Nadirashvili, N. Nigam, and I. Polterovich, *How large can the first eigenvalue be on a surface of genus two?*, Arxiv: math. SP/0509398.
[359] D. Jakobson and N. Nadirashvili, *Eigenfunctions with few critical points*, J. Differential Geom. **53(1)** (1999), 177–182.

[360] L. C. Jeffrey and F. C. Kirwan, *Localization and the quantization conjecture*, Topology **36(3)** (1997), 647-693.

[361] L.Z. Ji and R. MacPherson, *Geometry of compactifications of locally symmetric spaces*, Ann. Inst. Fourier (Grenoble) **52(2)** (2002), 457–559.

[362] M. Ji and G.Y. Wang, *Minimal surfaces in Riemannian manifolds*, Mem. Amer. Math. Soc. **104(495)** (1993).

[363] V. Jones, *Hecke algebra representations of braid groups and link polynomials*, Ann. of Math. (2) **126(2)** (1987), no. 2, 335–388.

[364] K. Jörgens, *Über die Losungen der Differentialgleichung* $rt - s^2 = 1$, Math. Ann. **127** (1954), 130–134.

[365] J. Jorgenson and A. Todorov, *Enriques surfaces, analytic discriminants, and Borcherds's* Φ *function*, Comm. Math. Phys. **191(2)** (1998), 249–264.

[366] J. Jost, *Embedded minimal surfaces in manifolds diffeomorphic to the three-dimensional ball or sphere*, J. Differential Geom. **30(2)** (1989), 555–577.

[367] J. Jost, *Convex functionals and generalized harmonic maps into spaces of nonpositive curvature*, Comment. Math. Helv. **70(4)** (1995), 659–673.

[368] J. Jost and S.-T. Yau, *Harmonic mappings and Kähler manifolds*, Math. Ann. **262(2)** (1983), 145–166.

[369] J. Jost and S.-T. Yau, *Harmonic maps and group representations*, Differential geometry, 241–259, Pitman Monogr. Surveys Pure Appl. Math., **52**, Longman Sci. Tech., Harlow, 1991.

[370] J. Jost and S.-T. Yau, *Applications of quasilinear PDE to algebraic geometry and arithmetic lattices*, Algebraic geometry and related topics, 169–193, Conf. Proc. Lecture Notes Algebraic Geom., I, Internat. Press, Cambridge, MA, 1993.

[371] J. Jost and S.-T. Yau, *Harmonic maps and superrigidity*, Differential geometry: partial differential equations on manifolds, 245–280, Proc. Sympos. Pure Math., **54**, Part 1, Amer. Math. Soc., Providence, RI, 1993.

[372] J. Jost and S.-T. Yau, *Harmonic maps and rigidity theorems for spaces of nonpositive curvature*, Comm. Anal. Geom. **7(4)** (1999), 681–694.

[373] D.D. Joyce, *Compact 8-manifolds with holonomy* Spin(7), Invent. Math. **123(3)** (1996), 507–552.

[374] D.D. Joyce, *Compact Riemannian 7-manifolds with holonomy* G_2, I, II, J. Differential Geom. **43(2)** (1996), 291–328, 329–375.

[375] D. Joyce, *A new construction of compact 8-manifolds with holonomy* Spin(7), J. Differential Geom. **53(1)** (1999), 89–130.

[376] D. Joyce, *Singularities of special Lagrangian fibrations and the SYZ conjecture*, Comm. Anal. Geom. **11(5)** (2003), 859–907.

[377] A.B. Katok, *Ergodic properties of degenerate integrable Hamiltonian systems*, Izv. Akad. Nauk SSSR. **37** (1973) (Russian), Math. USSR-Izv. **7** (1973) 535-571.

[378] X. Katz and C.-C. Liu, *Enumerative geometry of stable maps with Lagrangian boundary conditions and multiple covers of the disc*, Adv. Theor. Math. Phys. **5(1)** (2001), 1–49.

[379] J.L. Kazdan and F.W. Warner, *Existence and conformal deformation of metrics with prescribed Gaussian and scalar curvatures*, Ann. of Math. (2) **101** (1975), 317–331.

[380] D.A. Kazhdan, *On the connection of the dual space of a group with the structure of its closed subgroups* (Russian), Funkcional. Anal. i Prilozhen. **1** (1967), 71–74.

[381] L. Keen, *Collars on Riemann surfaces*, Discontinuous groups and Riemann surfaces (Proc. Conf., Univ. Maryland, College Park, Md., 1973), 263–268. Ann. of Math. Studies, **79**, Princeton Univ. Press, Princeton, NJ, 1974.

[382] J. Kim, C. LeBrun, and M. Pontecorvo, *Scalar-flat Kähler surfaces of all genera*, J. Reine Angew. Math. **486** (1997), 69–95.

[383] J.R. King, *The currents defined by analytic varieties*, Acta Math. **127(3-4)** (1971), 185–220.

[384] S. Klainerman and F. Nicolò, *Peeling properties of asymptotically flat solutions to the Einstein vacuum equations*, Classical Quantum Gravity **20(14)** (2003), 3215–3257.
[385] S. Klainerman and S. Selberg, *Bilinear estimates and applications to nonlinear wave equations*, Commun. Contemp. Math. **4(2)** (2002), 223–295.
[386] W. Klingenberg, *Lectures on closed geodesics*, Third edition, Mathematisches Institut der Universität Bonn, Bonn, 1977.
[387] S. Kobayashi, *On compact Kähler manifolds with positive definite Ricci tensor*, Ann. of Math. (2) **74** (1961) 570–574.
[388] S. Kobayashi, *Intrinsic metrics on complex manifolds*, Bull. Amer. Math. Soc. **73** (1967) 347–349.
[389] S. Kobayashi and T. Ochiai, *Meromorphic mappings onto compact complex spaces of general type*, Invent. Math. **31(1)** (1975), 7–16.
[390] K. Kodaira, *On a differential-geometric method in the theory of analytic stacks*, Proc. Nat. Acad. Sci. USA **39** (1953), 1268–1273.
[391] K. Kodaira, *On Kähler varieties of restricted type (an intrinsic characterization of algebraic varieties)*, Ann. of Math. (2) **60** (1954), 28–48.
[392] J. Kollár, *Effective base point freeness*, Math. Ann. **296(4)** (1993), 595–605.
[393] M. Kontsevich, *Enumeration of rational curves via torus actions*, The moduli space of curves (Texel Island, 1994), 335–368, Progr. Math., **129**, Birkhäuser, Boston, 1995.
[394] M. Kontsevich, *Feynman diagrams and low-dimensional topology*, First European Congress of Mathematics, Vol. **II** (Paris, 1992), 97–121, Progr. Math., **120**, Birkhäuser, Basel, 1994.
[395] M. Kontsevich, *Homological algebra of mirror symmetry*, Proceedings of the International Congress of Mathematicians, Vol. **1**, **2** (Zürich, 1994), 120–139, Birkhäuser, Basel, 1995.
[396] N. Korevaar, *An easy proof of the interior gradient bound for solutions to the prescribed mean curvature equation*, Proc. Sympos. Pure Math., **45**, Part 2, 81–89, Amer. Math. Soc., Providence, RI, 1986.
[397] N. Korevaar, *Upper bounds for eigenvalues of conformal metrics*, J. Differential Geom. **37(1)** (1993), 73–93.
[398] N. Korevaar, R. Mazzeo, F. Pacard, and R. Schoen, *Refined asymptotics for constant scalar curvature metrics with isolated singularities*, Invent. Math. **135(2)** (1999), 233–272.
[399] N. Korevaar and R. Schoen, *Sobolev spaces and harmonic maps for metric space targets*, Comm. Anal. Geom. **1(3-4)** (1993), 561–659
[400] A. Kovalev, *Twisted connected sums and special Riemannian holonomy*, J. Reine Angew. Math. **565** (2003), 125–160.
[401] P. Kramer, *Group actions on compact hyperbolic manifolds and closed geodesics*, Symmetry in physics, 113–124, CRM Proc. Lecture Notes, **34**, Amer. Math. Soc., Providence, RI, 2004.
[402] P.B. Kronheimer, and T.S. Mrowka, *Embedded surfaces and the structure of Donaldson's polynomial invariants*, J. Differential Geom. **41(3)** (1995), 573–734.
[403] M. Kruskal, *The birth of the soliton*, Nonlinear evolution equations solvable by the spectral transform (Internat. Sympos., Accad. Lincei, Rome, 1977), 1–8, Res. Notes in Math., **26**, Pitman, Boston, Mass.-London, 1978.
[404] V. Kulikov, *Epimorphicity of the period mapping for surfaces of type $K3$*, Uspehi Mat. Nauk **32(4)** (1977), 257–258.
[405] M. Kuranishi, *Strongly pseudoconvex CR structures over small balls.* I. *An a priori estimate*, Ann. of Math. (2) **115(3)** (1982), 451–500.
[406] M. Kuranishi, *Strongly pseudoconvex CR structures over small balls.* II. *A regularity theorem*, Ann. of Math. (2) **116(1)** (1982), 1–64.

[407] M. Kuranishi, *Strongly pseudoconvex CR structures over small balls.* III. *An embedding theorem*, Ann. of Math. (2) **116(2)** (1982), 249–330.

[408] F. Lalonde and D. McDuff, *The geometry of symplectic energy*, Ann. of Math. (2) **141** (1995), 349-371.

[409] H.B. Lawson, Jr., *Local rigidity theorems for minimal hypersurfaces*, Ann. of Math. (2) **89** (1969), 187C197.

[410] H.B. Lawson, Jr., *Compact minimal surfaces in* S^3, 1970 Global Analysis (Proc. Sympos. Pure Math., Vol. XV, Berkeley, Calif., 1968), 275–282, Amer. Math. Soc., Providence, RI.

[411] H.B. Lawson and J. Simons, *On stable currents and their application to global problems in real and complex geometry*, Ann. of Math. (2) **98** (1973), 427–450.

[412] H.B. Lawson Jr. and S.-T. Yau, *Compact manifolds of nonpositive curvature*, J. Differential Geometry **7** (1972), 211–228.

[413] H.B. Lawson Jr. and S.-T. Yau, *Scalar curvature, non-abelian group actions, and the degree of symmetry of exotic spheres*, Comment. Math. Helv. **49** (1974), 232–244.

[414] C.R. LeBrun, *Explicit Self-Dual Metrics on* $\mathbb{CP}_2\#\cdots\#\mathbb{CP}_2$, J. Differential Geom. **34** (1991), 223–253.

[415] C. LeBrun, *Ricci curvature, minimal volumes, and Seiberg-Witten theory*, Invent. Math. **145(2)** (2001), 279–316.

[416] J.-H. Lee and N.C. Leung, *Geometric structures on* G_2 *and* $Spin(7)$*-manifolds*, Arxiv: math.DG/0202045.

[417] Y.-I. Lee, *Embedded special Lagrangian submanifolds in Calabi-Yau manifolds*, Comm. Anal. Geom. **11(3)** (2003), 391–423.

[418] W. Lerche, C. Vafa, and N. Warner, *Chiral rings in* $N = 2$ *superconformal theories*, Nuclear Phys. B **324(2)** (1989), 427–474.

[419] N.C. Leung, *Einstein type metrics and stability on vector bundles*, J. Differential Geom. **45(3)** (1997), 514–546.

[420] N.C. Leung, *Lagrangian submanifolds in hyperKähler manifolds, Legendre transformation*, J. Differential Geom. **61(1)** (2002), 107–145.

[421] N.C. Leung, *Topological quantum field theory for Calabi-Yau threefolds and* G_2*-manifolds*, Adv. Theor. Math. Phys. **6(3)** (2002), 575–591.

[422] N.C. Leung, *Mirror symmetry without corrections*, Comm. Anal. Geom. **13(2)** (2005), 287–331.

[423] N.C. Leung and X.W. Wang, *Intersection theory of coassociative submanifolds in* G_2*-manifolds and Seiberg-Witten invariants*, Arxiv: math.DG/0401419.

[424] N.C. Leung, S.-T. Yau, and E. Zaslow, *From special Lagrangian to Hermitian-Yang-Mills via Fourier-Mukai transform*, Winter School on Mirror Symmetry, Vector Bundles and Lagrangian Submanifolds (Cambridge, MA, 1999), 209–225, AMS/IP Stud. Adv. Math., **23**, Amer. Math. Soc., Providence, RI, 2001.

[425] J.S. Li, *Kloosterman-Selberg zeta functions on complex hyperbolic spaces*, Amer. J. Math. **113(4)** (1991), 653–731.

[426] J. Li, *Algebraic geometric interpretation of Donaldson's polynomial invariants*, J. Differential Geom. **37(2)** (1993), 417–466.

[427] J. Li, *The geometry of moduli spaces of vector bundles over algebraic surfaces*, Proceedings of the International Congress of Mathematicians, Vol. **1**, **2** (Zürich, 1994), 507–516, Birkhäuser, Basel, 1995.

[428] J. Li, *Stable morphisms to singular schemes and relative stable morphisms*, J. Differential Geom. **57(3)** (2001), 509–578.

[429] J. Li, *A degeneration formula of GW-invariants*, J. Differential Geom. **60(2)** (2002), 199–293.

[430] J. Li, C.C. M. Liu, K.F. Liu and J. Zhou, *A Mathematical Theory of the Topological Vertex Authors*, arXiv:math.AG/0408426.

[431] J. Li and G. Tian, *Virtual moduli cycles and Gromov-Witten invariants of algebraic varieties*, J. Amer. Math. Soc. **11(1)** (1998), 119–174.

[432] J. Li and G. Tian, *Virtual moduli cycles and Gromov-Witten invariants of general symplectic manifolds*, Topics in symplectic 4-manifolds, 47–83, First Int. Press Lect. Ser., **I**, Internat. Press, Cambridge, MA, 1998.

[433] J. Li and S.-T. Yau, *Hermitian-Yang-Mills connection on non-Kähler manifolds*, Mathematical aspects of string theory (San Diego, Calif., 1986), 560–573, Adv. Ser. Math. Phys., **1**, World Sci. Publishing, Singapore, 1987.

[434] J. Li and S.-T. Yau, *The Existence of Supersymmetric String Theory with Torsion*, arXiv:hep-th/0411136.

[435] J. Li, S.-T. Yau, and F.Y. Zheng, *A simple proof of Bogomolov's theorem on class* VII_0 *surfaces with* $b_2 = 0$, Illinois J. Math. **34(2)** (1990), 217–220.

[436] P. Li, *A lower bound for the first eigenvalue of the Laplacian on a compact manifold*, Indiana Univ. Math. J. **28(6)** (1979), 1013–1019.

[437] P. Li, *Harmonic sections of polynomial growth*, Math. Res. Lett. **4(1)** (1997), 35–44.

[438] P. Li, R. Schoen, and S.-T. Yau, *On the isoperimetric inequality for minimal surfaces*, Ann. Scuola Norm. Sup. Pisa Cl. Sci. (4) **11(2)** (1984), 237–244.

[439] P. Li and G. Tian, *On the heat kernel of the Bergmann metric on algebraic varieties*, J. Amer. Math. Soc. **8(4)** (1995), 857–877.

[440] P. Li and J.P. Wang, *Counting dimensions of L-harmonic functions*, Ann. of Math. (2) **152(2)** (2000), 645–658.

[441] P. Li and J.P. Wang, *Complete manifolds with positive spectrum*, II, J. Differential Geom. **62(1)** (2002), 143–162.

[442] P. Li and S.-T. Yau, *Estimates of eigenvalues of a compact Riemannian manifold*, Geometry of the Laplace operator, 205–239, Proc. Sympos. Pure Math., **XXXVI**, Amer. Math. Soc., Providence, RI, 1980.

[443] P. Li and S.-T. Yau, *A new conformal invariant and its applications to the Willmore conjecture and the first eigenvalue of compact surfaces*, Invent. Math. **69(2)** (1982), 269–291.

[444] P. Li and S.-T. Yau, *On the Schrödinger equation and the eigenvalue problem*, Comm. Math. Phys. **88(3)** (1983), 309–318.

[445] P. Li and S.-T. Yau, *On the parabolic kernel of the Schrödinger operator*, Acta Math. **156(3-4)** (1986), 153–201.

[446] T.J. Li, *Symplectic 4-manifolds with Kodaira dimension zero*, preprint.

[447] T.J. Li and A. Liu, *General wall crossing formula*, Math. Res. Lett. **2(6)** (1995), 797–810

[448] T.J. Li and A. Liu, *Symplectic structure on ruled surfaces and a generalized adjunction formula*, Math. Res. Lett. **2(4)** (1995), 453–471.

[449] B.H. Lian, K.F. Liu, and S.-T. Yau, *Mirror principle*, I, Asian J. Math. **1(4)** (1997), 729–763.

[450] B.H. Lian, K.F. Liu, and S.-T. Yau, *Mirror principle*, II, Sir Michael Atiyah: a great mathematician of the twentieth century. Asian J. Math. **3(1)** (1999), 109–146.

[451] B.H. Lian, K.F. Liu, and S.-T. Yau, *Mirror principle*, III, Asian J. Math. **3(4)** (1999), 771–800.

[452] B.H. Lian, K.F. Liu, and S.-T. Yau, *Mirror principle*, IV, Surveys in differential geometry, 475–496, Surv. Differ. Geom., **VII**, Int. Press, Somerville, MA, 2000.

[453] A. Lichnerowicz, *Spineurs harmoniques*, C.R. Acad. Sci. Paris, **257** (1963), 7–9.

[454] A. Lichnerowicz, *Magnetohydrodynamics: waves and shock waves in curved space-time*, Mathematical Physics Studies, **14**, Kluwer Academic Publishers Group, Dordrecht, 1994.

[455] C.S. Lin, *The local isometric embedding in* R^3 *of 2-dimensional Riemannian manifolds with nonnegative curvature*, J. Differential Geometry **21** (1985), 213–230.

[456] C.S. Lin, *The local isometric embedding in* R^3 *of two-dimensional Riemannian manifolds with Gaussian curvature changing sign cleanly*, Comm. Pure Appl. Math. **39** (1986) 867–887.

[457] F. H. Lin, *Asymptotic behavior of area-minimizing currents in hyperbolic space*, Comm. Pure Appl. Math. **42(3)** (1989), 229–242.

[458] F.H. Lin, *Nodal sets of solutions of elliptic and parabolic equations*, Comm. Pure Appl. Math. **44(3)** (1991), 287–308.

[459] F.H. Lin, *Gradient estimates and blow-up analysis for stationary harmonic maps*, Ann. of Math. (2) **149(3)** (1999), 785–829.

[460] F.H. Lin, *Mapping problems, fundamental groups and defect measures*, Acta Math. Sin. (Engl. Ser.) **15(1)** (1999), 25–52.

[461] H. Lindblad and I. Rodnianski, *Global existence for the Einstein vacuum equations in wave coordinates*, Comm. Math. Phys. **256(1)** (2005), 43–110.

[462] A.K. Liu, *Some new applications of general wall crossing formula, Gompf's conjecture and its applications*, Math. Res. Lett. **3(5)** (1996), 569–585.

[463] C.-C. Liu, *Moduli of J-Holomorphic Curves with Lagrangian Boundary Conditions and Open Gromov-Witten Invariants for an* S^1*-Equivariant Pair*, Arxiv: math.SG/0210257.

[464] C.C.M. Liu and S.-T. Yau, *Positivity of Quasilocal Mass*, Phys. Rev. Lett. **90(23)** (2003), 231102 4 pp.

[465] C.C.M. Liu and S.-T. Yau, *Positivity of quasi-local mass*, II, J. Amer. Math. Soc. **19(1)** (2006), 181–204.

[466] G. Liu and G. Tian, *Floer homology and Arnold conjecture*, J. Differential Geom. **49(1)** (1998), 1–74.

[467] K.F. Liu, *On modular invariance and rigidity theorems*, J. Differential Geom. **41(2)** (1995), 343–396.

[468] K.F. Liu, *On elliptic genera and theta-functions*, Topology **35(3)** (1996), 617–640.

[469] K.F. Liu, *Heat kernel and moduli space*, Math. Res. Lett. **3(6)** (1996), 743–762.

[470] K.F. Liu, *Heat kernel and moduli spaces*, II, Math. Res. Lett. **4(4)** (1997), 569–588.

[471] K.F. Liu, X.F. Sun, and S.-T. Yau, *Canonical metrics on the moduli space of Riemann surfaces*, I, J. Differential Geom. **68(3)** (2004), 571–637.

[472] K.F. Liu, X.F. Sun, and S.-T. Yau, *Canonical Metrics on the Moduli Space of Riemann Surfaces*, II, J. Differential Geom. **69(1)** (2005), 163–216.

[473] J.C. Loftin, *Affine spheres and convex* $\mathbb{RP}^n$*-manifolds*, Amer. J. Math. **123(2)** (2001), 255–274.

[474] J. Loftin, S.-T. Yau, and E. Zaslow, *Affine Manifolds, SYZ Geometry, and the "Y" Vertex Authors*, arXiv:math.DG/0405061.

[475] J. Lohkamp, *Curvature h-principles*, Ann. of Math. (2) **142(3)** (1995), 457–498.

[476] E. Looijenga, L^2*-cohomology of locally symmetric varieties*, Compositio Math. **67(1)** (1988), 3–20.

[477] S.Y. Lu and S.-T. Yau, *Holomorphic curves in surfaces of general type*, Proc. Nat. Acad. Sci. USA **87(1)** (1990), 80–82.

[478] P. Lu and G. Tian, *The complex structure on a connected sum of* $S^3 \times S^3$ *with trivial canonical bundle*, Math. Ann. **298(4)** (1994), 761–764.

[479] Z.Q. Lu, *On the lower order terms of the asymptotic expansion of Tian-Yau-Zelditch*, Amer. J. Math. **122** (2000), 235–273.

[480] A. Lubotzky, R. Phillips, and P. Sarnak, *Ramanujan graphs*, Combinatorica **8(3)** (1988), 261–277.

[481] H. Luo, *Geometric criterion for Gieseker-Mumford stability of polarized manifolds*, J. Differential Geom. **49(3)** (1998), 577–599.

[482] W. Luo, Z. Rudnick, and P. Sarnak, *On Selberg's eigenvalue conjecture*, Geom. Funct. Anal. **5(2)** (1995), 387–401.

[483] G. Lusztig, *Novikov's higher signature and families of elliptic operators*, J. Differential Geometry **7** (1972), 229–256.
[484] E. Lutwak, *Selected affine isoperimetric inequalities. Handbook of convex geometry*, Vol. A, B, 151–176, North-Holland, Amsterdam, 1993.
[485] E. Lutwak, D. Yang, and G.Y. Zhang, *L_p affine isoperimetric inequalities*, J. Differential Geom. **56**(**1**) (2000), 111–132.
[486] M. Lübke, *Stability of Einstein-Hermitian vector bundles*, Manuscripta Math. **42**(**2-3**) (1983), 245–257.
[487] T. Lyons and D. Sullivan, *Function theory, random paths and covering spaces*, J. Differential Geom. **19**(**2**) (1984), 299–323.
[488] X.N. Ma and G. Marinescu, *Genegalized Bergman kernels on symplectic manifolds*, C.R. Math. Acad. Asi. Paris **339**(**7**) (2004), 493–498; the full version: math.DG/0411559.
[489] T. Mabuchi, *K-energy maps integrating Futaki invariants*, Tohoku Math. J. (2) **38**(**4**) (1986), 575–593.
[490] G.A. Margulis, *Discrete groups of motions of manifolds of nonpositive curvature*, Transl., I. Ser., Am. Math. Soc. **109** (1977), 33–45.
[491] G.A. Margulis, *Explicit constructions of graphs without short cycles and low density codes*, Combinatorica **2**(**1**) (1982), 71–78.
[492] J. Marsden and A. Weinstein, *Reduction of symplectic manifolds with symmetry*, Rep. Mathematical Phys. **5**(**1**) (1974), 121–130.
[493] T. Matsusaka, *Polarized varieties with a given Hilbert polynomial*, Amer. J. Math. **94** (1972), 1027–1077.
[494] Y. Matsushima, *Sur la structure du group d'homeomorphismes analytiques d'une certaine varietie Kaehleriennes*, Nagoya Math. J., **11** (1957), 145–150.
[495] Y. Matsushima, *On the first Betti number of compact quotient spaces of higher-dimensional symmetric spaces*, Ann. of Math. (2) **75** (1962), 312–330.
[496] D. McDuff, *Examples of symplectic structures*, Invent. Math. **89** (1987), 13-36.
[497] D. McDuff, *The structure of rational and ruled symplectic 4-manifolds*, J. Amer. Math. Soc. **3**(**3**) (1990), 679–712.
[498] D. McDuff. *Singularities and positivity of intersections of J-holomorphic curves*, Progr. Math., **117**, Holomorphic curves in symplectic geometry, 191–215, Birkhäuser, Basel, 1994.
[499] H.P. McKean, Jr and I.M. Singer, *Curvature and the eigenvalues of the Laplacian*, J. Differential Geometry **1**(**1**) (1967), 43–69.
[500] C. McMullen, *Iteration on Teichmüller space*, Invent. Math. **99**(**2**) (1990), 425–454.
[501] C. McMullen, *Riemann surfaces and the geometrization of 3-manifolds*, Bull. Amer. Math. Soc. (N.S.) **27**(**2**) (1992), 207–216.
[502] C.T. McMullen, *The moduli space of Riemann surfaces is Kähler hyperbolic*, Ann. of Math. (2) **151**(**1**) (2000), 327–357.
[503] W.H. Meeks, III and H. Rosenberg, *The uniqueness of the helicoid*, Ann. of Math. (2) **161**(**2**) (2005), 727–758.
[504] W.H. Meeks, L. Simon, and S.-T. Yau, *Embedded minimal surfaces, exotic spheres, and manifolds with positive Ricci curvature*, Ann. of Math. (2) **116**(**3**) (1982), 621–659.
[505] W.H. Meeks and S.-T. Yau, *Topology of three-dimensional manifolds and the embedding problems in minimal surface theory*, Ann. of Math. (2) **112**(**3**) (1980), 441–484.
[506] W.H. Meeks and S.-T. Yau, *The equivariant Dehn's lemma and loop theorem*, Comment. Math. Helv. **56**(**2**) (1981), 225–239.
[507] W.H. Meeks and S.-T. Yau, *The existence of embedded minimal surfaces and the problem of uniqueness*, Math. Z. **179**(**2**) (1982), 151–168.

[508] W.H. Meeks and S.-T. Yau, *The classical Plateau problem and the topology of three-dimensional manifolds. The embedding of the solution given by Douglas-Morrey and an analytic proof of Dehn's lemma*, Topology **21**(**4**) (1982), 409–442.

[509] V. Mehta and A. Ramanathan, *Restriction of stable sheaves and representations of the fundamental group*, Invent. Math. **77**(**1**) (1984), 163–172.

[510] A. d. Melas, *On the nodal line of the second eigenfunction of the Laplacian in R^2*, J. Differential Geom. **35**(**1**) (1992), 255–263.

[511] R. Melrose, *The inverse spectral problem for planar domains*, Instructional Workshop on Analysis and Geometry, Part I (Canberra, 1995), 137–160, Proc. Centre Math. Appl. Austral. Nat. Univ., **34**, Austral. Nat. Univ., Canberra, 1996.

[512] M.J. Micallef, *Stable minimal surfaces in Euclidean space*, J. Differential Geom. **19**(**1**) (1984), 57–84.

[513] M.J. Micallef and J.D. Moore, *Minimal two-spheres and the topology of manifolds with positive curvature on totally isotropic two-planes*, Ann. of Math. (2) **127**(**1**) (1988), 199–227.

[514] M.J. Micallef and M.Y. Wang, *Metrics with nonnegative isotropic curvature*, Duke Math. J. **72**(**3**) (1993), 649–672.

[515] J.H. Michael and L.M. Simon, *Sobolev and mean-value inequalities on generalized submanifolds of R^n*, Comm. Pure Appl. Math. **26** (1973), 361–379.

[516] M.L. Michelsohn, *Kähler manifolds with vanishing first Chern class*, Seminar on Differential Geometry, 359–361, Ann. of Math. Stud., **102**, Princeton Univ. Press, Princeton, NJ, 1982.

[517] M.L. Michelsohn, *On the existence of special metrics in complex geometry*, Acta Math. **149**(**3-4**) (1982), 261–295.

[518] J. Milnor, *Morse theory*, Based on lecture notes by M. Spivak and R. Wells, Annals of Mathematics Studies, **51**, Princeton University Press, Princeton, NJ, 1963.

[519] J. Milnor, *Eigenvalues of the Laplace operator on certain manifolds*, Proc. Nat. Acad. Sci. USA **51** (1964), 542.

[520] Y. Miyaoka, *On the Chern numbers of surfaces of general type*, Invent. Math. **42** (1977), 225–237.

[521] N. Mok, *Courbure bisectionnelle positive et variétés algébriques affines*, C.R. Acad. Sci. Paris Sér. I Math. **296**(**11**) (1983), 473–476.

[522] N. Mok, *An embedding theorem of complete Kähler manifolds of positive bisectional curvature onto affine algebraic varieties*, Bull. Soc. Math. France **112** (1984), 197–250.

[523] N. Mok, *The uniformization theorem for compact Kähler manifolds of nonnegative holomorphic bisectional curvature*, J. Differential Geom. **27**(**2**) (1988), 179–214.

[524] N. Mok, Y.T. Siu, and S.-T. Yau, *The Poincar-Lelong equation on complete Kähler manifolds*, Compositio Math. **44**(**1-3**) (1981), 183–218.

[525] N. Mok, Y.T. Siu, and S.K. Yeung, *Geometric superrigidity*, Invent. Math. **113**(**1**) (1993), 57–83.

[526] N. Mok and S.-T. Yau, *Completeness of the Kähler-Einstein metric on bounded domains and the characterization of domains of holomorphy by curvature conditions*, Proc. Sympos. Pure Math., **39**, 41-59, Amer. Math. Soc., Providence, RI, 1983.

[527] N. Mok and J.Q. Zhong, *Curvature characterization of compact Hermitian symmetric spaces*, J. Differential Geom. **23**(**1**) (1986), 15–67.

[528] N. Mok and J.Q. Zhong, *Compactifying complete Kähler-Einstein manifolds of finite topological type and bounded curvature*, Ann. of Math. (2) **129**(**3**) (1989), 427–470.

[529] J. Morgan, *On Thurston's uniformization theorem for three-dimensional manifolds*, The smith conjecture (New York, 1979), 37–115, Pure Appl. Math., **112**, Academic press, Orlando, FL, 1984.

[530] S. Mori, *Projective manifolds with ample tangent bundles*, Ann. of Math. (2) **110**(**3**) (1979), 593–606.

[531] G.D. Mostow, *Strong rigidity of locally symmetric spaces. Annals of Mathematics Studies*, **78**, Princeton University Press, Princeton, NJ; University of Tokyo Press, Tokyo, 1973.

[532] M. Mulase, *KP equations, strings, and the Schottky problem*, Algebraic analysis, Vol. II, 473–492, Academic Press, Boston, MA, 1988.

[533] W. Müller, *Analytic torsion and R-torsion of Riemannian manifolds*, Adv. in Math. **28**(**3**) (1978), 233–305.

[534] D. Mumford, J. Fogarty, and F. Kirwan, *Geometric invariant theory*, Third edition. Ergebnisse der Mathematik und ihrer Grenzgebiete (2) [Results in Mathematics and Related Areas (2)], **34**, Springer-Verlag, Berlin, 1994.

[535] A.M. Nadel, *Multiplier ideal sheaves and Kähler-Einstein metrics of positive scalar curvature*, Ann. of Math. (2) **132**(**3**) (1990), 549–596.

[536] J. Nash, *C^1 isometric imbeddings*, Ann. of Math. (2) **60** (1954), 383–396.

[537] J. Nash, *The imbedding problem for Riemannian manifolds*, Ann. of Math. (2) **63** (1956), 20–63.

[538] R. Neel and D. Stroock, *Analysis of the cutlocus via the heat kernel*, Surveys in Differential Geomtry, **IX**, International Press, 337-350.

[539] L. Ni, *Ancient solution to Kahler-Ricci flow*, arXiv:math.DG/0502494.

[540] L. Nirenberg, *The Weyl and Minkowski problems in differential geometry in the large*, Commun. Pure Appl. Math. **6** (1953), 337–394.

[541] S. P. Novikov, *Multivalued functions and functionals. An analogue of the Morse theory*, Dokl. Akad. Nauk SSSR **260**(**1**) (1981), 31–35.

[542] K. O'Grady, *The irreducible components of moduli spaces of vector bundles on surfaces*, Invent. Math. **112**(**3**) (1993), 585–613.

[543] Y.-G Oh, *Floer cohomology of Lagrangian intersections discs and pseudo-holomorphic curves with Lagrangian boundary conditions*, I, II, Comm. Pure Appl. Math. **46** (1992–1993), 949–1012.

[544] Y.-G. Oh, *Floer cohomology of Lagrangian intersections and pseudo-holomorphic disks*, III, Arnol'd-Givental conjecture. The Floer memorial volume, 555–573, Progr. Math., **133**, Birkhäuser, Basel, 1995.

[545] S. Osher, *A level set formulation for the solution of the Dirichlet problem for Hamilton-Jacobi equations*, SIAM J. Math. Anal. **24**(**5**) (1993), 1145–1152.

[546] R. Osserman, *A proof of the regularity everywhere of the classical solution to Plateau's problem*, Ann. of Math. (2) **91** (1970), 550–569.

[547] J. Otal, *Thurston's hyperbolization of Haken manifolds*, Surveys in differential geometry, Vol. III (Cambridge, MA, 1996), 77–194, Int. Press, Boston, MA, 1998

[548] P. Ozsváth and Z. Szabó, *The symplectic Thom conjecture*, Ann. of Math. (2) **151**(**1**) (2000), 93–124.

[549] T.H. Parker and J.G. Wolfson, *Pseudo-holomorphic maps and bubble trees* (English. English summary), J. Geom. Anal. **3**(**1**) (1993), 63–98.

[550] C.K. Peng and C.L. Terng, *Minimal hypersurfaces of spheres with constant scalar curvature*, Seminar on minimal submanifolds, Ann. of Math. Stud., **103**, Princeton Univ. Press, Princeton, NJ, 1983.

[551] G. Perelman, *The entropy formula for the Ricci flow and its geometric applications*, arXiv:math.DG/0211159.

[552] G. Perelman, *Ricci flow with surgery on three-manifolds*, arXiv:math.DG/0303109.

[553] G. Perelman, *Finite extinction time for the solutions to the Ricci flow on certain three-manifolds*, arXiv:math.DG/0307245.

[554] L. Pestov and G. Uhlmann, *The boundary distance function and the Dirichlet-to-Neumann map*, Math. Res. Lett. **11**(**2-3**) (2004), 285–297.

[555] D.H. Phong and J. Sturm, *On stability and the convergence of the Kähler-Ricci flow*, arXiv:math.DG/0412185

[556] D.H. Phong and J. Sturm, *The Monge-Ampère operator and geodesics in the space of Kähler potentials*, arXiv:math.DG/0504157.

[557] J.T. Pitts, *Existence and regularity of minimal surfaces on Riemannian manifolds*, Mathematical Notes, **27**, Princeton University Press, Princeton, NJ; University of Tokyo Press, Tokyo, 1981.

[558] J.T. Pitts and J.H. Rubinstein, *Equivariant minimax and minimal surfaces in geometric three-manifolds*, Bull. Amer. Math. Soc. (N.S.) **19(1)** (1988), 303–309.

[559] I.I. Pjateckiĭ-Šapiro and I.R. Šafarevič, *Torelli's theorem for algebraic surfaces of type* K3 (Russian), Izv. Akad. Nauk SSSR Ser. Mat. **35** (1971) 530–572; (English Translation) Math. USSR Izvestija **5** (1971) 547–588.

[560] A.V. Pogorelov, *On the question of the existence of a convex surface with a given sum principal radii of curvature*, Usp. Mat. Nauk **8** (1953), 127–130.

[561] A.V. Pogorelov, *On the isometric immersion of a two-dimensional Riemannian manifold homeomorphic to a sphere into a three-dimensional Riemannian space* (Russian), Dokl. Akad. Nauk SSSR **139** (1961), 818–820.

[562] A.V. Pogorelov, *Isometric immersion in the large of a two-dimensional Riemannian manifold into a three-dimensional one* (Russian), Dokl. Akad. Nauk SSSR **137** (1961), 282–283.

[563] A.V. Pogorelov, *On the improper convex affine hyperspheres*, Geometriae Dedicata **1**(**1**) (1972), 33–46.

[564] A.V. Pogorelov, *Existence of a closed convex hypersurface with prescribed curvature*, Dokl. Akad. Nauk SSSR **274(1)** (1984), 28–31.

[565] G. Pólya and G. Szegö, *Isoperimetric Inequalities in Mathematical Physics*, Annals of Mathematics Studies, **27**, Princeton University Press, Princeton, NJ, 1951.

[566] J. Porti, *Geometric structures on three-dimensional manifolds*, (Catalan) First Conference on Mathematics(Bellaterra, 1998), Butl. Soc. Catalana Mat. **14**(**1**) (1999), 99–114.

[567] D.R. Ray and I.M. Singer, *Analytic torsion for complex manifolds*, Ann. of Math. (2) **98**(1973), 154–177.

[568] M. Reid, *The moduli space of 3-folds with $K = 0$ may nevertheless be irreducible*, Math. Ann. **278(1-4)** (1987), 329–334.

[569] S.-S. Roan and S.-T. Yau, *On Ricci flat 3-fold*, Acta Math. Sinica (N.S.) **3**(**3**) (1987), 256–288.

[570] D.C. Robinson, *Uniqueness of the Kerr Black hole*, Phys. Rev. Lett., **34** (1975), 905-906.

[571] J. Rosenberg, *C^*-algebras, positive scalar curvature and the Novikov conjecture*, Publ. IHES **58** (1983), 197–212.

[572] J. Ross and R.P. Thomas, *An obstruction to the existence of constant scalar curvature Kähler metricsmath*, arXiv:math.DG/0412518.

[573] J. Ross and R.P. Thomas, *A study of the Hilbert-Mumford criterion for the stability of projective varieties*, arXiv:math.AG/0412519.

[574] H.L. Royden, *Report on the Teichmüler metric*, Proc. Nat. Acad. Sci. USA **65** (1970), 497–499.

[575] W.D. Ruan, *Canonical coordinates and Bergmann metrics*, Comm. Anal. Geom. **6**(**3**) (1998), 589–631.

[576] W. D. Ruan, *Generalized special Lagrangian torus fibration for Calabi-Yau hypersurfaces in toric varieties*, I, II, III, Arxiv: math.DG/0303114, math.DG/0303278, math.DG/0309450.

[577] W.D. Ruan, *The Fukaya category of symplectic neighborhood of a non-Hausdorff manifold*, math.SG/0602119

[578] Y.B. Ruan, *Symplectic topology on algebraic 3-folds*, J. Differential Geom. **39** (1994), 215-227.

[579] Y.B. Ruan, *Topological sigma model and Donaldson-type invariants in Gromov theory*, Duke Math. J. **83(2)** (1996), 461–500.

[580] Y.B. Ruan and G. Tian, *A mathematical theory of quantum cohomology*, J. Differential Geom. **42(2)** (1995), 259–367.

[581] J. Sacks and K. Uhlenbeck, *The existence of minimal immersions of* 2*-spheres*, Ann. of Math. (2) **113(1)** (1981), 1–24.

[582] L. Saloff-Coste, *A note on Poincaré, Sobolev, and Harnack inequalities*, Internat. Math. Res. Notices **2** (1992), 27–38.

[583] L. Saloff-Coste, *Analysis on Riemannian co-compact covers*, Surveys in Differential Geometry, **IX**, 351-385. International Press, Somerville, MA, 02143.

[584] L. Saper, *On the cohomology of locally symmetric spaces and of their compactifications*, Current developments in mathematics, 2002, 219–289, Int. Press, Somerville, MA, 2003.

[585] L. Saper and M. Stern, L_2*-cohomology of arithmetic varieties*, Ann. of Math. (2) **132(1)** (1990), 1–69.

[586] P. Sarnak, *Additive number theory and Maass forms*, Lecture Notes in Math., **1052**, 286–309, Springer, Berlin, 1984.

[587] P. Sarnak, *Spectra of hyperbolic surfaces*, Bull. Amer. Math. Soc. (N.S.) **40(4)** (2003), 441–478.

[588] W. Schmid, *Variation of Hodge structure: the singularities of the period mapping*, Invent. Math. **22** (1973), 211–319.

[589] M. Schmidt, *A proof of the Willmore conjecture*, math.DG/0203224

[590] R. Schoen, *Conformal deformation of a Riemannian metric to constant scalar curvature*, J. Differential Geom. **20(2)** (1984), 479–495.

[591] R. Schoen, L. Simon, and S.-T. Yau, *Curvature estimates for minimal hypersurfaces*, Acta Math. **134(3-4)** (1975), 275–288.

[592] R. Schoen and K. Uhlenbeck, *A regularity theory for harmonic maps*, J. Differential Geom. **17(2)** (1982), 307–335.

[593] R. Schoen and K. Uhlenbeck, *Boundary regularity and the Dirichlet problem for harmonic maps*, J. Differential Geom. **18(2)** (1983), 253–268.

[594] R. Schoen and J. Wolfson, *Minimizing area among Lagrangian surfaces: the mapping problem*, J. Differential Geom. **58(1)** (2001), 1–86.

[595] R. Schoen, S. Wolpert, and S.-T. Yau, *Geometric bounds on the low eigenvalues of a compact surface*, Geometry of the Laplace operator (Proc. Sympos. Pure Math., Univ. Hawaii, Honolulu, Hawaii, 1979), 279–285, Proc. Sympos. Pure Math., **XXXVI**, Amer. Math. Soc., Providence, RI, 1980.

[596] R. Schoen and S.-T. Yau, *On univalent harmonic maps between surfaces*, Invent. Math. **44(3)** (1978), 265–278.

[597] R. Schoen and S. T. Yau, *Existence of incompressible minimal surfaces and the topology of three-dimensional manifolds with nonnegative scalar curvature*, Ann. of Math. (2) **110(1)** (1979), 127–142.

[598] R. Schoen and S.-T. Yau, *On the proof of the positive mass conjecture in general relativity*, Comm. Math. Phys. **65(1)** (1979), 45–76.

[599] R. Schoen and S.-T. Yau, *Positivity of the total mass of a general space-time*, Phys. Rev. Lett. **43(20)** (1979), 1457–1459.

[600] R. Schoen and S.-T. Yau, *On the structure of manifolds with positive scalar curvature*, Manuscripta Math. **28(1-3)** (1979), 159–183.

[601] R. Schoen and S.-T. Yau, *Compact group actions and the topology of manifolds with nonpositive curvature*, Topology **18(4)** (1979), 361–380.

[602] R. Schoen and S.-T. Yau, *Proof of the positive mass theorem*, II, Comm. Math. Phys. **79(2)** (1981), 231–260.

[603] R. Schoen and S.-T. Yau, *Complete three-dimensional manifolds with positive Ricci curvature and scalar curvature*, Seminar on Differential Geometry, 209–228, Ann. of Math. Stud., **102**, Princeton Univ. Press, Princeton, NJ, 1982.

[604] R. Schoen and S.-T. Yau, *The existence of a black hole due to condensation of matter*, Comm. Math. Phys. **90(4)** (1983), 575–579.

[605] R. Schoen and S.-T. Yau, *Conformally flat manifolds, Kleinian groups and scalar curvature*, Invent. Math. **92(1)** (1988), 47–71.

[606] R. Schoen and S.-T. Yau, *Lectures on differential geometry*, Lecture notes prepared by Wei Yue Ding, Kung Ching Chang [Gong Qing Zhang], Jia Qing Zhong and Yi Chao Xu; Translated from the Chinese by Ding and S.Y. Cheng, Preface translated from the Chinese by Kaising Tso, Conference Proceedings and Lecture Notes in Geometry and Topology, **I**, International Press, Cambridge, MA, 1994.

[607] N. Seiberg and E. Witten, *Electric-magnetic duality, monopole condensation, and confinement in $N = 2$ supersymmetric Yang-Mills theory*, Nuclear Phys. B **426(1)** (1994), 19–52.

[608] N. Seiberg and E. Witten, *Monopoles, duality and chiral symmetry breaking in $N = 2$ supersymmetric QCD*, Nuclear Phys. B **431(3)** (1994), 484–550.

[609] P. Seidel and R. Thomas, *Braid group actions on derived categories of coherent sheaves*, Duke Math. J. **108(1)** (2001), 37–108.

[610] J. Shatah and M. Struwe, *Regularity results for nonlinear wave equations*, Ann. of Math. (2) **138(3)** (1993), 503–518.

[611] W.X. Shi, *Complete noncompact Kähler manifolds with positive holomorphic bisectional curvature*, Bull. Amer. Math. Soc. (N.S.) **23(2)** (1990), 437–440.

[612] W.X. Shi, *Ricci flow and the uniformization on complete noncompact Kähler manifolds*, J. Differential Geom. **45(1)** (1997), 94–220.

[613] W.X. Shi, *A uniformization theorem for complete Kähler manifolds with positive holomorphic bisectional curvature*, J. Geom. Anal. **8(1)** (1998), 117–142.

[614] Y.G. Shi and L.F. Tam, *Positive mass theorem and the boundary behaviors of compact manifolds with nonnegative scalar curvature*, J. Differential Geom. **62(1)** (2002), 79–125.

[615] B. Shiffman, *Complete characterization of holomorphic chains of codimension one*, Math. Ann. **274(2)** (1986), 233–256.

[616] T. Shioya and T. Yamaguchi, *Collapsing three-manifolds under a lower curvature bound*, J. Differential Geom. **56(1)** (2000), 1–66.

[617] T. Shioya and T. Yamaguchi, *Volume collapsed three-manifolds with a lower curvature bound*, math.DG/0304472

[618] L. Simon, *Theorems on the regularity and singularity of minimal surfaces and harmonic maps*, Geometry and global analysis (Sendai, 1993), 111–145,

[619] L. Simon, *Existence of Willmore surfaces*, Miniconference on geometry and partial differential equations (Canberra, 1985), 187–216, Proc. Centre Math. Anal. Austral. Nat. Univ., **10**, Austral. Nat. Univ., Canberra, 1986.

[620] L. Simon, *Theorems on regularity and singularity of energy minimizing maps*, Lectures in Mathematics ETH Zürich. Birkhäuser Verlag, Basel, 1996.

[621] C.T. Simpson, *Constructing variations of Hodge structure using Yang-Mills theory and applications to uniformization*, J. Amer. Math. Soc. **1(4)** (1988), 867–918.

[622] C.T. Simpson, *Moduli of representations of the fundamental group of a smooth projective variet*, I. Inst. Hautes Études Sci. Publ. Math. **79** (1994), 47–129.

[623] C.T. Simpson, *Moduli of representations of the fundamental group of a smooth projective variety*, II, Inst. Hautes Études Sci. Publ. Math. **80** (1994), 5–79.

[624] I.M. Singer, B. Wong, S.-T. Yau, and S.-T.S. Yau, *An estimate of the gap of the first two eigenvalues in the Schrödinger operator*, Ann. Scuola Norm. Sup. Pisa Cl. Sci. (4) **12(2)** (1985), 319–333.

[625] Y.T. Siu, *Complex-analyticity of harmonic maps and strong rigidity of compact Kähler manifolds*, Ann. of Math. **112** (1980), 73–111.

[626] Y.T. Siu, *A simple proof of the surjectivity of the period map of $K3$ surfaces*, Manuscripta Math. **35**(**3**) (1981), 311–321.

[627] Y.T. Siu, *Every $K3$ surface is Kähler*, Invent. Math. **73**(**1**) (1983), 139–150.

[628] Y.T. Siu, *Effective very ampleness*, Invent. Math. **124**(**1-3**) (1996), 563–571.

[629] Y.T. Siu, *Invariance of plurigenera*, Invent. Math. **134**(**3**) (1998), 661–673.

[630] Y.T. Siu, *Very ampleness part of Fujita's conjecture and multiplier ideal sheaves of Kohn and Nadel*, Complex analysis and geometry (Columbus, OH, 1999), 171–191, Ohio State Univ. Math. Res. Inst. Publ., **9**, de Gruyter, Berlin, 2001.

[631] Y.T. Siu, *Some recent transcendental techniques in algebraic and complex geometry*, Proceddings of the International Congress of Mathematicians, Vol. **I** (Beijing, 2002), 439–448.

[632] Y.T. Siu and S.-T. Yau, *Complete Kähler manifolds with nonpositive curvature of faster than quadratic decay*, Ann. of Math. (2) **105**(**2**) (1977), 225–264.

[633] Y.T. Siu and S.-T. Yau, *Compact Kähler manifolds of positive bisectional curvature*, Invent. Math. **59**(**2**) (1980), 189–204.

[634] Y.T. Siu and S.-T. Yau, *Compactification of negatively curved complete Kähler manifolds of finite volume*, Seminar on Differential Geometry, 363–380, Ann. of Math. Stud., **102**, Princeton Univ. Press, Princeton, NJ, 1982.

[635] I. Smith, R.P. Thomas, and S.-T. Yau, *Symplectic conifold transitions*, J. Differential Geom. **62**(**2**) (2002), 209–242.

[636] K. Smoczyk, *Longtime existence of the Lagrangian mean curvature flow*, Calc. Var. Partial Differential Equations **20**(**1**) (2004), 25–46.

[637] K. Smoczyk and M.T. Wang, *Mean curvature flows of Lagrangians submanifolds with convex potentials*, J. Differential Geom. **62**(**2**) (2002), 243–257.

[638] J.A. Smoller, A.G. Wasserman, and S.-T. Yau, *Existence of black hole solutions for the Einstein-Yang/Mills equations*, Comm. Math. Phys. **154**(**2**) (1993), 377–401.

[639] J.A. Smoller, A.G. Wasserman, S.-T. Yau, and J.B. McLeod, *Smooth static solutions of the Einstein/Yang-Mills equations*, Comm. Math. Phys. **143**(**1**) (1991), 115–147.

[640] M. Stern, *L^2-index theorems on locally symmetric spaces*, Invent. Math. **96**(**2**) (1989), 231–282.

[641] S. Stolz, *Simply connected manifolds of positive scalar curvature*, Ann. of Math. (2) **136**(**3**) (1992), 511–540.

[642] A. Strominger, *Superstrings with torsion*, Nuclear Phys. B **274**(**2**) (1986), 253–284.

[643] A. Strominger, S.-T. Yau, and E. Zaslow, *Mirror symmetry is T-duality*, Nuclear Phys. B **479**(**1-2**) (1996), 243–259.

[644] D. Sullivan, *A generalization of Milnor's inequality concerning affine foliations and affine manifolds*, Comment. Math. Helv. **51**(**2**) (1976), 183–189.

[645] D. Sullivan, *Cycles for the dynamical study of foliated manifolds and complex manifolds*, Invent. Math. **36** (1976), 225–255.

[646] D. Sullivan, *Infinitesimal computations in topology*, Inst. Hautes Études Sci. Publ. Math. **47** (1977), 269–331.

[647] D. Sullivan, *The Dirichlet problem at infinity for a negatively curved manifold*, J. Differential Geom. **18**(**4**) (1983), 723–732.

[648] D. Sullivan, *Closed string operators and Feynman graphs*, to appear.

[649] T. Sunada, *Riemannian coverings and isospectral manifolds*, Ann. of Math. (2) **121**(**1**) (1985), 169–186.

[650] Z. Szabó, *Locally non-isometric yet super isospectral spaces*, Geom. Funct. Anal. **9**(**1**) (1999), 185–214.

[651] G. Szegö, *Inequalities for certain eigenvalues of a membrane of given area*, J. Rational Mech. Anal. **3** (1954), 343–356.

[652] G. Talenti, *Best constant in Sobolev inequality*, Ann. Mat. Pura Appl. (4) **110** (1976), 353–372.

[653] T. Tao, *Low-regularity global solutions to nonlinear dispersive equations*, Surveys in analysis and operator theory (Canberra, 2001), 19–48, Proc. Centre Math. Appl. Austral. Nat. Univ., **40**, Austral. Nat. Univ., Canberra, 2002.

[654] D. Tataru, *On global existence and scattering for the wave maps equation*, Amer. J. Math. **123**(**1**) (2001), 37–77.

[655] C.H. Taubes, *Self-dual Yang-Mills connections on non-self-dual* 4*-manifolds*, J. Differential Geom. **17**(**1**) (1982), 139–170.

[656] C.H. Taubes, *Self-dual connections on* 4*-manifolds with indefinite intersection matrix*, J. Differential Geom. **19**(**2**) (1984), 517–560.

[657] C.H. Taubes, S^1 *actions and elliptic genera*, Comm. Math. Phys. **122**(1989), no. 3, 455–526.

[658] C.H. Taubes, *Casson's invariant and gauge theory*, J. Differential Geom. **31**(**2**) (1990), 547–599.

[659] C.H. Taubes, *The existence of anti-self-dual conformal structures*, J. Differential Geom. **36**(**1**) (1992), 163–253.

[660] C.H. Taubes, *The Seiberg-Witten invariants and symplectic forms*, Math. Res. Lett. **1**(**6**) (1994), 809–822.

[661] C.H. Taubes, SW $\Rightarrow$ Gr*: from the Seiberg-Witten equations to pseudo-holomorphic curves*, J. Amer. Math. Soc. **9**(**3**) (1996), 845–918.

[662] C.H. Taubes, *Counting pseudo-holomorphic submanifolds in dimension* 4, J. Differential Geom. **44**(**4**) (1996), 818–893.

[663] C.H. Taubes, Gr $\Longrightarrow$ SW*: from pseudo-holomorphic curves to Seiberg-Witten solutions*, J. Differential Geom. **51**(**2**) (1999), 203–334.

[664] C.H. Taubes, GR = SW*: counting curves and connections*, J. Differential Geom. **52**(**3**) (1999), 453–609.

[665] C.H. Taubes, *Seiberg Witten and Gromov invariants for symplectic* 4*-manifolds*, Edited by Richard Wentworth. First International Press Lecture Series, **2**, International Press, Somerville, MA, 2000.

[666] C.L. Terng, *A higher dimension generalization of the sine-Gordon equation and its soliton theory*, Ann. of Math. (2)**111**(**3**) (1980), 491–510.

[667] C.L. Terng, *Recent progress in submanifold geometry. Differential geometry: partial differential equations on manifolds*, 439–484, Proc. Sympos. Pure Math., **54**, Part 1, Amer. Math. Soc., Providence, RI, 1993.

[668] C.L. Terng and K. Uhlenbeck, *Poisson actions and scattering theory for integrable systems*, Surv. Differ. Geom., **V**, 315–402, Int. Press, Boston, MA, 1998.

[669] C.L. Terng and K. Uhlenbeck, *Bäcklund transformations and loop group actions*, Comm. Pure Appl. Math. **53**(2000), no. 1, 1–75.

[670] R.P. Thomas, *Moment maps, monodromy and mirror manifolds*, Symplectic geometry and mirror symmetry (Seoul, 2000), 467–498, World Sci. Publishing, River Edge, NJ, 2001.

[671] R.P. Thomas and S.-T. Yau, *Special Lagrangians, stable bundles and mean curvature flow* (English. English summary), Comm. Anal. Geom. **10**(**5**) (2002), 1075–1113.

[672] G. Thorbergsson, *A survey on isoparametric hypersurfaces and their generalizations*, Handbook of differential geometry, Vol. **I**, 963–995, North-Holland, Amsterdam, 2000.

[673] W.P. Thurston, *Three-dimensional manifolds, Kleinian groups and hyperbolic geometry*, Bull. Amer. Math. Soc. (N.S.) **6**(**3**) (1982), 357–381.

[674] W. Thurston, *Three-dimensional Geometry and Topology* (Silvio Levy, ed.), Princeton Math. Ser., **35**, Princeton Univ. Press, 1997.

[675] G. Tian, *Smoothness of the universal deformation space of compact Calabi-Yau manifolds and its Petersson-Weil metric*, Mathematical aspects of string theory (San

Diego, Calif., 1986), 629–646, Adv. Ser. Math. Phys., **1**, World Sci. Publishing, Singapore, 1987.

[676] G. Tian, *On a set of polarized Kähler metrics on algebraic manifolds*, J. Differential Geom. **32**(**1**) (1990), 99–130.

[677] G. Tian, *Calabi's conjecture for complex surfaces with positive first Chern class*, Invent. Math. **101**(**1**) (1990), 101–172.

[678] G. Tian, *Kähler-Einstein metrics with positive scalar curvature*, Invent. Math. **130**(**1**) (1997), 1–37.

[679] G. Tian, *Gauge theory and calibrated geometry*, I, Ann. of Math. (2) **151**(**1**) (2000), 193–268.

[680] G. Tian and S.-T. Yau, *Three-dimensional algebraic manifolds with $C_1 = 0$ and $\chi = -6$*, Mathematical aspects of string theory (San Diego, Calif., 1986), 543–559, Adv. Ser. Math. Phys., **1**, World Sci. Publishing, Singapore, 1987.

[681] G. Tian and S.-T. Yau, *Complete Kähler manifolds with zero Ricci curvature*, I, J. Amer. Math. Soc. **3**(**3**) (1990), 579–609.

[682] G. Tian and S.-T. Yau, *Complete Kähler manifolds with zero Ricci curvature*, II, Invent. Math. **106**(**1**) (1991), 27–60.

[683] A.N. Todorov, *Applications of the Kähler-Einstein-Calabi-Yau metric to moduli of $K3$ surfaces*, Invent. Math. **61**(**3**) (1980), 251–265.

[684] A.N. Todorov, *The Weil-Petersson geometry of the moduli space of* SU$(n \geq 3)$ *(Calabi-Yau) manifolds*, I, Comm. Math. Phys. **126**(**2**) (1989), 325–346.

[685] A.E. Treibergs, *Entire spacelike hypersurfaces of constant mean curvature in Minkowski space*, Invent. Math. **66**(**1**) (1982), 39–56.

[686] N.S. Trudinger, *Remarks concerning the conformal deformation of Riemannian structures on compact manifolds*, Ann. Scuola Norm. Sup. Pisa (3) **22** (1968), 265–274.

[687] N.S. Trudinger and X.J. Wang, *The Bernstein problem for affine maximal hypersurfaces*, Invent. Math. **140**(**2**) (2000), 399–422.

[688] K.K. Uhlenbeck, *Removable singularities in Yang-Mills fields*, Comm. Math. Phys. **83**(**1**) (1982), 11–29.

[689] K.K. Uhlenbeck, *Connections with L^p bounds on curvature*, Comm. Math. Phys. **83**(**1**) (1982), 31–42.

[690] K. Uhlenbeck, *Harmonic maps into Lie groups: classical solutions of the chiral model*, J. Differential Geom. **30**(**1**) (1989), 1–50.

[691] K. Uhlenbeck and S.-T. Yau, *On the existence of Hermitian-Yang-Mills connections in stable vector bundles*, Comm. Pure Appl. Math. **39**(**S**) (1986), S257–S293.

[692] C. Vafa, *Topological mirrors and quantum rings*, Essays on mirror manifolds, 96–119, Internat. Press, Hong Kong, 1992.

[693] N.Th. Varopoulos, L. Saloff-Coste, and T. Coulhon, *Analysis and geometry on groups*, Cambridge Tracts in Mathematics, **100**, Cambridge University Press, Cambridge, 1992.

[694] E. Viehweg, *Quasi-projective moduli for polarized manifolds*, Ergebnisse der Mathematik und ihrer Grenzgebiete (3) [Results in Mathematics and Related Areas (3)], **30**, Springer-Verlag, Berlin, 1995.

[695] M. Vigué-Poirrier and D. Sullivan, *The homology theory of the closed geodesic problem*, J. Differential Geometry **11**(**4**) (1976), 633–644.

[696] C. Voisin, *A counterexample to the Hodge conjecture for Kähler varieties*, Int. Math. Res. Not. **20** (2002), 1057–1075.

[697] C. Voisin, *On the homotopy types of compact Kähler and complex projective manifolds*, Invent. Math. **157**(**2**) (2004), 329–343.

[698] C. Voisin, *On the homotopy types of Kaehler manifolds and the birational Kodaira problem*, Arxiv: math.AG/0410040.

[699] T. Wan, *Constant mean curvature surface, harmonic maps, and universal Teichmüller space*, J. Differential Geom. **35(3)** (1992), 643–657.
[700] C.P. Wang, *Some examples of complete hyperbolic affine 2-spheres in R^3*, Global differential geometry and global analysis (Berlin, 1990), 271–280, Lecture Notes in Math., **1481**, Springer, Berlin, 1991.
[701] C.L. Wang, *On the incompleteness of the Weil-Petersson metric along degenerations of Calabi-Yau manifolds*, Math. Res. Lett. **4(1)** (1997), 157–171.
[702] H.C. Wang, *Complex parallisable manifolds*, Proc. Amer. Math. Soc. **5** (1954), 771–776.
[703] M.T. Wang, *A fixed point theorem of discrete group actions on Riemannian manifolds*, J. Differential Geom. **50(2)** (1998), 249–267.
[704] M.T. Wang, *Generalized harmonic maps and representations of discrete groups*, Comm. Anal. Geom. **8(3)** (2000), 545–563.
[705] M.T. Wang, *Mean curvature flow of surfaces in Einstein four-manifolds*, J. Differential Geom. **57(2)** (2001), 301–338.
[706] M.T. Wang, *Deforming area preserving diffeomorphism of surfaces by mean curvature flow*, Math. Res. Lett. **8(5-6)** (2001), 651–661.
[707] M.Y. Wang and W. Ziller, *Existence and nonexistence of homogeneous Einstein metrics*, Invent. Math. **84(1)** (1986), 177–194.
[708] X.J. Wang and X.H. Zhu, *Kähler-Ricci solitons on toric manifolds with positive first Chern class*, Adv. Math. **188(1)** (2004), 87–103.
[709] X.W. Wang, *Balance point and stability of vector bundles over a projective manifold*, Math. Res. Lett. **9(2-3)** (2002), 393–411.
[710] A. Weil, *Sur les théorémes de de Rham*, Comment. Math. Helv. **26** (1952), 119–145.
[711] H. Weyl, *Das asymptotische Verteilungsgesetz der Eigenwerte linearer partieller Differentialgleichungen* (mit einer Anwendung auf die Theorie der Hohlraumstrahlung) (German), Math. Ann. **71(4)** (1912), 441–479.
[712] S. Weinberger, *Aspects of the Novikov conjecture, in Geometric and Topological Invariants of Elliptic Operators*, J. Kaminker, editor, 1990, 281–297, AMS, Providence, RI.
[713] A. Weinstein, *On the volume of manifolds all of whose geodesics are closed*, J. Differential Geometry **9** (1974), 513–517.
[714] N. Wickramasekera, *A rigidity theorem for stable minimal hypercones*, J. Differential Geom. **68(3)** (2004), 433–514.
[715] N. Wickramasekera, *On the singularities and a Bernstein property of immersed stable minimal hypersurfaces*, Calc. Var. Partial Differential Equations **22(1)** (2005), 1–20.
[716] E. Witten, *A new proof of the positive energy theorem*, Comm. Math. Phys. **80(3)** (1981), 381–402.
[717] E. Witten, *Supersymmetry and Morse theory*, J. Differential Geom. **17(4)** (1982), 661–692.
[718] E. Witten, *Holomorphic Morse inequalities*, Algebraic and differential topology–global differential geometry, 318–333, Teubner-Texte Math., **70**, Teubner, Leipzig, 1984.
[719] E. Witten, *Elliptic genera and quantum field theory*, Comm. Math. Phys. **109(4)** (1987), 525–536.
[720] E. Witten, *On the structure of the topological phase of two-dimensional gravity*, Nuclear Phys. B **340(2-3)** (1990), 281–332.
[721] E. Witten, *Monopoles and four-manifolds*, Math. Res. Lett. **1(6)** (1994), 769–796.
[722] E. Witten and S.-T. Yau, *Connectedness of the boundary in the AdS/CFT correspondence*, Adv. Theor. Math. Phys. **3(6)** (1999), 1635–1655.
[723] M. Wolf, *The Teichmuller theory of harmonic maps*, J. Differential Geom. **29(2)** (1989), 449–479.

[724] S.A. Wolpert, *Geometry of the Weil-Petersson completion of Teichmüller space*, Surveys in differential geometry, Vol. **VIII**, 357–393, Int. Press, Somerville, MA, 2003.

[725] S.A. Wolpert, *Disappearance of cusp forms in special families*, Ann. of Math. (2) **139** (1994), 239–291.

[726] H. Wu, *Normal families of holomorphic mappings*, Acta Math. **119** (1967), 193–233.

[727] C.T. Yang, *Odd-dimensional wiedersehen manifolds are spheres*, J. Differential Geom. **15**(**1**) (1980), 91–96.

[728] C.T. Yang, *Smooth great circle fibrations and an application to the topological Blaschke conjecture*, Trans. Amer. Math. Soc. **320**(**2**) (1990), 507–524.

[729] C.T. Yang, *Any Blaschke manifold of the homotopy type of* $C\mathrm{P}^n$ *has the right volume*, Pacific J. Math. **151**(**2**) (1991), 379–394.

[730] P.C. Yang and S.-T. Yau, *Eigenvalues of the Laplacian of compact Riemann surfaces and minimal submanifolds*, Ann. Scuola Norm. Sup. Pisa Cl. Sci. (4) **7**(**1**) (1980), 55–63.

[731] S.-T. Yau, *On the fundamental group of compact manifolds of non-positive curvature*, Ann. of Math. (2) **93** (1971), 579–585.

[732] S.-T. Yau, *Harmonic functions on complete Riemannian manifolds*, Comm. Pure Appl. Math. **28** (1975), 201–228.

[733] S.-T. Yau, *Isoperimetric constants and the first eigenvalue of a compact Riemannian manifold*, Ann. Sci. École Norm. Sup. (4) **8**(**4**) (1975), 487–507.

[734] S.-T. Yau, *Some function-theoretic properties of complete Riemannian manifold and their applications to geometry*, Indiana Univ. Math. J. **25**(**7**) (1976), 659–670.

[735] S.-T. Yau, *Calabi's conjecture and some new results in algebraic geometry*, Proc. Nat. Acad. Sci. USA **74**(**5**) (1977), 1798–1799.

[736] S.-T. Yau, *On the Ricci curvature of a compact Kähler manifold and the complex Monge-Ampère equation*, I, Comm. Pure Appl. Math. **31**(**3**) (1978), 339–411.

[737] S.-T. Yau, *The role of partial differential equations in differential geometry*, Proceedings of the International Congress of Mathematicians (Helsinki, 1978), 237–250, Acad. Sci. Fennica, Helsinki, 1980.

[738] S.-T. Yau, *A general Schwarz lemma for Kähler manifolds*, Amer. J. Math. **100**(**1**) (1978), 197–203.

[739] S.-T. Yau, *Problem section*, Seminar on Differential Geometry, pp. 669–706, Ann. of Math. Stud., **102**, Princeton Univ. Press, Princeton, NJ, 1982.

[740] S.-T. Yau, *Compact three-dimensional Kähler manifolds with zero Ricci curvature*, Symposium on anomalies, geometry, topology (Chicago, Ill., 1985), 395–406, World Sci. Publishing, Singapore, 1985.

[741] S.-T. Yau, *Nonlinear analysis in geometry*, Monographies de L'Enseignement Mathématique, **33**, Série des Conférences de l'Union Mathématique Internationale, **8**, L'Enseignement Mathématique, Geneva, 1986.

[742] S.-T. Yau, *A review of complex differential geometry*, Proc. Symp. Pur Math. **52** (1991), 619-625.

[743] S.-T. Yau, *Open problems in Geometry*, Proc. Symp. Pure Math. **54** (1993), 1–28.

[744] S.-T. Yau, *A splitting theorem and an algebraic geometric characterization of locally Hermitian symmetric spaces*, Comm. Anal. Geom. **1**(**3-4**) (1993), 473–486.

[745] S.-T. Yau, *An application of eigenvalue estimate to algebraic curves defined by congruence subgroups*, Math. Res. Lett. **3**(**2**) (1996), 167–172.

[746] S.-T. Yau, *A note on the distribution of critical points of eigenfunctions*, Collection of papers on geometry, analysis and mathematical physics, 173–175, World Sci. Publishing, River Edge, NJ, 1997.

[747] S.-T. Yau, *An estimate of the gap of the first two eigenvalues in the Schrödinger operator. Lectures on partial diffcrcntial equations*, 223–235, New Stud. Adv. Math., **2**, Int. Press, Somerville, MA, 2003.

[748] S.-T. Yau and W.H. Meeks, *The equivariant loop theorem for three-dimensional manifolds and a review of the existence theorems for minimal surfaces*, The Smith conjecture (New York, 1979), 153–163, Pure Appl. Math., **112**, Academic Press, Orlando, FL, 1984.

[749] R.G. Ye, *Gromov's compactness theorem for pseudo holomorphic curves*, Trans. Amer. Math. Soc. **342**(**4**) (1994), 671–694.

[750] R.G. Ye, *Foliation by constant mean curvature spheres on asymptotically flat manifolds*, Geometric analysis and the calculus of variations, 369–383, Internat. Press, Cambridge, MA, 1996.

[751] G.L. Yu, *Novikov conjecture for groups with finite asymptotic dimension*, Ann. of Math. (2) **147**(**2**) (1998), 325–355.

[752] S. Zelditch, *Szegö kernels and a theorem of Tian*, Int. Math. Res. Not. **6**(1998), 317-331.

[753] S. Zelditch, *Spectral determination of analytic bi-axisymmetric plane domains*, Geom. Funct. Anal. **10**(**3**) (2000), 628–677.

[754] S.W. Zhang, *Heights and reductions of semi-stable varieties*, Compositio Math. **104**(**1**) (1996), 77–105.

[755] J.Q. Zhong and H.C. Yang, *On the estimate of the first eigenvalue of a compact Riemannian manifold*, Sci. Sinica Ser. A **27**(**12**) (1984), 1265–1273.

[756] A. Zinger, *Enumeration of genus-two curves with a fixed complex structure in* $\mathbb{P}^2$ *and* $\mathbb{P}^3$, arXiv:math.SG/0201254.

[757] A. Zinger, *Enumerative vs. Invariants and obstruction bundles*, arXiv:math.SG/0201255.

[758] P. Zograf, *A spectral proof of Rademacher's conjecture for congruence subgroups of the modular group*, J. Reine Angew. Math. **414** (1991), 113–116.

[759] K. Zuo, *Representations of fundamental groups of algebraic varieties*, Lecture Notes in Mathematics, **1708**, Springer-Verlag, Berlin, 1999.

Department of Mathematics, Harvard University, Cambridge, MA 02138
E-mail address: `yau@math.harvard.edu`

Distributions in algebraic dynamics

Shou-Wu Zhang

CONTENTS

0. Introduction

The complex dynamic system is a subject to study iterations on $\mathbb{P}^1$ or $\mathbb{P}^N$ with respect to complex topology. It originated from the study of Newton method and the three body problem in the end of 19th century and was highly developed in 20th century. It is a unique visualized subject in pure math because of the beautiful and intricate pictures of the Julia sets generated by computer. The subject of this paper, algebraic dynamics, is a subject to study iterations under Zariski topology and is still in its infancy. If the iteration is defined over a number field, then we are in the situation of arithmetical dynamics where the Galois group and heights will be involved. Here we know very little besides very symmetric objects like abelian varieties and multiplicative groups.

The development of arithmetical dynamics was initiated by Northcott in his study of heights on projective space [**47**], 1950. He showed that the set of rational preperiodic points of any endomorphism of $\mathbb{P}^N$ of degree ≥ 2 is always finite. The modern theory of canonical heights was developed by Call and Silverman in [**11**]. Their theory generalized earlier notions of Weil heights on projective spaces and Néron-Tate heights on abelian varieties. Thus many classical questions about abelian varieties and multiplicative groups can be asked again for the dynamical system, such as the size of rational points, preperiodic points, and their distributions. See §4.1.5 and §4.1.6 for some standard conjectures, such as Lehmer's conjecture and Lang's conjecture. In [**62**], we developed a height theory for subvarieties and an intersection theory for *integrable adelic metrized line bundles*, based on Gillet and Soulé's intersection theory [**26**]. Thus many questions about points can be asked again for subvarieties. Two questions we considered in [**62**] are the Manin-Mumford conjecture and the Bogomolov conjecture. See Conjecture 1.2.1 and 4.1.1.

This note in a large sense is an extension of our previous paper. Our first goal is to provide a broad background in Kähler geometry, algebraic geometry, and measure theory. Our second goal is to survey and explain the new developments. The following is a detailed description of the contents of the paper.

In §1, we will give some basic definitions and examples of dynamics in Kähler geometry and algebraic geometry, and study the Zariski properties of preperiodic points. Our dynamic Manin-Mumford conjecture says that a subvariety is preperiodic if and only if it contains many preperiodic points. One question we wish to know (but don't yet) is about the positivity of a canonical (1,1) class on the moduli of cycles on a Kähler variety.

In §2, we will study the classification problem about Kähler dynamics. In surfaces, the problem can be completely solved. In the smooth projective case, we will prove that the dynamics can only be either a quotient of a complex torus or uniruled. In the general case, we will give a factorization result with respect to rational connectedness.

In §3, we will study the measure theoretic properties of dynamics. We will first construct invariant metrics and measures on bundles and subvarieties. We will conjecture several properties about these invariant measures: they can be obtained by iterations of smooth measures, or by probability measures of backward orbits of general points. We also conjecture that the Kobayashi pseudo-metrics vanish. We will prove some special of these properties using the works of Yau [**59**] and Briend-Duval [**9**].

In §4, we will study dynamics over number fields. We will first propose a dynamic Bogomolov conjecture and an equidistribution conjecture for dynamically generic small points. Following Chambert-Loir [**16**], we can make an equidistrubution conjecture on Berkovich's *p-adic analytic spaces.* Finally, we will prove that the equidistribution conjecture and Bogomolov conjecture are essentially equivalent to each other using a recent work of Yuan [**60**] on arithmetic bigness.

What should be, but is not, discussed in this article. Because of limitations of our time and knowledge, many interesting and important topics will not be treated in this article.

- First is the "real theory of dynamics". We prove some properties about the distribution of backward orbits but we say nothing about the forward orbits. Also we have zero knowledge about support of the canonical measure (which is actually crucial in our arithmetic theory). We have to learn from "real or p-adic experts of the dynamical system" about what we should do in the next step. We refer to Katok and Hasselblatt's book [**32**] for dynamics on manifolds, Milnor's book [**41**] for $\mathbb{P}^1$, Sibony's article [**54**] for $\mathbb{P}^N$, and Dinh-Sibony [**21**] for general complex variety respectively.

- The second topic is about the dynamics of correspondences and automorphisms of positive entropy. There are many beautiful examples that have been discovered and studied. For classification and complex theory of automorphisms of surfaces, in particular $K3$ surfaces, we refer to Cantat [**15, 14**] and McMullen [**40, 39**]. For arithmetic theory, we refer to the work of Autissier [**2**] for Hecke correspondences, Silverman [**55**] and Mazur [**38**] for involutions on $K3$ surfaces, and Kawaguchi [**33**] for some generalizations.
- The last topic is about the moduli of dynamical system. We will discuss the classification problem for which variety to have a dynamical system, and construction of dynamical system for moduli of subvarieties, but we will not study all polarized endomorphisms on a fixed variety. As the moduli of abelian varieties play a fundamental role in modern number theory and arithmetic geometry, it will be an interesting question to construct some interesting moduli spaces for dynamics. We refer to Silverman's paper [**56**] for the moduli of dynamics on $\mathbb{P}^1$.

Acknowledgments. This work grew out of a talk given at a memorial conference for S.S. Chern at Harvard University. I would like to thank S.-T. Yau for this assignment so that I have the motivation to learn a lot of mathematics from the web in order to give a fair background on arithmetical dynamics. I would also like to thank Yuefei Wang and Yunping Jiang for explaining to me some *common sense* of the general dynamical system, and Antoine Chmbert-Loir, Xander Faber, Curtis McMullen, Bjorn Poonen, Nessim Sibony, Joe Silverman for their helpful comments on an early version of this paper. Finally, I would like to thank Xander Faber, Johan de Jong, Kathy O'Neil, and Xinyi Yuan for their patience in listening to my lectures during the initial preparation of the note. This work has been supported by the National Science Foundation of the USA and the Core Research Group of the Chinese Academy Sciences.

1. Kähler and algebraic dynamics

In this section, we will first give some basic definitions of polarized dynamics in Kähler geometry and algebraic geometry, and some basic categorical constructions, such as the fiber product and quotients. Then we will propose our first major conjecture: a dynamic Manin-Mumford conjecture. Finally we will list some examples, including abelian varieties, projective spaces, and the Chow variety of 0-cycles. The main tools in this section are Serre's theorem on a Kähler analogue of the Weil conjecture, Deligne's theory on intersections of line bundles, and a conjectured Kähler analogue of the positivity of Deligne's pairing on the Chow variety.

1.1. Endomorphisms with polarizations.

Kählerian dynamical system. Let us first recall some definitions about Kähler manifolds. See [**27**] for details. Recall that a *Kähler manifold* is a complex manifold X with a differential form ω of type $(1,1)$ such that $d\omega = 0$ and that locally if we write

$$\omega = i \sum h_{i,j} dz_i \wedge d\bar{z}_j$$

then $(h_{i,j})$ is a positive definite hermitian matrix. The form ω here is called a *Kähler form* and its class

$$[\omega] \in H^{1,1}(X, \mathbb{R}) := H^{1,1}(X, \mathbb{C}) \cap H^2(X, \mathbb{R})$$

is called a *Kähler class.*

By a *Kähler variety* X with a *Kähler form* ω we mean an analytic variety which admits a finite map $f : X \to M$ to a Kähler manifold M with a Kähler form η such that $f^*\eta = \omega$.

Let $\phi : X \longrightarrow X$ be an endomorphism of a compact Kähler variety. Then ϕ acts on $H^{1,1}(X, \mathbb{R})$ by the pull-back ϕ^*. We say that ϕ is *polarizable* by a Kähler class ξ if

$$\phi^*\xi = q\xi$$

for some integer $q > 1$. A *polarized Kähler dynamical system* is by definition a triple (X, ϕ, ξ) as above. The number $\dim X \cdot \log q$ is called the *entropy* of the dynamical system, and $\log q$ is called the *entropy slope.* One immediate fact about polarized endomorphisms is the following:

LEMMA 1.1.1. *Let $\phi : X \longrightarrow X$ be a polarized endomorphism. Then ϕ is finite with degree* $\deg \phi = q^{\dim X}$.

PROOF. Indeed, for any subvariety Y in X, one has the formula

$$\deg(\phi|_Y) \int_{\phi(Y)} \omega^{\dim Y} = \int_Y \phi^*\omega^{\dim Y} = q^{\dim Y} \int_Y \omega^{\dim Y} \neq 0.$$

Here $\deg(\phi|Y)$ is defined to be 0 if $\dim \phi(Y) < \dim Y$. The above equation implies that $\deg(\phi|_Y) \neq 0$. Taking $Y = \phi^{-1}(x)$, we get that Y is finite. Thus ϕ is finite. Taking $Y = X$, we get that $\deg(\phi) = q^{\dim X}$. □

A deep property of it is the following Kähler analogue of Weil's conjecture about eigenvalues of ϕ^* on cohomology:

THEOREM 1.1.2 (Serre [**53**]). *Let $\phi : X \to X$ be a polarizable endomorphism of degree q^n. Then the eigenvalues of ϕ^* on each cohomology $H^i(X, \mathbb{R})$ have absolute value $q^{i/2}$.*

PROOF. Consider the cup product

$$H^i(X,\mathbb{C}) \times H^{2n-i}(X,\mathbb{C}) \longrightarrow H^{2n}(X,\mathbb{C}) \simeq \mathbb{C}.$$

Here the last map is given by integration. Let ξ be a Kähler class such that $\phi^*\xi = q\xi$. Notice that ξ^n is a generator of $H^{2n}(X,\mathbb{C})$. So ϕ^* on $H^{2n}(X,\mathbb{C})$ is given by multiplication by q^n. Now let g denote the endomorphism on $H^*(X,\mathbb{C}) = \oplus_i H^i(X,\mathbb{C})$ that has restriction $q^{-i/2}\phi^*$ on $H^i(X,\mathbb{C})$. Then the above product is invariant under g, and so is the class ξ. Now we use the Hard Lefshetz theorem ([**27**], page 122) to give a decomposition of $H^i(X,\mathbb{C})$. For $i \le n$, let $P_i(X)$ denote the kernel of the map

$$H^i(X,\mathbb{C}) \longrightarrow H^{2n-i+2}(X,\mathbb{C}), \quad \alpha \longrightarrow \xi^{n-i+1}\wedge\alpha.$$

Then $H^*(X,\mathbb{C})$ is a direct sum of $\xi^j P_i$ with $i \le n$, $i+2j \le 2n$. Obviously, this decomposition is invariant under the action by g, and so it suffices to show that the eigenvalues of g on P_i have absolute value 1. Moreover, by the Hodge index theorem (or Hodge and Riemann bilinear relations, [**27**], page 123) the pairing on P_i defined by

$$P_i \times P_i \longrightarrow \mathbb{C}, \quad (\alpha,\beta) = \int \alpha C(\beta)\kappa^{n-i}$$

is positively definite. Here C is an operator on $H^*(X,\mathbb{C})$ such that on the Hodge component $H^{p,q}$ with $p+q=i$, it is given by

$$\alpha \mapsto (-1)^{(n-i)(n-i-1)/2}\sqrt{-1}^{p-q}\bar{\alpha}.$$

It is easily checked that g is unitary with respect to this pairing. It follows that g has eigenvalues with norm 1. Thus the eigenvalues of ϕ^* have norm $q^{i/2}$ on $H^i(X,\mathbb{C})$. □

Endomorphisms with positive entropy. We say that a finite endomorphism ϕ of a compact Kähler variety has *positive entropy* if there is a semi-positive class $\xi \in H^{1,1}(X,\mathbb{R})$ such that

$$\phi^*\xi = q\xi$$

with $q > 1$. The notion of "positive entropy" here is equivalent to the same notion in the topological sense and to the statement that ϕ^* on $H^{1,1}(X,\mathbb{R})$ has an eigenvalue of absolute value greater than 1. See Dinh-Sibony's paper [**22**] for details. The proof of Corollary 2.2 of that paper also shows that ϕ^* on $H^p(X,\mathbb{Z})$ has eigenvalues with absolute values bounded by the p-th power of the absolute value of the eigenvalues on $H^{1,1}(X,\mathbb{R})$. Notice that all eigenvalues on $H^i(X,\mathbb{Z})$ are algebraic integers with product a positive integer. Thus if ϕ does not have positive entropy then all of its eigenvalues on $H^*(X,\mathbb{Z})$ are roots of unity. Thus $(\phi^*)^N$ for some fixed N has eigenvalues equal to 1. This implies in particular that ϕ is the identity on $H^{2n}(X,\mathbb{Z})$. Thus ϕ is a biholomorphic map. We conclude that *if ϕ has zero entropy then ϕ is an automorphism.* Notice that this statement is true for a dynamical system on a compact manifold. See Theorem 8.3.1 in Katok-Hasselblatt's book [**32**].

The category of dynamical system. We can define a *morphism* $f : \phi \to \psi$ of two endomorphisms $\phi : X \longrightarrow X$ and $\psi : Y \to Y$ as usual by a morphism $f : X \to Y$ such that $f \circ \phi = \psi \circ f$:

$$\begin{array}{ccc} X & \xrightarrow{\phi} & X \\ \downarrow{\scriptstyle f} & & \downarrow{\scriptstyle f} \\ Y & \xrightarrow{\psi} & Y \end{array}$$

If ϕ and ψ are polarized by classes ξ and η with the same entropy slope, then ϕ is polarized by all positive classes $\xi + cf^*\eta$ where $c \in \mathbb{R}$. Especially, if f is finite and ψ is polarized by η, then ϕ is polarized by $f^*\eta$.

If f is proper and flat with relative dimension d, and ϕ is polarized by a class ξ with entropy slope $\log q$, then we claim that ψ is polarized by the form

$$\eta := \int_{X/Y} \xi^{d+1}$$

with the same entropy slope $\log q$ as ϕ provided that η *is a Kähler class on* Y. See Conjecture 1.2.3 and Remark 1.2.4. below. Indeed, for any point $y \in Y$, in the diagram

$$\begin{array}{ccc} X_y & \xrightarrow{\phi_y} & X_{\psi(y)} \\ \downarrow & & \downarrow \\ y & \longrightarrow & \psi(y) \end{array}$$

the morphism ϕ_y has degree q^d as $\int_{X_y} \xi^d > 0$, and

$$\deg(\phi_y) \int_{X_{\psi(y)}} \xi^d = \int_{X_y} \phi^* \xi^d = q^d \int_{X_y} \xi^d.$$

It follows that

$$(\psi^*\eta)(y) = \int_{X_{\psi(y)}} \xi^{d+1} = \frac{1}{\deg \phi_y} \int_{X_y} \phi^* \xi^{d+1} = q \int_{X_y} \xi^{d+1} = q\eta.$$

Especially, if f is finite and flat, then ϕ is polarized if and only if ψ is polarized. One application is the normalization $f : \widetilde{X} \longrightarrow X$: obviously ϕ induces an endomorphism $\widetilde{\phi}$ of $\widetilde{X}$ which is polarized by the class $f^*\xi$.

We say two endomorphisms $\phi, \psi : X \longrightarrow X$ are *equivalent* if there are positive numbers m, n such that $\phi^m = \psi^n$. We will mainly study the properties of endomorphisms depending only on their equivalence classes. Thus it makes sense to define the *entropy class* for the equivalence class of an endomorphism ϕ to be

$$\mathbb{Q} \log(\deg \phi)$$

as a $\mathbb{Q}$-line in $\mathbb{R}$.

Notice that the product of two polarized endomorphisms may not be polarizable. If we allow to replace them by equivalent ones, then a sufficient condition is that they have the same entropy class. More precisely, let $\phi : X \longrightarrow X$ and $\psi : Y \longrightarrow Y$ be two endomorphisms of compact Kähler varieties polarized by $\omega \in H^{1,1}(X)$ and $\eta \in H^{1,1}(Y)$. The following two statements are equivalent:

(1) the endomorphism

$$\phi \times \psi : X \times Y \longrightarrow X \times Y$$

is polarizable by $\pi_1^*\omega + \pi_2^*\eta$ where π_i are projections from $X \times Y$ to X and Y;

(2) ϕ and ψ have the same entropy slope.

If $f_1 : \phi_1 \to \psi$ and $\phi_2 \to \psi$ are two morphism from two dynamic systems X_1, X_2 to a variety Y, then we can form the fiber product

$$\phi_1 \times_Y \phi_2 : X_1 \times_Y X_2 \longrightarrow X_1 \times_Y X_2.$$

If they have the same entropy slope, then the product is again a polarized dynamical system in an obvious way.

Algebraic dynamical system. We now consider an endomorphism $\phi : X \longrightarrow X$ of projective varieties. We may define *algebraic polarization* by replacing (1,1)-classes by line bundles. Let $\mathrm{Pic}\,(X)$ denote the group of line bundles on X which is isomorphic to $H^1(X, \mathcal{O}_X)$ and let $\mathrm{Pic}^{\,0}(X)$ denote the subgroup of line bundles which are algebraically equivalent to 0, and let $NS(X)$ denote the quotient $\mathrm{Pic}\,(X)/\mathrm{Pic}^{\,0}(X)$ which is called the *Neron-Severi group.* Then the exact sequence

$$0 \longrightarrow \mathbb{Z} \longrightarrow \mathcal{O}_X \longrightarrow \mathcal{O}_X^\times \longrightarrow 0$$

induces the following natural isomorphisms:

$$\mathrm{NS}\,(X) \simeq H^{1,1}(X, \mathbb{C}) \cap H^2(X, \mathbb{Z}), \quad \mathrm{Pic}^{\,0}(X) \simeq H^1(X, \mathcal{O}_X)/H^1(X, \mathbb{Z}).$$

Recall that a line bundle $\mathcal{L}$ is ample if some positive power $\mathcal{L}^m$ is isomorphic to the pull-back of the hyperplane section bundle for some embedding $i : X \longrightarrow \mathbb{P}^N$. By Kodaira's embedding theorem, $\mathcal{L}$ is ample if and only if its class in $NS(X) \subset H^{1,1}(X, \mathbb{R})$ is a Kähler class.

Let $\phi : X \longrightarrow X$ be an endomorphism of a projective variety. Then ϕ acts on $\mathrm{Pic}\,(X)$. We say that ϕ is *polarizable* by a line bundle $\mathcal{L}$ (resp. $\mathbb{R}$-line bundle $\mathcal{L} \in \mathrm{Pic}\,(X) \otimes \mathbb{R}$) if

$$\phi^*\mathcal{L} \simeq \mathcal{L}^q$$

for some $q > 0$.

An endomorphism $\phi : X \longrightarrow X$ of projective variety polarized by a line bundle $\mathcal{L}$ will be polarized by an *integral* Kähler class: we just take

$$\xi = c_1(\mathcal{L}) \in H^{1,1}(X, \mathbb{Z}).$$

We want to show the converse is true:

PROPOSITION 1.1.3. *Let $\phi : X \longrightarrow X$ be an endomorphism of smooth projective variety with a polarization by a Kähler class ξ such that ξ is integral, and that $\phi^*\xi = q\xi$ with q integral and > 1. Then there are line bundles $\mathcal{L}$ with class ξ such that*

$$\phi^*\mathcal{L} \simeq \mathcal{L}^q.$$

PROOF. Let $\mathrm{Pic}^{\xi}(X)$ denote the variety of line bundles on X with class ξ. Then we have a morphism

$$\lambda : \quad \mathrm{Pic}^{\xi}(X) \longrightarrow \mathrm{Pic}^{0}(X), \quad \lambda(\mathcal{L}) = \phi^*\mathcal{L} \otimes \mathcal{L}^{-q}.$$

Notice that $\mathrm{Pic}^{\xi}(X)$ is a principal homogenous space of $\mathrm{Pic}^{0}(X)$. The induced homomorphism on H^1's is an endomorphism on $H^1(X, \mathbb{Z})$ given by $\lambda := \phi^* - q$. By Proposition 1.1.2, all eigenvalues of ϕ^* on $H_1(X, \mathbb{Z})$ have eigenvalues with absolute values $q^{1/2}$. It follows from the assumption that λ is finite and thus surjective. In particular we have an $\mathcal{L} \in \mathrm{Pic}^{\xi}(X)$ such that $\lambda(\mathcal{L}) = 0$. In other words $\phi^*\mathcal{L} = \mathcal{L}^q$. □

Category of algebraic dynamical system. In the same manner as in Kähler case, we may define the morphism $f : \phi \to \psi$ between two endomorphisms of projective varieties X, Y. If ϕ and ψ are both polarized by line bundles $\mathcal{L}$ and $\mathcal{M}$ with the same entropy slope, then ϕ is also polarized by any positive class of the form $\mathcal{L} \otimes f^*\mathcal{M}^n$. If f is flat of relative dimension d, and ϕ is polarized by a line bundle $\mathcal{L}$, then ψ is polarized by the following Deligne's pairing ([**19**], See also [**63**]):

$$\int_{X/Y} \mathcal{L}^{\langle d+1 \rangle} := \langle \mathcal{L}, \dots, \mathcal{L} \rangle.$$

For convenience to reader, let us recall the definition. Let $\pi : \quad Z \longrightarrow C$ be a flat family of projective varieties of pure relative dimension d. Let $\mathcal{L}_0, \dots \mathcal{L}_d$ be line bundles on Z. The Deligne pairing $\langle \mathcal{L}_0, \dots, \mathcal{L}_d \rangle$ is a line bundle on C which is locally generated by a symbol $\langle \ell_0, \dots \ell_d \rangle$ modulo a relation, where ℓ_i are sections of $\mathcal{L}_i$ such that their divisors $\mathrm{div}(\ell_i)$ have empty intersection on fibers of f. The relation is given as follows. If α is a function, and i is an index between 0 and d such that $\mathrm{div}(\alpha)$ has disjoint intersection $Y := \prod_{j \neq i} \mathrm{div}(\ell_j)$, then Y is finite over C, and

$$\langle \ell_0, \dots, \alpha\ell_i, \dots, \ell_d \rangle = \mathrm{N}_Y(\alpha) \langle \ell_0, \dots, \ell_d \rangle.$$

Here $\mathrm{N}_Y(\alpha)$ is the usual norm map $\mathrm{N}_Y : \pi_*\mathcal{O}_Y \longrightarrow \mathcal{O}_C$.

We may also define the polarized product or fiber product for polarized endomorphisms with the same entropy slope in the same manner as in Kähler case.

1.2. Preperiodic subvarieties. Let $\phi : X \longrightarrow X$ be an endomorphism of Kähler variety with a polarization. Let Y be an analytic subvariety of X. We say that Y is *periodic* if for some $k > 0$, $\phi^k(Y) = Y$, and *preperiodic* if for some m, $\phi^m(Y)$ is periodic. Equivalently, Y is preperiodic if the orbits of Y under ϕ are finite. When X is projective, it shown by Fakhruddin

([**23**], Corollary 2.2) that for some $\ell, m \geq 1$ such that the system $(X, \phi, \mathcal{L}^m)$ can be extended to a dynamic system of $\mathbb{P}^N$, where $N = \dim \Gamma(X, \mathcal{L}^m)$.

If Y is periodic, say $\phi^k(Y) = Y$. Then the restriction of ϕ^k on Y is still polarized with entropy slope $k \log q$. The aim of our paper is to study the distribution properties of the set Prep (X) of preperiodic points of X in various topology. In this section we mainly focus on Zariski topology. Our first basic conjecture is the following:

CONJECTURE 1.2.1 (Dynamic Manin-Mumford). *A subvariety Y of X is preperiodic if and only if $Y \cap \operatorname{Prep}(X)$ is Zariski dense in Y.*

Dynamic topology. For a better understanding of the nature of the dynamic Manin-Mumford conjecture, it is helpful to introduce the following so-called *dynamic topology* on a dynamical system (ϕ, X, ξ) in which all closed sets are preperiodic subvarieties. To see it is really a topology, we check that the intersections of preperiodic subvarieties are still preperiodic. In this topology, the set of minimal subvarieties are exactly the set of preperiodic points. The conjecture 1.2.1 is equivalent to the following two statements:

(1) The preperiodic points on any preperiodic subvariety are Zariski dense; when X is projective, this is actually a Fakhruddin [**23**], Theorem 5.1.
(2) On Prep (X), dynamic topology = Zariski topology.

The Zariski closure of preperiodic points in a preperiodic subvariety is again preperiodic. Thus for the first statement it suffices to consider *periodic points in the periodic subvariety.*

CONJECTURE 1.2.2. *Let Y be a periodic subvariety of dimension r: $\phi^m Y = Y$ for some $m > 0$. Then as $k \longrightarrow \infty$,*

$$\#\{y \in Y, \phi^{km}(x) = x\} = q^{rkm}(1 + o(1)).$$

By Serre's Theorem 1.1.2, the conjecture is true if Y is smooth, polarizable, and if most of the fixed points have multiplicity one. Indeed, in this case without loss of generality we may simply assume that $Y = X$ and that $a = 1$. For any fixed point x of ϕ^k the multiplicity $m_k(x)$ is defined to be the length of the dimension of the maximal quotient of the local $\mathcal{O}_{X,x}$ where the action of $(\phi^k)^*$ is trivial:

$$m_k(x) := \dim_{\mathbb{C}} \mathcal{O}_{X,x}/((\phi^k)^* - 1)\mathcal{O}_{X,x}.$$

We define $m_k(x) = 0$ if x is not a fixed point of ϕ^k. Then by Lefshetz fixed point theorem ([**27**], page 421), the left hand side is

$$\sum_x m_k(x) = \sum_i (-1)^i \operatorname{tr}((\phi^k)^* : H^i(X, \mathbb{C})). \tag{1.2.1}$$

By Theorem 1.1.2, the right hand has estimate q^{kn} if ϕ is *polarizable.*

A consequence of Conjecture 1.2.2. is that *the set of pre-periodic points of X is countable.* This is the true for general preperiodic subvarieties proved Corollary 1.2.7.

Dynamical systems of subvarieties. In the following, we will introduce some dynamical system on the Chow variety. Notice that the Chow variety is not of finite type; it is a union of a countably many subvarieties of finite type. Later on, we will construct some dynamic systems on subvarieties of Chow variety of finite type which are conjectured to be the Zariski closure of periodic subvarieties.

Let us start with a compact Kähler variety X with a Kähler class ξ. Let $\mathcal{C}(X)$ denote the variety of cycles on X with pure dimension [**4**]. Then $\mathcal{C}(X)$ is a union of countably many Kähler varieties. We call $\mathcal{C}(X)$ the Chow variety of X, as when X is projective, $\mathcal{C}(X)$ is simply the usual Chow variety of X.

We may equip $\mathcal{C}(X)$ with the structure of a Kähler variety as follows. Let

$$(i, \pi) : \mathcal{Z}(X) \longrightarrow X \times \mathcal{C}(X)$$

be the universal family of cycles. For each d between 0 and n let π_d : $\mathcal{Z}_d(X) \to \mathcal{C}_d(X)$ denote the moduli of cycles of pure dimension d. Then for any Kähler class ξ of X, we define

$$\eta_d := \int_{\mathcal{Z}_d(X)/\mathcal{C}_d(X)} (i^*\xi)^{d+1} \in H^{1,1}(\mathcal{C}_d(X)).$$

CONJECTURE 1.2.3. The class η_d is a Kähler class on $\mathcal{C}_d(X)$.

REMARKS 1.2.4.

(1) This conjecture implies that for any flat morphism of compact Kähler manifold $f : X \to Y$ and any Kähler class ξ on X, the class $\eta = \int_{X/Y} \xi^{d+1}$ is a Kähler on Y. Indeed, in this case we have an embedding $Y \longrightarrow \mathcal{C}_d(X)$ for d the relative dimension of f.

(2) The conjecture is true when both X and Y are projective varieties and when ξ, the first Chern class of an ample line bundle on X. Indeed, in this case replacing $\mathcal{L}$ by a power we may assume that $\mathcal{L} = i^*\mathcal{O}(1)$ for some embedding $X \longrightarrow \mathbb{P}^n(\mathbb{C})$; then $\eta = j^*\mathcal{O}(1)$ for some embedding $j : Y \longrightarrow \mathbb{P}^N(\mathbb{C})$. See [**63**].

If $\phi : X \longrightarrow X$ is an endomorphism polarized by a positive class ξ then ϕ induces an endomorphism

$$\phi_d : \mathcal{C}_d(X) \longrightarrow \mathcal{C}_d(X), \qquad \phi_d(Z) = \phi_*(Z) = \deg(\phi|_Z) \cdot \phi(Z).$$

It clear that ϕ_d is polarized by η_d:

$$\phi^*(\eta_d) = q^{d+1} \cdot \eta_d.$$

In the following, we want to construct countably many subvarieties $\mathcal{C}(d, \gamma, k)$ of $\mathcal{C}_d(X)$ of finite type and endomorphisms $\phi_{d,\gamma,k}$ with polarizations such that every periodic subvariety is represented by points in $\mathcal{C}(d, \gamma, k)$.

First we may decompose $\mathcal{C}_d(X)$ further as a union of closed subvarieties $\mathcal{C}(d, \delta)$ representing cycles with degree δ.

Recall that for an integral subvariety Z,

$$\phi_*(Z) = \deg \phi|_Z \cdot \phi(Z).$$

The class $\gamma' := [\phi(Z)]$. Compute the degree to obtain

$$\deg(\phi(Z)) = q^d \frac{\deg(Z)}{\deg \phi|_Z}.$$

If furthermore Z is fixed by some positive power ϕ^k of ϕ, then the above implies that

$$\deg \phi^k|_Z = q^{kd}.$$

For each positive integer k, let $\mathcal{C}(d, \delta, k)$ denote the subvariety of $\mathcal{C}(d, \delta)$ of cycles Z of degree δ such that

$$q^{-kd\ell}(\phi_*^{k\ell} Z), \qquad \ell = 1, 2, \ldots$$

are all integral. Then we can define an endomorphism

$$\phi_{d,\delta,k} : \mathcal{C}(d, \delta, k) \longrightarrow \mathcal{C}(d, \delta, k), \qquad Z \longrightarrow q^{-kd}(\phi_*^k Z).$$

PROPOSITION 1.2.5. *The endomorphisms $\phi_{d,\gamma,k}$ are all polarized with respect to the bundle η_d with entropy slope $k \log q$:*

$$\phi_{d,\delta,k}^* \eta_d = q^k \cdot \eta_d.$$

PROOF. By integration over fibers over Z for the form ξ^{d+1}, we have

$$\phi_{d,\delta,k}^* \eta_d = \frac{q^{(d+1)k}}{q^{dk}} \cdot \eta_d = q^k \eta_d. \tag{1.2.2}$$

□

In view of Conjecture 1.2.1 for $\mathcal{C}(d, \delta, k)$ we have the following:

CONJECTURE 1.2.6. The variety $\mathcal{C}(d, \delta, k)$ is the Zariski closure of points in $\mathcal{C}(X)$ representing periodic cycles Y of X such that the following identities hold:

$$\dim Y = d, \quad \deg Y = \delta, \quad \phi^k(Y) = Y.$$

When X is projective, this conjecture is a theorem of Fakhruddin [**23**].

Notice that each periodic subvariety represents a fixed point in some $\phi_{d,\gamma,k}^*$. Thus they are finite in each $\mathcal{C}(d, \gamma, k)$. In other words, the set of preperiodic subvarieties of X is countable.

COROLLARY 1.2.7. *Let $\phi : X \longrightarrow X$ be an endomorphism of a compact Kähler manifold with a polarization. Then the set of preperiodic subvarieties of X is countable.*

From the known example, it seems that all irreducible preperiodic subvarieties of X have the bounded geometry, i.e., lie in a finite union of components of the Chow variety $\mathcal{C}(X)$. The following is a reformulation of the question:

QUESTION 1.2.8. Does there exist a number δ such that

$$\int_Y \xi^d \le \delta,$$

for any irreducible preperiodic subvariety Y of dimension d?

If X is polarized by line bundles, then we may replace the above integrals by Deligne's pairing in §1.1. Thus we will naturally define line bundles on $\mathcal{C}_d(X)$ denoted by

$$\mathcal{L}^{\langle d+1\rangle} := \langle i^*\mathcal{L}, \dots, i^*\mathcal{L}\rangle = \text{ Deligne pairing of } d+1 \ \mathcal{L}\text{'s}$$

whose Chern class is equal to

$$\int_{\mathcal{Z}_d(X)/\mathcal{C}_d(X)} c_1(\mathcal{L})^{d+1}.$$

1.3. Examples. In this subsection we want to give some examples of endomorphisms with polarizations.

Complex torus. Our first example is the complex torus $X = \mathbb{C}^n/\Lambda$ where Λ is a lattice in $\mathbb{C}^n$ with Kähler class

$$\xi = i\sum dz_i \wedge d\bar{z}_i,$$

and ϕ is given by multiplication by an integer $m > 1$. Then we have

$$\phi^*\xi = q\xi, \qquad q = m^2.$$

In this case the preperiodic points are exactly the torsion points:

$$\text{Prep}\,(X) = \Lambda \otimes \mathbb{Q}/\Lambda.$$

The conjecture 1.2.2 is trivial: the set of fixed points by ϕ^k is the set of torsion points $X[m^k - 1]$ which has cardinality

$$(m^k - 1)2n = q^{nk} + O(q^{(n-1)k}).$$

The preperiodic subvarieties are translations of abelian subvarieties by torsion points. When X is an abelian variety, the dynamic Manin-Mumford conjecture is the original Mumford-Manin conjecture proved firstly by Raynaud [**52**]: *Let Y be a subvariety of X which is not a translate of an abelian subvariety. Then all the torsion points on Y are included in a proper subvariety.* There are other proofs ([**64, 18**]), but all of them uses heavily the algebraic property (or even arithmetic property) of X. Thus they can't be generalized directly to the general complex torus.

Projective spaces. Let $X = \mathbb{P}^n$, and $\phi : X \longrightarrow X$ be any map of degree $d > 1$ defined by $n + 1$-homogenous polynomials of degree q with no non-trivial common zeros. By Fakhruddin's result ([**23**],Corollary 2.2), any polarized dynamic system is a subsystem of a certain system on $\mathbb{P}^n$. Conjecture 1.2.2 looks easy but I don't know how to prove it. In the simplest case $\phi(x_0, \dots x_n) = (x_0^m, \dots, x_n^m)$ where $m \ne \pm 1$, the preperiodic points are exactly the points where the coordinates x_i are either 0 or roots of unity. This is a multiplicative group analogous situation of abelian varieties: $\mathbb{P}^n$

is the union of multiplicative groups defined by the vanishing of some coordinates. The Manin-Mumford conjecture is true! See Lang [**36**], page 207, Ihara-Serre-Tate for $n = 2$, and Laurent [**37**] for the general case. On each multiplicative group, the conclusions are the same as in the abelian varieties case.

The next nontrivial work is when $X = \mathbb{P}^1 \times \mathbb{P}^1$, $\phi = (\phi_1, \phi_2)$. The conjecture is true when Julia sets of ϕ_i are very different ([**42**]). *Any curve C in $\mathbb{P}^1 \times \mathbb{P}^1$ which is neither horizontal nor vertical contains at most finitely many preperiodic points.*

Weighted projective spaces. Fix an $n+1$-tuple of positive integers $r = (r_0, \dots, r_n)$. Then we have an action of $\mathbb{C}^\times$ on $\mathbb{C}^{n+1} \setminus \{0\}$ by

$$(z_0, \dots, z_n) \mapsto (t^{r_0} z_0, \dots, t^{r_n} z_n), \qquad (t \in \mathbb{C}^\times).$$

The quotient is called a called a weighted projective space and denoted by $\mathbb{P}^n_r$. Notice that $\mathbb{P}^n_r$ is a projective space and can be defined by Proj $\mathbb{C}[z_0, \dots, z_n]_r$ where $\mathbb{Z}[z_0, \dots, z_n]_r$ is the graded algebra $\mathbb{Z}[z_0, \dots, z_n]$ with weighted degree $\deg z_i = r_i$. Any endomorphism of $\mathbb{P}^n_r$ is again given by homogenous polynomials with nontrivial zeros and with the same degree, say q, and is polarized by the the bundle $\mathcal{O}(1)$. Notice that $\mathbb{P}^n_r$ is in general a singular variety and is a quotient of $\mathbb{P}^n$ by the diagonal action the product of roots of r_i-th roots of unity:

$$\mu_{r_0} \times \cdots \mu_{r_n}.$$

Thanks to N. Sibony who showed this example to me!

Dynamical projective bundles. Let $\phi: \quad X \longrightarrow X$ be an endomorphism of compact Kähler variety polarized by a Kähler class ξ of entropy slope $\log q$. Let $\mathcal{L}_i$ $(i = 1, \dots, n)$ be line bundles on X such that

$$\psi_i: \quad \phi^* \mathcal{L}_i \simeq \mathcal{L}_i^q.$$

Define a vector bundle $\mathcal{V}$ as follows:

$$\mathcal{V} = \mathcal{L}_0 \oplus \mathcal{L}_1 \oplus \cdots \oplus \mathcal{L}_{n_1}$$

and define Y to be the corresponding projective bundle:

$$Y = \mathbb{P}(\mathcal{V}).$$

Then ψ_i induces embeddings of vector bundles

$$\phi^* \mathcal{V} \longrightarrow \mathrm{Sym}^q \mathcal{V}.$$

Thus we have an endomorphism

$$f: \mathbb{P}(\mathcal{V}) \longrightarrow \mathbb{P}(\mathcal{V})$$

such that

$$f^* \mathcal{O}_{\mathbb{P}(\mathcal{V})}(1) = \mathcal{O}_{\mathbb{P}_X(\mathcal{V})}(q).$$

Then f is polarizable by bundles $c_1(\mathcal{O}_{\mathbb{P}(\mathcal{V})}(1)) + m\xi$ which is positive on $\mathbb{P}(\mathcal{V})$ when $m \gg 0$. We don't know if Conjecture 1.2.1 is true or not on $Y = \mathbb{P}_X(\mathcal{V})$ if it is already true on X. A typical example is when $X = A$ is an abelian variety, $\phi = [n]$ for some $n > 1$, $q = n^2$, and $\mathcal{L}_i$ $(i = 1, \dots, n)$ are line bundles such that $\mathcal{L}_i$ are ample and symmetric:

$$[-1]^*\mathcal{L}_i \simeq \mathcal{L}_i.$$

Another case is when $\mathcal{L}_i$ are torsion bundles. Then we will have $[n]^*\mathcal{L}_i \simeq \mathcal{L}_i^{n^2}$ if $\mathcal{L}_i^{n-1} = \mathcal{O}_X$ for all i. This case corresponds to the almost split semiabelian variety. The conjecture 1.2.1 is true by Chambert-Loir [**17**].

Chow variety of 0*-cycles.* Let $\phi : X \longrightarrow X$ be an endomorphism with a polarization. Let δ be a positive number. Then the Chow variety $\mathcal{C}(0, \delta)$ of zero cycles of degree δ has an endomorphism ϕ_0 which is polarized by classes η_0. Recall that η_0 is defined as $\pi_* i^* \xi$, where (i, π) is the embedding of universal δ-cycles $\mathcal{Z}(0, \delta) \longrightarrow \mathcal{C}(0, \delta) \times X$.

Here is situation of curves: If $X = \mathbb{P}^1$, then $\mathcal{C}_\delta = \mathbb{P}^\delta$. If C is an elliptic curve, then $\mathcal{C}_d$ is a $\mathbb{P}^{\delta-1}$ bundle over E.

2. Classifications

In the following we want to discuss some classification problems for the dynamical system. We will first study the first Chern class for *smooth dynamics* and classify them when the first Chern class vanishes using a result of Beauville. Then we show that the smooth dynamics is uniruled in the remaining case using results of Miyaoka-Mori on a criterion on uniruledness and Bouchson-Demailly-Paum-Peternell on a criterion on pseudo-effectiveness. Using a result of Miyaoka-Mori and Campana, we will also give a fiberation decomposition with respect to the rational connectedness for general dynamics. Finally we give a full classification for which surface admits a polarized endomorphism using work of Fujimoto and Nakayama.

2.1. Positivity of the first Chern class. First notice that for any dynamical system X, the canonical class can't be positive when X is smooth.

PROPOSITION 2.1.1 (Fakhruddin [**23**], Theorem 4.2 for X projective). *Let $\phi : X \longrightarrow X$ be an endomorphism of a compact Kähler manifold with a polarization by a class ξ. Let K_X be the canonical class of X. Let R_ϕ be the ramification divisor of ϕ. Then the following statements hold:*

(1)
$$(1-q)\xi^{n-1} \cdot K_X = \xi^{n-1} \cdot R_\phi.$$

(2) *The Kodaira dimension of X is ≤ 0.*
(3) *If $c(X) = -c_1(K_X) = 0$ in $H^{1,1}(X, \mathbb{Z})$, then X has an etale cover by complex torus:*

$$X \simeq T/G, \qquad T = \mathbb{C}^n/\Lambda,$$

where Λ is a full rank Lattice in $\mathbb{C}^n$, and G is a finite group acting on T without fixed points. Moreover the endomorphism ϕ is induced by a linear endomorphism $\widetilde{\phi}$ on $\mathbb{C}^n$ as a $\mathbb{C}$-vector space.

PROOF. By definition of ramification divisor,

$$K_X = \phi^* K_X + R_\phi.$$

Thus

$$\begin{aligned}\xi^{n-1} \cdot R_\phi &= \xi^{n-1} \cdot K_X - \xi^{n-1} \cdot \phi^* K_X \\ &= \xi^{n-1} \cdot K_X - q^{1-n} \phi^*(\xi^{n-1} \cdot K_X) \\ &= \xi^{n-1} \cdot K_X - q \cdot \xi^{n-1} \cdot K_X = (1-q)\xi^{n-1} \cdot K_X.\end{aligned}$$

This proves the first part of the proposition.

If the Kodaira dimension of X is positive, then some multiple of K_X is effective and nonzero; then $\xi^{n-1} K_X > 0$. As R_ϕ is effective and $q > 0$, we thus have a contradiction! So we have proved the second part.

If $c_1(X) = -c_1(K_X) = 0$, then both sides are zero in the equation in Part 1. Thus $R_\phi = 0$ and ϕ is unramified. In this case ϕ is induced from an unramified automorphism $\widetilde{\phi}$ of the universal covering $\widetilde{X}$. Now we apply a theorem of Beauville ([**5**], Theorem 1, page 759) that $\widetilde{X}$ is isomorphic to $\mathbb{C}^k \times M$ where M is a simply connected Kähler manifold and the pull-back $\widetilde{\xi}$ on $\widetilde{X}$ of ξ is a sum $\widetilde{\xi} = \kappa + \eta$, where κ is a flat Kähler class on $\mathbb{C}^k$,

$$\kappa = \sqrt{-1} \sum_{j=1}^{k} a_j dz_j d\bar{z}_j,$$

and η is a class on M. Since all morphisms from M to $\mathbb{C}^k$ are constant, the morphism $\widetilde{\phi}$ has the form

$$\widetilde{\phi}(z, m) = (u(z), v_z(m)).$$

Here $u(z)$ is an automorphism of $\mathbb{C}^k$ and v_z is an automorphism of M for each given $z \in \mathbb{C}^k$. Now we apply the equality $\widetilde{\phi}^* \widetilde{\xi} = q\widetilde{\xi}$ to conclude that $v_z^* \eta = q\eta$. As M is compact and $q > 1$ and η is positive, this is impossible unless M is a point. Thus we have shown that $\widetilde{X} = \mathbb{C}^k$. By the same theorem in [**5**], X is then an unramified quotient of a complex torus $T = \mathbb{C}^n/\Lambda$ and Λ is invariant under the action by $\widetilde{\phi}$.

To show the last statement, we need only to show that the induced endomorphism ψ on $T = \mathbb{C}^n/\Lambda$ has a fixed point. We will use the Lefshetz fixed point theorem. Notice that ψ is again polarized. Thus the eigenvalues λ_i of ψ^* on $H^1(T, \mathbb{Z})$ are greater than 1 by Theorem 1.1.2. Notice that $H^i(X, \mathbb{Z})$ is the i-th exterior power of $H^1(X, \mathbb{Z})$, and thus the eigenvalues of ψ^* are given by $\lambda_{j_1} \cdots \lambda_{j_i}$ of i distinct elements in λ_i's. Thus the number of fixed points with multiplicity is given by

$$\sum_i (-1)^i \mathrm{tr}\,(\psi^*, H^i(X, \mathbb{Z})) = \sum_{j_1, \ldots, j_k} (-\lambda_{j_1}) \cdots (-\lambda_{j_k}) = \prod (1 - \lambda_i).$$

As $|\lambda_i| > 1$, the above number is nonzero. Thus ψ has a fixed point. After changing coordinates, we may assume that ψ and $\widetilde{\phi}$ fix the origin. □

2.2. Uniruledness. By Proposition 2.1.2, the classification of *smooth* dynamical systems is reduced to the case where X has Kodaira dimension $-\infty$. If $n = \dim X = 1$, then X is simply $\mathbb{P}^1$. Later on we will discuss the case of surfaces. In general, it is conjectured that a Kähler manifold with Kodaira dimension $-\infty$ is always uniruled, i.e., covered by rational curves. The conjecture is true for projective varieties of dim ≤ 3 by Mori [**44**], and non-algebraic Kähler manifolds of dimensions ≤ 3 with possibly the exception of simple threefolds by Peternell [**48**]. In the following we want to prove the uniruledness in the smooth and projective case using a ruledness criterion of Miyaoka-Mori [**43**] and a pseudo-effectiveness criterion of Bouchson, Demailly, Paum, and Peternell [**51**].

PROPOSITION 2.2.1. *Let $\phi : X \longrightarrow X$ be an endomorphism of a projective manifold with a polarization by line bundles $\mathcal{L}$. Assume that $c_1(X) \neq 0$. Then X is uniruled.*

PROOF. Let ξ be the corresponding Kähler class of $\mathcal{L}$. By the first part of Proposition 2.1.1, $\xi^{n-1}K_X \leq 0$. If $K_X \cdot \xi^{n-1} < 0$, then since $\mathcal{L}$ is ample, K_X has negative intersection with *strongly movable curves* as in [**51**]. By Theorem 0.2 in [**51**], K_X is not *pseudo-effective*, i.e., $c_1(K_X)$ is not in the closure of the cone in $H^{1,1}(X, \mathbb{R})$ generated by effective divisors. By Corollary 0.3 in [**51**], X is uniruled.

It remains to treat the case where $\xi^{n-1}K_X = 0$. Since $K_X \neq 0$, this case can't happen by the following proposition. □

PROPOSITION 2.2.2. *Let X be a projective variety of dimension n, and let $\xi_i \in \mathrm{NS}\,(X)$ be divisor classes such that*

(1) *ξ_1 is pseudo-effective;*
(2) *ξ_i $(i > 1)$ are ample;*
(3) $\prod_i \xi_i = 0$.

Then $\xi_1 = 0$.

PROOF. There is nothing needed to prove if $n \leq 1$. If $n = 2$, we will use Hodge index theorem: since $\xi_1 \cdot \xi_2 = 0$, one has $\xi_1^2 \leq 0$, and the equality holds only when $\xi_1 = 0$. On the other hand, we may take N a positive integer such that $N\xi_2 + \xi_1$ is ample and thus has non-negative intersection with ξ_1 as ξ_1 is pseudo-effective. Thus we have

$$\xi_1^2 = (\xi_1 + N\xi_2)\xi_1 \geq 0.$$

Combining with Hodge index theorem, we have $\xi_1 = 0$.

Now we assume that $n \geq 3$ and we want to reduce to the case $n = 2$ and to use the following Lefshetz Theorem in hyperplane section ([**27**], page 156): Let D_i be smooth divisors D_i $(i = 3, \dots n)$ representing positive multiples of ξ_i such that the partial products $Y_k := \prod_{i=k+1}^n D_i$ are smooth subvarieties of X of dimension k. Then for each k, the restriction map

$$H^2(Y_k, \mathbb{Q}) \longrightarrow H^2(Y_{k-1}, \mathbb{Q})$$

is an isomorphism when $k \geq 4$, and injective when $k = 3$.

By induction we can show that the restriction of ξ_1 on Y_k is pseudo-effective. It is sufficient to show that any effective divisor A of Y_{k+1} will have the restriction $[A] \cdot Y_k$ represented by an effective divisor. Indeed, write $A = B + mD_k$ with B properly intersecting D_k, and then $[A] \cdot Y_k$ will be represented by $[B \cdot D_k] + m[D'_k]$ where D'_k is some effective representative of ξ_k on Y_k, which always exists as ξ_k is ample.

Now on the surface Y_2, ξ_1 and ξ_2 satisfy the conditions of the Proposition, so we must have that $\xi_1 = 0$ on Y_2. Now we apply the Lefshetz Theorem on the hyperplane section to conclude that $\xi_1 = 0$ on X. □

REMARK 2.2.3. The above proposition can be considered as a supplement to Theorem 2.2 in [**51**] which says that a class $\alpha \in \mathrm{NS}\,(X)_{\mathbb{R}}$ is pseudo-effective if and only if it is in the dual of the cone SME(X) of strongly movable curves. Our proposition just says that the pairing of α on SME(X) is strictly positive if $\alpha \neq 0$.

Rationally connected factorization. Let us discuss some factorization results of Miyaoka and Mori [**43**] (see also Campana [**12**]). Let X be a projective variety. Then their result says that there is a rational morphism $f : X \longrightarrow Y$ classifying the rational connected components, i.e., the following conditions hold:

(1) f is dominated with rationally connected fiber;
(2) there is a Zariski open subset X^* over which f is regular and proper;
(3) for a general point x of X, the fiber of f over x is the set $R(x)$ of points y which can be connected to x by a finite chain of rational curves. Here "general" means outside of a countably many proper subvarieties.

We may pick up a canonical $f : X \longrightarrow Y$ as follows: let Y be the Zariski closure of points $[R(x)]$ in the Chow variety $\mathcal{C}(X)$ corresponding to the general points of X. Let

$$(\rho, \pi) : \widetilde{X} \longrightarrow X \times Y$$

the universal family of cycles parameterized by Y. Then the morphism $\rho : \widetilde{X} \longrightarrow X$ is birational. We define $f = \pi \circ i^{-1}$ as a rational morphism

$$\begin{array}{ccc} \widetilde{X} & \xrightarrow{\ \rho\ } & X \\ {\scriptstyle \pi}\downarrow & \swarrow{\scriptstyle f} & \\ Y & & \end{array}$$

If X has an endomorphism $\phi : X \longrightarrow X$ with polarization by ϕ, then ϕ takes rational curves to rational curves, and thus takes $R(x)$ to $R(\phi(x))$. In other words, ϕ induces an endomorphism ψ on Y and an endomorphism $\widetilde{\phi}$

on $\widetilde{X}$ with commutative diagrams

$$\begin{array}{ccc} \widetilde{X} & \xrightarrow{\widetilde{\phi}} & \widetilde{X} \\ \downarrow{\scriptstyle \pi} & & \downarrow{\scriptstyle \pi} \\ Y & \xrightarrow{\psi} & Y \end{array} \qquad \begin{array}{ccc} \widetilde{X} & \xrightarrow{\widetilde{\phi}} & \widetilde{X} \\ \downarrow{\scriptstyle \rho} & & \downarrow{\scriptstyle \rho} \\ X & \xrightarrow{\phi} & X \end{array} .$$

PROPOSITION 2.2.4. *Both endomorphisms $\widetilde{\phi}$ and ψ are polarizable.*

PROOF. Let $\mathcal{M}$ Deligne's pairing on Y:

$$\mathcal{M} := \int_{\widetilde{X}/Y} \rho^* \mathcal{L}^{\langle d+1 \rangle}$$

where d is the relative dimension of π. Then $\mathcal{M}$ is an ample line bundle and $\psi^*\mathcal{M} = \mathcal{M}^q$. In other words ψ is polarized by $\mathcal{M}$. For the polarization of $\widetilde{\phi}$, we notice that $\rho^*\mathcal{L}$ is ample on each fiber of π with property

$$\widetilde{\phi}^* \rho^* \mathcal{L} = \rho^* \phi^* \mathcal{L} = \rho^* \mathcal{L}^q.$$

Thus for some positive number N,

$$\widetilde{\mathcal{L}} := \rho^* \mathcal{L} \otimes \pi^* \mathcal{M}^N$$

will be ample with the property

$$\widetilde{\phi}^* \widetilde{\mathcal{L}} = \widetilde{\mathcal{L}}^q.$$

Thus we obtain a polarization for $\widetilde{\phi}$. □

REMARKS 2.2.5. Here are some obvious questions about the classifications of the general dynamical system:

(1) extend Proposition 2.1.1 and 2.2.1 to general Kähler variety. We may replace them by a bi-rationally equivalent dynamic system if they are helpful;
(2) classify the dynamical system into two extreme cases: the non-uniruled case, and the rationally connected case. It is not true that every rationally connected variety carries an endomorphism of degree ≥ 2. For example, Beauville [**6**] showed that any smooth hypersurface in projective space with dimension ≥ 2 and degree ≥ 3 does not admit any endomorphism of degree ≥ 2.

REMARK 2.2.6. For algebraic endomorphism $\phi : X \longrightarrow X$ with a polarization, we are in the opposite situation of general type: X and the finite étale coverings do not admit a rational map to a positive dimensional variety of general type. See Harris-Tschinkel [**31**] and Campana [**13**] for a detailed discussion of the geometry and arithmetic of these *varieties of special type* in contrast to varieties of general type.

2.3. Dynamic surfaces. In the following we would like to classify the dynamic systems on surfaces.

PROPOSITION 2.3.1. *Let $\phi : X \longrightarrow X$ be an endomorphism of a Kähler surface. Then ϕ is polarizable if and only if X is one of the following types:*

(1) *complex torus;*
(2) *hyperelliptic surfaces, i.e., the unramified quotients of the product of two genus 1 curves;*
(3) *toric surfaces, i.e., the completions of $\mathbb{G}_m^2$ with extending action by $\mathbb{G}_m^2$;*
(4) *a ruled surface $\mathbb{P}_C(\mathcal{E})$ over an elliptic curve such that either*
   (a) *$\mathcal{E} = \mathcal{O}_C \oplus \mathcal{M}$ with $\mathcal{M}$ torsion or of positive degree;*
   (b) *$\mathcal{E}$ is not decomposable and has odd degree.*

PROOF. By a result of Fujimoto and Nakayama ([**25**], Theorem 1.1), the only non-algebraic Kähler surfaces admitting endomorphisms of degree ≥ 2 are complex tori. So we will only consider algebraic ones. By Proposition 2.1.1, we need only consider unramified quotients of abelian surfaces and algebraic surfaces with negative Kodaira dimension. So we have the first two cases listed above, plus rational surfaces and irrational ruled surfaces.

By a result of Noboru Nakayama ([**46**], Theorem 3), a rational surface X has an endomorphism ϕ of degree ≥ 2 if and only if it is toric. We may take ϕ to be the "square morphism" on X, i.e., the morphism on X satisfies the equation $\phi(tx) = t^2\phi(x)$ for any $t \in \mathbb{G}_m^2$ and $x \in X$. For polarization, we may simply take $\mathcal{L}$ to be the divisor of the complement of $\mathbb{G}_m^2$ in X:

$$\phi^*\mathcal{L} = \mathcal{L}^2.$$

It remains to work on $\mathbb{P}^1$-bundle $\pi : X \longrightarrow C$ over a curve of genus $\neq 0$. We will use an idea of Nakayama ([**46**], proof of Proposition 5). We need to check when such an X has a polarizable endomorphism ϕ. Notice that any such ϕ will take rational curves to rational curves. Thus ϕ will dominate an endomorphism g of C:

$$\begin{array}{ccc} X & \xrightarrow{\phi} & X \\ \downarrow{\scriptstyle \pi} & & \downarrow{\scriptstyle \pi} \\ C & \xrightarrow{g} & C \end{array}$$

Let $g^*X = X \times_g C$, then ϕ is the decomposition $X \xrightarrow{\alpha} g^*X \xrightarrow{\beta} X$, α is a morphism over C, and β is the projection. Let $\mathcal{L}$ be an ample line bundle on X such that

$$\phi^*\mathcal{L} \simeq \mathcal{L}^q. \tag{2.3.1}$$

It follows that α has degree q. Since $\deg \phi = q^2$, it follows also that $\deg \beta = q$, and that C must have genus 1.

LEMMA 2.3.2. *Let $g : C \longrightarrow C$ be a morphism of curve of genus 1 of degree $q > 1$. Then any endomorphism $\phi : X \longrightarrow X$ of ruled surface* *over C is induced by a homomorphism $g : C \longrightarrow C$ of degree q, and a homomorphism of vector bundles*

$$g^*\mathcal{E} \longrightarrow \operatorname{Sym}^q\mathcal{E} \otimes \mathcal{N}$$

with $\mathcal{N}$ a line bundle on C. Moreover, ϕ is polarized if and only if $\deg\mathcal{N} = 0$.

PROOF. The first statement is well known. It remains to study when such ϕ is polarizable. Let $\mathcal{L}_0$ be the $\mathcal{O}(1)$-bundle corresponding to $\mathcal{E}$. Then we can write

$$\mathcal{L} = \mathcal{L}_0^m \otimes \pi^*\mathcal{N}_0.$$

Here $\mathcal{N}_0$ is some line bundle on C. The equation $\phi^*\mathcal{L} = \mathcal{L}^q$ is equivalent to

$$(\phi^*\mathcal{L}_0 \otimes \mathcal{L}_0^{-q})^m = \pi^*(g^*\mathcal{N}_0 \otimes \mathcal{N}_0^{-q}).$$

This equality shows that $\phi^*\mathcal{L}_0 \otimes \mathcal{L}_0^{-q}$ has degree 0 on all fibers. It follows that for some bundle $\mathcal{N}$ on C,

$$\phi^*\mathcal{L}_0 \simeq \mathcal{L}_0^q \otimes \pi^*\mathcal{N}, \qquad g^*\mathcal{N}_0 \simeq \mathcal{N}_0^q \otimes \mathcal{N}^m. \tag{2.3.2}$$

Since $\deg g = q$, the second equation gives $\deg\mathcal{N} = 0$.

Conversely, if ϕ is induced by a homomorphism as in the lemma with $\deg\mathcal{N} = 0$, then we may find a line bundle $\mathcal{N}_0$ on C of degree 1 such that $g^*\mathcal{N}_0 \simeq \mathcal{N}_0^q \otimes \mathcal{N}$. Then we can check that $\mathcal{L} := \mathcal{L}_0 \otimes \pi^*\mathcal{N}_0$ will give the right polarization for ϕ. □

After being twisted by a line bundle on C, any vector bundle of rank 2 on C is one of the following three types:

(1) there is a splitting, $\mathcal{E} \simeq \mathcal{O}_C \oplus \mathcal{M}$ with $\deg\mathcal{M} \geq 0$;

(2) there is a non-split exact sequence

$$0 \longrightarrow \mathcal{O}_C \longrightarrow \mathcal{E} \longrightarrow \mathcal{O}_C \longrightarrow 0;$$

(3) there is a non-split exact sequence

$$0 \longrightarrow \mathcal{O}_C \longrightarrow \mathcal{E} \longrightarrow \mathcal{M} \longrightarrow 0$$

where $\deg\mathcal{M} = 1$.

In case (1), since C has genus 1 there is a point O such that $\mathcal{M} = \mathcal{O}_C(d \cdot O)$ if $d = \deg\mathcal{M} > 0$. We give C an elliptic curve structure such that O is the unit element. Let $a \geq 2$ be a fixed integer. Then the multiplication by a gives

$$[a]^*\mathcal{M} \simeq \mathcal{M}^b, \qquad b = \begin{cases} a^2, & \text{if } \deg\mathcal{M} \neq 0, \\ a, & \text{if } \deg\mathcal{M} = 0. \end{cases}$$

If $\mathcal{M}$ is torsion of order t, then we may take $a = t+1$, and we can replace b by a^2. Thus in the case that either $\deg\mathcal{M} > 0$ or $\mathcal{M}$ is torsion, we have a morphism of bundles on C:

$$[a]^*\mathcal{E} \simeq \mathcal{O}_C \oplus \mathcal{M}^{a^2} \longrightarrow \operatorname{Sym}^{a^2}\mathcal{E}.$$

This induces a morphism $\phi : X \longrightarrow X$ such that ϕ is compatible with multiplication $[a]$ on C. By the lemma, this homomorphism has polarization.

We want to show that when $\mathcal{M}$ is a non-torsion degree 0 bundle, any homomorphism $\phi : X \longrightarrow X$ is not polarizable. Otherwise, we will have a morphism $g : C \longrightarrow C$ of degree $q > 1$ and a homomorphism of vector bundles:

$$g^*\mathcal{E} \longrightarrow \mathrm{Sym}^q\mathcal{E} \otimes \mathcal{N}$$

where $\mathcal{N}$ is of degree 0. Let x be the section of $\mathcal{E}$ corresponding to the embedding $1 \in \mathcal{O}_C \subset \mathcal{E}$. Then the above equation gives

$$\mathcal{O}_C \oplus g^*\mathcal{M} \longrightarrow \sum_{i=0}^{q} x^{q-i}\mathcal{M}^i \otimes \mathcal{N}.$$

Since $\mathcal{M}$ is not torsion, the bundles in the right hand side are not isomorphic to each other. Since all of them have degree 0, we have i and j such that the above homomorphism is given by two isomorphisms:

$$\mathcal{O}_C \simeq \mathcal{M}^i \otimes \mathcal{N}, \qquad g^*\mathcal{M} = \mathcal{M}^j \otimes \mathcal{N}.$$

Since this homomorphism defines morphism $X \longrightarrow X$, one must have that $\{i, j\} = \{0, q\}$. Thus in any case, we have

$$g^*\mathcal{M} = \mathcal{M}^{\pm q}.$$

Let O be a fixed point of g. Then we may consider C as an elliptic curve with origin O. Write $\mathcal{M} = \mathcal{O}(P - O)$. Then $g^*\mathcal{M} = \mathcal{O}([g^\vee O] - [O])$ and $\mathcal{M}^q = \mathcal{O}([qP] - O)$, where $g^\vee$ is the conjugate of g: $gg^\vee = \deg g = q$. Thus the above equation gives

$$g^\vee P = \pm qP, \qquad (g^\vee \mp q)P = 0.$$

This implies again that P is torsion. Thus we have a contradiction.

In case (2), the bundle $g^*\mathcal{E}$ is still non-split so it is isomorphic to $\mathcal{E}$. In other words g^*X is isomorphic to X. Indeed, the extension

$$0 \longrightarrow \mathcal{O}_C \longrightarrow \mathcal{E} \longrightarrow \mathcal{O}_C \longrightarrow 0$$

is given by a nonzero element t in $H^1(C, \mathcal{O}_C)$. The $g^*\mathcal{E}$ will correspond to g^*t in $H^1(C, \mathcal{O}_C)$. As $H^1(C, \mathcal{O}_C) \simeq k$, $g^*t = at$ with $a \in k^\times$, there is a homomorphism $g^*\mathcal{E} \simeq \mathcal{E}$ over C.

Thus, ϕ induces (and is induced by) a C-endomorphism of X of degree $q > 1$. Now we want to apply a result of Silverman about the moduli space of endomorphisms of $\mathbb{P}^1$ [**56**]. For a positive integer d, let Rat_d denote the space of endomorphisms of $\mathbb{P}^1$. We only take the base $\mathbb{C}$ here. Then Rat_d has a natural action by $\mathrm{Aut}\,(\mathbb{P}^1)$ as follows:

$$(h, f) \mapsto f \circ h \circ f^{-1}, \quad h \in Rat_d,\ f \in \mathrm{Aut}\,(\mathbb{P}^1).$$

Let $\mathrm{SL}_2 \longrightarrow \mathrm{Aut}\,(\mathbb{P}^1)$ be the Mobius transformation. Then by Theorem 1.1 and Theorem 3.2 in [**56**], the quotient $M_d := Rat_d/\mathrm{SL}_2$ exists as an affine variety over $\mathbb{C}$. In order to apply this to our situation, we give a slightly different interpretation of M_d: M_d is a *fine* moduli space of triples (V, h, ℓ) where V is a vector space of dimension 2, ϕ is an endomorphism of the projective line $\mathbb{P}(V)$, and ℓ is an isomorphism $\det V \simeq \mathbb{C}$.

In case (2), since we have an isomorphism $\ell : \det \mathcal{E} \simeq \mathcal{O}_X$, the morphism α induces an morphism $C \longrightarrow M_q$. As C is projective and M_q is projective, we must have that this morphism is constant. Thus $\mathcal{E}$ must be a trivial vector on C. We get a contradiction!

In case (3), we claim that $X \simeq \mathrm{Sym}^2 C$; thus, X has an endomorphism by multiplication by 2. First there is a section O so that $\mathcal{M} = \mathcal{O}_C(O)$. In this way, C becomes an elliptic curve with origin O. Let us consider the following maps

$$\begin{array}{ccccc} C\times C & \xrightarrow{[2]\times[2]} & C\times C & & (x,y) \quad . \\ f\downarrow & & \downarrow f & & \downarrow \\ \mathrm{Sym}^2 C & \xrightarrow{\phi} & \mathrm{Sym}^2 C & & (x)+(y) \\ \pi\downarrow & & \downarrow\pi & & \downarrow \\ C & \xrightarrow{[2]} & C & & x+y \end{array}$$

Let p_1, p_2 be two projections of $C \times C$ to C. Let $\mathcal{N} = \pi_1^*\mathcal{O}_C(O) + \pi_2^*\mathcal{O}(O)$ be a line bundle on $C \times C$, and $\mathcal{L}$ the descent bundle of $\mathcal{N}$ on $\mathrm{Sym}^2 C$. The bundle $\mathcal{L}$ is ample, with fiber on $(P) + (Q)$ canonically isomorphic to $\mathcal{O}(O)|_P \otimes \mathcal{O}(O)|_Q$ up to the order of tensor product. The multiplication on $C \times C$ by an integer a induces an endomorphism ϕ on $\mathrm{Sym}^2 C$. As $[a]^*\mathcal{N} \simeq \mathcal{N}^{a^2}$, $\phi^*\mathcal{L} \simeq \mathcal{L}^{a^2}$.

We claim the following:

(1) *$\mathcal{L}$ has degree 1 on the fiber of π, thus $\mathcal{V} := \pi_*\mathcal{L}$ is a rank 2 bundle on C;*

(2) *$\Gamma(\mathcal{L}) = \Gamma(\mathcal{M})$ is one dimensional, say generated by ℓ, such that $\mathrm{div}\,(\ell) = s^*C$, where $s(P) = (P) + (O)$;*

(3) *$s^*\mathcal{L} \simeq \mathcal{O}(O)$.*

This claim implies that $X \simeq \mathbb{P}(\mathcal{V})$ and that $\mathcal{V}$ is not decomposable and fits in an exact sequence:

$$0 \longrightarrow \mathcal{O}_C \longrightarrow \mathcal{V} \longrightarrow \mathcal{O}_C(O) \longrightarrow 0.$$

This of course implies that $\mathcal{V} \simeq \mathcal{E}$ and $X \simeq \mathrm{Sym}^2 C$.

It remains to prove our claim. For item (1) we need to check the degree of $\mathcal{L}$ on the fiber $\mathrm{Sym}^2(C \times C)_O$ over $O \in C$. The pull-back of this fiber on $C \times C$ is the image of the following map

$$\delta : C \longrightarrow C \times C, \qquad p \mapsto (p, -p).$$

It follows that

$$\begin{aligned} \mathcal{L} \cdot \mathrm{Sym}^2(C)_0 &= \frac{1}{2}\pi^*\mathcal{L} \cdot \pi^*\mathrm{Sym}^2(C)_0 = \frac{1}{2}\mathcal{N} \cdot \delta(C) \\ &= \frac{1}{2}\deg \delta^*(\mathcal{N}) = \frac{1}{2}\deg \mathcal{O}(O) \otimes [-1]^*\mathcal{O}(O) = 1. \end{aligned}$$

For item (2) we see that $\Gamma(\mathcal{N})$ is equal to the symmetric part of $\Gamma(\mathcal{L})$. It is easy see that

$$\Gamma(\mathcal{N}) = p_1^*\Gamma(\mathcal{O}(O)) \otimes p_2^*\Gamma(\mathcal{O}(O)) = \mathbb{C}p_1^*\alpha \otimes p_2^*\alpha$$

where α is the canonical section of $\mathcal{O}(O)$ with divisor O. It is clear that the section $p_1^*\alpha \otimes \pi_2^*\alpha$ is symmetric and thus descends to section ℓ on $\mathcal{L}$ with divisor div $(\ell) = s^*C$. For the last part, for any point $P \in C$, we see that

$$s^*\mathcal{L} \simeq \mathcal{O}(O)|_O \otimes \mathcal{O}(O) \simeq \mathcal{O}(O).$$

This completes the proof of the claim. □

3. Canonical metrics and measures

We will fix an endomorphism $\phi : X \longrightarrow X$ of a Kähler variety with a polarization by a Kähler class ξ. Our aim in this section is to study the distributional properties of the set Prep (X) of all preperiodic points on X. We will first construct a canonical current ω to represent ξ. The class ω is integrable in the sense that the restriction of ω^d on any subvariety Y of dimension d defines a measure. By a result of Bedford-Taylor and Demailly [**20**], the support of the measure is Zariski dense. Then we conjecture some properties about this invariant measure. First of all, this measure can be obtained from any smooth measure by iterations. Second, this measure can be constructed from the probability measures of the backward images of a general point. We will prove some of these properties in the special cases using the work of Yau [**59**] and Briend-Duval [**9**]. Some of our results follow from some very general results of Dinh-Sibony [**21**], Corollary 5.4.11 and Theorem 5.4.12. We present here a self-contained treatment for the simplicity. Also our treatment is completely global and thus easily extended to p-adic Berkovich spaces.

Finally, with hope to initiate a dynamic Nevanlinna theory of holomorphic curves, we construct a canonical order function on a Kähler dynamical system. As an application we will show that the Fatou set is Kobayashi hyperbolic.

One question remains unsolved: the positivity of a canonical current on the Chow variety.

3.1. Canonical forms and currents. First we will try to find canonical representatives for the classes in $H^{1,1}(X)$. Let $\mathcal{Z}^{1,1}(X)$ denote the space of ∂ and $\bar\partial$ closed currents on X which have the form $\omega + \frac{\partial\bar\partial}{\pi i} g$ with ω smooth and g continuous. Then there is a class map

$$c: \quad \mathcal{Z}^{1,1}(X) \longrightarrow H^{1,1}(X).$$

Notice that ϕ^* acts on both spaces and this class map is a homomorphism of ϕ^*-modules. The kernel of c is the space of forms $\frac{\partial\bar\partial}{\pi i} g$ for continuous functions g.

PROPOSITION 3.1.1. *The class map c of ϕ^*-modules has a unique section, i.e., there is a unique ϕ^*-subspace $\mathcal{H}^{1,1}(X)$ of $\mathcal{Z}^{1,1}(X)$ such that c induces an isomorphism*

$$\mathcal{H}^{1,1}(X) \simeq H^{1,1}(X).$$

The space $\mathcal{H}^{1,1}(X)$ is called the space of canonical forms. Moreover, if ξ is an eigenclass of ϕ^* with eigenvalue λ which is represented by a smooth form ω_0 then the canonical lifting is the limit

$$\omega := \lim_{k\to\infty} (\lambda^{-1}\phi^*)^k \omega_0.$$

PROOF. Let $C(X)$ denote the space of continuous functions on X. Then we have an exact sequence

$$0 \longrightarrow \mathbb{C} \longrightarrow C(X) \longrightarrow \mathcal{Z}^{1,1}(X) \longrightarrow H^{1,1}(X) \longrightarrow 0$$

where the map $C(X) \longrightarrow \mathcal{Z}^{1,1}(X)$ is given by $\frac{\partial\bar\partial}{\pi i}$. Let $P(T)$ be the characteristic polynomial of ϕ^* acting on $H^{1,1}(X)$. We want to show that for $P(\phi^*)$ is invertible over $C(X)$. In this way, we may take $\mathcal{H}^{1,1}(X) = \ker P(\phi^*)$. In other words, every element κ in $H^{1,1}(X)$ has a lifting η such that $P(\phi^*)\eta = 0$. Indeed, if η_0 is one lifting of κ in $\mathcal{Z}^{1,1}(X)$ then $P(\phi^*)\eta_0$ is in the image of $C(X)$. Thus we have a $g \in C(X)$ such that

$$P(\phi^*)\eta_0 = \frac{\partial\bar\partial}{\pi i} g. \tag{3.1.1}$$

It is easy to see that κ has a lifting in the kernel of $P(\phi^*)$ with the following form:

$$\eta = \eta_0 + \frac{\partial\bar\partial}{\pi i} P(\phi^*)^{-1} g. \tag{3.1.2}$$

It remains to show that $P(\phi^*)$ is invertible over $C(X)$. We write $P(T) = \prod_i (T - \lambda_i)$ where λ_i are eigenvalues of ϕ^* on $H^{1,1}(X)$. By Theorem 1.1.2, all $|\lambda_i| > 1$. It follows that $\lambda_i^{-1}\phi^*$ is a compact operator on $C(X)$. Indeed,

$$\|\lambda_i^{-1}\phi^*\alpha\|_{\sup} \le |\lambda_i|^{-1}\|\alpha\|_{\sup}, \qquad \alpha \in C(X).$$

It follows that $P(\phi^*)$ has an inverse on $C(X)$:

$$P(\phi^*)^{-1}\alpha := \prod_i \sum_k -\lambda_i(\lambda_i^{-1}\phi^*)^k \alpha.$$

This proves the first part of the proposition.

For the second part, let g be a smooth function such that

$$(1 - \lambda^{-1}\phi^*)\omega_0 + \frac{\partial\bar\partial}{\pi i} g = 0.$$

Then

$$(\lambda^{-1}\phi^*)^j\omega_0 - (\lambda^{-1}\phi^*)^{j+1})\omega_0 + \frac{\partial\bar\partial}{\pi i}(\lambda^{-1}\phi^*)^j g = 0.$$

Add the above equality from $j = 1$ to $j = k - 1$ to obtain

$$(\lambda^{-1}\phi^*)^k\omega_0 = \omega_0 + \frac{\partial\bar\partial}{\pi i}\sum_{j=1}^{k}(\lambda^{-1}\phi^*)^j g.$$

It is easy to see from this expression that $(\lambda^{-1}\phi^*)^k\omega_0$ has a limit as the canonical lifting of ξ:

$$\omega = \omega_0 + \frac{\partial\bar\partial}{\pi i}(1 - \lambda\phi^*)^{-1}g.$$

□

We have an analogue for algebraic polarizations. Let $\phi : X \longrightarrow X$ be an endomorphism with a polarization. Let $\widehat{\mathrm{Pic}}\,(X)$ denote the group of (continuously) metrized line bundles on X. Then we have a class map

$$\gamma : \widehat{\mathrm{Pic}}\,(X) \longrightarrow \mathrm{Pic}\,(X).$$

Again ϕ^* acts on both groups and this map is a homomorphism of ϕ^*-modules.

Proposition 3.1.2. *The class map γ has a projective section, i.e., there is a unique ϕ^*-submodule $\overline{\mathrm{Pic}}\,(X)$ of $\widehat{\mathrm{Pic}}\,(X)$ such that the map γ induces an exact sequence*

$$0 \longrightarrow \mathbb{R} \longrightarrow \overline{\mathrm{Pic}}\,(X) \longrightarrow \mathrm{Pic}\,(X) \longrightarrow 0.$$

Here $\mathbb{R}$ maps $r \in \mathbb{R}$ to the metrized line bundle $(\mathcal{O}_X, \|1\| = e^{-r})$. The metrics in $\overline{\mathrm{Pic}}\,(X)$ are called canonical metrics.

Proof. By Proposition 3.1.1, for any line bundle $\mathcal{L}$ there is a unique metric up to a constant with curvature in $\mathcal{H}^{1,1}(X)$. □

Remarks 3.1.3. The proof of the above proposition applies to $H^{p,p}(X)$ and Green's currents for codimension p-cycles if we can show that $\lambda^{-1}\phi^*$ is compact on the space $C^{p-1,p-1}(X)$ of continuous (p,p)-forms on X. For example, if ϕ is polarized by a Kähler class ξ which has a lifting ω which is *continuous and positive pointwise*, then we may equip $C^{p-1,p-1}(X)$ with norm by ω. In this way we have

$$\|\phi^*\alpha\|(x) = q^{p-1}\|\alpha\|(\phi(x)), \qquad \alpha \in C^{p,p}(X).$$

Then by Theorem 1.1.2, the eigenvalue λ on $H^{p,p}(X, \mathbb{C})$ has absolute value q^p and again

$$\|\lambda^{-1}\phi^*\alpha\|_{\sup} \le q^{-1}\|\alpha\|_{\sup}.$$

In the following we want to study the volume forms defined by polarizations by ξ. Let ω denote its canonical form in $\mathcal{H}^{1,1}(X)$. If ω is a continuous form then we will have a volume form

$$d\mu_Y = \omega|_Y^{\dim Y}/\mathrm{vol}\,(Y).$$

Here

$$\mathrm{vol}\,(Y) = \xi^{\dim Y} \cdot [Y] = \int_Y \omega_0^{\dim Y}.$$

Only require ω is to be a current, and the above definition does not make sense. In the following we will use the limit process to show that the above definition still gives a measure. Let's study a slightly more general situation.

Let $d = \dim Y$ and pick up d classes $\eta_1, \dots, \eta_d$ so that

(1) η_i are semi-positive;
(2) $\phi^* \eta_i = \lambda_i \eta_i$ with $\lambda > 1$.

Let ω_{i0} be semipositive, smooth forms for η_i. Let $\omega_{ik} = \lambda_i^{-k} (\phi^*)^k \omega_{i0}$.

PROPOSITION 3.1.4. *With notation as above the following hold:*

(1) *ω_{ik} converges to the canonical lifting ω_k of η_k as a current,*
(2) *the volume form $\omega_{1k} \cdots \omega_{dk} \delta_Y$ is weakly convergent with a limit measure*

$$\omega_1 \cdots \omega_d \delta_Y := \lim_{k \to \infty} \omega_{1k} \cdots \omega_{dk} \delta_Y$$

on Y which is independent of the choice of initial forms ω_{i0}.

Integrable forms and metrics. We want to show that the proposition follows a more general theory about integrable metrics [**62**]. More precisely, a class $\omega = \omega_0 + \frac{\partial\bar\partial}{\pi i} g \in \mathcal{Z}^{1,1}$ is called *semi-positive* if $g = \lim_n g_n$ is the limit in $C(X)$ of a sequence of smooth functions g_n such that $\omega_0 + \frac{\partial\bar\partial}{\pi i} g_n$ are smooth positive forms. A class ω is called *integrable* if ω is the difference $\omega_1 - \omega_2$ of two semi-positive classes. Let $\mathcal{S}^{1,1}(X)$ denote the space of integrable forms. A function g is called a *Green's function* if there is a divisor $D = \sum_i a_i D_i$ on X with real coefficients such that g is continuous on $X \setminus D$ with logarithmic singularity near D: if locally D_i is defined by equations $f_i = 0$ near a point x, then g has an asymptotic formula near x:

$$g = \sum a_i \log |f_i| + h \tag{3.1.3}$$

where h is a continuous function. Let $\mathcal{G}(X)$ denote the space of Green's functions.

PROPOSITION 3.1.5. *Let X be a compact Kähler variety and let Y be a subvariety of dimension d. There is a unique integration pairing*

$$\mathcal{G}(Y) \times \mathcal{S}^{1,1}(X)^d \longrightarrow \mathbb{C},$$

$$(g, \omega_1, \dots, \omega_d) \mapsto \int_Y g \omega_1 \cdots \omega_d$$

such that the following properties are verified:

(1) *the pairing is linear in each variable;*
(2) *if each ω_i is semi-positive and is a limit of smooth forms ω_{ik} on X, then the above pairing is the limit of the usual integral pairings of smooth forms.*

PROOF. Write $\omega_i = \lim \omega_{ik}$ with ω_{ik} smooth and positive. Thus we have $\omega_{ik} = \omega_{i0} + \frac{\partial\bar\partial}{\pi i} h_{ik}$ with h_{ik} smooth and convergent to h_i. First we want to show that the functional

$$g \mapsto \int_Y g\omega_{1k}\cdots\omega_{dk}$$

is convergent on the restriction of $g \in \mathcal{G}^\infty(X)$ on Y, the space of functions whose local asymptotic formula (3.1.3) has smooth h. Let g be a smooth function on X; the difference of the integrations is given by

$$\begin{aligned}&\int_Y g(\omega_{1k}\cdots\omega_{dk} - \omega_{1\ell}\cdots\omega_{d\ell})\\ &= \int_Y g\sum_{i=1}^d \omega_{1k}\cdots\omega_{i-1,k}(\omega_{i,k} - \omega_{i,\ell})\omega_{i+1,\ell}\cdots\omega_{d,\ell}.\end{aligned}$$

From our expression of ω_{ik},

$$\omega_{ik} - \omega_{ij} = \frac{\partial\bar\partial}{\pi i}(h_{ik} - h_{i\ell}).$$

It follows that

$$\begin{aligned}&\int_Y g\cdot(\omega_{1k}\cdots\omega_{dk} - \omega_{1\ell}\cdots\omega_{d\ell})\\ &= \int_Y \sum_{i=1}^d \omega_{1k}\cdots\omega_{i-1,k}(h_{i,k} - h_{i,\ell})\omega_{i+1,\ell}\cdots\omega_{d,\ell}\frac{\partial\bar\partial}{\pi i}g.\end{aligned}$$

Since g is smooth, we have a formula

$$\frac{\partial\bar\partial}{\pi i}g = \sum a_i\delta_{D_i} - \alpha$$

where α is a smooth (1,1)-form. Let M be a positive number such that

$$\omega'_{i,0} := \omega_{i0} - \frac{1}{M}\alpha$$

is positive point wise. Then the above sum can be written as

$$\begin{aligned}&\int_Y g\cdot(\omega_{1k}\cdots\omega_{dk} - \omega_{1\ell}\cdots\omega_{d\ell})\\ &= \sum_j a_j\int_{D_j}\sum_{i=1}^d \omega_{1k}\cdots\omega_{i-1,k}(h_{i,k} - h_{i,\ell})\\ &\quad + M\int_Y\sum_{i=1}^d \omega_{1k}\cdots\omega_{i-1,k}(h_{i,k} - h_{i,\ell})\omega_{i+1,\ell}\cdots\omega_{d,\ell}(\omega'_{i0} - \omega_{i0}).\end{aligned}$$

Replacing $h_{i,k} - h_{i,\ell}$ by its L^∞-norm and $\omega'_{i0} - \omega_{i0}$ by $\omega'_{i0} + \omega_{i0}$, we have the following estimate:

$$\left| \int_Y g \cdot (\omega_{1k} \cdots \omega_{dk} - \omega_{1\ell} \cdots \omega_{d\ell}) \right|$$
$$\leq \sum_i \left(\sum_j |a_j| (\eta_1 \cdots \widehat{\eta_i} \cdots \eta_d [D_i]) + 2M(\eta_1 \cdots \eta_d [Y]) \right) \|h_{ik} - h_{i\ell}\|_{\sup},$$

where η_i are the classes of ω_i in $H^{1,1}(X, \mathbb{R})$. This shows that $\omega_{1k} \cdots \omega_{dk} \delta_Y$ converges as a distribution, say $\omega_1 \cdots \omega_d \delta_Y$. To show this limit can be extended into a continuous Green's function, we need only consider the continuous function $g \in C(X)$, or equivalently show that the limits is actually a measure. It suffices to show the following:

(1) the functional on $C^\infty(X)$, $g \longrightarrow \int_Y g\omega_1 \cdots \omega_d$ is continuous with respect to the supreme norm and
(2) the restriction of $C^\infty(X)$ on Y is dense in $C(Y)$.

The first property is clear since $\omega_{1k} \cdots \omega_{1,d}$ is semi-positive with volume $\eta_1 \cdots \eta_d|_Y$. For a smooth function f on X:

$$\left| \int_Y g \cdot \omega_1 \cdots \omega_d \right| = \lim_k \left| \int_Y g\omega_{1k} \cdots \omega_{dk} \right| \leq \|g\|_{\sup} (\eta_1 \cdots \eta_d [Y]).$$

For the second property, we use Stone-Weierstrass theorem: $C^\infty(X)$ is dense in $C(X)$ which is surjective on $C(Y)$ by restriction map.

Finally, we want to show the independence on ω_{ik}. This can be done by the same argument as above. Indeed, let ω'_{ik} be different smooth and positive forms convergent to ω_i, which induce a differential sequence of forms $\omega'_{1k} \cdots \omega'_{dk}$. The same argument as above can be used to show that

$$\left| \int_Y g \cdot (\omega'_{1k} \cdots \omega'_{dk} - \omega_{1k} \cdots \omega_{dk}) \right| \leq C \sum_i \|\alpha_{ik}\|_{\sup}$$

where C is a constant depending only on f, and α_{ik} are smooth functions such that

$$\omega'_{ik} - \omega_{ik} = \frac{\partial \bar{\partial}}{\pi i} \alpha_{ik}.$$

It is easy to show that $\alpha_{ik} \longrightarrow 0$. Thus two limits are same. □

For any open connected subset (in complex topology) U of a subvariety Y of X of dimension d and the current $\omega_1 \cdots \omega_d$ defined by integrable forms $\omega_1, \ldots, \omega_d$, the support of $\omega_1 \cdots \omega_d$ is defined as the smallest closed subset $\mathrm{Supp}_U(\omega_1 \cdots \omega_d)$ of Y in complex topology such that

$$\int_Y f\omega_1 \cdots \omega_d = 0$$

whenever $f \in C_0(U)$ vanishes on $\mathrm{Supp}_U(\omega_1 \cdots \omega_d)$. When X is projective, the above proposition shows that $\mathrm{Supp}_Y(\omega_1 \cdots \omega_d)$ is not included in any proper subvariety. Otherwise, $\mathrm{Supp}_Y(\omega_1 \cdots \omega_d)$ will be included in the support of an effective divisor D. Then we can take g to be the Green's function for D. The integral will be infinite! This contradicts our proposition.

For general Kähler variety, Chambert-Loir pointed to the following result of Bedford-Taylor and Demailly:

THEOREM 3.1.6 (Bedford-Taylor-Demailly).
The set $\mathrm{Supp}_U(\omega_1 \cdots \omega_d)$ *is either empty or Zariski dense in* U.

PROOF. When Y is smooth, this is simply a result of Bedford-Taylor and Demailly [**20**], Corollary 2.3. In the general case, let $\pi : \widetilde{Y} \longrightarrow Y$ be a resolution of singularity. Then we can define the pull-back of forms in $\mathcal{Z}^{1,1}$ by the usual way: if $\omega = \omega_0 + \frac{\partial\bar\partial}{\pi i} g$ then

$$\pi^*\omega = \pi^*\omega_0 + \frac{\partial\bar\partial}{\pi i} g \circ \pi.$$

If ω is integrable, then it is easy to show that $\pi^*\omega$ is integrable. For any continuous function f on Y, and any integrable currents $\omega_1, \dots, \omega_d$, it is easy to check that

$$\int_{\widetilde{Y}} \pi^* f \cdot \pi^*\omega_1 \cdots \pi^*\omega_d = \int_Y f \cdot \omega_1 \cdots \omega_d.$$

It follows that

$$\pi^{-1}\mathrm{Supp}_U(\omega_1 \cdots \omega_d) \subset \mathrm{Supp}_{\pi^{-1}U}(\pi^*\omega_1 \cdots \pi^*\omega_d).$$

Thus we are reduced to the smooth case. □

REMARK 3.1.7. The same proof as in Proposition 3.1.5 can be used to show the following weaker form of Theorem 3.1.6: *Assume that* X *is projective; then the measure* $\omega_1 \cdots \omega_d|_Y$ *on* Y *does not support on any proper subvariety.* Let D be a any divisor of Y, and g be a Green's function for D, i.e., a function on Y with logarithmic singularity such that

$$\frac{\partial\bar\partial}{\pi i} g = \delta_D - h$$

where h is a smooth $(1,1)$ form on Y. We need to show that the integral

$$\int_Y \omega_1 \cdots \omega_d$$

makes sense and is finite, which then implies that the support of the measure is not supported on D.

Metrics on Chow varieties. In the following we want to introduce the canonical forms or metrics for the Chow varieties and show their compatibility with the induced endomorphism ϕ_* and $\phi_{d,\delta,k}$. One basic question in this theory is about the fiber pairing of integrable metrics on X.

QUESTION 3.1.8. Let X be a compact Kähler variety. How could one construct an integration pairing

$$\mathcal{S}^{1,1}(X)^{d+1} \longrightarrow \mathcal{S}^{1,1}(\mathcal{C}_d(X)),$$

$$(\omega_0, \ldots, \omega_d) \mapsto \int_{\mathcal{Z}_d(X)/\mathcal{C}_d(X)} \omega_0 \cdots \omega_d?$$

When X is algebraic, this question has a positive answer, see [**63**]. Mimicking what has been done in the projective case, our first step to answer this question is to restrict to smooth forms ω_i' in the same class of ω_i and try to show that the above integral defines some integral forms. If $\omega_i = \omega_i' + \frac{\partial\bar{\partial}}{\pi i} h_i$ then we compute the difference formally by:

$$\begin{aligned}
&\int_{\mathcal{Z}_d(X)/\mathcal{C}_d(X)} \omega_0 \cdots \omega_d - \int_{\mathcal{Z}_d(X)/\mathcal{C}_d(X)} \omega_0 \cdots \omega_d \\
&= \int_{\mathcal{Z}_d(X)/\mathcal{C}_d(X)} \sum_{i=0}^{d} \omega_1 \cdots \omega_{i-1}(\omega_i - \omega_i')\omega_{i+1}' \cdots \omega_d' \\
&= \int_{\mathcal{Z}_d(X)/\mathcal{C}_d(X)} \sum_{i=0}^{d} \omega_1 \cdots \omega_{i-1} \frac{\partial\bar{\partial}}{\pi i} h_i \omega_{i+1}' \cdots \omega_d' \\
&= \frac{\partial\bar{\partial}}{\pi i} \int_{\mathcal{Z}_d(X)/\mathcal{C}_d(X)} h_i \sum_{i=0}^{d} \omega_1 \cdots \omega_{i-1} \omega_{i+1}' \cdots \omega_d'.
\end{aligned}$$

Our Proposition 3.1.5 shows the last integral is well defined at each point. But then one has to prove that this integral defines a continuous function on $\mathcal{C}_d(X)$.

Let $\phi : X \longrightarrow X$ be an endomorphism of a compact Kähler variety with a polarization by a positive class ξ which is represented by a canonical form ω. If Question 3.1.8 has a positive answer, then we will have canonical forms ω_d on the varieties $\mathcal{C}_d(X)$ compatible with the action of ϕ^*. Of course for the variety $\mathcal{C}(d, \delta, k)$ which is of finite type, we will have the usual theory of canonical metrics.

3.2. Equidistribution of backward orbits. Fix an endomorphism $\phi : X \longrightarrow X$ of a compact Kähler variety with polarization. Let $d\mu_0$ be a continuous probability measure. Define $d\mu_k$ by the following inductive formula

$$d\mu_k = \frac{\phi^* d\mu_{k-1}}{\deg \phi}.$$

More precisely, for any continuous function f on X,

$$\int_X f d\mu_k d\mu_k = (\deg \phi)^{-k} \int_X \phi_*^k f d\mu_0.$$

Here $\phi_*^k f$ is a function defined by

$$\phi_*^k f(x) = \sum_{\phi^k(y)=x} f(y)$$

where the sum is over pre-images of x with multiplicity. Notice that $\phi_*^k(f)$ is a bounded on X and continuous on a Zariski open subset of X, and thus is measurable with respect to any continuous measure. The following is a simple consequence of a result of Yau:

THEOREM 3.2.1. *Let $\phi : X \longrightarrow X$ be an endomorphism of a compact Kähler manifold with polarization ξ. Then $d\mu_k$ converges to the canonical measure on X:*

$$\lim_{k\to\infty} d\mu_k = \omega^n/(\xi^n[X])$$

where ω is the canonical form for the class ξ and $n = \dim X$.

PROOF. Notice that for a continuous function f on X,

$$\int f d\mu_k = (\deg\phi)^{-k}\int \phi_*^k(f)d\mu_0$$

where $\phi_*^k(f)$ is defined such that

$$\phi_*^k(f)(x) = \sum_{\phi^k(y)=x} f(x)$$

where the sum is over the preimage of x under ϕ^k with multiplicity. It is easy to check that $\phi_*^k(f)$ is bounded and continuous on a Zariski open subset of X. So the above integral makes sense.

As every continuous measure is a strong limit of smooth volume forms, we may assume that $d\mu_0$ is a smooth volume form. By a theorem of Yau [**59**], $d\mu_0$ on X is induced from a unique class ω_0 in ξ by formula

$$d\mu_0 = \omega_0^n/(\xi^n[X]).$$

Now we can apply Proposition 3.1.4. □

We would like to conjecture that this theorem is true without assumption on smoothness of X:

CONJECTURE 3.2.2. Let $\phi : X \longrightarrow X$ be an endomorphism of a (possibly singular) compact Kähler variety with a polarization ξ. Let $d\mu_0$ be a continuous probability measure on X. Define $d\mu_k$ by inductive formula

$$d\mu_k = \frac{\phi^* d\mu_{k-1}}{\deg\phi}.$$

Then $d\mu_k$ is convergent to the probability measure $d\mu_X$ of the form

$$\omega^n/(\xi^n[X]), \qquad (n = \dim X.)$$

Let $\phi : X \longrightarrow X$ be an endomorphism of a compact Kähler variety with a polarization ξ. Let x be a point in X. For any positive integer k, let's define a probability measure the k-th preimage:

$$\mu_{x,k} = (\deg \phi)^{-k} \sum_{\phi^k(y)=x} \delta_y$$

where the sum is over the k-th preimage with multiplicity.

By this conjecture, for almost all p the $\mu_{p,n}$ is convergent to the canonical measure $d\mu$. Indeed, for any continuous probability measure $d\mu_0$, the $d\mu_n$ in Theorem 3.1.6 can be written as

$$\int_X f d\mu_n = \int_X \left\{ \int_X f(x) d\mu_{p,n}(x) \right\} d\mu(p).$$

In a more precise way we would like to make the following conjecture:

CONJECTURE 3.2.3. Let x be a point of X and let Y be the Zariski closure of the complete orbit, i.e., Y is the minimal subvariety of X containing all $\phi^{-k}(x)$ and $\phi^k(x)$ for all k. Then $\mu_{p,n}$ is weakly convergent to the canonical measure $\omega^{\dim Y}/\mathrm{vol}\,(Y)$ on Y, where $\omega^{\dim Y}$ is the canonical measure on Y defined by the canonical form ω on Y.

It seems that a more natural subvariety than Y for the conjecture is the backward limit of x: i.e., the minimal subvariety Y' containing $\phi^{-k}(x)$ for k sufficiently large. We claim that $Y = Y'$. Indeed, for each $k \geq 0$ let Y_k be the Zariski closure of the union $\cup_{i \geq k} \phi^{-k}(x)$. Then Y_k is a decreasing sequence of closed subvarieties of X. Thus $Y = Y_k = \cap Y_k$ for k sufficiently large. It is easy to see that Y is periodic, so $Y = Y'$.

When $X = \mathbb{P}^N$,the conjecture is a theorem by recent work of Briend-Duval [**9**] and Briend-Cantat-Shishikura [**10**]. Indeed by Briend-Cantat-Shishikura, the full orbit of any $x \in \mathbb{P}^N$ is always a finite union of linear subspace, and thus isomorphic to $\mathbb{P}^k$ if ϕ is replaced by a power. Then by Briend-Duval $\mu_{m,x}$ is equidistributed on $\mathbb{P}^k$.

3.3. Hyperbolicity and holomorphic curves. Let $\phi : X \longrightarrow X$ be an endomorphism of a compact Kähler variety with a polarization. As discussed in §2.3, X should have a fibration over a variety Y of Kodaira dimension 0 with rationally connected fibers. Thus X should be *special* in the sense of Campana [**13**]. By Campana, Conjecture 9.2 in [**13**], this should be equivalent to *hyperbolically special*:

CONJECTURE 3.3.1. Let $\phi : X \longrightarrow X$ be an endomorphism of a compact Kähler variety with a polarization. Then the pseudo-metric d_X on X vanishes.

Here the *pseudo-metric* is defined in Kobayashi's hyperbolic geometry [**34**]. We can only prove the vanishing of the pseudo-distance under Conjecture 3.2.3:

PROPOSITION 3.3.2. *Let $\phi : X \longrightarrow X$ be an endomorphism of a compact Kähler variety with a polarization. Assume Conjecture* 3.2.1 *for the endomorphism $\phi \times \phi$ for $X \times X$. Then the pseudo-distance vanishes on X.*

PROOF. For two positive number $a > 0$, let $T(a)$ be the open subset of $X \times X$ of points (x, y) with the pseudo-distance satisfying the inequality $d(x, y) > a$. Since the pseudo-distance is decreasing under ϕ:

$$d(\phi x, \phi y) \leq d(x, y),$$

we see that $\phi^{-1}T(a) \subset T(a)$.

If $T(a)$ is not empty, then it has a non-empty interior, and thus supports continuous probability measure $d\mu_0$. By our assumption, the limit $\deg \phi^{-k}\phi^{*k} d\mu_0$ converges to the canonical measure on $X \times X$. Notice that the canonical measure $d\mu_X \times d\mu_X$ on $X \times X$ is the product measure on X's. Thus the support of this measure contains the support of $d\mu_X$ by diagonal map $X \to X \times X$. It follows that $T(a)$ contains the diagonal elements. This is a contradiction as the distance of $d(x, x) = 0$ for any $x \in X$. So we have shown that $d(x, y) = 0$ for all $x, y \in X$. □

Combined with Theorem 3.2.1, we can prove the conjecture for endomorphisms with polarizations:

COROLLARY 3.3.3. *Let $\phi : X \longrightarrow X$ be an endomorphism of a compact Kähler manifold with a polarization by a Kähler form. Then the pseudo-distance vanishes on X everywhere.*

One consequence of Conjecture 3.3.1 is the vanishing pseudo-volume form of Kobayashi which is apparently easy to prove:

PROPOSITION 3.3.4. *The Kobayashi pseudo-measure vanishes.*

PROOF. Let Φ_X denote the Kobayashi pseudo-volume form. Then $\phi^*\Phi_X \leq \Phi_X$. Taking integration of both sides we obtain

$$\deg \phi \int \Phi \leq \int \Phi.$$

It follows that $\int \Phi = 0$ or $\Phi = 0$ almost everywhere. □

Another consequence is the existence of many non-degenerate holomorphic maps $\mathbb{C} \longrightarrow X$ by work of Brody (see Lang [**36**], Theorem 2.2). Fix a Kähler form ω_0 on X which induces a metric on X. From the proof of Theorem 2.2. in [**36**], one sees that for any point $x \in X$ and any tangent vector $v \in T_{X,x}$ of norm 1, there is a holomorphic map $f : \mathbb{C} \longrightarrow X$ such that

$$df_0(\partial/\partial z) = v, \qquad \|df_z\| \leq 1, \quad \forall z \in \mathbb{C}.$$

Here df_z is the map $T_{\mathbb{C},z} \longrightarrow T_{X,f(x)}$ between two tangent spaces with norms induced by the Euclidean norm on $\mathbb{C}$ and the norm ω_0 on X.

Holomorphic curves. Fix a Kähler dynamical system (X, ϕ, ξ). Let ω be the canonical form. The remainder of this section is devoted to proving the following:

THEOREM 3.3.5. *Let C be an affine complex curve; i.e., the complement of finitely many points in a Riemann surface $\bar{C}$. There is no nonconstant holomorphic map $\psi : C \longrightarrow X$ such that $\psi^*(\omega) = 0$.*

REMARK 3.3.6. Let F be the complement of the support of ω. As in Sibony [**54**] Theorem 1.6.5, one may show that F is the *Fatou* set of ϕ, i.e., the set of points p which has neighborhood U such that $f|_U$ is equicontinuous. Thus our theorem shows that F is Brody hyperbolic, which is equivalent to Kobayashi hyberbolic as F is the complement of a compact set.

Our main idea for the proof is to use Nevanlinna theory of holomorphic curves. We refer to Griffiths [**28**] for basic theory. Let's fix a holomorphic map $p : \bar{C} \longrightarrow \mathbb{P}^1$ such that $p^{-1}\mathbb{C} \subset C$. Replacing C by $p^{-1}\mathbb{C}$ we may simply assume that C has a finite and flat morphism to *the affine line* $\mathbb{C}$.

Order function. For each positive number r, let C_r denote the set of points $p \in C$ such that $|p(z)| \le r$. Let ω_0 be a closed $(1,1)$ form in the class. Then we can define the *order function* (or *characteristic function*) for C_r as follows:

$$T_{\omega_0}(C_r) = \int_0^r \frac{dt}{t} \int_{C_t} \psi^* \omega_0.$$

If ω_0' is in the same class as ω_0, i.e., there is a real function $g \in C^\infty(X)$ such that $\omega_0' = \omega_0 + \frac{\partial\bar\partial}{\pi i} g$, then

$$\begin{aligned} T_{\omega_0'}(C_r) - T_{\omega_0}(C_r) &= \int_0^r \frac{dt}{t} \int_{C_t} \frac{\partial\bar\partial}{\pi i} \psi^* g \\ &= \frac{1}{\pi i} \int_0^r \frac{dt}{t} \int_{\partial C_t} \bar\partial(\psi^* g) = \frac{1}{\pi i} \int_0^r \frac{dt}{t} \int_{\partial \mathbb{C}_t} \bar\partial(p_* \psi^* g). \end{aligned}$$

Now we use polar coordinate $z = te^{i\theta}, \bar z = te^{-i\theta}$, and

$$\bar\partial = \frac{\partial}{\partial \bar z} \otimes d\bar z = \frac{1}{2}\left(\frac{\partial}{\partial t} + \frac{i}{t}\frac{\partial}{\partial \theta}\right) \otimes (dt - itd\theta).$$

It follows that

$$\begin{aligned} T_{\omega_0'}(C_t) - T_{\omega_0}(C_t) &= \frac{-1}{2\pi} \int_0^r dt \int_0^{2\pi} \left(\frac{\partial}{\partial t} + \frac{i}{t}\frac{\partial}{\partial \theta}\right) p_* \psi^* g d\theta \\ &= \frac{-1}{2\pi} \int_0^r dt \int_0^{2\pi} \frac{\partial}{\partial t} p_* \psi^* g d\theta \\ &= \frac{-1}{2\pi} \int_0^{2\pi} (p_* \psi^* g(t, \theta) - p_* \psi^* g(0, \theta)) d\theta. \end{aligned}$$

In summary we have the formula:

$$T_{\omega_0'}(C_r) - T_{\omega_0}(C_r) = \frac{-1}{2\pi} \int_{\partial C_r} \psi^* g p^* d\theta + \psi^* g(p^* 0). \tag{3.3.1}$$

In terms of supreme norm, we have

$$|T_{\omega_0'}(C_r) - T_{\omega_0}(C_r)| \le 2\deg p \cdot \|g\|_{\sup}.$$

First Nevanlinna Inequality. Notice that every ω_0 is induced from a metric $\|\cdot\|_0$ on the line bundle $\mathcal{L}$ on X. Let s be a nonzero rational section of $\mathcal{L}$. Assume that $D = \operatorname{div} s$ intersects C properly and $\psi p^{-1}(0)$ is disjoint with D. Then

$$\omega = \delta_D + \frac{\partial\bar\partial}{\pi i}\log\|s\|_0$$

where δ_D is the Dirac distribution of D, and $\frac{\partial\bar\partial}{\pi i}$ is computed in the sense of distributions. The same formal computation gives

$$\begin{aligned} T_{\omega_0}(C_r) &= \frac{1}{\pi}\int_0^t \frac{dt}{t}\int_{C_r}\left(\delta_D + \frac{\partial\bar\partial}{\pi i}\log\|s\|_0\right) \\ &= \frac{1}{\pi}\int_0^r (C_t\cdot D)\frac{dt}{t} - \int_{\partial C_r}\psi^*\log\|s\|_0\frac{d\theta}{2\pi} + \log\|s\|_0(p^*0). \end{aligned}$$

Define the *counting function* and *proximity function* by

$$N_D(C_r) = \frac{1}{\pi}\int_0^r (C_t\cdot\psi^*D)\frac{dt}{t}$$

and

$$m_s(\partial C_t) = -\int_{\partial C_r}\psi^*\log\|s\|_0\frac{d\theta}{2\pi} + \log\|s\|_0(p^*0).$$

Then the identity we just proved is the First Main Theorem in Nevanlina theory:

$$T_{\omega_0}(C_r) = N_D(C_r) + m_s(\partial C_r). \tag{3.3.2}$$

If s is regular, then last two terms are bounded from below. Thus we obtain the Nevanlina inequality:

$$N_D(C_r) \le T_\omega(C_r) + O(1). \tag{3.3.3}$$

One consequence of this inequality is that C is algebraic if

$$T_\omega(C_r) = O(\log r).$$

Indeed, if this equality holds, then by (3.3.3), $C\cdot f^*D$ will be bounded by some positive number e. It follows that C is an algebraic curve of degree at most e. The converse is also true, see Griffiths [**28**]. In fact, assume that $\mathcal{L}$ is ample on the Zariski closure Y of $f(C)$, and let $Y \longrightarrow \mathbb{P}^m$ be the embedding defined by $\mathcal{L}$. Then with respect to a Fubini-Study metric on $\mathbb{P}^m$ and its dual metric on $\mathbb{P}^{m*}$, there is a Crofton formula ([**28**], pp. 22-23):

$$\int_{\mathbb{P}^{m*}} N_D(C_r)dD = T_{\omega_0}(C_r).$$

Proof of Theorem 3.3.5. Assume ω_0 is semi-positive and apply this to $\omega' = d^{-k}\phi^*\omega_0$. Then we can define the order function for the canonical form ω. Write $\omega = \omega_0 + \frac{\partial\bar{\partial}}{\pi i}g$. As $\psi^*\omega = 0$, the above inequality implies the following

$$|T_{\omega_0}(C_r)| \leq 2\deg p\|g\|_{\sup}.$$

Thus the order function of ψ is finite. It follows that ψ can be extended to a holomorphic map $\bar{\phi} : \bar{C} \longrightarrow X$. In this case ω will be supported on a finite number of points. This is impossible as there is no continuous function g such that $\frac{\partial\bar{\partial}}{\pi i}g$ gives a Dirac measure on points, see Theorem 3.1.6.

4. Arithmetic dynamics

In this section we will study (polarized) endomorphisms defined over a number field. We will first propose a conjecture about the equidistribution of the Galois orbits of a dynamically generic sequence of preperiodic points. This equidistribution conjecture implies the dynamic Manin-Mumford conjecture assuming the Zariski density conjecture of support of canonical measures. Then we will define the canonical heights and generalize these conjectures to *small points*. Following Chambert-Loir, we also propose an equidistrubution conjecture on Berkovich's *p-adic analytic spaces*. Finally, we will prove an equidistribution theorem for Zariski generic sequences of small points on any variety using the recent work of Yuan on arithmetic bigness. One immediate consequence is the equivalence of the dynamic Bogomolov conjecture and the dynamic equidistrubution conjecture.

4.1. Preperiodic points and small points. Let $\phi : X \longrightarrow X$ be an endomorphism of a projective variety with a polarization defined over a number field K. In other words, this means that X and ϕ are defined over K, and that there is an ample line bundle $\mathcal{L} \in \mathrm{Pic}\,(X) \otimes \mathbb{R}$ such that $\phi^*\mathcal{L} = \mathcal{L}^q$ for some integer $q > 1$.

The set Prep (X) of preperiodic points is defined over $\bar{K}$. Moreover, by a theorem of Northcott [**47**], for each number D, the set of preperiodic points x with degree $\deg(x) := [K(x) : K] \leq D$ is finite. Let Γ denote the absolute Galois group over K: $\Gamma = \mathrm{Gal}\,(\bar{K}/K)$. Then Γ acts on the set Prep (X) of preperiodic points on X.

Let us fix an embedding $\bar{K} \subset \mathbb{C}$ and write $X(\mathbb{C})$ for complex points via this embedding. Let $d\mu$ be the probability measure on $X(\mathbb{C})$ defined by the Chern class c_1 of the bundle $\mathcal{L}_{\mathbb{C}}$ constructed in §3.1.1. Notice that $d\mu$ is the invariant measure on X; i.e., the probability measure $d\mu$ defined on X such that

$$\phi^*d\mu = \deg\phi \cdot d\mu.$$

CONJECTURE 4.1.1 (Equidistribution of dynamically generic preperiodic points). Let x_i be a sequence of preperiodic points on X such that no infinite subsequence is supported in a proper preperiodic subvariety. Then the

Galois orbits of x_i are equidistributed with respect to the canonical measure $d\mu$ on $X(\mathbb{C})$.

More precisely, for $x \in \mathrm{Prep}\,(X)$, define probability measure

$$\mu_{\Gamma x} := \frac{1}{\deg x} \sum_{y \in \Gamma x} \delta_y.$$

Then the conjecture says the probability measures $\delta_{\Gamma x_n}$ converge weakly to the invariant measure $d\mu$ in the following sense: for a continuous function f on $X(\mathbb{C})$,

$$\frac{1}{\deg(x_n)} \sum_{y \in \Gamma x_n} f(y) \longrightarrow \int_{X(\mathbb{C})} f(x) d\mu(x)$$

as $n \longrightarrow \infty$.

Consequences. In the following let us give some consequences of the conjecture. The first consequence is the dynamic Manin-Mumford Conjecture 1.2.1.

PROPOSITION 4.1.2. *Conjecture* 4.1.1 $\Rightarrow$ *Conjecture* 1.2.1.

PROOF. Let Y be a subvariety containing a Zariski dense subset of preperiodic points $\mathrm{Prep}\,(X) \cap Y$ of ϕ. There is nothing to prove if $\dim Y = 0$. Otherwise $\mathrm{Prep}\,(X) \cap Y$ is an infinite set. Let Z be the intersection of all preperiodic subvarieties defined over K of X containing Y. Then Z itself is preperiodic. As the set of all algebraic subvarieties of X is countable, we may list the set of preperiodic proper subvarieties of Z in a sequence $Z_1, \ldots, Z_n, \cdots$. Using induction, we can define a sequence of points $x_n \in \mathrm{Prep}\,(X) \cap Y$ such that $x_n \notin \cup_{i=1}^n Z_i$. In this way, the sequence x_n will have finite intersection with any proper preperiodic subvariety of Z. Now we apply Conjecture 4.1.1 to conclude that the Galois orbits of x_n converge to the invariant measure on Z. As all Galois orbits lie in Y, the invariant measure $d\mu$ must be supported on Y. Now by Theorem 3.1.6, $Y = Z$. In other words, Y is preperiodic. □

Topological interpretation. To understand the nature of the equidistribution conjecture, let's introduce the following topologies:

(1) *Zariski topology*: the topology defined by algebraic equations;
(2) *dynamic topology*: the topology defined by preperiodic subvarieties;
(3) *archimedean topology*: the topology defined by inequalities of usual continuous functions on $X(\mathbb{C})$;
(4) *distributional topology*: Let $\mathrm{Meas}(X(\mathbb{C}))$ denote the set of probability measures on $X(\mathbb{C})$ with weak topology; let's define a map

$$\mathbb{D} : X \longrightarrow \mathrm{Meas}(X(\mathbb{C})), \qquad y \mapsto d\mu_{\bar{y}(\mathbb{C})}.$$

Here $y \in X$ is a point of *scheme* X, and $\bar{y}$ denotes the Zariski closure. Then the distributional topology is the induced topology on X from $\mathbb{D}$. Theorem 3.1.6 implies that the map $\mathbb{D}$ is injective.

We have the following order of fineness of the topologies:

$$\text{Distributional} > \text{Archimedean} > \text{Zariski} > \text{Dynamic}.$$

On Prep (X), Conjecture 4.1.1 says that the dynamic topology equals the distributional topology while Conjecture 1.2.1 says that the Zariski topology equals the dynamic topology.

What is proved?

(1) If X is an abelian variety, and $\phi = [m]$ is the multiplication by a positive number $m > 1$, the conjecture was proved in [**64**] and [**58**].
(2) If $X = \mathbb{P}^n$ and ϕ is the morphism $(x_0, ..., x_m) \longrightarrow (x_0^m, \ldots, x_n^m)$, the conjecture was proved by Bilu [**8**].
(3) If X is a compactification of an almost split semi-abelian variety, then the conjecture was proved by Chambert-Loir [**17**].

Height machinery. In the following, we want to introduce the concept of heights and state a conjecture about small points.

Let K be a number field. For each place v of K let $|\cdot|_v$ be the normalized places: for any haar measure dx_v on K_v,

$$d(ax)_v = |a|_v dx_v.$$

For a point $x = (x_0, \ldots, x_n) \in \mathbb{P}^n(\bar{\mathbb{Q}})$, the *naive height* $h_{naive}(x)$ of x is defined by

$$h_{naive}(x) = \frac{1}{[K : \mathbb{Q}]} \sum_v \log \max(|x_0|_v, \ldots, |x_n|_v).$$

It can be shown that the definition $h(x)$ can be extended to $\mathbb{P}^n(\bar{\mathbb{Q}})$. If x is a rational point represented by an $(n+1)$-tuple of integers $(x_0, \ldots, x_n)$ with no common divisor, then

$$h_{naive}(x) = \log \max(|x_0|_\infty, \ldots, |x_n|_\infty).$$

If we define the complexity $c(x)$ of x as the maximum number of digits of x_i, which measures the time spent to write a number down, then

$$h_{naive}(x) - c(x) \log 10$$

is bounded on the set of rational points of $\mathbb{P}^n$. A basic property of h_{naive} is the following Northcott Theorem: for any given numbers D and H, the set of points in $\mathbb{P}^n$ with height $\leq H$ and degree $\leq D$ is finite.

Let X be a projective variety over a number field. Let $\mathcal{L}$ be a very ample line bundle. Then there is an embedding $i : X \to \mathbb{P}_K^N$ such that $i^*\mathcal{O}(1) \simeq \mathcal{L}$. In this way we obtain a height function

$$h_\mathcal{L} : \quad X(\bar{\mathbb{Q}}) \longrightarrow \mathbb{R}, \quad h_\mathcal{L}(x) = h_{naive}(i(x)).$$

Here are some standard properties of $h_\mathcal{L}$:

(1) $h_\mathcal{L}$ up to a bounded function on $X(\bar{\mathbb{Q}})$ does not depend on the choice of the embedding.

(2) $h_{\mathcal{L}}$ is additive in the sense that for two very ample line bundles $\mathcal{L}$ and $\mathcal{M}$,

$$h_{\mathcal{L}\otimes\mathcal{M}} = h_{\mathcal{L}} + h_{\mathcal{M}} + O(1).$$

(3) $h_{\mathcal{L}}$ is bounded from below:

$$h_{\mathcal{L}}(x) \geq C.$$

(4) (Northcott Theorem) for any positive numbers D, H, the set

$$\left\{x \in X(\bar{\mathbb{Q}}), \quad \deg(x) < D, \quad h_{\mathcal{L}}(x) < H\right\}$$

is finite.

(5) For two very ample line bundles $\mathcal{L}$ and $\mathcal{M}$, the equality $h_{\mathcal{L}} = h_{\mathcal{M}} + O(1)$ implies that $\mathcal{L} \otimes \mathcal{M}^{-1}$ is a torsion line bundle.

From these properties, we can extend the definition of height function to every $\mathcal{L} \in \mathrm{Pic}\,(X) \otimes \mathbb{R}$ by linearity. In other words, we have an embedding from the $\mathbb{R}$-vector space of $\mathbb{R}$-bundles into the space of functions on $X(\bar{\mathbb{Q}})$ modulo bounded functions:

$$\mathrm{Pic}\,(X) \otimes \mathbb{R} \longrightarrow \mathrm{Funct}(X(\bar{\mathbb{Q}}), \mathbb{R})/O(1).$$

This is a homomorphism of $\mathbb{R}$-vector spaces and such that the Northcott theorem is true if $\mathcal{L}$ is numerically positive in $\mathrm{Pic}\,(X) \otimes \mathbb{R}$. Indeed, if $\mathcal{M} \in \mathrm{Pic}\,(X)$ is a very ample line bundle in the usual sense, then there is a positive number ϵ such that $\mathcal{L} - \epsilon\mathcal{M}$ is positive. Thus the Northcott Theorem for $\mathcal{M}$ implies that for $\mathcal{L}$.

Now let us go back to our situation: let $\phi : X \longrightarrow X$ be an endomorphism of a projective variety over a number field K with a polarization by an $\mathbb{R}$-line bundle $\mathcal{L} \in \mathrm{Pic}\,(X) \otimes \mathbb{R}$. Fix a height function $h_{\mathcal{L}}$. Then we can define the *canonical height* $\widehat{h}_{\mathcal{L}}$ on $X(\bar{\mathbb{Q}})$ by the following formula:

$$\widehat{h}_{\mathcal{L}}(x) := \lim_{k\longrightarrow\infty} \frac{h_{\mathcal{L}}(\phi^k(x))}{q^k}. \tag{4.1.1}$$

The height $\widehat{h}_{\mathcal{L}}$ defined in this way does not depend on the choice of height function $h_{\mathcal{L}}$: one may characterize $\widehat{h}_{\mathcal{L}}$ as the unique height function for $\mathcal{L}$ such that

$$\widehat{h}_{\mathcal{L}}(\phi(x)) = q\widehat{h}_{\mathcal{L}}(x).$$

Then $\widehat{h}_{\mathcal{L}}$ has the following properties:

PROPOSITION 4.1.3. *For any $x \in X(\bar{\mathbb{Q}})$, $\widehat{h}_{\mathcal{L}}(x) \geq 0$. Moreover, $\widehat{h}_{\mathcal{L}}(x) = 0$ if and only if x is a torsion point.*

PROOF. Since each $\mathcal{L}$ is ample, we may choose an initial height function $h_{\mathcal{L}}$ with positive values. In this way, the above definition formula 4.1.1 implies that $\widehat{h}_{\mathcal{L}} \geq 0$. If $\widehat{h}_{\mathcal{L}}(x) = 0$, then

$$\widehat{h}_{\mathcal{L}}(\phi^k(x)) = 0 \qquad \forall k.$$

In other words, every point in the forward orbit of x has height 0. By Northcott's theorem, this orbit must be finite. Thus x must be preperiodic.

Conversely, if x is preperiodic, say, $\phi^k(x) = \phi^\ell(x)$ with $k \neq \ell$, then $\widehat{h}_{\mathcal{L}}(\phi^k(x)) = \widehat{h}_{\mathcal{L}}(\phi^\ell(X))$, or $q^k\widehat{h}_{\mathcal{L}}(x) = q^\ell h_{\mathcal{L}}(x)$. Thus $\widehat{h}_{\mathcal{L}}(x) = 0$. □

Now we apply the Northcott theorem to give:

COROLLARY 4.1.4. *For any number D, the set of preperiodic points on X with degree bounded by D is finite.*

REMARK 4.1.5. There are some conjectures concerning the lengths of orbits of preperiodic points and heights of non-torsion points:

(1) If x is non-preperiodic then a Lehmer type conjecture states that

$$\widehat{h}_{\mathcal{L}}(x) \geq c \cdot \deg(x)^{-1}.$$

(2) If x is a preperiodic point, then the length $\mathrm{ord}\,(x)$ of the orbit of x should have the bound

$$\mathrm{ord}\,(x) \leq c \deg(x).$$

(3) (Morton-Silverman [**45**]) For all positive integers D, N, d with $d > 2$, there exists an integer $\kappa(D, N, d)$ such that for each number field of degree D over $\mathbb{Q}$, and each finite endomorphism $\psi : \mathbb{P}^N_k \longrightarrow \mathbb{P}^N_k$ of degree d^N, the number of preperiodic points ψ in $\mathbb{P}^N(k)$ is less than or equal to $\kappa(D, N, d)$. By Fakhruddin [**23**], this conjecture implies the corresponding uniform on torsion points on abelian varieties.

See Fakhruddin [**23**] for some other interesting questions about rational points on a dynamical system.

Rational points. From the classification in §2, one sees that for a polarized endomorphism $\phi : X \longrightarrow X$, X is either a quotient of an abelian variety or a rationally connected variety. Thus it is conjectured that for some finite extension F of K, the set of rational points $X(F)$ is Zariski dense in X. Here we would like to conjecture that this is already true for some orbits:

CONJECTURE 4.1.6. Let $\phi : X \longrightarrow X$ be an endomorphism of a projective variety defined over a number field K with a polarization. Then there is a point $x \in X(\bar{K})$ such that the forward orbit

$$\{x, \phi(x), \dots, \phi^k(x), \dots\}$$

is Zariski dense in X.

Small points. In the following we want to extend the dynamic Manin-Mumford Conjecture 1.2.1 and the equidistribution conjecture for small points to

CONJECTURE 4.1.7 (Dynamic Bogomolov conjecture). A subvariety Y of X is preperiodic if and only if for any Zariski open subvariety U of Y, and for any positive number $\epsilon > 0$, there is a point $x \in U(\bar{\mathbb{Q}})$ such that $\widehat{h}_{\mathcal{L}}(x) < \epsilon$.

CONJECTURE 4.1.8 (Equidistribution of dynamically generic small points). Fix a complex place of K. Let x_i be a sequence of points on X such that no infinite subsequence is included in a proper preperiodic subvariety, and that $\lim_{i\to\infty} \widehat{h}_{\mathcal{L}}(x_i) = 0$. Then the Galois orbits of x_i are equidistributed with respect to the canonical measure on X.

What is proved?

(1) If X is an abelian variety, then both conjectures were proved in [**64**] and [**58**].
(2) If X is the multiplicative group then Conjecture 4.1.7 was proved in [**61**], and the equidistribution Conjecture 4.1.8 was proved by Bilu [**8**].
(3) If X is an almost split semiabelian group, then the Bogomolov conjecture was proved by Chambert-Loir [**17**].

4.2. Metrized line bundles and heights of subvarieties. In this subsection, we want to associate to every metrized line bundle a height function. By an integral and hermitian model of $(X, \mathcal{L})$ we mean a triple $\bar{L} := (\widetilde{X}, \widetilde{\mathcal{L}}, \|\cdot\|_v)$ where

(1) $\widetilde{X}$ is an projective and flat scheme over $\operatorname{Spec} \mathcal{O}_K$ with generic fiber X;
(2) $\widetilde{\mathcal{L}}$ is a line bundle on $\widetilde{X}$;
(3) $\|\cdot\|_v$ is a collection of metrics of $\mathcal{L}_v := \mathcal{L}_{K,v} \otimes \mathbb{C}$ on $X_v = X \otimes_v \mathbb{C}$ for each embedding $v : K \longrightarrow \mathbb{C}$. We assume that for an embedding v and its complex conjugate v^c,

$$\|\ell\|_v(x) = \|\ell^c\|_{v^c}(x^c).$$

Here the bar map is the complex conjugation

$$\begin{array}{ccc} \mathcal{L}_v & \xrightarrow{c} & \mathcal{L}_{v^c} \\ \downarrow & & \downarrow \\ X_v & \xrightarrow{c} & X_{v^c} \end{array}$$

Let x be a point in $X(\bar{\mathbb{Q}})$. Then the Zariski closure $\bar{x}$ of x has a normalization $f : \operatorname{Spec}\mathcal{O}_F \to \widetilde{X}$ where $\mathcal{O}_F$ is the ring of integers of some number field F over K. The invertible $\mathcal{O}_F$-module $\mathcal{N} := f^*\mathcal{L}$ is equipped with a hermitian metric on $\mathcal{N} \otimes_\sigma \mathbb{C}$ for each embedding $\sigma : F \to \mathbb{C}$. Then we define the degree of $\bar{\mathcal{L}}$ on $\bar{x}$ by

$$\deg_{\bar{\mathcal{L}}} \bar{x} = \log \frac{\#\mathcal{N}/n\mathcal{O}_F}{\prod_{v|\infty} \|n\|_v}$$

where n is any nonzero element in $\mathcal{N}$. One can show that

$$h_{\bar{\mathcal{L}}}(x) = \frac{1}{[F:\mathbb{Q}]} \deg_{\bar{\mathcal{L}}}(\bar{x}).$$

One immediate advantage of using metrized line bundles is to extend the definition of heights to arbitrary cycles of $\widetilde{X}$ by using the intersection theory of Gillet and Soulé [**26**]. Indeed, on a subvariety Y of X over K of dimension d, the intersection number of the metrized line bundles $\bar{\mathcal{L}}_1, \dots, \bar{\mathcal{L}}_d$ can be computed by induction:

$$\widehat{c}_1(\bar{\mathcal{L}}_1) \cdots \widehat{c}_1(\bar{\mathcal{L}}_d) \cdot Y = \widehat{c}_1(\bar{\mathcal{L}}_1) \cdots \widehat{c}_1(\bar{\mathcal{L}}_{d-1}) \cdot [\operatorname{div} s]$$
$$- \sum_{v|\infty} \int_{X_v(\mathbb{C})} \log \|s\|_v c_1(\bar{\mathcal{L}}_{1,v}) \cdots c_1(\bar{\mathcal{L}}_{d-1,v}),$$

where s is a nonzero section of $\mathcal{L}_d$ on Y.

If Y is a closed subvariety of X, then the height $h_{\bar{\mathcal{L}}}(Y)$ is defined by the formula

$$h_{\bar{\mathcal{L}}}(Y) = \frac{\deg_{\bar{\mathcal{L}}}(\widetilde{Y})}{(\dim Y + 1) \deg_{\mathcal{L}}(Y)}$$

where $\widetilde{Y}$ is the Zariski closure of Y in $\widetilde{X}$.

Integrable metrized line bundles [**62**]. Consider a projective variety X over $\operatorname{Spec} K$. For a line bundle $\mathcal{L}$ on X and an integral model $(\widetilde{X}, \widetilde{\mathcal{L}})$ of $(X, \mathcal{L}^e)$ over $\operatorname{Spec} \mathcal{O}_K$, one can define an adelic metric $\|\cdot\|_{\widetilde{\mathcal{L}}} = \{\|\cdot\|_v, v \in \mathcal{S}\}$ on $\mathcal{L}$, where e is a positive integer, $\mathcal{S}$ is the set of places of K, and $\|\cdot\|_v$ is a metric on $\mathcal{L} \otimes_K \overline{K}_v$ on $X(\overline{K}_v)$. The metrics constructed this way are called *smooth* metrics. The metric so obtained will be invariant under the Galois group $\operatorname{Gal}(\bar{K}_v/K_v)$.

Let $\mathcal{L}_1, \dots, \mathcal{L}_d$ $(d = \dim X + 1)$ be line bundles on X. For each positive integer n, let $(\widetilde{X}_n, \widetilde{\mathcal{L}}_{1,n}, \dots, \widetilde{\mathcal{L}}_{d,n})$ be an arithmetic model of $(X, \mathcal{L}_1^{e_{1,n}}, \dots, \mathcal{L}_d^{e_{d,n}})$ on $\operatorname{Spec} \mathcal{O}_K$. Assume for each i that $(\mathcal{L}, \|\cdot\|_{\widetilde{\mathcal{L}}_{i,n}})$ converges to an adelic metrized line bundle $\overline{\mathcal{L}}_i$. One might ask whether the number

$$c_n = \frac{c_1(\widetilde{\mathcal{L}}_{1,n}) \cdots c_1(\widetilde{\mathcal{L}}_{d,n})}{e_{1,n} \quad \cdots \quad e_{d,n}}$$

in Gillet-Soulé's intersection theory converges or not.

We showed in [**62**] that c_n converges if all $\widetilde{\mathcal{L}}_{i,n}$ are relatively semipositive, and that $\lim_{n\to\infty} c_n$ depends only on $\overline{\mathcal{L}}_i$. We say that an adelic line bundle $\overline{\mathcal{L}}$ is *integrable* if $\overline{\mathcal{L}} \xrightarrow{\sim} \overline{\mathcal{L}}_1 \otimes \overline{\mathcal{L}}_2^{-1}$ with $\overline{\mathcal{L}}_i$ semipositive. It follows that Gillet-Soulé's theory can be extended to integrable metrized line bundles. Some theorems such as the Hilbert-Samuel formula, the Nakai-Moishezon theorem, and the successive minima inequality remain valid on semi-smooth metrized line bundles.

Admissible metrized line bundles [**62**]. Let $\phi : X \to X$ be an endomorphism of a projective variety with a polarization over K. Thus we have an ample line bundle and an isomorphism $\alpha : \mathcal{L}^q \simeq \phi^*\mathcal{L}$ with $q > 1$. Using Tate's argument, in [**62**] we constructed a unique integrable metric $\|\cdot\|$ on $\mathcal{L}$ such that

$$\|\cdot\|^q = \alpha^* \phi^* \|\cdot\|.$$

Now for any effective cycle Y of X of pure dimension, we can define an (absolute) height

$$h_{\mathcal{L}}(Y) = \frac{c_1(\overline{\mathcal{L}}|_Y)^{\dim Y+1}}{(\dim Y + 1)\deg_{\mathcal{L}}(Y)}.$$

The height $h_{\mathcal{L}}$ can be characterized by the property that

$$h_{\mathcal{L}}(\phi(Y)) = q h_{\mathcal{L}}(Y).$$

As Tate did, $h_{\mathcal{L}}$ can be defined without an admissible metric. Some situations are studied by Philippon [**49**], Kramer [**35**], Call and Silverman [**11**], and Gubler [**29**]. In this case, if Y is preperiodic: the orbit

$$\{Y, f(Y), f^2(Y), \ldots\}$$

is finite, then $h_{\mathcal{L}}(Y) = 0$.

We showed in [**62**] that the Bogomolov conjecture is equivalent to the following converse:

Conjecture 4.2.1. Let $\phi : X \longrightarrow X$ be an endomorphism of a projective variety over a number field K with a polarization by an ample line bundle $\mathcal{L}$. Then $h_{\mathcal{L}}(Y) = 0$ if and only if Y is preperiodic.

This is a theorem [**61**] for the case of multiplicative group. A consequence is the generalized Lang's conjecture which claims that if Y is not preperiodic then the set of preperiodic points in Y is not Zariski dense. Lang's conjecture is proved by Laurent [**37**] and by Raynaud [**52**] for abelian varieties.

Measures on Berkovich spaces [**30**] [**16**]. Fix a place v of K. Then there is a v-adic analytic space X_v^{an} – *Berkovich space* [**7**]. If v is complex this is usual $X_v(\mathbb{C})$; if v is real this is $X_v(\mathbb{C})/\{1, c\}$ where c is the complex conjugation on $X_v(\mathbb{C})$. For v a finite place, we have an embedding of topology space $|X_v| \longrightarrow X_v^{\mathrm{an}}$ with dense image, where $|X_v|$ denotes the set of closed points on X_v with v-adic topology, or equivalently, the set of Galois orbits of $X(\bar{K}_v)$ under $\mathrm{Gal}(\bar{K}_v/K_v)$. Moreover the metrized line bundles on some model of X_v over $\mathcal{O}_{K_v}$ will induce some *continuously* metrized line bundles on X_v^{an} whose restriction on $|X_v|$ is the usual metrized line bundles constructed as above. Thus we will have the notion of integrable metrized line bundles.

A continuous function f on X_v^{an} is called *smooth* if its restriction on $|X_v|$ is the logarithm of a smooth metric $\|\cdot\|_v$ at v of $\mathcal{O}_{X_v^{\mathrm{an}}}$ defined by an integral model:

$$f = \log \|1\|.$$

By the work of Gubler [**30**], the smooth functions are dense in the continuous functions on X_v^{an}. In other words, let $C^\infty(|X_v|)$ denote the space of smooth functions on $|X_v|$ which may not be closed under multiplication, and let $R(X_v)$ be the ring of functions on $|X_v|$ generated by smooth functions with supremum norm. Then $C^\infty(|X_v|)$ is dense in $R(X_v)$ and $X^{\mathrm{an}}{}_v$ as a topological space is the spectrum of $R(X_v)$:

$$X_v^{\mathrm{an}} = \mathrm{Hom}_{\mathrm{cont}}(R(X_v), \mathbb{R}).$$

Here the right hand side stands for continuous homomorphisms of $\mathbb{R}$-algebras. In other words, X is the unique compact space such that $C(X_v^{\mathrm{an}}) = R(|X_v|)$.

Now we consider the situation of a polarized dynamical system $(X, \phi, \mathcal{L})$. For any subvariety Y of X_v of dimension d and integrable metrized line bundles $\bar{\mathcal{L}}_1, \ldots, \bar{\mathcal{L}}_d$, Chambert-Loir [**16**] defined the measure

$$c_1(\bar{\mathcal{L}}_1) \cdots c_1(\bar{\mathcal{L}}_d) \delta_{Y_v^{\mathrm{an}}}$$

supported on the image $Y^{\mathrm{an}}{}_v$ which has the usual properties as in the archimedean case in §3.1

For example, for a subvariety Y of X over K of dimension d and adelic metrized line bundles $\bar{\mathcal{L}}_1, \ldots, \bar{\mathcal{L}}_d$, one can compute the intersection by induction:

$$\begin{aligned} \widehat{c}_1(\bar{\mathcal{L}}_1) \cdots \widehat{c}_1(\bar{\mathcal{L}}_d) \cdot Y = {} & \widehat{c}_1(\bar{\mathcal{L}}_1) \cdots \widehat{c}_1(\bar{\mathcal{L}}_{d-1}) \cdot [\operatorname{div} s] \\ & - \sum_v \int_{X_v^{\mathrm{an}}} \log \|s\|_v c_1(\bar{\mathcal{L}}_{1,v}) \cdots c_1(\mathcal{L}_{d-1,v}) \delta_{Y_v^{\mathrm{an}}} \end{aligned}$$

where s is a nonzero section of $\mathcal{L}_d$ on Y.

In this case, the construction of Chambert-Loir as above gives the canonical measures on X_v^{an} for each integral subvariety δ_Y for each embedding $v : K \longrightarrow \mathbb{C}_p$:

$$d\mu_{Y,v} := c_1(\bar{\mathcal{L}}_v)^{\dim Y} \cdot \delta_{Y_v^{\mathrm{an}}} / \deg_{\mathcal{L}}(X).$$

One can show an analogue of Proposition 3.1.5, such the support of the measure is any Z_v^{an} for any subvariety Z of Y.

We want to propose a generalization of the equidistribution conjecture:

CONJECTURE 4.2.2 (v-adic Equidistribution of dynamically generic small points). Let $\phi : X \longrightarrow X$ be an endomorphism of a projective variety over a number field K with a polarization by an ample line bundle $\mathcal{L}$. Let v be a place of K. Fix an embedding $v : K_v \longrightarrow \mathbb{C}_p$ and write X_v^{an} for the induced analytic space. Let x_i be a sequence of points on X such that no infinite subsequence is included in a proper preperiodic subvariety, and that $\lim_{i \to \infty} \widehat{h}_{\mathcal{L}}(x_i) = 0$. Then the Galois orbits of x_i are equidistributed with respect to the canonical measure on X_v^{an}.

REMARK 4.2.3. We would like to consider an *adelic version* of the above equidistribution. Let $\phi : X \longrightarrow X$ be an endomorphism of a projective variety over a number field K with a polarization by an ample line bundle $\mathcal{L}$. Let S be a finite set of places K. For each place $v \in S$, fix an embedding $v : K_v \longrightarrow \mathbb{C}_p$ and write X_v^{an} for the induced analytic space. Let $d\mu_v$ denote the probability measure on X_v^{an}. Let x_i be a sequence of points on X such that no infinite subsequences are included in a proper preperiodic subvariety, and that $\lim_{i \to \infty} \widehat{h}_{\mathcal{L}}(x_i) = 0$. Then we want to conjecture that the Galois orbits of x_i are equidistributed with respect to the canonical measure on $\prod X_v^{\mathrm{an}}$, with respect to the product measures $d\mu_v$.

In one special case where S is the set of places over a prime p, we may reduce the conjecture for the dynamical system on $\mathrm{Res}_{K/\mathbb{Q}}(X)$ at the place p. Indeed, by definition $\mathrm{Res}_{K/\mathbb{Q}}(X)(L) = X(K \otimes L)$. Thus its fiber over p is given by $X \otimes_K (K \otimes \mathbb{Q}_p) = \prod_{v|p} X_v$.

REMARK 4.2.4. We still have some topological interpretation of the above conjecture by introducing the distributional topology on X_v^{an}. We will still have a conjecture that the support of $d\mu_v$ is Zariski dense. Also we have the analogue of Proposition 4.1.2: the v-adic equidistribution conjecture plus the density of the support of $(d\mu_v)$ will imply the "if" part of Conjecture 1.2.1.

REMARK 4.2.5. In [**50**], Szpiro and Tucker gave a formula for canonical heights for the dynamical system on $\mathbb{P}^1$ by working on successive blow-ups.

4.3. A generic equidistribution theorem. In this section we want to show that the Bogomolov conjecture is equivalent to the equidistribution conjecture. This is actually a consequence of the following equidistribution theorem for Galois orbits of *generic* sequences of small points:

THEOREM 4.3.1 (Equidisdibution for Zariski generic small points [**57**], [**64**], [**16**], [**60**]). *Let X be a projective variety over a number field K of dimension n. Let v be a place of K. Let $\bar{\mathcal{L}} = (\mathcal{L}, \|\cdot\|_v)$ be a metrized line bundle on X such that the following hold:*

(1) *$\mathcal{L}$ is ample,*
(2) *$\|\cdot\|_v$ is semipositive,*
(3) *$h_{\bar{\mathcal{L}}}(X) = 0$.*

Let x_n be a sequence of points on X such that $\lim h_{\bar{\mathcal{L}}}(x_n) = 0$ and that no infinite subsequence of x_n is included in a proper subvariety. Then the Galois orbits of x_n are equidistributed in X_v^{an} with respect to the measure

$$d\mu := c_1(\mathcal{L}_v, \|\cdot\|_v)^n / \deg_{\mathcal{L}}(X).$$

COROLLARY 4.3.2. *The dynamic Bogomolov Conjecture* 4.1.7 *is equivalent to the equidistribution conjecture* 4.1.8 *and* 4.2.2.

PROOF. By a standard trick, we need only show that any infinite subsequence x_{i_n} contains another infinite subsequence whose Galois orbits are equidistributed. With x_{i_n} replaced by a subsequence we may assume the following:

(1) the Zariski closure Y of $\{\Gamma x_{i_n}, n = 1, \dots\}$ is an integral subvariety of X;
(2) no infinite subsequence of x_{i_n} is included into a proper subvariety of Y.

By the Bogomolov conjecture, Y is a preperiodic subvariety of X. By the assumption of the conjecture 4.2.2, $X = Y$. Now Theorem 4.3.1 gives the equidistribution of Galois orbits. □

Sketch of proof of Theorem 4.3.1. The theorem was first proved in Szpiro-Ullmo-Zhang [**57**] when X_K is smooth and the curvature of $\bar{\mathcal{L}}$ is smooth and positive point-wise on $X(\mathbb{C})$, and extended in [**64**] when $X_{\mathbb{C}}$ is a subvariety of a smooth variety Y and $\bar{\mathcal{L}}_{\mathbb{C}}$ is the restriction of a metrized line bundle $\mathcal{M}$ with smooth and positive curvature point-wise. Then Chambert-Loir [**16**] further extended all of these results to v-adic Berkovich spaces. The general case stated here is due to Yuan [**60**] as a consequence of his theorem of *arithmetic bigness of line bundles*:

THEOREM 4.3.3 (Yuan [**60**]). *Let $\bar{\mathcal{L}}_1$ and $\bar{\mathcal{L}}_2$ be two arithmetically ample line bundles on X such that*

$$c_1(\bar{\mathcal{L}}_1)^{n+1} - (n+1)c_1(\bar{\mathcal{L}}_1)^n \cdot c_1(\bar{\mathcal{L}}_2) > 0.$$

Then the bundle $\bar{\mathcal{L}}_1 \otimes \bar{\mathcal{L}}_2^{-1}$ is big in the following sense:

$$\log \# \left\{ s \in \Gamma(X, (\mathcal{L}_1 \otimes \mathcal{L}_2^{-1})^k) : \|s\|_{\sup} \leq 1 \right\} \geq ck^{n+1} + o(k^{n+1})$$

where $\|\cdot\|_{\sup} = \sup_v \|\cdot\|_{v,\sup}$ is the superum norm over all places and c is a positive number independent of k.

Let f be a semi-smooth function on X_v^{an}. For each $t \in \mathbb{R}$, let $\bar{\mathcal{L}}_t$ denote the metrized line bundle $(\mathcal{L}, \|\cdot\|_t)$ with adelic metric $\|\cdot\|_t = \|\cdot\| e^{-ft}$ (which differs from $\|\cdot\|$ only at v) . Since f is smooth, we have a line bundle $\mathcal{O}_X(f)$ on a model of X induces the metric on $\mathcal{O}_X$ such that the induced metric has property $\|1\|_v = e^f$ and $\|1\|_w = 1$ for $w \neq v$. It is easy to see that there are two semi-ample line bundles $\bar{\mathcal{M}}_1$ and $\bar{\mathcal{M}}_1$ such that

$$\mathcal{O}(f) = \bar{\mathcal{M}}_1 \otimes \bar{\mathcal{M}}_2^{-1}.$$

Now we have the expression

$$c_1(\bar{\mathcal{L}}_t) = c_1(\bar{\mathcal{L}} \otimes \bar{\mathcal{M}}_1^t) - c_1(\bar{\mathcal{M}}_2^t).$$

Now we apply Yuan's bigness theorem to the two line bundles on the right hand side. The quantity in Yuan's theorem is

$$\begin{aligned}
&(c_1(\bar{\mathcal{L}}) + tc_1(\bar{\mathcal{M}}_1))^{n+1} - (n+1)(c_1(\bar{\mathcal{L}}) + tc_1(\bar{\mathcal{M}}))^n (t\bar{c}_1(\mathcal{M}_2)) \\
&= tc_1(\bar{\mathcal{L}})^n (c_1(\bar{\mathcal{M}}_1) - c_1(\bar{\mathcal{M}}_2)) + O(t^2) \\
&= (n+1)tc_1(\bar{\mathcal{L}})^n \cdot c_1(\mathcal{O}(f)) + O(t^2) \\
&= (n+1)t \int fc_1(\bar{\mathcal{L}})^n + O(t^2).
\end{aligned}$$

Thus $\int fc_1(\bar{\mathcal{L}})^n > 0$ will imply that $\bar{\mathcal{L}}_t$ is big. In this case there is a section s of $\bar{\mathcal{L}}_t^k$ with norm ≤ 1. Now let us use this section to compute the heights of $x \in X(\bar{K})$ when x not in the divisor of s,

$$h_{\bar{\mathcal{L}}_t}(x) = -\frac{1}{k}\frac{1}{\deg x} \sum_v \sum_{\sigma : K(x) \longrightarrow \bar{K}_v} \log \|s\|_v(\sigma(x)) \geq 0.$$

On the other hand, we can compute the height by the additivity

$$\bar{\mathcal{L}}_t = \bar{\mathcal{L}} + O(tf)$$

and Section 1 for $O(tf)$ with norm e^{-tf}:

$$\begin{aligned} h_{\bar{\mathcal{L}}_t}(x) &= h_{\bar{\mathcal{L}}}(x) + h_{O(tf)}(x) \\ &= h_{\bar{\mathcal{L}}}(x) + \frac{t}{\deg x} \sum_{\sigma:K(x)\longrightarrow \bar{K}_v} f(\sigma(x)). \end{aligned}$$

Combining both expressions we obtain that $\int f c_1(\bar{\mathcal{L}})^n > 0$ implies that for x in a Zariski dense subset,

$$\frac{t}{\deg x} \sum_{\sigma:K(x)\longrightarrow \bar{K}_v} f(\sigma(x)) \geq -h_{\bar{\mathcal{L}}}(x).$$

Now we apply this inequality to x_n in Theorem 4.3.1 to obtain that

$$\lim_{n\longrightarrow\infty}\inf \frac{t}{\deg x_n} \sum_{\sigma:K(x_n)\longrightarrow \bar{K}_v} f(\sigma(x_n)) \geq 0.$$

For arbitrary f, we may replace f by $f - \int f d\mu + \epsilon$ (which has positive integral $\epsilon > 0$) to obtain the following unconditional inequality

$$\lim_{n\longrightarrow\infty}\inf \frac{1}{\deg x} \sum_{\sigma:K(x_n)\longrightarrow \bar{K}_v} f(v(x_n)) \geq \int f d\mu.$$

We may replace f by $-f$ in the above expression to obtain

$$\lim_{n\longrightarrow\infty}\sup \frac{1}{\deg x} \sum_{\sigma:K(x_n)\longrightarrow \bar{K}_v} f(\sigma(x_n)) \leq \int f d\mu.$$

Thus we have shown that

$$\lim_{n\longrightarrow\infty} \frac{1}{\deg x} \sum_{\sigma:K(x_n)\longrightarrow \bar{K}_v} f(\sigma(x_n)) = \int f d\mu.$$

REMARK 4.3.4. When $\dim X = 1$, Theorem 4.3.1 has been proved by A. Chambert-Loir in [**17**] using a bigness type result of Autissier [**1**] for arithmetic surfaces. In the spacial case where $X = \mathbb{P}^1$, and $\mathcal{L}$ is equipped with the canonical metric induced from an endomorphism ϕ of degree > 1, Theorem 4.3.1 has been proved by two different groups of people: M. Baker and R. Rumely [**3**] and C. Favre and J. Rivera-Letelier in [**24**].

References

[1] P. Autissier, *Points entiers sur les surfaces arithmétiques*, J. Reine Angew. Math. **531** (2001).

[2] P. Autissier, *Dynamique des correspondances algebriques et hauteurs*, Int. Math. Res. Not. **69** (2004).

[3] M. Baker and R. Rumely, *Equidistribution of small points, rational dynamics, and potential theory*, arXiv: math.NT/0407426.

[4] D. Barlet, *Espace analytique réduit des cycles analytiques complexes compacts d'un espace analytique complexe de dimension finie*, in 'Fonctions de plusieurs variables complexes, II', Lecture Notes in Math., **482**, 1–158. Springer, Berlin, 1975.

[5] A. Beauville, *Variétés kähleriennes dont la première classe de Chern est nulle*, Journal of Differential Geometry **87** (1983).

[6] A. Beauville, *Endomorphisms of hypersurfaces and other manifolds*, Internat. Math. Res. Notices **1** (2001), 53–58.

[7] V.G. Berkovich, *Spectral theory and analytic geometry over non-Archimedean fields*, Mathematical Surveys and Monographs, **33**, American Mathematical Society, Providence, RI, 1990.

[8] Y. Bilu, *Limit distribution of small points on algebraic tori*, Duke Mathematical Journal **89(3)** (1997).

[9] J.-V. Briend and J. Duval, *Deux caractérisations de la mesure d'équilibre d'un endomorphisme de* $\mathbf{P}^k(\mathbf{C})$, Publ. Math. Inst. Hautes Études Sci. **93** (2001), 145–159.

[10] J.-Y. Briend, S. Cantat, and M. Shishikura, *Linearity of the exceptional set for maps of* $\mathbb{P}^k(\mathbb{C})$, Math. Ann. **330(1)** (2004), 39–43.

[11] G. Call and J.H. Silverman, *Canonical heights on varieties with morphisms*, Compositio Math. **89(2)** (1993), 163–205.

[12] F. Campana, *Connexité rationnelle des variétés de Fano*, Ann. Sci. école Norm. Sup. (4) **25** (1992), 539–545.

[13] F. Campana, *Orbifolds, special varieties and classification theory*, Ann. Inst. Fourier (Grenoble) **54(3)** (2004), 499–630.

[14] S. Cantat, *Dynamique des automorphismes des surfaces* $K3$, Acta Math. **187(1)**, 1–57.

[15] S. Cantat, *Dynamique des automorphismes des surfaces projectives complexes*, C.R. Acad. Sci. Paris Sér. I Math. **328(10)** (1999), 901–906.

[16] A. Chambert-Loir, *Mesures et équidistribution sur les espaces de Berkovich*, arXiv: math.NT/0304023 v3, 12 Nov 2004.

[17] A. Chambert-Loir, *Points de petite hauteur sur les variétés semi-abéliennes*, Ann. Sci. école Norm. Sup. (4) **33(6)** (2000), 789–821.

[18] S. David and P. Philippon, *Minorations des hauteurs normalisées des sous-variétés de variétés abéliennes*, in 'Number theory', Contemp. Math., **210**, Amer. Math. Soc., 1998.

[19] P Deligne, *Le déterminant de la cohomologie*, in 'Current trends in arithmetical algebraic geometry' (Arcata, Calif., 1985), Contemp. Math., **67**, 93–177, Amer. Math. Soc., Providence, RI, 1987.

[20] J.-P. Demailly, *Mesures de monge-ampére et caractérisation géométrique des variétés algébriques affines*, Mém. Soc. Math. France (N.S.), 1–125, 1985.

[21] T. Dinh and N. Sibony, *Dynamique des endomorphismes holomorphes*, preprint, 2002.

[22] T. Dinh and N. Sibony, *Groupes commutatifs d'automorphismes d'une variété kählérienne compacte*, Duke Math. J. **123(2)** (2004), 311–328.

[23] N. Fakhruddin, *Questions on self maps of algebraic varieties*, J. Ramanujan Math. Soc. **18(2)** (2003), 109–122.

[24] C. Favre and J. Rivera-Letelier, *Equidistribution quantitative des points de petite hauteur sur la droite projective*, arXiv:math. NT/0407471.

[25] Y. Fujimoto and N. Nakayama, *Compact complex surfaces admitting non-trivial surjective endomorphisms*, Tohoku Mathematical Journal **57(3)** (2005), 395–426.
[26] H. Gillet and C. Soulé, *Arithmetic intersection theory*, Inst. Hautes Etudes Sci. Publ. Math. **72** (1991).
[27] P. Griffiths and J. Harris, *Principles of algebraic geometry*, Wiley-Interscience, 1978.
[28] P. Griffiths, *Entire holomorphic mappings in one and several complex variables.* Princeton University Press, 1976.
[29] W. Gubler. *Höhentheorie, with an appendix by jurg kramer*, Math. Ann. **298(3)** (1994), 427–455.
[30] W. Gubler, *Local heights of subvarieties over non-archimedean fields*, J. Reine Angew. Math. **498** (1998), 61–113.
[31] J. Harris and Y. Tschinkel, *Rational points on quartics*, Duke Math. J. **104(3)** (2000), 477–500.
[32] A. Katok and B. Hasselblatt, *Introduction to the Mordern Theory of Dynamical Systems*, Cambridge University Press, 1995.
[33] S. Kawaguchi, *Canonical heights, invariant currents, and dynamical systems of morphoisms associated to line bundles*, math.NT/0405006.
[34] S. Kobayashi, *Intrinsic distances, measures and geometric function theory*, Bull. Amer. Math. Soc. **82(3)** (1976), 357–416.
[35] J. Kramer, *On a generalization of the Neron-Tate height on abelian varieties*, in 'Abelian varieties' (Egloffstein, 1993), 199–209, de Gruyter, Berlin, 1995.
[36] S. Lang, *Fundamentals of Diophantine geometry*, Springer-Verlag, 1983.
[37] M. Laurent, *Équations diophantiennes exponentielles*, Inventiones Mathematicae **78(2)** (1984), 299–327.
[38] B. Mazur, *The topology of rational points*, J. EXp. Math. **1** (1992), 35–46.
[39] C. McMullen, *Dynamics on blowups of the projective plane.*
[40] C. McMullen, *Dynamics on K3 surfaces: Salem numbers and siegel disks*, J. reine angen. Math. **545** (2002), 201–233.
[41] J. Milnor, *Dynamics in one complex variable*, Annals of Mathematics Studies, **160**, Princeton University Press, Princeton, NJ, 3 edition, 2006.
[42] A. Mimar, PhD thesis, 1995.
[43] Y. Miyaoka and S. Mori, *A numerical criterion for uniruledness*, Ann. of Math. **124(1)** (1986), 65–69.
[44] S. Mori, *Flip theorem and the existence of minimal models for 3-folds*, J. Amer. Math. Soc. **1(1)** (1988), 117–253.
[45] P. Morton and J.H. Silverman, *Rational periodic points of rational functions*, Internat. Math. Res. Notices.
[46] N. Nakayama, *Ruled surfaces with non-trivial surjective endomorphism*, Kyushu Journal of Mathematics (2), 2002.
[47] D.G. Northcott, *Periodic points on an algebraic variety*, Ann. of Math. (2) **51** (1950), 167–177.
[48] T. Peternell, *Towards a Mori theory on compact Kähler threefolds* III, Bull. Soc. Math. France **129(3)** (2001), 339–356.
[49] P. Philippon, *Sur des hauteurs alternatives*, I, Math. Ann. **289(2)** (1991), 255–283.
[50] J. Pineiro, L. Szpiro, and T. Tucker. *Mahler measure for dynamical systems on $\mathbb{P}^1$ and intersection theory on a singular arithmetic surface*, Geometric Methods in Algebra and Number Theory, **235**, 2004.
[51] M. Paun, S. Bouchson, J.-P. Demailly and T. Peternell, *The pseudo-effective cone of a compact Kähler manifold and varieties of negative kodaira dimension*, arXiv:math.AG/0405285 v1, 14 May 2004.
[52] M. Raynaud, *Sous-variétés d'une variété abélienne et points de torsion*, in 'Arithmetic and Geometry' (J. Coates and S. Helgason, editors), **1**, Birkauser, 1983.

[53] J.-P. Serre, *Analogues kählériens de certaines conjectures de Weil*, Annales of Mathematics **71** (1960), 392–394.
[54] N. Sibony, *Dynamique des applications rationnelles de* $\mathbb{P}^k$, in 'Dynamique et géométrie complexes', Panor. Synthéses, **8**, 97–185, Soc. Math. France, Paris, 1999.
[55] J.H. Silverman, *Rational points on* $K3$ *surfaces: a new canonical height*, Invent. Math. **105**(**2**) (1991), 347–373.
[56] J.H. Silverman, *The space of rational maps on* $\mathbb{P}^1$, Duke Math. J. **94**(**1**) (1998), 41–77.
[57] L. Szpiro, E. Ullmo, and S. Zhang, *Équirépartition des petits points*, Invent. Math. **127**(**2**) (1997).
[58] E. Ullmo, *Positivité et discrétion des points algébriques des courbes*, Ann. of Math. (2) **147**(**1**) (1998).
[59] S.-T. Yau, *On the Ricci curvature of a compact Käler manifold and the complex Monge-Ampere equation*, I, Comm. Pure Appl. Math. **31**(**3**) (1978), 339–411.
[60] X. Yuan, Ph.D. thesis, in preparation.
[61] S.-W. Zhang, *Positive line bundles on arithmetic varieties*, J. Amer. Math. Soc. **8**(**1**) (1995).
[62] S.-W. Zhang, *Small points and adelic metrics*, Journal of Algebraic Geometry **4**(**2**) (1995).
[63] S.-W. Zhang, *Heights and reductions of semi-stable varieties*, Compositio Math. **104**(**1**) (1996), 77–105.
[64] S.-W. Zhang, *Equidistribution of small points on abelian varieties*, Ann. of Math. (2) **147**(**1**) (1998).

Columbia University, 2990 Broadway, 509 Mathematics Buidling, NY, NY 10027
E-mail address: szhang@cpw.math.columbia.edu